food microbiology
AN INTRODUCTION

food microbiology
AN INTRODUCTION

Thomas J. Montville
and Karl R. Matthews

Department of Food Science, Cook College
Rutgers, The State University of New Jersey
New Brunswick, New Jersey

ASM PRESS

WASHINGTON, D.C.

Address editorial correspondence to ASM Press, 1752 N St. NW, Washington, DC
20036-2904, USA

Send orders to ASM Press, P.O. Box 605, Herndon, VA 20172, USA
Phone: (800) 546-2416 or (703) 661-1593
Fax: (703) 661-1501
E-mail: books@asmusa.org
Online: www.asmpress.org

Library of Congress Cataloging-in-Publication Data

Montville, Thomas J.
  Food microbiology : an introduction / Thomas J. Montville and Karl Matthews.
    p. cm.
  Includes index.
  ISBN 1-55581-308-9 (hardcover)
  1. Food—Microbiology. I. Matthews, Karl. II. Title.

  QR115.M625 2005
  664′.001′579—dc22

                         2004017880

10 9 8 7 6 5 4 3 2 1

Cover and interior design: Susan Brown Schmidler

Cover photo: Scanning electron micrograph showing *Listeria monocytogenes* cells in
the early stage of biofilm formation attached to stainless steel. Courtesy of Amy
Wong, University of Wisconsin—Madison.

*We dedicate this book to the thousands of scientists who
made the discoveries that are now presented as fact,
to all the scientists and regulators who use this knowledge
to ensure the safety of the food supply,
and to the advancement of food microbiology*

Contents

SECTION *II* Gram-Negative Foodborne Pathogenic Bacteria   83

SECTION III Gram-Positive Foodborne Pathogenic Bacteria 157

20

Molds   269

21

Viruses and Prions   289

SECTION *V* Control of Microorganisms in Food   299

22

Antimicrobial Chemicals   301

23

Biologically Based Preservation and Probiotic Bacteria   311

Preface

Food microbiology is an exciting field that reaches into every home and supports a multibillion dollar food industry. This book provides a taste of its complexity and challenge. The safety of food requires more than memorization of microbiological minutiae. It calls for critical thinking, innovative approaches, and healthy skepticism. We have tried to foster these skills so that today's students will be able to solve tomorrow's problems.

We would have never attempted to write a textbook on such a wide and complex topic as food microbiology "from scratch." Fortunately, ASM Press had published an advanced text for researchers, graduate students, and professors who needed the most up-to-date and in-depth treatment of food microbiology. *Food Microbiology: Fundamentals and Frontiers* (FMFF) was written by an army of subject area experts who presumed that the reader had a working knowledge of microbiology, biochemistry, and genetics. The success of the two editions of that book gave us the courage (and the resource) to write a food microbiology textbook for undergraduates. *Food Microbiology: an Introduction* is the child of the "big book." We have rewritten the experts' chapters to make them accessible to an undergraduate with a semester of microbiology and no biochemistry. In some cases, this meant adding foundational material; in others, it entailed deleting details that only an expert needs to know. In all cases, we have tried to write in a style, at a level, and in language appropriate for undergraduates. To enhance its utility as a textbook, we have added case studies, chapter summaries, questions for critical thinking, and a glossary.

The book is divided into five major sections. The first section covers the foundational material, describing how bacteria grow in food, how the food affects their growth, control of microbial growth, spores, detection, and microbiological criteria. Instructors may choose to use the other four sections in virtually any order. The gram-negative and gram-positive foodborne pathogens are covered in sections II and III, respectively. Section IV contains chapters on beneficial microbes and spoilage organisms. Molds are covered both as spoilage organisms and as potential toxin producers. Viruses may cause more than half of all foodborne illnesses, so the student should become familiar with their properties and control. Prions are not bacteria, molds, or viruses. In fact, they are not "microbes" at all; however, they are a major biological concern to the public and food safety experts.

The first reported case of mad cow disease in the United State was reported as we finished preparing the manuscript. Section V covers the chemical, biological, and physical methods of controlling foodborne microbes and closes by examining industrial and regulatory strategies for ensuring food safety.

Although only our names appear on the cover of this book, many people have made important contributions to it. First and foremost, we acknowledge the subject experts who wrote the chapters of *FMFF* from which this book is drawn. Each of these scientists is acknowledged by name in the appropriate chapter(s). Any errors, omissions, or oversimplifications fall on our shoulders, not on theirs. We thank the editors of *FMFF* for recruiting the subject experts and giving their blessing to this project. The reader should thank the students who reviewed each chapter for level and depth of coverage, writing style, and "what an undergraduate could be expected to know," in addition to grammar and usage. We have already thanked them and thank them again. Their names are Marcelo Bonnet, Rebecca Dengrove, Siobain Duffy, Megha Gandhi, Glynis Kolling, Wendy M. Iwanyshyn, Jennifer McEntire, Rebecca I. Montville, June Oshiro, Hoan-Jen Pang, Ethan Solomon, and Sarah Smith-Simpson. They helped to make this text "student friendly."

THOMAS J. MONTVILLE
KARL R. MATTHEWS

1

The Trajectory of Food Microbiology

LEARNING OBJECTIVES

The information in this chapter will help the student to:

- increase his or her awareness of the antiquity of microbial life and the newness of food microbiology as a scientific field
- appreciate how fundamental discoveries in microbiology still influence the practice of food microbiology
- understand the origins of food microbiology and thus anticipate its forward path

INTRODUCTION

Where should we start? A former president of the American Society for Microbiology defined *microbiology* as an artificial subdiscipline of biology based on size. This suggests that basic biological principles hold true, and are often discovered, in the field of microbiology. *Food microbiology* is a further subdivision of microbiology. It studies microbes that grow in food and how food environments influence microbes. In some ways, food microbiology has changed radically in the last 20 years. The number of recognized foodborne pathogens has doubled. "Safety through end product testing" has given way to the "safety by design" provided by Hazard Analysis Critical Control Points. Genetic and immunological probes have replaced biochemical tests and reduced testing time from days to minutes. In other ways, food microbiology is still near the beginning. Louis Pasteur would find his pipettes in a modern laboratory. Robert Petri would find his plates (albeit plastic rather than glass). Christian Gram would find all the reagents required for his stain. Food microbiologists still study only the microbes that we can see under the microscope and that grow on agar media in petri dishes. However, experts suggest that only 1% of all the bacteria in the biosphere can be detected by cultural methods.

This chapter's discussion of microbes per se sets the stage for a historical review of food microbiology. The bulk of it concerns bacteria. Less is known about viruses; prions were discovered relatively recently. The chapter ends with some thoughts about future developments in the field.

WHO'S ON FIRST?

Let there be no doubt about it—the microbes were here first. If the earth came into being at 12:01 a.m. of a 24-h day, microbes would appear at dawn and remain the only living things until well after dusk. Around 9 p.m.,

Authors' note

"The invention of the plastic petri dish set microbiology back 20 years."
—*J. L. Smith*

3

larger animals would appear, and a few seconds before midnight, humans would appear. The microbes were here first, they cohabit the planet with us, and they will be here after we are gone. Bacteria live in airless bogs, in thermal vents, in boiling geysers, in us, and in foods. We are lucky that they are here, for microbes form the foundation of the biosphere. We could not exist without them. Photosynthetic bacteria fix carbon into usable forms and make much of our oxygen. *Rhizobium* species fix the air's elemental nitrogen into ammonia that can be used for a variety of life processes. Degradative enzymes allow ruminants to digest cellulose. Microbes recycle dead organisms into their basic components, which can be used again and again. Microbes in our intestines aid in digestion, produce vitamins, and prevent colonization by pathogens. For the most part, microbes are our friends.

Life is not sterile. We live in a microbial world. Microbes can never be (nor should they be) conquered, once and for all. The food microbiologist can only create foods that microbes don't "like," manipulate the growth of microbes that are in food, kill them, or exclude them by physical barriers.

FOOD MICROBIOLOGY, PAST AND PRESENT

From the dawn of humanity until ~10,000 years ago, humans were hunter-gatherers. Humans were lucky to have enough food. There was neither surplus nor a settled place to store it. Preservation was not an issue. With the shift to agricultural societies, storage, spoilage, and preservation became important challenges. The first preservation methods were undoubtedly accidental. Sun-dried, salted, or frozen foods did not spoil. In the classic "turning lemons into lemonade" style, early humans learned that "spoiled" milk could be acceptable, or even desirable, if viewed as fermented. Fermenting food became an organized activity around 4000 B.C. (Table 1.1). Breweries and bakeries sprang up long before the idea of yeast was conceived.

Humans remained ignorant of microbes for thousands of years. It was not until the late 1600s that Antonie van Leeuwenhoek (Fig. 1.1) used a

Table 1.1 Significant events in the history of food microbiology[a]

Decade	Event
~4000 B.C.	Fermentation of food becomes an organized activity.
1670s	van Leeuwenhoek observes "little animals" in rain water; "microbiology" is born.
1760s	Spallanzani's experiments with boiled beef strike a blow against spontaneous generation.
1800s	Nicolas Appert invents the canning process. Amazingly, this is still a mainstay of food processing 200 years later.
1810s	P. Dumond patents the tin can, making Appert's life much easier.
1850s	Appert and R. Chevallier are issued a patent for steam sterilization (retort). The use of steam under pressure increases processing temperatures, decreases processing times, and radically improves the quality of canned food.
	Louis Pasteur demonstrates that living organisms cause lactic and alcoholic fermentations.
1860s	Pasteur disproves spontaneous generation. Life can come only from other life.
	Joseph Lister develops the concept of antiseptic practice. Convincing surgeons to wash their hands saves thousands of lives.

Table 1.1 *(continued)*

Decade	Event
1880s	Robert Koch postulates bacteria as causative agents of disease. To this day, Koch's postulates remain the "gold standard" for proving that a bacterium causes a disease.
	Christian Gram invents the Gram stain.
	R. J. Petri invents the petri dish. Petri worked in Koch's laboratory, where the usefulness of agar was also discovered.
	A. A. Gartner isolates *Salmonella enteritidis* from a food-poisoning outbreak. A century later, salmonellae are still the leading cause of death among foodborne microbes.
1890s	Pasteurization of milk begins in the United States.
1900s	The Food and Drug Act is passed in response to Upton Sinclair's exposé of the meat industry in *The Jungle*.
1920s	U.S. Public Health Commission publishes methods to prevent botulism.
	Alexander Fleming discovers antibiotics. Less than 100 years later, multiply antibiotic-resistant pathogens threaten possible epidemics.
	Clarence Birdseye introduces frozen foods to the retail marketplace.
1930s	G. M. Dack confirms that *S. aureus* makes a toxin.
	Widespread introduction of home refrigerators.
	The Food, Drug, and Cosmetic Act strengthens regulation over foods.
	Viruses are discovered.
1940s	First freeze-dried food.
	The supermarket replaces assorted shops as the place to buy food.
1950s	Watson and Crick discover the structure of DNA. Rosalind Franklin played a large but uncredited role.
	Research on food irradiation begins.
1960s	C. Duncan and D. Strong demonstrate that perfringens food poisoning is caused by a toxin.
	The role of fungal toxins is discovered when "turkey X disease" breaks out in poultry.
1970s	S. Cohen discovers genetic recombination in bacteria.
	L. McKay reports plasmids in gram-positive bacteria.
	Introduction of "good manufacturing practices" for low-acid foods.
	Invention of monoclonal antibodies lays the foundation for mass-produced antibody-based tests.
1980s	First recognized outbreak of listeriosis.
	Escherichia coli O157:H7 is first recognized as a pathogen.
	First genetic probe for detection of salmonellae.
	Polymerase chain reaction is invented.
	Prions are discovered.
1990s	Irradiation is approved for pathogen control in meat and poultry.
	"Mad cow" crisis hits the United Kingdom.
	Hazard Analysis Critical Control Points required by U.S. Department of Agriculture.
2000s	Irradiation is approved for shell eggs.
	Regulatory concerns about bioterroristic contamination of food lead to new laws on facility registration, product traceability, and prenotification of imports.
	First report of mad cow disease in United States.

*a*Compiled from T. D. Brock, *Milestones in Microbiology: 1546 to 1940* (ASM Press, Washington, D.C., 1999); P. Hartman, *in* M. P. Doyle, L. R. Beuchat, and T. J. Montville (ed.), *Food Microbiology: Fundamentals and Frontiers*, 2nd ed. (ASM Press, Washington, D.C., 2001), and other sources.

Figure 1.1   Antonie van Leeuwenhoek.

crude microscope (a replica of which is shown in Fig. 1.2) to see small living things in pond water. *Microbiology was born!*

Nonetheless, it took another 200 years to prove that microbes exist and cause fermentative processes. In the mid-1700s, Lazzaro Spallanzani showed that boiled meat placed in a sealed container did not spoil. Advocates of spontaneous generation, however, argued that air is needed for life and that the air was sealed out. It took another 100 years for Pasteur's elegant swan neck flask experiment (see Box 24.1) to replicate Spallanzani's experiment in a way that allowed access to air but not microbes. During this interval, Nicolas Appert (Fig. 1.3) discovered that foods would not spoil if they were heated in sealed containers. Thus, canning was invented without any knowledge of microbiology. Indeed, it was not until the 1900s that mathematical bases for canning processes were developed.

Robert Koch was a giant of microbiology. In the late 1800s, he established criteria for proving that a bacterium caused a disease. To satisfy Koch's postulates, one had to (i) isolate the suspect bacterium in pure culture from the diseased animal, (ii) expose a healthy animal to the bacterium and make it sick, and (iii) reisolate the bacterium from the newly infected animal. This approach was so brilliant that it is still used today to prove that a disease has a microbial origin. Microbiology still emphasizes the study of single bacterial species in pure culture. Unfortunately, most microbes cannot be cultured and exist in nature as community members, not in pure culture. Two other innovations from Koch's laboratory are still with us. Robert Petri, an assistant in Koch's laboratory, invented the petri dish as a minor modification of Koch's plating method. Workers in the Koch laboratory were often frustrated by the gelatin used in the plating method. Gelatin would not solidify when the temperature was too warm and was dissolved by certain organisms (which produced enzymes that degrade gelatin). Walter Hesse, a physician who had joined the Koch laboratory to study infectious bacteria in the air, shared this frustration with his wife, Fanny. Frau Hesse had been using agar to thicken jams and jellies for years and suggested that they try agar in the laboratory. You know the rest of the story.

The first half of the 20th century was marked by the discovery of the traditional foodborne pathogens (*Salmonella* species, *Clostridium botulinum*,

Figure 1.2   Replica of van Leeuwenhoek's microscope.

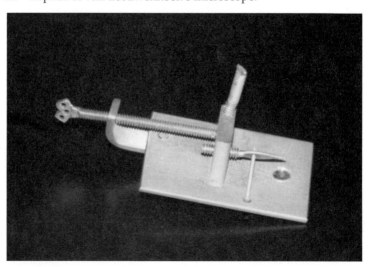

Figure 1.3 Stamp honoring Nicolas Appert.

Staphylococcus aureus, Bacillus cereus, and *Clostridium perfringens*) and their association with certain illnesses. Viruses were first crystallized and associated with disease in the 1930s. In the middle of the 20th century, biology made a quantum leap forward; James Watson and Francis Crick discovered the structure of DNA, giving birth to the era of molecular genetics (Fig. 1.4). This has led to better understanding of how bacteria cause illness, revolutionary genetic and antibody-based detection methods for bacteria, genetic "fingerprints" as epidemiological tools, and the ability to genetically alter fermentation organisms to improve their industrial characteristics. The full impact of molecular biology has yet to hit food microbiology.

The movement away from massive end product testing to safety by design also started in this era. The adoption of "good manufacturing practices" provided manufacturers with procedures that should yield safe products. Hazard Analysis Critical Control Points codified this into a safety assurance system. Food irradiation was approved as a kill step in the processing of raw poultry and meat. Risk assessment provides the basis for more sophisticated regulatory approaches based on the end result (fewer sick people) rather than dictating processes that give fewer microbes.

Figure 1.4 The discovery of DNA's double helix marked a quantum leap in the history of microbiology.

TO THE FUTURE AND BEYOND

My mentor once told me, "If you stare into the crystal ball too long, you end up with shards of glass in your face." It is hard to predict the future. However, it is safe to say that there will be greater demand for more, safer foods; that innovations in medical microbiology will impact food microbiology in ~15 years; and that entirely new concepts will alter the course of food microbiology.

About one-third of the world's food supply is lost to spoilage. A global population that will double in your lifetime cannot afford this loss. If everyone is to eat, that level of spoilage must be reduced. Hunger hovers over humanity and may come to dwarf safety as an issue. Fermentation, a low-energy "appropriate technology," may become more important as a preservation method. Microbes themselves may become a food source. (Yeast was touted as "single-cell protein" in the 1970s). Probiotic bacteria may help people maintain health through their diet. Perhaps, just as pesticides have been cloned into plants to kill insects, antimicrobials will be cloned into foods to make them kill bacteria.

While 30% of the world's population is worried about getting *enough* food, the 10% of the world's population that worries about "overnutrition" demands *safer* food. Zero risk is impossible, and the cost of safety increases exponentially as one approaches zero. How safe is safe enough? When should a society shift resources from increased safety to increased availability? The requirements of international trade mandate the harmonization of safety standards. First- and third-world countries will need similar sanitary standards.

Scientific innovations usually appear first in medical technology and impact the food system later. This can be seen in the adaptation of rapid and automated methods, food-processing facilities increasingly adapting the methods of pharmaceutical houses, and increased emphasis on molecular understanding. Look at what's hot in the biomedical area today, and you will see the food microbiology of tomorrow.

This chapter highlights the scientific limits of microbiology. They are starting to fall. The causes of half of the cases of foodborne illness are unknown. Could they be caused by viruses? Could they be due to the microbes that are there but are unculturable? Will new molecular techniques for studying these bacteria allow us to understand and control them? The first 200 years of microbiology have been devoted to the study of bacteria as if they were noninteracting ball bearings. We now know that they "talk" to each other using chemical signals and sensors. This *quorum sensing* helps microbial communities decide when they will take action (for example, to mount an infectious attack). If bacteria talk to each other, what are they saying? If we knew, could we prevent spoilage and disease by scrambling the signal, drowning the signal out with other biochemical noise, or inactivating the receivers? What about the 99% of bacteria that remain undiscovered? What is in their gene pool? How might we use those genes and their products to make food better and safer? What are the roles of prions and viral agents in foodborne disease?

The future of food microbiology may be its most exciting part. It is the part that you will live in and maybe even create. To prepare for it, gain a foundational understanding of the field today, learn how to think critically and creatively, and learn to love these little creatures that are so small but so smart.

Summary

- Microbes were the first living things on earth and remain the foundation of the biosphere.
- The fundamental biological discoveries of the 1800s, such as the disproof of spontaneous generation, Koch's postulates, and pure culture of microbes, continue to influence the field.
- Molecular biology and genetics have begun to impact food microbiology and will shape its future.
- Food microbiology is a relatively young field of study.

Suggested reading

Brock, T. D. 1999. *Milestones in Microbiology: 1546 to 1940.* ASM Press, Washington, D.C.

Hartman, P. 2001. The evolution of food microbiology, p. 3–12. *In* M. P. Doyle, L. R. Beuchat, and T. J. Montville (ed.), *Food Microbiology: Fundamentals and Frontiers,* 2nd ed. ASM Press, Washington, D.C.

Needham, C., M. Hoagland, K. McPherson, and B. Dodson. 2000. *Intimate Strangers: Unseen Life on Earth.* ASM Press, Washington, D.C.

Questions for critical thought _____

1. What would have happened if there had been spores in Pasteur's swan neck flask?

2. Can you think of another experiment that would disprove spontaneous generation? How would food safety be affected if there were spontaneous generation?

3. Extend Table 1.1 fifty years into the future. Enter what you think might happen.

4. Discuss three historical discoveries in microbiology that still influence food microbiology.

5. It is said that if Pasteur walked into a food microbiology laboratory today, he could be comfortably at work in ~15 min. Is this true? Is it good?

6. Why did the invention of the plastic petri dish set food microbiology back twenty years?

2

Factors That Influence Microbes in Foods*

LEARNING OBJECTIVES

The information in this chapter will enable the student to:

- distinguish among different methods of culturing foodborne microbes and choose the correct technique for a given application
- recognize how intrinsic and extrinsic factors are used in controlling microbial growth
- quantitatively predict the size of microbial populations using the equations that describe the kinetics of microbial growth
- qualitatively describe how altering the variables in the equations for the kinetics of microbial growth alters growth patterns
- relate biochemical pathways to energy generation and metabolic products of foodborne bacteria

INTRODUCTION

Never say that you are *just* a food microbiologist. Food microbiology is a specialty area in microbiology. Just as a heart surgeon is more than a surgeon, a food microbiologist is a microbiologist with a specialty. Food microbiologists must understand microbes. They must understand complex food systems. Finally, they must integrate the two to solve microbiological problems in complex food ecosystems. Even the question "How many bacteria are in that food?" has no simple answer. The first part of this chapter examines the question of how we detect and quantify bacteria in food. We present the concept of culturability and ask if it really reflects viability. Even though injured or otherwise nonculturable cells cannot form colonies on petri dishes, they do exist and can cause illness.

The second part of the chapter presents internal and environmental factors that control bacterial growth. Ecology teaches that the *interaction* of factors determines which organisms can or cannot grow in an environment. The use of multiple environmental factors (i.e., pH, salt concentration, temperature, etc.) to inhibit microbial growth is called *hurdle technology*. Hurdle technology is becoming an increasingly popular preservation strategy.

The kinetics of microbial growth are covered in the third part of the chapter. All four phases of the microbial growth curve are important to food microbiologists. If the lag phase can be extended beyond the normal shelf life, the food is microbially safe. If the growth rate can be slowed so

---

*This chapter is based on one written by Thomas J. Montville and Karl R. Matthews for *Food Microbiology: Fundamentals and Frontiers,* 2nd ed. It has been adapted and augmented with material on prions for use in an introductory text.

Mechanisms of energy generation

Anaerobes cannot liberate all of the energy in a substrate. They typically make one or two ATP molecules by burning a mole of glucose. This process is called *substrate level phosphorylation*. Strict anaerobes are killed by air.

Aerobes can burn glucose completely in the presence of oxygen. They generate 34 ATP molecules from a mole of glucose through a process called *respiration*. Strict aerobes cannot grow in the absence of oxygen.

Facultative anaerobes can generate energy by both substrate level phosphorylation and respiration. They can grow in a wider variety of environments than strict aerobes or strict anaerobes.

that the population size is too small to cause illness, the product is safe. If the rate of the death phase can be controlled, it can be used to ensure safety, as in the case of cheddaring cheese made from unpasteurized milk.

The fourth part of the chapter introduces the physiology and metabolism of foodborne microbes. Because this book does not assume a background in biochemistry, the biochemistry associated with foodborne microbes is introduced. The biochemical pathways used by lactic acid bacteria determine the characteristics of fermented foods. Spoilage is also caused by a product(s) of a specific pathway(s). Lactic acid makes milk sour; carbon dioxide causes cans to swell. The ability of bacteria to use different biochemical pathways that generate different amounts of energy (in the form of adenosine triphosphate [ATP]) governs their ability to grow under adverse conditions. Organisms grown in the presence of oxygen (aerobically) produce more energy (ATP) than those grown in the absence of oxygen (anaerobically) (Box 2.1). Thus, organisms such as *Staphylococcus aureus* can withstand hostile environments better under aerobic conditions than under anaerobic conditions.

FOOD ECOSYSTEMS, HOMEOSTASIS, AND HURDLE TECHNOLOGY

Foods are ecosystems composed of the environment and the organisms that live in it. The food environment is composed of intrinsic factors inherent to the food (i.e., pH, water activity, and nutrients) and extrinsic factors external to it (i.e., temperature, gaseous environment, and the presence of other bacteria). Intrinsic and extrinsic factors can be manipulated to preserve food. When applied to microbiology, *ecology* was defined by the International Commission on Microbial Specifications for Foods as "the study of the *interactions* between the chemical, physical, and structural aspects of a niche and the composition of its specific microbial population." "Interactions" emphasizes the dynamic complexity of food ecosystems.

Foods can be heterogeneous on a micrometer scale. Heterogeneity and gradients of pH, oxygen, nutrients, etc., are key ecological factors in foods. Foods may contain multiple microenvironments. This is well illustrated by the food poisoning outbreaks in aerobic foods caused by the strict anaerobe *Clostridium botulinum*. *C. botulinum* growth in potatoes, sautéed onions, and cole slaw has caused botulism outbreaks. The oxygen in these foods is driven out during cooking and diffuses back in so slowly that most of the product remains anaerobic.

CLASSICAL MICROBIOLOGY AND ITS LIMITATIONS

Limitations of Detection and Enumeration Methods
All methods based on the plate count and pure-culture microbiology have the same limitations. The plate count assumes that every cell forms one colony and that every colony originates from only one cell. The ability of a given cell to grow to the size of a macroscopic colony depends on many things. These include the cell's physiological state, the growth medium used (and the presence or absence of selective agents), and the incubation temperature.

Figure 2.1 illustrates the concepts of plate counting, qualitative detection, and most-probable-number (MPN) methodologies. Consider a 50-ml

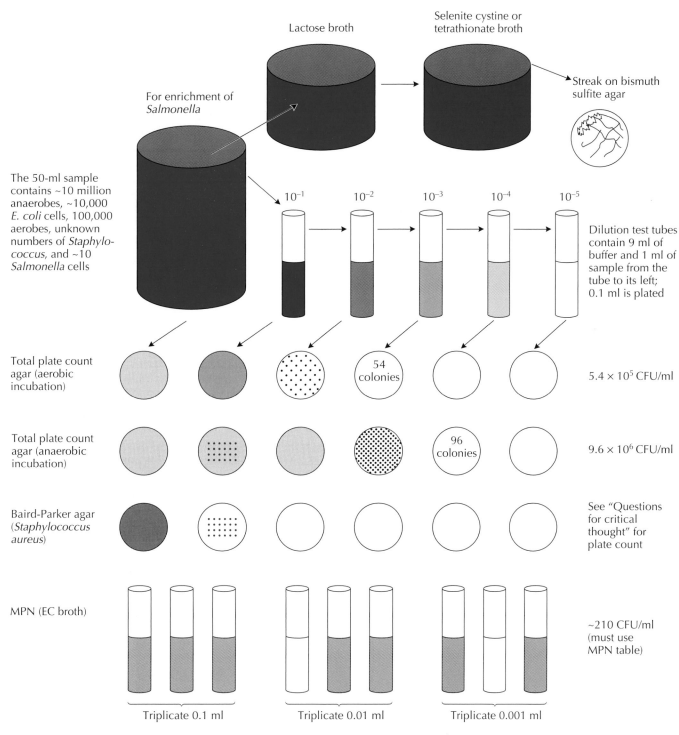

Figure 2.1 Methods of enumerating bacteria. The sample (which represents a 10^0 dilution) is diluted through a series of 10-fold dilutions prior to enumeration. To enumerate the bacteria, 0.1 ml from each dilution tube is plated on an agar plate with appropriate medium. After 24- to 48-h incubations, the colonies are counted and multiplied by the dilution factor to give the microbial load of the original sample on a per-milliliter basis. The color intensity indicates the culture density. If the number of cells present is expected to be low, the MPN technique is used. In this case, the sample is diluted into triplicate series of test tubes containing the media, and the pattern of positive tubes is used to determine the MPN from an MPN table (Table 2.2).

BOX 2.2

Powers of 10

Because microbial populations are so large, microbiologists usually speak in powers of 10. The answer to the question "How many bacteria are in that culture?" would probably be "10^8 or 10^9." This suggests that there is not much difference between 100 million and 1 billion microbes. Usually there is not. Try to remember these powers of 10:

$$
\begin{aligned}
100 &= 10^2 \\
1,000 &= 10^3 \\
100,000 &= 10^5 \\
1,000,000 &= 10^6 \\
1,000,000,000 &= 10^9
\end{aligned}
$$

When bacteria are being killed, a similar vocabulary is used. A 1-log-unit reduction is equal to a 10-fold reduction (see table): 90% of the population has been killed. This may sound effective, but if the initial number is 10^6, 10^5 are still left. That's a lot! For practical purposes, reductions of <1 log unit have no real significance. Regulatory agencies frequently require an "x log" reduction in microbial populations. "x" is product and process dependent, equaling 7 for pasteurized milk and 12 for low-acid canned food.

Log unit reduction	% Reduction
1	90
2	99
3	99.9
4	99.99
5	99.999
6	99.9999
7	99.99999
12	99.9999999999

sample of milk that contains 10^6 colony-forming units (CFU) of anaerobic bacteria per ml, 10^5 aerobic bacteria, a significant number of *S. aureus* (a pathogen) organisms, some *Escherichia coli* organisms, and a few salmonellae (Box 2.2). (A CFU ideally would be a single cell, but in reality it could be a chain of 10 or a clump of 100.) Since the bacteria are either too numerous to be counted directly or too few to be found easily, samples have to be diluted or enriched.

Plate Counts

Plate counts are used to quantify bacterial populations of >250 CFU/ml (for liquids) or >2,500 CFU/ml (for solids, which must be diluted 1:10 in a liquid to be pipettable). The food sample is first homogenized 1:10 (wt/vol) in a buffer to give a 10-fold dilution. This is further diluted through a series of 10-fold dilutions to give 10-, 100-, 1,000-, 10,000-fold, etc., dilutions and spread onto an agar medium. After incubation at 35°C, the average number of colonies per plate (on plates having between 25 and 250 colonies) is determined and multiplied by the dilution factor to give the number of bacteria in the original sample. This sounds simple, right? Wrong. At one time, the procedure described above was referred to as the *total plate count*. However, it counts only those bacteria that can grow on that agar medium at 35°C. The anaerobes, which are present in higher numbers, are not detected because they cannot grow in the presence of air. Bacteria that grow only below 15°C might be good predictors of refrigerated milk's shelf life. They are not detected either. *S. aureus,* which can make you sick, would be lost in the crowd of harmless bacteria. Variations of the standard aerobic plate count consider these issues. Plate counts can be incubated at different temperatures or under different atmospheres to increase their specificities for certain types of bacteria.

Table 2.1 Some selective media used in food microbiology

Bacterium	Medium	Selective agent(s)
S. aureus	Baird-Parker agar[a]	Lithium chloride, glycine, tellurite
Listeria species	LPM	Lithium chloride, phenyl-ethanol, antibiotic
Lactic acid bacteria	deMann-Rogosa-Sharp (MRS)	Low pH, surfactant
Campylobacter species	Abeyta-Hunt-Bark agar	Microaerobic incubation, antibiotics

[a]The highest honor for a microbiologist is to have an organism named after oneself. Not far behind, and much easier to attain, is naming a medium after oneself.

Selective, or Differential, Media

The concept of selective media is relatively simple. Some ingredient of the medium helps (in the case of an unusual sugar used as a carbon source) the growth of the target bacteria or inhibits (in the cases of antibiotics, salt, surfactants, etc.) the growth of bacteria other than the target bacteria. Some examples of selective media are given in Table 2.1. In the case of Baird-Parker agar (Fig. 2.2), lithium chloride, glycine, and tellurite allow *S. aureus* to grow while suppressing a larger number of other organisms. In addition, the *S. aureus* colonies are a distinctive black color. However, target organisms that are injured may be killed by the selective agent. In addition, the selective media are not entirely selective. For example, *Listeria monocytogenes* can grow on *Lactobacillus* deMann-Rogosa-Sharp (MRS) medium. Thus, the identities of suspect pathogens isolated on selective media must be confirmed by biochemical or genetic methods.

Differential media are those designed to make a colony of the target organism look different and stand out from the background of other organisms. For example, *S. aureus* colonies are black with a precipitation zone on Baird-Parker agar. *L. monocytogenes* has a blue sheen on McBride's agar and turns Fraiser broth black.

Figure 2.2 *S. aureus* forms black colonies when grown on Baird-Parker agar.

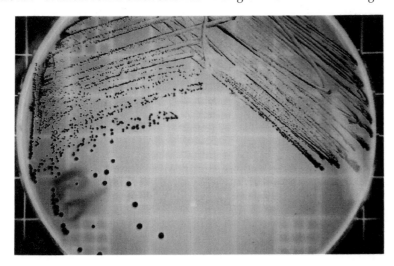

Most-Probable-Number Methods

MPN methods are used to estimate low numbers of organisms in a sample (Fig. 2.1 and Table 2.2). They are used most commonly for *E. coli* and coliform bacteria. Selective agents are added to broth medium. One milliliter of sample or diluted sample is transferred to each of three test tubes of medium for each of at least three 10-fold serial dilutions. The samples are incubated, and the pattern of positive tubes is recorded. This pattern is used to derive the most probable number of bacteria from a statistical chart. The lower sensitivity limit for a three-tube MPN is <3.0 bacteria per g. This can be lowered to <0.96 bacterium per g by using 10 tubes at each dilution instead of 3, but this is rarely cost-effective.

Enrichment Techniques

There is zero tolerance (actually, 0/25 g) for *Salmonella*, *L. monocytogenes*, and *E. coli* O157:H7 in ready-to-eat foods. The good news is that this makes it unnecessary to count these pathogens. The bad news is that one must be able to find one salmonella organism hiding among a million other bacteria. This is done by coupling preenrichment medium with selective enrichment, followed by plating the bacteria on selective medium and biochemical or genetic confirmation (Fig. 2.1). The preenrichment in a nonselective medium allows injured pathogens to repair themselves, recover, and regain resistance to selective agents. The enrichment step uses selective agents to promote pathogen growth to high levels while inhibiting growth of the background organisms. Then it is possible to isolate the pathogen on a selective agar and confirm the identity of the colony using other methods.

Table 2.2 Three-tube MPN table for quantifying low levels of bacteria[a]

No. of positive tubes			MPN/g	Confidence limits		No. of positive tubes			MPN/g	Confidence limits	
0.10	0.01	0.001		Low	High	0.10	0.01	0.001		Low	High
0	0	0	<3.0		9.5	2	2	0	21	4.5	42
0	0	1	3.0	0.15	9.6	2	2	1	28	8.7	94
0	1	0	3.0	0.15	11	2	2	2	35	8.7	94
0	1	1	6.1	1.2	18	2	3	0	29	8.7	94
0	2	0	6.2	1.2	18	2	3	1	36	8.7	94
0	3	0	9.4	3.6	38	3	0	0	23	4.6	94
1	0	0	3.6	0.17	18	3	0	1	38	8.7	110
1	0	1	7.2	1.3	18	3	0	2	64	17	180
1	0	2	11	3.6	38	3	1	0	43	9	180
1	1	0	7.4	1.3	20	3	1	1	75	17	200
1	1	1	11	3.6	38	3	1	2	120	37	420
1	2	0	11	3.6	42	3	1	3	160	40	420
1	2	1	15	4.5	42	3	2	0	93	18	420
1	3	0	16	4.5	42	3	2	1	150	37	420
2	0	0	9.2	1.4	38	3	2	2	210	40	430
2	0	1	14	3.6	42	3	2	3	290	90	1,000
2	0	2	20	4.5	42	3	3	0	240	42	1,000
2	1	0	15	3.7	42	3	3	1	460	90	2,000
2	1	1	20	4.5	42	3	3	2	1,100	180	4,100
2	1	2	27	8.7	94	3	3	3	>1,100	420	

[a]Source: http://www.cfsan.fda.gov/~ebam/bam-a2.html#tables. For three tubes with 0.1-, 0.01-, and 0.001-g inocula, the MPN per gram and 95% confidence intervals are shown.

Factors That Influence Microbes in Foods **17**

Table 2.3 Influence of thermal history and enumeration protocols on experimentally determined $D_{55°C}$ values for *L. monocytogenes*

	$D_{55°C}$ value (min)			
	TSAY medium		McBride's medium	
Atmosphere	+ Heat shock	− Heat shock	+ Heat shock	− Heat shock
Aerobic	18.7	8.8	9.5	6.6
Anaerobic	26.4	12.0	No growth	No growth

[a]+, with; −, without.

Values derived from plate counts are also influenced by these factors. Table 2.3 illustrates these points by providing $D_{55°C}$ values (the time required to kill 90% of the population at 55°C) for *L. monocytogenes* with different thermal histories (heat shocked for 10 min at 80°C or not heat shocked) on selective (McBride's) or nonselective (tryptic soy agar with yeast extract [TSAY]) medium under an aerobic or anaerobic atmosphere. Cells that have been heat shocked and recovered on TSAY medium are about four times more heat resistant (as shown by the larger D values) than cells which have not been heat shocked and recovered on McBride's medium under aerobic conditions. Injured cells and cells that are viable but nonculturable (VNC) pose additional problems for food microbiologists. While they cannot form colonies and be detected, they can make people sick.

Injury

Microbes may be injured rather than killed by *sublethal* levels of heat, radiation, acid, or sanitizers. Injured cells are less resistant to selective agents or have increased nutritional requirements. Injury is a complex process influenced by time, temperature, the concentration of the injurious agent, the strain of the target pathogen, the experimental methodology, and other factors. For example, a standard sanitizer test may suggest that a sanitizer *kills* listeria cells. However, cells might be recovered by using listeria repair broth. Injured cells die as the exposure time and sanitizer concentration increase.

Data that illustrate injury are given in Fig. 2.3. Cells subjected to mild stress are plated on a rich nonselective medium and on a selective medium containing 6% salt. The difference between the populations of cells able to form colonies on the two media represents injured cells. (If 10^7 CFU of a population/ml are found on the selective medium and 10^4 CFU/ml can grow on the nonselective medium, then 10^3 CFU/ml are injured.) Media for recovery of injured cells are often specific for a given bacteria or stress condition.

Injury is important to food safety for several reasons. (i) If injured cells are classified as dead during heat resistance determinations, the effect of heating will be overestimated and the resulting heat process will be ineffective. (ii) Injured cells that escape detection at the time of postprocessing sampling may repair themselves before the food is eaten and cause illness. (iii) The selective agent may be a common food ingredient, such as salt or organic acids or even suboptimal temperature. For example, *S. aureus* cells injured by acid during sausage fermentation can grow on tryptic soy agar but not on tryptic soy agar +7.5% salt. These injured cells can repair themselves in sausage if held at 35°C (but not if held at 5°C) and then grow and produce enterotoxin.

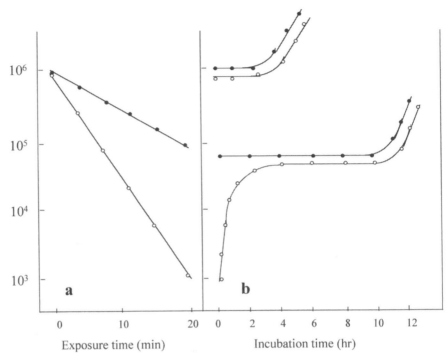

Figure 2.3 Data indicative of injury and repair. When bacteria are plated on selective (○) or nonselective (●) medium during exposure to some stressor (e.g., heat [a]), the decrease in CFU on a nonselective medium represents the true lethality, while the difference between the values obtained the two media is defined as injury. During repair (b), resistance to selective agents is regained, and the value obtained on the selective medium approaches that on the nonselective medium. Unstressed controls are shown at the top of panel b. (Source: M. P. Doyle, L. R. Beuchat, and T. J. Montville [ed.], *Food Microbiology: Fundamentals and Frontiers,* 2nd ed. [ASM Press, Washington, D.C., 2001].)

Repair is the process by which cells recover from injury. Repair requires RNA and protein synthesis and is often made evident by lengthening of the lag phase. Environmental factors influence the extent and rate of repair. The microbial stability of cured luncheon meat is due to the extended lag period required for the repair of spoilage organisms.

Viable but Nonculturable
Salmonella, Campylobacter, Escherichia, Shigella, and *Vibrio* species can exist in a state in which they are viable but cannot be cultured by normal methods. The differentiation of vegetative cells into a dormant VNC state is a survival strategy for many nonsporulating bacteria. During the transition to the VNC state, rod-shaped cells shrink and become small spherical bodies. These are totally different from bacterial spores. It takes from 2 days to several weeks for an entire population of vegetative cells to become VNC.

Although VNC cells cannot be cultured, their viability can be determined through microscopic methods. Fluorescent molecular probes are used to determine the integrity of the bacterial membrane. Bacteria with intact cell membranes stain fluorescent green, but bacteria with damaged membranes stain fluorescent red. Dyes can also be used to identify VNC cells, since respiring cells reduce the colorless soluble dye to a colored insoluble compound that can be seen under the microscope. Powerful new methods for detecting VNC cells are being developed as our understanding of bacteria at the molecular level increases.

Because the VNC state is most often induced by nutrient limitation in aquatic environments, it might appear irrelevant to nutrient-rich food environments. However, changes in the salt concentration, exposure to sanitizers, and shifts in temperature can also cause the VNC state. When *Vibrio vulnificus* (10^5 CFU/ml) is shifted to refrigeration temperatures, it becomes nonculturable (<0.04 CFU/ml) but is still lethal to mice. The bacteria resuscitate in the mice and can be isolated postmortem using culture methods. Foodborne pathogens can become VNC when shifted to a cold temperature in nutritionally rich media. This has chilling implications for the safety of refrigerated foods.

Our increased awareness of the VNC state should lead to the reexamination of our concept of viability, our dependence on enrichment culture to isolate pathogens, and our reliance on culture methods to monitor microbes in the environment. We do not understand the mechanisms of VNC formation, the mechanisms that make VNC cells resistant to stressors, and what event signals the cell to resuscitate. The relationship between viability and culturability needs to be better understood.

Intrinsic Factors That Influence Microbial Growth

Characteristics of the food itself are called *intrinsic* factors. These include naturally occurring compounds that influence microbial growth, compounds added as preservatives, the oxidation-reduction potential, water activity, and pH. Most of these factors are covered separately in the chapters on physical and chemical methods of food preservation.

The influence of pH on gene expression is a relatively new area. The expression of genes governing proton transport, amino acid degradation, adaptation to acidic or basic conditions, and even virulence can be regulated by the external pH. Cells sense changes in acidity (pH) through several mechanisms:

1. Organic acids enter the cell only in the protonated (undissociated) form. Once inside the cell, they dissociate, releasing H^+ (the proton) into the cell cytoplasm. The cell may sense the accumulation of the resulting anions, which cannot pass back across the cell membrane. Additionally, they may sense the released protons, which cause the cytoplasm to become more acidic.

2. The change in pH results in a change in the transmembrane proton gradient (i.e., the difference in pH between the inside and outside of the cell), which can also serve as a sensor to start or stop energy-dependent reactions.

3. The changes in acidity inside the cell may result in the protonation or deprotonation of amino acids in proteins. This may alter the secondary or tertiary structures of the proteins, changing their functions and signaling the change in pH to the cell.

Cells must maintain their intracellular pH (pH_i) above some critical pH_i at which intracellular proteins become denatured. *Salmonella enterica* serovar Typhimurium has three progressively more stringent mechanisms to maintain a pH_i that supports life. These three mechanisms are the homeostatic response, the acid tolerance response (ATR), and the synthesis of acid shock proteins.

At a pH outside (pH_o) of >6.0, salmonella cells adjust their pH_i through the *homeostatic* response. The homeostatic response maintains the pH_i by increasing the activity of proton pumps to expel more protons from the cytoplasm. The homeostatic mechanism is always on and functions in the presence of protein synthesis inhibitors.

The ATR is triggered by a pH_o of 5.5 to 6.0. This mechanism is sensitive to protein synthesis inhibitors; there are at least 18 ATR-induced proteins. ATR appears to involve the membrane-bound ATPase proton pump and maintains the pH_i at >5.0 at pH_o values as low as 4.0. The loss of ATPase activity caused by gene disruption mutations or metabolic inhibitors abolishes the ATR but not the pH homeostatic mechanism.

The synthesis of acid shock proteins is the third way that cells regulate pH_i. The synthesis of these proteins is triggered by a pH_o of 3.0 to 5.0. They constitute a set of regulatory proteins distinct from the ATR proteins.

Extrinsic Factors That Influence Microbial Growth

Extrinsic factors are *external* to the food. Temperature and gas composition are the main extrinsic factors influencing microbial growth. The influence of temperature on microbial growth and physiology is huge. While the influence of temperature on the growth rate is obvious and is covered in some detail here, the influence of temperature on gene expression is equally important. Cells grown at refrigeration temperature do not just grow more slowly than those grown at room temperature, they express different genes and are physiologically different. Later chapters provide details about how temperature regulates the expression of genes governing traits ranging from motility to virulence in specific organisms.

A rule of thumb in chemistry suggests that reaction rates double with every 10°C increase in temperature. This simplifying assumption is valid for bacterial growth rates only over a limited range of temperatures (Fig. 2.4). Above the optimal growth temperature, growth rates decrease rapidly. Below the optimum, growth rates also decrease, but more gradually. Bacteria can be classified as psychrophiles, psychrotrophs, mesophiles, and thermophiles according to how temperature influences their growth.

Figure 2.4 Relative growth rates of bacteria at different temperatures. (Source: M. P. Doyle, L. R. Beuchat, and T. J. Montville [ed.], *Food Microbiology: Fundamentals and Frontiers,* 2nd ed. [ASM Press, Washington, D.C., 2001].)

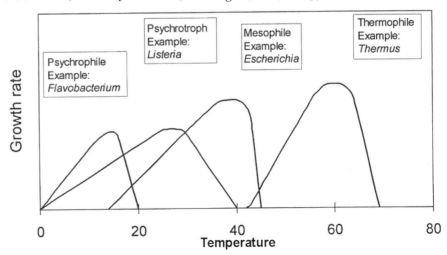

Both *psychrophiles* and *psychrotrophs* grow, albeit slowly, at 0°C. True psychrophiles have optimum growth rates at 15°C and cannot grow above 25°C. Psychrotrophs, such as *L. monocytogenes* and *C. botulinum* type E, have an optimum growth temperature of ~25°C and cannot grow at >40°C. Because these foodborne pathogens, and even some mesophilic *S. aureus* strains, can grow at <10°C, conventional refrigeration is inadequate to ensure the safety of a food.

The growth temperature influences a cell's thermal resistance. *L. monocytogenes* cells preheated at 48°C have increased heat resistance. Holding listeria cells at 48°C for 2 h in sausages doubles their heat resistance at 64°C. This thermotolerance is maintained for 24 h at 4°C. Subjecting *E. coli* O157:H7 to sublethal heating at 46°C increases its thermal resistance at 60°C by 50%. Shock proteins synthesized in response to one stress may provide cross-protection against other stresses. For example, cells exposed to acid may become more heat resistant.

Homeostasis and Hurdle Technology

Consumer demands have decreased the use of intrinsic factors, such as acidity and salt, as the sole means of inhibiting microbes. Many food products use multiple-hurdle technology to inhibit microbial growth. Instead of setting one environmental parameter to the extreme limit for growth, hurdle technology "deoptimizes" several factors. For example, limiting the amount of available water to a water activity (a_w, the equilibrium relative humidity of the product's atmosphere divided by 100) of <0.85 *or* a limiting pH of 4.6 prevents the growth of foodborne pathogens. Hurdle technology might obtain similar inhibition at pH 5.2 *and* an a_w of 0.92.

Hurdle technology assaults multiple homeostatic processes (Box 2.3). For example, the intracellular pH must be maintained within narrow limits. This is done by using energy to pump out protons, as described above. In low-a_w environments, cells must use energy to accumulate compatible solutes. Membrane fluidity must be maintained through homeoviscous adaptation, another energy-requiring process. The expenditure of energy to maintain homeostasis is fundamental to life. When cells channel the energy needed for growth into maintenance of homeostasis, their growth is inhibited. When the energy demands of homeostasis exceed the cell's energy-producing capacity, the cell dies.

Growth Kinetics

Growth curves showing the lag, exponential, stationary, and death phases of a culture are normally plotted as the number of cells on a log scale, or log_{10} cell number versus time. These graphs represent the state of microbial populations rather than individual microbes. Thus, both the lag phase and the stationary phase of growth represent periods when the growth rate equals the death rate to produce no net change in cell numbers. Food microbiology is the only area of microbiology where all four phases of the microbial growth curve are important. Microbial inhibitors can extend the lag phase, decrease the growth rate, decrease the size of the stationary phase population, and increase the death rate.

During the lag phase, cells adjust to their new environment by turning genes on or off, replicating their genes, and in the case of spores, differentiating into vegetative cells (see chapter 3). The lag phase duration depends on the temperature, the inoculum size (larger inocula usually have shorter

BOX 2.3

Hurdle theory and interactions of inhibitors

Food preservation can be thought of as a race between the human and the microbe to see who gets to eat first (see cartoon). Some of the preservative hurdles, such as pH 4.6 for *C. botulinum*, are just too big. They can stop a microbe when used alone. However, more often, combinations of preservatives are used. These may take so much of the microbe's energy that it slows down or even stops.

Food microbiologists agree that the interactions among preservatives are of utmost importance. These interactions are difficult to envision and hard to teach. The interaction of temperature, pH, and water activity on the growth of *L. monocytogenes* is illustrated more quantitatively in the graph. Under optimal conditions of 30°C, pH 6.5, and an a_w of 0.99 (plot a), the population grows rapidly to high levels. Dropping the pH to 5.5 (plot b) slows down the organism a little. Dropping the temperature by 15°C (plot c) has a greater, but still not very large, effect. When the pH and the temperature are both dropped (plot d), there is a much greater effect. At this point, the addition of a third hurdle ($a_w = 0.95$) (plot e) makes a radical improvement, and it takes the population 15 rather than 5 days to reach maximum size.

Courtesy of Donna Curran.

The data for this figure were taken from the U.S. Department of Agriculture pathogen-modeling program. This interactive program is wonderful for demonstrating interactions against a variety of foodborne pathogens, and it can be downloaded from http://www.ars.errc/mfs/PATHOGEN.htm.

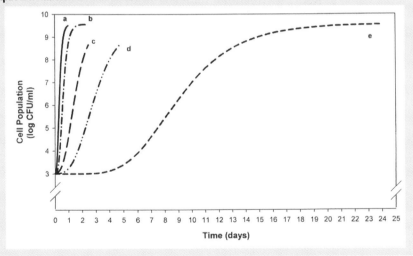

lag phases), and the physiological history of the organism. If actively growing cells are transferred into an identical fresh medium at the same temperature, the lag phase may vanish. These factors can also be manipulated to extend the lag phase to the point where some other quality attribute of the food (such as proteolysis or browning) becomes unacceptable. Foods are considered microbially safe if the food spoils before pathogens become an issue. However, "spoiled" is a subjective and culturally biased concept. It is safer to produce conditions that prevent cell growth regardless of time (such as reduction of the pH to <4.6 to inhibit botulinal growth).

Table 2.4 Representative specific growth rates and doubling times of microorganisms

Organism	Conditions	μ (h^{-1})	t_d (h)
Bacteria	Optimal	2.3	0.3
	Limited nutrients	0.20	3.46
	Psychrotroph, 5°C	0.023	30
Molds	Optimal	0.1–0.3	6.9–20

During the log, or exponential, phase of growth, bacteria reproduce by binary fission. One cell divides into two cells that divide into four cells that divide into eight cells, etc. Food microbiologists often use doubling times (t_d) as the constant to describe the rate of logarithmic growth. Doubling times, which are also referred to as generation times (t_{gen}), are inversely related to the specific growth rate (μ), as shown in Table 2.4.

Equations can be used to calculate the influence of different variables on a food's final microbial load. The number of organisms (N) at any time (t) is proportional to the initial number of organisms (N_0).

$$N = N_0 e^{\mu t} \tag{1}$$

Thus, decreasing the initial microbial load 10-fold will reduce the number of cells at any time by 10-fold. Because μ and t are in the power function of the equation, they have a greater effect on N than does N_0. Consider a food where N_0 is 1×10^4 CFU/g and μ is 0.2 h^{-1} at 37°C. After 24 h, N would be 1.2×10^6 CFU/g. Reducing the initial number 10-fold reduces the number after 24 h 10-fold to 1.2×10^5 CFU/g. However, reducing the temperature from 37 to 7°C has a more profound effect. If the growth rate decreases twofold with every 10°C decrease in temperature, then μ will be decreased eightfold to 0.025 h^{-1} at 7°C. When equation 1 is solved using these values (i.e., $N = 10^4 e^{0.025 \times 24}$), N at 24 h is 1.8×10^4 CFU/g. Both time and temperature have much greater influence over the final number of cells than does the initial microbial load.

How long it will take a microbial population to reach a certain level can be determined from the following equation:

$$2.3 \log(N/N_0) = \mu \Delta t \tag{2}$$

Consider the case of ground meat manufactured with an N_0 of 10^4 CFU/g. How long can it be held at 7°C before reaching a level of 10^8 CFU/g? According to equation 2, $t = [2.3\log(10^8/10^4)]/0.025$, or 368 h.

Food microbiologists frequently use doubling times to describe growth rates of foodborne microbes. The relationship between t_d and μ is more obvious if equation 2 is written using natural logarithms (i.e., $\ln[N/N_0] = \mu \Delta t$) and solved for the condition where $t = t_d$ and $N = 2N_0$. Since the natural logarithm of 2 is 0.693, the solution is

$$0.693/\mu = t_d \tag{3}$$

Some typical specific growth rates and doubling times are given in Table 2.4.

MICROBIAL PHYSIOLOGY AND METABOLISM

Think about your room. Its natural state is disordered. It takes energy to combat the disorder. This is as it should be—it is a law of the universe. (Tell that to your mother, roommate, or partner the next time she or he

*Authors' note*_____
We have written this book assuming no coursework in biochemistry, but microbes cannot be understood without understanding some biochemistry. Consider this section your minimal primer in biochemistry.

complains about your messy room.) The second law of thermodynamics dictates that *all* things go to a state of maximum disorder in the absence of energy input. Since life is a fundamentally ordered process, all living things must generate energy to maintain their ordered state. Foodborne bacteria do this by oxidizing reduced compounds. Oxidation occurs only when the oxidation of one compound is coupled to the reduction of another. In the case of aerobic bacteria, the initial carbon source, for example, glucose, is oxidized to carbon dioxide, oxygen is reduced to water, and 34 ATP molecules are generated. Most of the ATP is generated through oxidative phosphorylation in the electron transport chain. In *oxidative phosphorylation,* energy is generated (in the same sense that money is made to be spent later; think of ATP as the cell's energy currency) when oxygen is used as the terminal electron acceptor. This drives the formation of a high-energy bond between free phosphate and adenosine diphosphate (ADP) to form ATP. Anaerobic bacteria, which do not use oxygen, must use an internal organic compound as an electron acceptor in a process of fermentation. They generate only 1 or 2 moles (mol) of ATP per mol of glucose used. The ATP is formed by *substrate-level phosphorylation,* and the phosphate group is transferred from a phosphorylated organic compound to ADP to make ATP.

Carbon Flow and Substrate-Level Phosphorylation

The Embden-Meyerhoff-Parnas Pathway Forms Two ATP Molecules from Six-Carbon Sugars

The Embden-Meyerhoff-Parnas (EMP) pathway (Fig. 2.5) is the most common pathway for glucose catabolism (glycolysis). The overall rate of glycolysis is regulated by the activity of phosphofructokinase. This enzyme converts fructose-6-phosphate to fructose-1,6-bisphosphate. Phosphofructokinase activity is regulated by the binding of adenosine monophosphate (AMP) or ATP to inhibit or stimulate (respectively) the phosphorylation of fructose-6-phosphate. Fructose-1,6-bisphosphate activates lactate dehydrogenase (see below) so that the flow of carbon to pyruvate is tightly linked to nicotinamide adenine dinucleotide (NAD) regeneration. This occurs when pyruvate is reduced to lactic acid, the only product of this homolactic fermentation. Two ATP molecules are formed in this pathway.

The Entner-Doudoroff Pathway Allows the Use of Five-Carbon Sugars and Makes One ATP Molecule

The Entner-Doudoroff pathway is an alternate glycolytic pathway that yields one ATP molecule per molecule of glucose and diverts one three-carbon unit to biosynthesis. In aerobes that use this pathway, such as *Pseudomonas* species, the difference between forming one ATP molecule by this pathway versus the two ATP molecules formed by the EMP pathway is inconsequential compared to the 34 ATP molecules formed from oxidative phosphorylation. In the Entner-Doudoroff pathway, glucose is converted to 2-keto-3-deoxy-6-phosphogluconate. The enzyme keto-deoxy phosphogluconate aldolase cleaves this to one molecule of pyruvate (directly, without the generation of an ATP molecule) and one molecule of 3-phosphoglyceraldehyde. The 3-phosphoglyceraldehyde then follows the same reactions of the EMP pathway. One ATP molecule is made by substrate-level phosphorylation using phosphoenol pyruvate to donate the phosphoryl group.

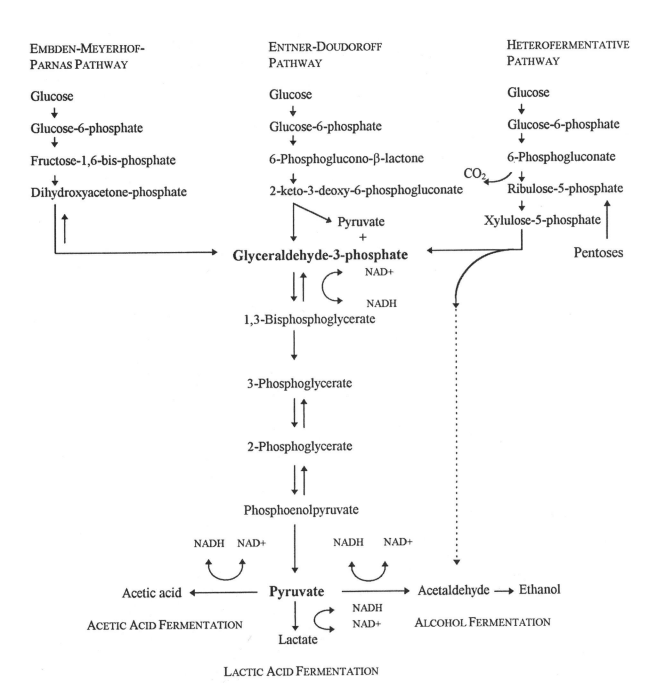

Figure 2.5 Major catabolic pathways used by foodborne bacteria. (Source: M. P. Doyle, L. R. Beuchat, and T. J. Montville [ed.], *Food Microbiology: Fundamentals and Frontiers*, 2nd ed. [ASM Press, Washington, D.C., 2001].)

Homolactic Catabolism Makes Only Lactic Acid

Homolactic bacteria in the genera *Lactococcus* and *Pediococcus*, and some *Lactobacillus* species, produce lactic acid as the sole fermentation product. The EMP pathway is used to produce pyruvate, which is then reduced by lactate dehydogenase, forming lactic acid and regenerating NAD.

Heterofermentative Catabolism Gives Several End Products

Heterofermentative bacteria, such as leuconostocs and some lactobacilli, have neither aldolases nor keto-deoxy phosphogluconate aldolase. The heterofermentative pathway is based on the catabolism of five-carbon sugars (pentoses). The pentose can be transported into the cell or made by decarboxylating hexoses. In either case, the pentose is converted to xylulose-5-phosphate, with ribulose-5-phosphate as an intermediate. The xylulose-5-phosphate is split into glyceraldehyde-3-phosphate and a two-carbon unit. The two-carbon unit can be converted to acetaldehyde, acetate, or ethanol. Although this pathway yields only one ATP molecule, it offers cells a competitive advantage by allowing them to utilize pentoses that homolactic organisms cannot use.

The Tricarboxylic Acid Cycle Links Glycolysis to Aerobic Respiration

The tricarboxylic acid (TCA) cycle links gylcolytic pathways to respiration. It generates reduced NAD (NADH) and reduced flavin adenine dinucleotide (FADH) as substrates for oxidative phosphorylation and makes some ATP through substrate level phosphorylation. With each turn of the TCA cycle, the reaction 2 pyruvate + 2 ADP + 2 FAD + 8 NAD → $6CO_2$ + 2 ATP + 2 FADH + 8 NADH occurs, and oxygen is used as the terminal electron acceptor to form water. The TCA cycle is used by all aerobes. Anaerobes may have some of the enzymes in the TCA cycle, but not enough to complete the full cycle.

The TCA cycle is also the basis for two industrial fermentations important to the food industry. The industrial fermentations for the acidulant citric acid and the flavor enhancer glutamic acid have similar biochemical bases. Both fermentations take advantage of impaired TCA cycles.

CONCLUSION

Microbial growth in foods is complex. It is governed by genetic, biochemical, and environmental factors. Developments in molecular biology and microbial ecology will change or deepen our perspective on the growth of microbes in foods. Some of these developments are detailed in this book. Other developments will unfold over the coming decades, perhaps accomplished by readers whose journey begins here.

Summary ─────────────────────────────────

- Food microbiology is a highly specialized subfield of microbiology.
- The amount of energy a microbe makes depends on the metabolic pathways it can use.
- The plate count determines how many of a specific organism can grow on a given medium under the incubation conditions used.
- Selective media allow low numbers of specific pathogens to be enumerated when they are present in a larger population of other bacteria.
- Enrichment allows very low numbers of specific pathogens to be detected in the presence of large numbers of other bacteria.
- The MPN method allows very low numbers of bacteria to be estimated from statistical tables.
- Injured cells may escape detection by culture methods but still cause illness when consumed.

- Viable but nonculturable cells are exactly that.
- Intrinsic factors, such as pH and water activity, and extrinsic factors, such as time and temperature, can be manipulated to control microbial growth.
- Hurdle technology challenges bacteria by several mechanisms to inhibit growth.
- The microbial growth curve consists of lag, log, stationary, and death phases of growth. Each phase is important to microbial food safety and can be manipulated.
- The equation $N = N_0 e^{\mu t}$ describes the exponential phase of microbial growth.
- Bacteria use different metabolic pathways to generate the energy they need to maintain an ordered state.

Suggested reading

Barer, M., and C. R. Harwood. 1999. Bacterial viability and culturability. *Adv. Microbiol. Physiol.* **41:**93–137.

International Commission on Microbiological Specifications for Foods. 1980. *Microbial Ecology of Foods,* vol. 1. *Factors Affecting Life and Death of Microorganisms.* Academic Press, New York, N.Y.

International Commission on Microbiological Specifications for Foods. 1980. *Microbial Ecology of Foods,* vol. 2. *Food Commodities.* Academic Press, New York, N.Y.

Kell, D. B., A. S. Kaprelyants, D. H. Weichart, C. R. Harwood, and M. R. Barer. 1998. Viability and activity in readily culturable bacteria: a review and discussion of the practical issues. *Antonie Leeuwenhoek* **73:**169–187.

Montville, T. J., and K. R. Matthews. 2001. Principles which influence microbial growth, survival, and death in foods, p. 13–32. *In* M. P. Doyle, L. R. Beuchat, and T. J. Montville (ed.), *Food Microbiology: Fundamentals and Frontiers,* 2nd ed. ASM Press, Washington, D.C.

Questions for critical thought

1. The plate count is based on the assumptions that one cell forms one colony and that every colony is derived from only one cell. For each assumption, list two situations that would violate the premise.
2. Given the data in Fig. 2.1, calculate the number of *S. aureus* organisms in the sample using the proper units.
3. Why *can't* you count, say, 3.2×10^4 CFU of *S. aureus*/ml in a population of 5.3×10^7 other bacteria using a standard plate count?
4. If the solid circles in the diagram below represent test tubes positive for growth, what would the MPN of the sample be?

0.1 ml	0.01 ml	0.001 ml	0.0001 ml	0.00001 ml
●●●	●●●	○○●	○○●	○○○

5. What is the difference between intrinsic and extrinsic factors in food? Name one intrinsic factor and one extrinsic factor that you think are very important for controlling microbes in food. Why do you think each is important? How does it work?
6. What three mechanisms do cells use to maintain pH homeostasis? Why do they need three? Loss of which mechanism would be worst for the cell? Why?

7. Consider equation 1. Your food product has an initial aerobic plate count of 3.8×10^4 CFU/g. After 1 week of refrigerated storage, the count reaches an unacceptable 3.8×10^6 CFU/g. You decide that an acceptable level of 3.8×10^5 CFU/g could be achieved if the initial number were lower. What starting level of bacteria would you need to achieve this goal? (Hint: put your calculator away and think about it.) In real life, how might a food manufacturer achieve this lower initial count?

8. Make a chart listing the three main catabolic pathways and, for each, the key regulatory enzyme, carbohydrate-splitting enzyme, amount of ATP generated, and final metabolic product.

9. Both anaerobic lactic acid bacteria and aerobic *Pseudomonas* spp. are naturally present in ground beef. In beef packed in oxygen-permeable cellophane, *Pseudomonas* is the main spoilage organism. In beef packed in oxygen-impermeable plastic film, lactic acid bacteria predominate. Why?

3 Spores and Their Significance*

LEARNING OBJECTIVES

The information in this chapter will help the student to:

- discuss how the modern canned food industry has developed to overcome the challenge of spore heat resistance
- define the traits of a low-acid canned food and explain how it must be processed
- identify, characterize, and differentiate among spore-forming bacteria that cause illness and spoilage
- understand the fundamental difference between a bacterial spore and a cell
- correlate the unique properties of spores with the challenge they present for food preservation
- draw a diagram of a spore and compare it to a diagram of a bacterial cell
- explain the physical and chemical basis for spore heat resistance
- compare the cycles of sporulation and germination with the gain and loss of spore resistance characteristics

INTRODUCTION

This chapter examines the spore. It is a unique life form of tremendous significance in food processing. Why spores are so important in the food industry and the steps used to control them are discussed first. Then, the structure and unique properties of the spore are described. Finally, the complex life cycle of a spore, which consists of *sporulation* (the conversion of a vegetative cell to a spore) and *germination* (the conversion of a spore to a vegetative cell), is presented.

SPORES IN THE FOOD INDUSTRY

Spore-forming bacteria and heat-resistant fungi pose specific problems for the food industry. Three species of sporeformers, *Clostridium botulinum*, *Clostridium perfringens*, and *Bacillus cereus*, are infamous for producing toxins. Other sporeformer species cause spoilage. Sporeformers causing foodborne illness and spoilage are particularly important in low-acid foods (equilibrium pH $\geq$ 4.6) packaged in cans, bottles, pouches, or other hermetically (vacuum) sealed containers ("canned" foods), which are processed by heat. Other sporeformers cause spoilage of high-acid foods (equilibrium pH < 4.6). Psychrotrophic sporeformers cause spoilage of refrigerated

---

*This chapter is based on the chapter written by Peter Setlow and Eric A. Johnson for *Food Microbiology: Fundamentals and Frontiers*, 2nd ed., and has been adapted for use in an introductory text.

foods. Fungi that produce heat-resistant ascospores cause spoilage of acidic foods and beverages.

During investigations of butyric acid fermentation in wines, Louis Pasteur discovered that spore-forming bacteria caused food spoilage. Pasteur isolated an organism he called *Vibrion butyrique,* which is probably what we now call *Clostridium butyricum*. Spores were discovered independently by Ferdinand Cohn and Robert Koch in 1876. Investigations by Pasteur and Koch linked microbial activity with food quality and safety. During investigation of the anthrax bacillus, *Bacillus anthracis,* the famous Koch's postulates were born. These proved that a disease is caused by a specific microorganism and led to the development of pure-culture techniques. Diseases and spoilage caused by sporeformers are associated with thermally processed foods, since heat kills vegetative cells but allows the survival and growth of spore-forming organisms.

In the late 1700s, Nicolas Appert invented the process of appertization. This process preserved food sterilized by heat in a hermetically sealed container. Appert believed that the elimination of air was responsible for the stability of canned foods. The empirical use of thermal processing gradually developed into modern-day thermal processing industries. In the late 1800s and early 1900s, scientists in the United States developed scientific principles to ensure the safety of thermally processed foods. As a result of these studies, thermal processing of foods in hermetically sealed containers became an important industry. S. C. Prescott and W. L. Underwood at the Massachusetts Institute of Technology and H. L. Russell at the University of Wisconsin found that spore-forming bacilli caused the spoilage of thermally processed clams, lobsters, and corn. The classic study by J. R. Esty and K. F. Meyer in California provided definitive values for the heat resistance of *C. botulinum* type A and B spores. This helped rescue the U.S. canning industry from its near demise due to botulism in commercially canned olives and other foods. Quantitative thermal processes, understanding of spore heat resistance, aseptic processing, and implementation of the Hazard Analysis Critical Control Point concept also grew out of the canned food industry.

Low-Acid Canned Foods

The U.S. Food and Drug Administration (FDA) and the U.S. Department of Agriculture (USDA) Food Safety Inspection Service define a low-acid canned food as one with a final pH of >4.6 and a water activity (a_w) of >0.85. The regulations regarding thermal processing of canned foods are described in the U.S. *Code of Federal Regulations* (21 CFR, parts 108 to 114). The USDA has regulatory control of foods that contain at least 3% raw red meat or 2% cooked poultry. The FDA regulates everything else.

Low-acid foods are packaged in hermetically sealed containers. These are usually cans or glass jars but can also be plastic pouches and other types of containers. These "cans" (as collectively defined in this chapter) containing low-acid food products must be processed by heat to achieve commercial sterility. *Commercial sterility* is a condition achieved by using heat that inactivates microorganisms of public health significance, as well as any other microorganisms that can grow in the food under normal nonrefrigerated conditions. Commercial sterility indicates a negligible level of microbial survival, as well as shelf stability. Preservation procedures, such as acidification or lowering a_w, can also be used to attain commercial sterility. These preservation techniques are often combined to reduce heat treatments. A description of the process for achieving commercial sterility and the facility, equipment, and formulations must be filed with the FDA or USDA prior to commercial processing. The processor must report process deviations or instances of spoilage.

In canned low-acid foods, the primary goal is to inactivate spores of *C. botulinum*, which has the highest heat resistance of microbial pathogens. The degree of heat treatment required depends on the class of food; its spore content, pH, and storage conditions; and other factors. For example, canned low-acid vegetables and uncured meats usually receive a 12D process (see below), or *botulinum cook*. Lesser heat treatments are applied to shelf-stable canned cured meats, as well as to foods with reduced a_w or other antimicrobial factors. Certain foods and ingredients, such as mushrooms, potatoes, spices, sugars, and starches, may contain high spore levels. The spore loads impact the process and must be monitored.

Researchers in the early and mid-1900s described thermal processes to prevent botulism and spoilage in canned foods. I. J. Pflug refined the semilogarithmic microbial-death model and designed a strategy to achieve the required heat process F_T value. (The F_T value, also referred to as F_o or F_R, is the equivalent process time in minutes at 121°C.) He also introduced the concept of the probability of a nonsterile unit on a one-container basis. This reasoning logically explains the traditional 12D term, the time required in a thermal process for a 12-log-unit reduction in *C. botulinum* spores. In thermal processing, two values are used to describe thermal inactivation. The D value is the time required for a 1-log-unit reduction in viability of a microbial population (a 1-log-unit reduction equates to killing 90% of the population). The z value is the temperature change required to change the D value by a factor of 10. The D value represents the resistance of an organism to a specific temperature. The z value represents the relative resistance of an organism to inactivation at different temperatures. The thermal process for the 10^{-11}-to-10^{-13} level of probability of a botulism incident occurring depends on the initial number of *C. botulinum* spores present in a container. This value can be quite high (for example, 10^4 for a container of mushrooms) or very low (such as 10^{-1} or less for a container of meat product). Pflug emphasized that the thermal-processing industry should prioritize process design to protect against (i) public health hazard (botulism) from *C. botulinum* spores, (ii) spoilage from mesophilic spore-forming organisms, and (iii) spoilage from thermophilic organisms in containers stored in warm environments. Generally, low-acid foods in hermetically sealed containers are heated to achieve 3 to 6 min at a temperature of 121°C (250°F) or equivalent at the coldest spot of the food. This ensures the inactivation of the most heat-resistant *C. botulinum* spores with a D value at 121°C ($D_{121°C}$) of 0.21 min and a z value of 10°C (18°F). Economic spoilage is avoided by achieving ~5D killing of mesophilic spores that typically have a $D_{121°C}$ of ~1 min. Foods distributed in warm climates require a particularly severe thermal treatment of ~20 min at 121°C to achieve a 5D killing of *Clostridium thermosaccharolyticum*, *Bacillus stearothermophilus*, and *Desulfotomaculum nigrificans*. These organisms have a $D_{121°C}$ of ~3 to 4 min. Such severe treatment decreases the nutrient content and sensory qualities, but it ensures a shelf-stable food.

Low-acid canned foods can receive aseptic heat treatments. In aseptic processing, (i) the product is commercially sterilized outside the container, (ii) presterilized containers are filled with the cooled product, and (iii) the containers receive an aseptic hermetic sealing in a sterile environment. This technology was initially used for commercial sterilization of milk and creams in the 1950s. It is now used for other foods, such as soups, eggnog, cheese spreads, sour cream dips, puddings, and high-acid products, such as fruit and vegetable drinks. Aseptic processing and packaging systems reduce energy, packaging, and distribution costs.

Bacteriology of Sporeformers of Public Health Significance

Three species of sporeformers, *C. botulinum*, *C. perfringens*, and *B. cereus*, cause foodborne illness and are covered extensively in other chapters. There have also been devastating incidences of intestinal anthrax caused by ingestion of contaminated raw or poorly cooked meat. However, *B. anthracis* is not considered a foodborne pathogen.

The principal microbial hazard in heat-processed foods and in minimally processed refrigerated foods is *C. botulinum*. The genus *Clostridium* consists of gram-positive, anaerobic, spore-forming bacilli that obtain energy by fermentation. The species *C. botulinum* is heterogeneous, but all botulinal strains produce a potent toxin. *C. botulinum* spores swell the mother sporangium, giving a "tennis racket" or club-shaped appearance (Fig. 3.1). The ability of some *C. botulinum* strains to grow at low temperatures has generated concern that refrigerated food products could cause botulism outbreaks.

C. perfringens is widespread in soils and in the intestinal tracts of humans and certain animals. *C. perfringens* can grow extremely rapidly in high-protein foods, such as meats that have been cooked to eliminate competitors and are inadequately cooled. This allows it to grow and to produce an enterotoxin that causes diarrheal disease. It also produces a variety of other extracellular toxins and enzymes, but these are mainly of significance in gas gangrene and diseases in animals. Its ubiquitous distribution in foods and food environments, its formation of resistant spores that survive cooking, and an extremely rapid growth rate in warm foods (6 to 9 min at 43 to 45°C) contribute to the ability of *C. perfringens* to cause foodborne illness.

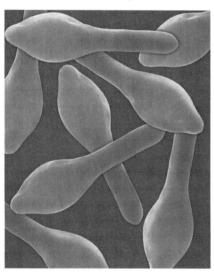

Figure 3.1 Scanning electron micrograph of *C. botulinum* vegetative cells swollen with terminal spores.

The heat resistance of *C. perfringens* spores varies among strains. In general, two classes of heat sensitivity are common. Heat-resistant spores have $D_{90°C}$ ($D_{194°F}$) values of 15 to 145 min and z values of 9 to 16°C (16 to 29°F), whereas heat-sensitive spores have $D_{90°C}$ values of 3 to 5 min and z values of 6 to 8°C (11 to 14°F). The spores of the heat-resistant class generally require a heat shock of 75 to 100°C (167 to 212°F) for 5 to 20 min in order to germinate. The spores of both classes may survive cooking of foods and may be stimulated by heat shock for germination during the heating procedures. Both classes cause diarrheal foodborne illness.

The genus *Bacillus* contains only two pathogenic species, *B. anthracis* and *B. cereus*. *B. cereus* can produce a heat-labile enterotoxin that causes diarrhea and a heat-stable toxin that causes emesis (vomiting) in humans. Generally, the organism must grow to very high numbers ($>10^6$/g of food) to cause human illness. *B. cereus* is closely related to *Bacillus megaterium*, *Bacillus thuringiensis*, and *B. anthracis*. *B. cereus* can be distinguished from these species by biochemical tests and the absence of toxin crystals. Other bacilli, including *Bacillus licheniformis*, *Bacillus subtilis*, and *Bacillus pumilis*, have caused foodborne disease outbreaks, primarily in the United Kingdom.

Spores of *B. cereus* are ellipsoidal and central to subterminal and do not distend the sporangium. Spore germination occurs over a range of 8 to 30°C (46.4 to 86°F). Spores from strains associated with food poisoning have a heat resistance of ~24 min ($D_{95°C\ [203°F]}$). Other strains have a wider range of heat resistances. The strains involved in food poisoning may have higher heat resistances and therefore be more apt to survive cooking.

Heat Resistance of *C. botulinum* Spores

Heat resistance of spores can vary greatly among species, and even among strains (Table 3.1). Group I *C. botulinum* type A and B strains produce extremely heat-resistant spores. They are the most important sporeformers for the public health safety of canned foods. Esty and Meyer in California carried out the classic investigation of their heat resistance because of commercial outbreaks of botulism in canned olives and certain other canned vegetables. They examined 109 type A and B strains at five heating temperatures over the range 100 to 120°C (212 to 248°F). They found that the inactivation rate is logarithmic between 100 and 120°C and that the inactivation rate depends on the spore concentration, the pH, and the heating medium. Esty and Meyer demonstrated that 0.15 M phosphate buffer (Sorensen's buffer), pH 7.0, gave the most consistent heat inactivation results, and their use of a standardized system allows comparisons of heat resistance by researchers today. The data of Esty and Meyer can be extrapolated to give a maximum $D_{121.1°C}$ value of 0.21 min for

Table 3.1   Heat resistance of sporeformers important in foods[a]

Organism	Approx *D* value (min) at 100°C
Spores of public health significance	
Group I *C. botulinum* types A and B	7–30
C. botulinum type E	0.01
B. cereus	3–200
C. perfringens	0.3–18
Mesophilic aerobes	
B. subtilis	7–70
B. licheniformis	13.5
B. megaterium	1
Bacillus polymyxa	0.1–0.5
Bacillus thermoacidurans	2–3
Thermophilic aerobes	
B. stearothermophilus	100–1,600
Bacillus coagulans	20–300
Mesophilic anaerobes	
Clostridium sporogenes	80–100
Thermophilic anaerobes	
D. nigrificans	≤480
C. thermosaccharolyticum	400

[a]Source: P. Setlow and E. A. Johnson, p. 33–70, *in* M. P. Doyle, L. R. Beuchat, and T. J. Montville (ed.), *Food Microbiology: Fundamentals and Frontiers,* 2nd ed. (ASM Press, Washington, D.C., 2001.)

C. botulinum type A and B spores in phosphate buffer. The canning industry uses $D_{121°C}$ as a standard in calculating process requirements. Proteolytic type F *C. botulinum* spores have a heat resistance of 12.2 to 23.2 min ($D_{98.9°C}$) and 1.45 to 1.82 min ($D_{110°C}$). This is much lower than that of type A spores. Spores of nonproteolytic type B and E *C. botulinum* have much lower heat resistance than proteolytic A and B strains. Type E spores have a $D_{70°C\ (158°F)}$ varying from 29 to 33 min and a $D_{80°C\ (176°F)}$ from 0.3 to 2 min, depending on the strain. The z value ranges from 13 to 15°F. The heat resistance of *C. botulinum* spores depends on environmental and recovery conditions. Heat resistance is markedly affected by acidity. Esty and Meyer found that spores have maximum resistance at pH 6.3 and 6.9, and resistance decreases markedly at pH values of <5 or >9. Increased levels of sodium chloride or sucrose and decreased a_w increase the heat resistance of *C. botulinum* spores. *C. botulinum* spores heated in oil are more resistant to heat than those heated in water. Sporulation of *C. botulinum* at higher temperatures results in spores with greater heat resistance, possibly through the formation of heat shock proteins.

    C. botulinum spores of groups I and II are more highly resistant to irradiation than most vegetative cells. It is probably not practical to inactivate them in foods by irradiation. *C. botulinum* spores have a *D* value of 2.0 to 4.5 kilograys. (Unfortunately the *D* values used in irradiation and thermal processing have different meanings. In thermal processing, it is the time at a given temperature required to reduce viability by 90%. In irradiation, it is the dose in kilograys required to reduce viability by 90%). *C. botulinum* spores are also highly resistant to ethylene oxide but are inactivated by halogen sanitizers and by hydrogen peroxide. Hydrogen peroxide is commonly used for sanitizing surfaces in aseptic packaging. Halogen sanitizers are used in cannery cooling waters.

Spoilage of Acid and Low-Acid Canned and Vacuum-Packaged Foods by Sporeformers

Thermally processed low-acid foods receive a heat treatment adequate to kill spores of *C. botulinum* but not sufficient to kill more heat-resistant spores. Acid and acidified foods with an equilibrium pH of ≤4.6 are not heated enough to inactivate all spores. Most species of sporeformers do not grow under acid conditions, and inactivation of all spores would decrease food quality and nutrition. Certain foods, such as cured meats and hams, are not heated enough to kill spores. They must be kept refrigerated for microbial stability. Nonpathogenic sporeformers cause spoilage in these foods.

    The spores naturally present in foods and the cannery environment contribute to spoilage problems. Dry ingredients, such as sugar, starches, flours, and spices, often contain high spore levels. Spore populations can also accumulate in a food-processing plant, such as thermophilic spores on heated equipment and saccharolytic clostridia in plants processing sugar-rich foods, such as fruits.

    The principal spoilage organisms and spoilage manifestations are presented in Table 3.2. The principal classes of sporeformers causing spoilage are thermophilic *flat-sour organisms* (organisms which sour the product without producing gas), thermophilic anaerobes not producing hydrogen sulfide, thermophilic anaerobes forming hydrogen sulfide, putrefactive anaerobes, facultative *Bacillus* mesophiles, butyric clostridia, lactobacilli, and heat-resistant molds and yeasts. *Alicyclobacillus acidoterrestris* and psychrotrophic clostridia can spoil fruit and meat products, respectively. These

Table 3.2 Spoilage of canned foods by sporeformers[a]

Type of spoilage or sporeformer	pH	Major sporeformer(s) responsible	Spoilage defects
Flat-sour	≥5.3	*Bacillus coagulans, B. stearothermophilus*	No gas; pH lowered. May have abnormal odor and cloudy liquor.
Thermophilic anaerobe	≥4.8	*C. thermosaccharolyticum*	The can swells and may burst. Anaerobe and products give sour, fermented, or butyric odor. Typical foods are spinach, corn.
Sulfide	≥5.3	*D. nigrificans, Clostridium bifermentans*	Hydrogen sulfide produced, giving rotten-egg odor. Iron sulfide precipitate gives blackened appearance. Typical foods are corn, peas.
Putrefactive anaerobe	≥4.8	*Clostridium sporogenes*	Plentiful gas. Disgusting putrid odor. pH often increased. Typical foods are corn, asparagus.
Psychrotrophic clostridia	≥4.6		Spoilage of vacuum-packaged chilled meats. Production of gas, off flavors and odors, discoloration.
Aerobic sporeformers	≥4.8	*Bacillus* spp.	Gas usually absent except for cured meats; milk is coagulated. Typical foods are milk, meats, beets.
Butyric	≥4.0	*C. butyricum, Clostridium tertium*	Gas; acetic and butyric odor. Typical foods are tomatoes, peas, olives, cucumbers.
Acid	≥4.2	*B. thermoacidurans*	Flat (*Bacillus*) or gas (butyric anaerobes). Off odors depend on organism. Common foods are tomatoes, tomato products, other fruits.
	<4	*A. acidoterrestris*	Flat spoilage with off flavors. Most common in fruit juices and acid vegetables and also reported to spoil iced tea.

[a]Source: P. Setlow and E. A. Johnson, p. 33–70, *in* M. P. Doyle, L. R. Beuchat, and T. J. Montville (ed.), *Food Microbiology: Fundamentals and Frontiers,* 2nd ed. (ASM Press, Washington, D.C., 2001).

organisms are controlled by monitoring raw foods entering the cannery to limit the initial spore load in a food product, thermal processing appropriate for subsequent storage and distribution conditions, cooling products rapidly, chlorination of cooling water, and good manufacturing practices.

Heat-resistant fungi cause spoilage of acidic foods, particularly fruit products. While heating for a few minutes at 60 to 75°C kills most filamentous fungi and yeasts, heat-resistant fungi produce thick-walled ascospores that survive heating at ≥85°C for 5 min. The most common heat-resistant spoilage fungi are *Byssochlamys, Neosartorya, Talaromyces,* and *Eupenicillium.* Certain heat-resistant fungi also produce toxic secondary metabolites called mycotoxins. To prevent spoilage of heat-treated foods, raw materials should be screened for heat-resistant fungi and strict good manufacturing practices and sanitation programs should be followed. Manipulation of the a_w and oxygen tension and the application of antimycotic agents can also prevent fungal growth.

SPORE BIOLOGY

Structure

The spore is released from the mother cell at the end of sporulation. It is biochemically, structurally, and physiologically different from the vegetative cell. The spore has seven layers: the exosporium, coat, outer membrane, cortex, germ cell wall, inner membrane, and core (Fig. 3.2). Many spore structures have no counterparts in the vegetative cell.

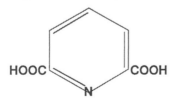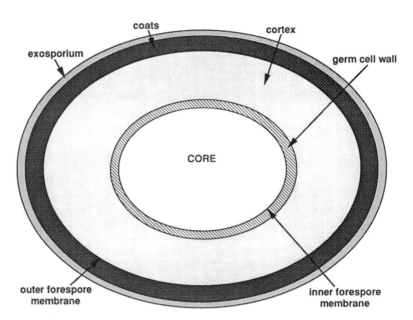

Figure 3.2 Structure of a dormant spore. The various structures are not drawn precisely to scale, especially the exosporium, whose size varies tremendously between spores of different species. The relative size of the germ cell wall is also generally smaller than shown. The positions of the inner and outer forespore membranes, between the core and the germ cell wall and between the cortex and coats, respectively, are also noted. (Source: P. Setlow and E. A. Johnson, p. 33–70, *in* M. P. Doyle, L. R. Beuchat, and T. J. Montville [ed.], *Food Microbiology: Fundamentals and Frontiers*, 2nd ed. [ASM Press, Washington, D.C., 2001].)

The outermost spore layer, the exosporium, varies in size between species. Underlying the exosporium are the spore coats. The spore coats protect the spore cortex from attack by lytic enzymes. Underlying the spore coats is the outer spore membrane. It helps keep the spore impermeable to small molecules.

The cortex is under the membrane. The peptidoglycan that makes up the cortex is structurally similar to cell wall peptidoglycan but with several differences. The spore peptidoglycan always contains pyridine-2, 6-dicarboxylic (dipicolinic acid [DPA]) which is not found in the vegetative-cell wall (Fig. 3.3). The spore cortex mechanically dehydrates the spore core and is responsible for much of the spore resistance (see below).

Between the cortex and the inner membrane is the germ cell wall. Its structure may be identical to that in vegetative cells. The next structure, the inner spore membrane, is a complete membrane. It is a very strong permeability barrier. This membrane's phospholipid content is similar to that of vegetative cells.

Finally, the spore core contains the spore's DNA, ribosomes, and most enzymes, as well as the deposits of DPA and divalent cations. There are many unique compounds in the dormant spore, including the large pool of SASP (small acid-soluble proteins), which constitutes 10 to 20% of the spore protein. Much of the SASP pool is bound to spore DNA. Spore cores have very low water content. While vegetative cells have ~4 g of water per g (dry weight), the spore core has only 0.4 to 1 g of water per g (dry weight). The core's low water content is responsible for spore dormancy and spore resistance. The spore cortex generates and maintains spore core dehydration, but the precise mechanism(s) involved is unknown.

Figure 3.3 Structure of DPA. Note that at physiological pH both carboxyl groups will be ionized. (Source: P. Setlow and E. A. Johnson, p. 33–70, *in* M. P. Doyle, L. R. Beuchat, and T. J. Montville [ed.], *Food Microbiology: Fundamentals and Frontiers*, 2nd ed. [ASM Press, Washington, D.C., 2001].)

Macromolecules

Spores are biochemically different from vegetative cells. Some proteins found in the spores are absent in the vegetative cell. The SASP play a major role in spore resistance. The binding of these proteins to DNA has striking effects on DNA properties, including providing resistance to chemical and enzymatic cleavage of the DNA backbone and altering the DNA's UV photochemistry. Not only do spores contain unique proteins, but many proteins present in vegetative cells are absent from spores. These include amino acid and nucleotide biosynthetic enzymes. The catabolic enzymes for using amino acid and carbohydrate are present in spores and cells. Spore DNA appears to be identical to cell DNA.

Small Molecules

The spore's small molecules, located in the cortex, are different from the cell's small molecules. The ions in the spore core are immobile, since there is no free water. The pH in the spore core is 1 to 1.5 units lower than that in a growing cell. In contrast to growing cells, spores have little, if any, of the common high-energy compounds found in cells, including deoxynucleoside triphosphates, ribonucleoside triphosphates, reduced pyridine nucleotides, and acyl coenzyme A (acyl-CoA) (Table 3.3).

Table 3.3 Small molecules in cells and spores of *Bacillus* species

Molecule	Content (mmol/g [dry weight]) in:	
	Cells[a]	Spores[b]
ATP	3.6	≤0.005
ADP	1	0.2
AMP	1	1.2–1.3
Deoxynucleotides	0.59[c]	<0.025[d]
NADH	0.35	<0.002[e]
NAD	1.95	0.11[e]
NADPH	0.52	<0.001[e]
NADP	0.44	0.018[e]
Acyl-CoA	0.6	<0.01[e]
CoASH[f]	0.7	0.26[e]
CoASSX[g]	<0.1	0.54[e]
3PGA[h]	<0.2	5–18
Glutamic acid	38	24–30
DPA	<0.1	410–470
Ca^{2+}		380–916
Mg^{2+}		86–120
Mn^{2+}		27–56
H^+	7.6–8.1[i]	6.3–6.9[h]

[a]Values for *B. megaterium* in mid-log phase.
[b]Values are the range from spores of *B. cereus*, *B. subtilis*, and *B. megaterium*.
[c]Value is the total of all four deoxynucleoside triphosphates.
[d]Value is the sum of all four deoxynucleotides.
[e]Values are for *B. megaterium* only.
[f]CoASH, free CoA.
[g]CoASSX, CoA in disulfide linkage to CoA or a protein.
[h]3PGA, 3-phosphoglyceric acid.
[i]Values are expressed as pH and are the range with *B. cereus*, *B. megaterium*, and *B. subtilis*.

Jurassic spores

Just how long can spores be dormant and still be viable? Think in geological time. The earth was formed ~6 billion years ago. Bacteria came on the scene ~3 billion years ago. Dinosaurs went extinct 250 million years ago. "Humans" are ~6 six million years old. Bacteria as a class clearly have staying power. Individual organisms also have staying power. Raúl Cano, a microbiologist at California Polytechnic University, resurrected spores that had been dormant for 25 million to 40 million years. *Bacillus* species have a symbiotic relationship with bees, which are frequently embedded in amber. Spores had been seen in these bees by using electron microscopes, and spore DNA had been isolated, but no one had ever determined whether the spores were still alive. Cano rigorously sterilized amber samples and all of the experimental material before pulverizing it in broth in which spores could normally grow. Grow they did! The spores produced bacilli. Analysis of their DNA showed that they were ancient and unrelated to any modern bacilli that could have been potential contaminants.

Cano, R. J., and M. K. Borucki. 1995. Revival and identification of bacterial spores in 24–40 million year old Dominican amber. *Science* **228:**1060–1064.

Dormancy

Spores are metabolically dormant. They have no detectable metabolism. The major cause of the spore's metabolic dormancy is undoubtedly its low water content, which precludes enzyme action. Dormancy is further demonstrated by enzyme-substrate pairs in the core, which are stable for months to years but which are degraded in the first 15 to 30 min of germination.

Resistance

The spore's dormancy helps it survive for extremely long periods in the absence of nutrients (Box 3.1). A second factor in long-term spore survival is the spore's extreme resistance to heat, radiation, chemicals, and desiccation. Spores are much more resistant than vegetative cells. Spore resistance is due to a variety of factors, such as spore core dehydration, SASP, and impermeability. Since different factors contribute to different types of spore resistance, it is not surprising that different types of spore resistance are gained at different times in sporulation. However, in recent years, the mechanisms of spore heat, ultraviolet (UV) light, and H_2O_2 resistance have been determined. The following discussion of spore resistance concentrates on *B. subtilis* because of the detailed mechanistic data available for the organism. However, factors involved in the resistance of *B. subtilis* spores are also involved in the resistance of other species.

Freezing and Desiccation Resistance

Some growing bacteria are killed during freezing, and more killing occurs during drying. The precise mechanism(s) of killing is not clear, but one cause may be DNA damage; freeze-drying cells can cause significant mutagenesis. Spores are resistant to multiple cycles of freeze-drying. A complete explanation for spore drying resistance is not yet available. However, the SASP contribute to spore drying resistance by preventing DNA damage.

Pressure Resistance

Spores are much more resistant to high pressures ($\geq$12,000 atmospheres) than are cells. Although spores are more resistant than cells to lower pressures, spores are killed more rapidly at lower pressures than at higher pressures. This apparent anomaly occurs because lower pressures promote spore germination; the germinated spores are then rapidly killed by the pressure treatment. Spore germination is not promoted by high pressures.

γ-Radiation Resistance

Spores are generally more resistant to γ-radiation than are vegetative cells. In the few organisms studied, γ-radiation resistance is gained 1 to 2 h before the acquisition of heat resistance. The precise factors involved in spore γ-radiation resistance are not known, although SASP are not involved. The low water content in the spore core may provide protection against γ-radiation. However, the amount of water in the spore core has not been correlated with spore γ-radiation resistance. Radiation presumably damages the spore DNA, but through an unknown mechanism.

UV Radiation Resistance

Spores are 7 to 50 times more resistant to UV than are vegetative cells. UV resistance is acquired 2 h before spores become heat resistant, in parallel with the synthesis of SASP. These proteins are essential for spore UV resistance. Core dehydration is not necessary for spore UV resistance.

Chemical Resistance

Spores are much more resistant than cells to many chemicals, pH extremes, and lytic enzymes, such as lysozyme. The resistance of spores to chemicals is acquired at different times in sporulation. For some compounds, spore coats play a role in chemical resistance, possibly by providing an initial barrier against attack. This is clearly true for lytic enzymes, as spores with coats removed are sensitive to lysozyme. The inability of most molecules to penetrate the spore core plays an important role in chemical resistance.

Heat Resistance

Heat resistance has huge implications for the food industry. It is probably the best-studied form of resistance. Spore heat resistance is truly remarkable; many spores withstand temperatures of 100°C for several minutes. Heat resistance is often quantified as a D_T value. This is the time in minutes at temperature D_T needed to kill 90% of a population. Generally, D values for spores at a temperature of $D_T + 40$°C are about equal to those for their vegetative-cell counterparts at temperature D_T. An often-overlooked feature of spore heat resistance is that the extended survival of spores at elevated temperatures is paralleled by even longer survival times at lower temperatures. Spore D values increase 4- to 10-fold for each 10°C decrease in temperature. Consequently, a spore with a D value at 90°C ($D_{90°C}$) of 30 min may have a $D_{20°C}$ value of many years.

We do not know what target(s) is damaged by heat to kill the spores. The target is probably not spore DNA. Spore heat killing is not associated with DNA damage or mutagenesis. Protein(s) may be the target of spore heat killing. Sublethal heat treatment can also injure spores. This damage can be repaired during spore germination and outgrowth. Several factors that modulate spore heat resistance are discussed below.

Sporulation Temperature

Elevated sporulation temperatures increase spore heat resistance. Indeed, spores of thermophiles are generally more heat resistant than spores of mesophiles. Since spore macromolecules are identical to cell macromolecules, spore macromolecules are not intrinsically heat resistant. Presumably, the total macromolecular content of spores from thermophiles is more heat stable than that from mesophiles, accounting for the higher heat resistance. However, spores of the same strain prepared at various temperatures are most heat resistant when prepared at the highest temperature.

In growing bacteria, adaptation to heat stress involves the proteins of the heat shock response. The levels of these proteins increase with increasing sporulation temperature. Although heat shock proteins could play some role in spore heat resistance, this appears not to be the case.

α/β-Type SASP

One striking finding about the killing of spores by wet heat is that neither general mutagenesis nor DNA damage occurs. This is despite the fact that the elevated temperatures would be expected to cause DNA damage. Therefore, spore DNA must be well protected against heat damage, and the thermal inactivation of spores must be due to some other mechanisms. The major cause of spore DNA protection against heat damage appears to be the saturation of spore DNA by α/β-type SASP. Consequently, $\alpha^- \beta^-$ spores of *B. subtilis* are more heat sensitive.

Dry heat affects spores differently than does heating in water. First, spores are much more resistant to dry heat than to aqueous heat. D values are 100- to 1,000-fold higher in dry versus hydrated spores. Second, spores have a high level of mutagenesis (~12% of survivors) when killed by dry heat. This mutagenesis damages spore DNA.

Spore Core Water Content and Heat Resistance

Low core water content is a major factor in spore heat resistance. The spore becomes more heat resistant as it is dehydrated. The spore cortex is essential for creating and maintaining the dehydrated state of the spore core. This is undoubtedly due to the cortex's ability to change its volume upon changes in ionic strength and/or pH. If an expansion in cortex volume were restricted to one direction, i.e., toward the spore core, mechanical action in the opposite direction would express core water.

There is good correlation between spore core water content and heat resistance over a 20-fold range of D values. However, at the extremes of core water contents, D values vary widely, presumably reflecting the importance of other factors, such as sporulation temperature and cortex structure. There is little, if any, free water in the core. Presumably, water-driven chemical reactions are inhibited by the spore core's low water content. This also contributes to heat resistance. Low water content stabilizes macromolecules, such as proteins, against denaturation by restricting their molecular motion.

THE CYCLE OF SPORULATION AND GERMINATION

Sporulation

Bacillus and *Clostridium* spp. make spores in response to environmental stress or nutrient depletion. The molecular biologies of sporulation and spore resistance have been extensively studied only in the genus *Bacillus*. Spores formed by *Alicyclobacillus*, *Bacillus*, *Clostridium*, *Desulfotomaculum*, and *Sporolactobacillus* spp. also pose problems in foods.

An unequal cell division is the first event in sporulating cells. This creates the smaller spore compartment and the larger mother cell compartment. As sporulation proceeds, the mother cell engulfs the forespore, resulting in a cell (the forespore) within a cell (the mother cell), both of which have a complete genome. Since the spore is formed within the mother cell, it is called an endospore.

Genes are expressed at specific times and places during sporulation. Some genes are expressed only in the mother cell, and others are expressed only in the spore. Gene expression is controlled by the ordered synthesis and activation of new sigma (specificity) factors for RNA polymerase. Many DNA binding proteins, both repressors and activators, also regulate gene expression during sporulation. As sporulation proceeds, there are striking morphological and biochemical changes in the developing spore. It becomes encased in two novel layers, a peptidoglycan layer (the spore cortex) and spore coats. The spore also accumulates a huge store ($\geq$10% of its dry weight) of DPA (Fig. 3.3), found only in spores, and a large number of divalent cations. The spore becomes both metabolically dormant and extremely resistant to heat, radiation, and chemicals.

Activation

Dormant spores can rapidly return to active metabolism by germinating (Fig. 3.4). Spores germinate more rapidly and completely if activated prior to exposure to a nutrient that induces germination (i.e., a germinant). There are many ways to activate spores. The most widely used agent is sublethal heat (e.g., 10 min at 80°C). The precise changes induced by spore activation are not clear. In some species, the activation is reversible. In most species, heat activation releases a small amount of the spore's DPA.

Germination

Spores lose most of their unique characteristics within minutes of encountering a germinant. They lose DPA and SASP. The spore's resistance is also lost in the first minutes of germination. During this time, active metabolism begins and synthesis of large molecules starts. Eventually, the germinated spore is converted back to a growing vegetative cell through the process of outgrowth.

Spore germination occurs during the first 20 to 30 min after mixing of spores and germinant. During this period, a resistant dormant spore with a cortex and a large pool of DPA, minerals, and SASP is transformed into a sensitive, actively metabolizing germinated spore in which the cortex and SASP have been degraded. These changes can occur in the absence of nutrients. However, further conversion of the germinated spore into a growing cell requires added nutrients.

*Authors' note*_____
If one enumerated a population of 850,000 spores per ml on agar medium, the plate count might be as low as 75,000. If the spores are heated for 10 minutes at 80°C (heat shocked), the count might increase to 775,000.

Figure 3.4   Cycle of sporulation, dormancy, activation, and outgrowth.

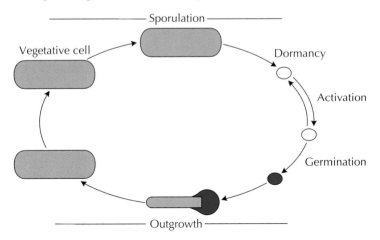

Spore germination can be triggered by a wide variety of compounds, including amino acids, sugars, salts, and DPA. How these compounds trigger spore germination is not clear. Metabolism of the germinant is not required. Indeed, inorganic salts germinate some spores. The stereospecificity exhibited by germinants (e.g., L-alanine is a germinant, whereas D-alanine often inhibits germination) suggests that some germinants interact directly with specific proteins.

In the first minutes of spore germination, protons and some divalent cations, Zn^{2+} in particular, are released. Release of DPA, loss of spore refractility, and cortex degradation follow. During germination, the spore excretes up to 30% of its dry weight, and the core increases its water content to that of the vegetative cell. These events clearly require large changes in the permeability of the inner spore membrane. However, neither the nature nor the mechanism of these changes is understood. The changes accompanying the initiation of spore germination may be extremely fast. An individual spore can lose refractility in as little as 30 s to 2 min. However, for a spore population, the time can be much longer. Individual spores initiate germination after widely different lag times.

RNA synthesis is initiated in the first few minutes of germination, using nucleotides stored in the spore or generated by the breakdown of preexisting spore RNA. Protein synthesis begins shortly after RNA synthesis. Amino acids produced from SASP breakdown support most protein synthesis in the first 20 to 30 min of germination.

Outgrowth

The transition from spore germination to spore outgrowth is not distinct. It is the time from ~25 min after the initiation of germination until the first cell division. Most processes during germination are supported from stored reserves. Outgrowth requires outside resources. During outgrowth, the spore regains the ability to synthesize amino acids, nucleotides, and other small molecules.

DNA replication is not generally initiated until at least 60 min after the start of germination. DNA repair can occur well before DNA replicative synthesis. Even in the first minute of germination, spores contain deoxynucleoside triphosphates. During spore outgrowth, the volume of the outgrowing spore continues to increase, requiring the synthesis of membrane and cell wall components.

Summary

- Spores are unique life forms that are resistant to many stresses.
- The heat resistance of spores in food is measured as their D value: the number of minutes at a given temperature required to kill 90% of the spores. The z value is the number of degrees that it takes to change the D value by a factor of 10.
- Low-acid canned foods are those with a pH of >4.6 and an a_w value of >0.85.
- Commercial sterility is achieved through the application of a 12D botulinum cook.
- Many sporeformers cause economic spoilage.
- Sporulation is the process by which a vegetative cell produces a spore.

- Germination and outgrowth enable the spore to resume life as a vegetative cell.
- The spore cortex is a unique form of peptidoglycan that contributes to spore resistance.
- DPA and SASP contribute to spore resistance.
- The thermal resistance of a spore crop is influenced by many factors, including its sporulation temperature.
- The relatively low water content of the spore core helps make it resistant.
- Sporulation, germination, and outgrowth are parts of the spore's life cycle.

Suggested reading

Driks, A. 2002. Overview: development in bacteria: spore formation in *Bacillus subtilis. Cell. Mol. Life Sci.* **59:**389–391.

Nicholson, W. L., P. Fajardo-Cavazo, R. Rebeil, T. A. Slieman, P. J. Riesenman, J. F. Law, and Y. Xue. 2002. Bacterial endospores and their significance in stress resistance. *Antonie Leeuwenhoek* **81:**27–32.

Setlow, P., and E. A. Johnson. 2001. Spores and their significance, p. 33–70. *In* M. P. Doyle, L. R. Beuchat, and T. J. Montville (ed.), *Food Microbiology: Fundamentals and Frontiers,* 2nd ed. ASM Press, Washington, D.C.

Questions for critical thought

1. You may remember that low-acid canned foods must receive a thermal process equivalent to an F_o of 2.4 min. This is based on a 12D reduction in *C. botulinum* spores, which have a $D_{250°F}$ of 0.2 min. Imagine two process deviations: in deviation A, the time is 10% less than it should be (i.e., the product receives 2.16 min at 250°F), while in deviation B, the temperature is 10% too low (i.e., the product receives 2.40 min at 225°F). Which process deviation represents a greater threat to public safety? Be quantitative in your answer (i.e., calculate the log reduction in *C. botulinum* spores that results from each deviation, assuming a z value of 18°F). Show all of your work.

2. Define z value. If an organism has a $D_{80°C}$ of 20 min, a $D_{95°C}$ of 2 min, and a $D_{110°C}$ of 0.2 min, what is its z value? (Hint: if you understand the concept, you won't need a calculator.)

3. If you had to choose, would you rather be a spore or a vegetative cell? Why?

4. What are two components found in spores but not vegetative cells? What are their functions?

5. Is it acceptable for cans of tomatoes to contain viable *C. botulinum* spores? Why or why not?

6. Explain in your own words what "gene expression is regulated in time and space" means.

7. Discuss how spore water content influences heat resistance. How does this integrate with other things you know about thermal lethality (i.e., wet versus dry heat, heat resistance at low a_w, etc.)?

4

Detection and Enumeration of Microbes in Food

LEARNING OBJECTIVES

The information in this chapter will enable the student to:
- discuss methods available for microbiological analysis of food
- compare methods of analysis, indicating advantages and disadvantages of each method
- detail procedures for collection and processing of food samples
- calculate the microbial load of a sample
- recognize the difference between conventional and rapid microbiological methods

---

INTRODUCTION

Microorganisms inhabit our bodies and our environment. Therefore, it is not surprising to find many types of microorganisms (yeasts, molds, viruses, bacteria, and protozoa) in food. Often, food microbiologists refer to microorganisms in food as "the good, the bad, and the ugly." Good microorganisms are those that are used in the production of food (e.g., yogurt, wine, and beer), bad microorganisms cause spoilage, and ugly microorganisms cause human illness. Determining the types and numbers of microorganisms in a food is an important aspect of food microbiology.

The *aerobic plate count* (APC) provides an estimation of the number of microorganisms in a food. As the name implies, strict anaerobes will not grow on the plate and therefore are not included in the APC. The APC changes—increases or decreases—during the processing, handling, and storage of foods. In foods that are held raw, such as refrigerated meats, the APC will increase during storage, whereas in dried or frozen foods, the APC will remain unchanged or decrease. Depending on the food, the APC can be as low as 10 to as high as 1,000,000 microorganisms per g.

The levels of microorganisms in products from the same origin can differ greatly. For instances the microbial load of ground meat is much greater than that of whole cuts of meat. During the handling of meat pieces, grinding, and packaging, bacteria may multiply or bacteria on equipment can be transferred to the product. The starting number of microorganisms associated with a food will have a significant impact on the number of microorganisms present in the finished product, even for foods that receive heat treatment. Poor-quality ingredients, poor sanitation, recontamination, and improper handling can result in high levels of microorganisms even in foods that were properly heat treated.

45

In order to determine if a product meets the microbial levels prescribed in specifications, guidelines, or standards, an estimation of the number of microorganisms in or on a food is needed. A rule of thumb is that as the microbial count increases, the quality of the food decreases. Of course there are exceptions to this rule; for instance, in fermented foods the level of microorganisms would be expected to increase. This chapter will focus on some of the methods for determining the microbial levels of foods. Note that for the most part microbiological analysis of food samples refers to determining the levels of bacteria.

SAMPLE COLLECTION AND PROCESSING

The methods used for sample collection and processing vary from food to food and for specific microorganisms. The Food and Drug Administration outlines the methods that they use in the *Bacteriological Analytical Manual,* and the Food Safety Inspection Service of the U.S. Department of Agriculture publishes the *Microbiology Laboratory Guidebook.* Regardless of the protocol used, a sample will only yield significant and meaningful information if it represents the mass of material being examined, if the method of collection used protects against microbial contamination, and if the sample is handled in a manner that prevents changes in microbial numbers between collection and analysis.

Care must be taken during sample collection to prevent the introduction of microbes into the food sample. If possible, individual containers of food should be submitted for analysis. However, this is not always practical, and therefore a representative sample must be collected using an aseptic technique. Instruments used for sample collection should be sterilized in the laboratory rather than at the place of sampling. Sterile containers (plastic bags or wide-mouth jars) should be labeled with the name of the food, the date collected, and other information that may be useful in the analysis of results and tracking of the sample.

Once the sample is collected, it should be analyzed as quickly as possible to prevent a change in the microbial population. However, this is not always possible, and in that case, the sample should be refrigerated and frozen samples should be kept frozen. Products that are normally refrigerated should not be frozen, since freezing may cause death or damage to some bacterial cells, producing incorrect results.

Prior to analysis, some preparation of the sample is generally required (Fig. 4.1). For meaningful results, the sample should be processed to produce a homogeneous suspension of bacteria so they can be pipetted. If the sample is solid, the food is generally mixed with a sterile diluent, such as Butterfield's buffered phosphate or 0.1% peptone water. If an appropriate buffer is not used, the bacteria may multiply in the dilution medium prior to being plated. To make a homogeneous suspension, the sample is added to a sterile plastic stomacher bag, diluent is added, and the sample is processed in a *stomacher* (a device that has two paddles that move rapidly back and forth against the sample). A typical-size food sample of 25 g is added to the sterile bag containing 225 ml of sterile diluent and mixed (placed in the stomacher), and a 1:10 dilution of the food and associated bacteria is obtained.

ANALYSIS

The APC, discussed above, provides a good estimate of the number of microorganisms associated with a sample. Drawbacks of surface plating are the *coalescence* of colonies (two or more colonies growing together) and the

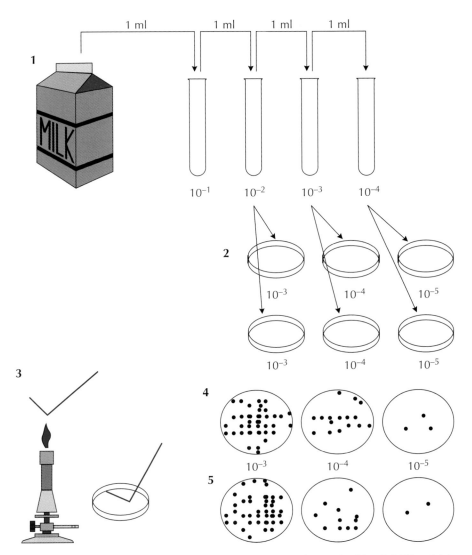

Figure 4.1 Preparation of sample for determination of microbial load. Milk, which initially contains millions of microbes, is diluted, and samples are spread plated so that individual colonies can be counted. Appropriate calculations are then made to determine the actual number of microbes in the milk. Step 1: since a liquid sample is being tested, a 1-ml aliquot can be removed and added to 9 ml of buffer for direct preparation of serial 10-fold dilutions. Step 2: 100 μl of each dilution is transferred to replicate plates. Step 3: a glass "hockey stick" is flame sterilized, allowed to cool, and used to spread the sample on the surface of the agar; the plate should be rotated to ensure distribution across the entire surface. Step 4: the plates are incubated under specified conditions. Step 5: dilutions yielding 30 to 300 colonies per plate are counted and expressed as numbers of CFU per milliliter of milk.

growth of spreaders. In the APC, the hardened agar surfaces must be dried thoroughly to prevent the spread of bacteria by moisture droplets. A 0.1-ml aliquot of the selected dilution of the sample is dispensed onto the surface of the agar and spread with the aid of "hockey sticks," or bent glass rods. For the standard plate count (SPC; also called a pour plate), a measured aliquot (1 ml) of the appropriate dilution is dispensed onto the surface of a sterile petri plate and sterile, melted, tempered (45°C) agar is added to the plate. The agar and sample are mixed, and the agar is allowed to solidify. The agar plate is then incubated, and isolated colonies known as colony-forming units (CFU) are counted. This method has many disadvantages, including the time required for setup, expense, and lack of accuracy.

A CFU in theory arises from a single bacterial cell; however, this may not always hold true. In all likelihood, colonies arise from clusters or chains of bacteria, resulting in an underestimation of the number of bacteria in a sample. Therefore, samples may need to be processed to separate the cells into individual reproductive units. This is achieved through homogenization of a sample and by making a series of 1:10 dilutions (Fig. 4.1). A 1:10 dilution of a 1:10 dilution is a 1:100 dilution. For samples with unknown microbial populations, an extensive (10^{-1} to 10^{-7}) series of dilutions may need to be made and plated. In general, standard 100-mm-diameter plates should contain between 30 and 300 colonies. Fewer or more colonies per plate can affect accuracy. The total number of bacteria is determined by multiplying the colony count by the dilution factor to yield the number of bacteria per gram of food.

The *roll tube* method is basically the same as the pour plate method, but screw-cap tubes are used in place of petri plates. Test tubes containing 2 to 4 ml of plate count agar or other appropriate medium are sterilized, and the melted agar is allowed to cool to 45°C (tempered). A 0.1-ml aliquot of the appropriate dilution of the sample is added, and the tube is immersed in cold water in a horizontal position and rolled. The agar will solidify, forming a thin layer on the inner wall of the tube. The tubes are then incubated upside down to prevent any condensation that may form from causing smearing of the colonies. A decided advantage of this method is the saving associated with the use of less plate count agar.

Mechanical devices are available to facilitate the determination of bacterial numbers in food samples. The *spiral plater* uses a robotic arm to dispense a sample in an Archimedes spiral onto the surface of a rotating agar plate. A continuously decreasing volume of the inoculum is deposited from near the center of the plate to the outside of the plate so that a concentration range of up to 10,000:1 is achieved on a single plate. The spiral plater compares favorably with the SPC. A specialized counting grid is used to count colonies on a plate, although the colonies are tightly packed together and may be difficult to count. The method is best suited for liquid samples, such as milk, since food particles may cause blocking of the dispenser. As with any method, there are advantages and disadvantages. The advantages include the use of fewer plates, fewer dilution bottles, and less agar and the ability to process a greater number of samples. The disadvantages include limitation of the types of samples that can be processed and difficulty in counting colonies.

Fluid samples (water or diluent) can be passed through membrane filters to collect bacteria. Membranes with a pore size of 0.45 μm are typically used. The filter can either be placed onto a plate containing a culture medium of choice or used in the direct microscopic count (DMC) assay. This *membrane filter technique* is useful for examination of water or other dilute liquid samples that can easily pass through the filter. Therefore, even low numbers of microbes can be detected, since large volumes (1 liter or more) can be passed through a single filter. The method can also be used to determine microbial numbers in air. Small amounts of viscous liquid samples, such as milk or juice, can be processed without clogging the membrane.

A number of commercial methods based on the membrane filter technique are available. Some of these methods are discussed in chapter 5. The *direct epifluorescent filter technique* uses fluorescent dyes to stain bacteria that are then counted using a fluorescence microscope. The number of cells per gram or milliliter is calculated by multiplying the average number per field by the microscope factor. This method has been used for estimating numbers of bacteria in milk, meat, and poultry products and on food contact surfaces. The *hydrophobic grid membrane filter* method is also used to estimate microbial numbers associated with a variety of foods. For this method, specially designed filters containing 1,600 grids that restrict microbial growth and colony size are used. Generally, 1 ml of a 1:10 sample homogenate is passed through the filter. Grids that contain colonies are counted, and a most-probable number is determined.

The DMC is faster than most methods, since no incubation period is required for cells to metabolize and grow (multiply). With this method, a smear of a food sample or a liquid sample (0.01 ml) is uniformly spread over a 1-cm^2 area on a microscope slide. Generally, liquid foods can be analyzed directly, but solid foods must be put into a suspension (1:10 dilution) before analysis. Once the sample is fixed, defatted, and stained, the cells are counted. A calibrated microscope must be used, since the diameter of the field to be examined must be known in order to calculate the number of organisms per gram. A stage micrometer can used to measure the diameter of the circle (on the slide described above) to the nearest 0.001 mm. The average number of cells or clumps is calculated on a per-field basis and divided by the area of the field to determine the number per square millimeter.

The DMC is perhaps the simplest and most rapid method for estimating bacterial numbers in a sample. Therefore, adjustments or changes in processing can be implemented immediately to correct any problems. The test has other advantages, including little processing of samples, limited equipment required, a slide(s) that can be maintained as a record, and the need for only a small amount of sample. The disadvantages include fatigue (eyes becoming tired from peering through a microscope) and detection of live and dead cells without the ability to distinguish live cells from dead ones, and the method has little or no value for foods with low microbial numbers.

Fluorescent dyes are now available that can be used to differentially stain live and dead bacteria. Using such stains, the total bacterial count and a live-dead count can be obtained. Since the stains are intensely fluorescent, they can be used in samples that may contain low numbers of bacteria. The disadvantages are that the fluorescent signal fades over time and food constituents may also stain or fluoresce, interfering with the ability to count the bacterial cells.

Metabolism-Based Methods

In general microbiology, metabolism of microorganism is used to determine starch hydrolysis, sugar fermentation, production of hydrogen sulfide and indole, or nitrate reduction. Measurements of microbial metabolism or production of metabolic products in foods can be used to estimate the bacterial population. Today many of these tests have been miniaturized and are the bases of a number of rapid test methods.

Microorganisms obtain energy through chemical reactions, i.e., oxidation-reduction reactions, where the energy source becomes oxidized while another compound is reduced. Oxidation-reduction reactions consist of electron transfers and can therefore be measured electrically with a potentiometer. The measured oxidation-reduction potential is known as the *redox potential*. The redox potential can also be determined with indicators and dyes. The addition of such compounds to metabolizing bacteria results in the transfer of electrons to the indicator and subsequent change in color.

Several oxidation-reduction indicators or dyes, including methylene blue, resazurin, and tetrazoliums, can be used. The rate at which the indicator shows a change in color is related to the metabolic rate of the microbial culture (sample). Basically, the greater the number of bacteria, the faster the color change occurs. Tests using these compounds are often referred to as *dye reduction tests,* but this is not strictly correct, since resazurin and tetrazoliums are not dyes but indicators.

Reductase tests can be used for a variety of products, and the results compare favorably with those of the SPC. However, for some food samples, including raw meat, the test may not be appropriate, since meat contains inherent high levels of reductive substances. The test is commonly used in the dairy industry for determining the microbial quality of raw milk. The advantage of reductase tests over the SPC is that they can be completed in a shorter time.

Surface Testing

This chapter has concentrated on methods designed to detect and enumerate microorganisms associated with food samples. However, to ensure that food contact surfaces are being maintained in a hygienic state, they should be tested regularly for microorganisms. The inherent problem with surface testing is the consistent removal of microorganisms from the test surface. A number of methods are available; however, each may be appropriate only in specified areas of a food-processing plant.

The *swab test* method is perhaps the most widely used method for microbiological examination of surfaces in food-processing facilities. The basics of the method include swabbing a given area with a moistened cotton or calcium alginate swab. The area to be examined can be defined through the use of templates that have predefined openings (usually 1 cm^2). The template should be sterilized prior to use. After the defined area is swabbed, the swab is returned to a test tube containing a suitable diluent and agitated to dislodge the bacteria. To facilitate the process, calcium alginate swabs can be used and dissolved with the addition of sodium hexametaphosphate to the diluent. The inoculated diluent can be used in both conventional and rapid microbiological assays. The SPC is generally used to determine the number of bacteria in the inoculated diluent. The swab test method is ideal for testing rough, uneven surfaces.

Other methods for examining food contact surfaces include direct contact of agar with a test surface and the use of sticky film or tape. In the *replicate organism direct area contact* method, specially designed petri plates are used so that the medium, when poured, produces a raised agar surface. The agar can then make direct contact with the test surface. Once the plate has been used to test the desired surface area, the lid is replaced and the plate is incubated. Selective media can be used to reduce the growth of bacteria that may spread across the surface of the plate. This method is not suitable for heavily contaminated or rough surfaces.

Sterile sticky tape is commercially available for sampling of surfaces from equipment to beef carcasses. The tape and dispenser are similar in appearance to a standard roll of adhesive tape used for sealing packages. The tape is withdrawn from the roll and held over a premeasured (in square centimeters) area located on the tape dispenser. The sticky surface of the tape is then pressed against the area to be tested, and then the tape is pressed onto the surface of an agar plate. The method is comparable to the swab test except on wooden surfaces.

A sponge system has been employed for examination of animal carcasses and food contact surfaces. Similar to other contact systems, a moistened sponge is wiped across the test area and then placed into a container (test tube, sterile plastic bag, etc.) containing an appropriate diluent. The container is agitated to dislodge the bacteria from the sponge, and an aliquot of the inoculated diluent is used in the SPC or other microbiological assay.

The methods outlined in this chapter are all commonly referred to as conventional methods. To facilitate the detection and enumeration of microorganisms from food or food-processing facilities, selective medium is commonly used and is commercially available. Selective medium may contain antibiotics or other inhibitors to permit the growth of select microorganisms. Some selective media contain specific substrates that are used by only certain microorganisms. A few of these media are discussed in chapter 5. Often, conventional methods are combined with rapid methods to decrease the time required for identification of microorganisms found in food.

Summary

- Samples must be collected using aseptic techniques and held in such a way as to prevent the growth of microorganisms.
- Food samples should be blended or homogenized prior to examination.
- The SPC is typically used to provide an estimate of the number of microorganisms in a sample.
- Surface test methods are essential in determining whether food contact surfaces are being maintained in a hygienic state.
- Rapid methods (miniaturized tests) based on conventional biochemical methods are now available, reducing the time required to identify a microorganism.

Suggested reading

Food and Drug Administration. 2001. *Bacteriological Analytical Manual Online.* http://vm.cfsan.fda.gov/~ebam/bam-mi.html.

Questions for critical thought

1. The DMC method has little or no value for foods with low microbial numbers. Why?
2. What is the difference between the surface and pour plate SPCs?
3. How can the buffer used for dilution of a sample influence the estimation of bacterial numbers associated with a sample?
4. The swab test and sticky-tape test are both methods for direct sampling of contact surfaces. List advantages and disadvantages for each.
5. Outline a protocol for the collection of a sample of ground beef from a 25-kg block. Ultimately, you want to perform an SPC.
6. The spiral plater is a convenient method for microbial analysis of a sample. Why?
7. Select a food and, using the *Bacterial Analytical Manual,* available through the Food and Drug Administration website, outline the methodology for determining total bacterial numbers and the tests required for detection of a food pathogen of concern in the selected food.

5

Rapid and Automated Microbial Methods

LEARNING OBJECTIVES

The information in this chapter will enable the student to:
- identify the various categories of rapid identification methods
- explain the basis of immunological, nucleic acid, and biochemical methods
- recognize when rapid methods are suitable to use
- understand the advantages and disadvantages associated with the use of rapid methods

INTRODUCTION

Microbiological analysis of food is not a "piece of cake." Foods are complex matrices of fats, carbohydrates, proteins, preservatives, and other chemicals. Foods also vary with respect to their physical natures: solid, dry, liquid, or semisolid. Collectively, these attributes can make it difficult to process a sample for microbiological analysis. Even if hurdles associated with processing of a sample are overcome, often foodborne pathogens are present at extremely low levels, further complicating the detection process. To decrease the time required to detect and identify target microbes in a food, an array of rapid microbiological methods have been developed.

Significant advances in the development of rapid methods have been made during the last decade, although rapid methods in some form have been available for more than 20 years. Initially, the medical community led these advances, but as a result of increased consumer awareness of the microbial safety of foods, food microbiologists have closed the gap. The task was made more difficult by the increased range of microorganisms now associated with food compared to 20 or 30 years ago. In the "old" days, food microbiologists tested for *Salmonella*, *Clostridium botulinum*, and *Staphylococcus aureus*; now *Listeria monocytogenes*, *Escherichia coli* O157:H7, *Campylobacter jejuni*, and *Vibrio parahaemolyticus* are also of concern. The change in the complexity of microorganisms associated with food is linked to many factors, including changes in processing and consumer preference and a global marketplace.

The basis for ensuring that safe and wholesome foods are available for the consumer is the testing of foods for pathogens and spoilage microorganisms. Conventional test methods (see chapter 4), although not lacking in sensitivity and cost-effectiveness, can be laborious and require several days before results are known. Products that are minimally processed have an inherently short shelf life, which prevents the use of many conventional methods that may require several days to complete, ultimately limiting product time on the shelf. Rapid assays are based on immunological, biochemical, microbiological, molecular, and serological methods for isolation, detection, enumeration, characterization, and identification.

The times required to complete different assays vary greatly. Indeed, "rapid" may imply seconds, minutes, hours, or even days. When considering whether an assay is rapid, all steps in the assay must be included, not just the time required to complete the rapid test itself. For instance, a rapid assay may require only 15 min to complete, but the food sample may need to be mixed with general growth medium (*preenrichment*) and incubated for several hours, with an aliquot then transferred to a selective enrichment medium and incubated before an aliquot is finally used in the rapid 15-min assay. Most rapid methods are adaptable to a wide range of foods, since they rely on culture methods to recover injured cells and increase the number of target cells.

The fast pace at which rapid methods are being developed precludes discussion of all available methods. In this chapter, the breadth of rapid methods available and the scientific principles of the methods used for detection of specific microbes in foods are examined. Existing rapid methods and those that show promise for the future but that are not yet commercially available are discussed.

SAMPLE PROCESSING

As mentioned above, food samples generally require some type of processing before a rapid method can be employed. The methods mentioned in this chapter are generally accepted for a broad range of foods, including milk, yogurt, apple juice, ground beef, and tomatoes. The composition of the food—the presence of fats, carbohydrates, and protein—can have a dramatic impact on successfully determining whether a target organism is present. The way that the food is processed, such as cooling, drying, heating, and addition of chemical additives, can also influence the ability to detect the presence of a target microorganism. In general, food-borne pathogens are present in low numbers in foods. Therefore, most rapid methods include enrichment, preenrichment, and/or selective enrichment to facilitate detection of the target pathogen.

Rapid tests have detection limits ranging from 10^2 to 10^5 colony-forming units (CFU)/g or 10^2 to 10^5 CFU/ml; therefore, enrichment increases the probability of detecting low levels of a target organism. The duration of the enrichment incubation varies considerably, based on the manufacturer's directions. Often, the period is relatively short, <12 h, but it can be as long as 24 h. The rapid test manufacturer may specify the medium to be used for enrichment, or a standard growth medium may be appropriate. Prior to selecting a rapid method, the assay protocol should be read carefully to determine whether an enrichment step is required, since this may dramatically increase the time needed to complete the assay.

REQUIREMENTS AND VALIDATION OF RAPID METHODS

Considerable effort goes into the development of rapid methods. Before rapid methods can be discussed, an understanding of how the accuracy and validity of a test is determined is necessary. To develop a reliable method, basic information about the target microbe is required. This facilitates the identification of unique features that can then be used for rapid detection. The cornerstone of any method is its *accuracy*. This consists of the *sensitivity*, or the ability of the assay to detect low numbers of the target microorganism, and the *specificity*, or its capability to differentiate the microbe of interest from other microorganisms. A *false-negative* result occurs when an assay fails to detect a target pathogen that is present upon culture. Similarly, a *false-positive* result occurs when the test system gives a positive result for a culture-negative sample. R. R. Beumer et al. give standard equations to calculate sensitivity and specificity rates: sensitivity = $(p \times 100)/(p$ + number of false negatives) and specificity = $(n \times 100)/(n$ + number of false positives), where p is the number of true positives and n is the number of true negatives. Without question, the assay must be as sensitive as possible and the detection limit must be as low as possible. For microorganisms that cause disease, the criterion is <1 cell per 25 g of food.

The intent in developing a rapid assay is to reduce the time required to obtain an accurate result. Conventional testing may require several days, whereas an ideal rapid test should provide results within an 8-h time frame, or a typical workday. In an ideal world, a rapid test would provide results nearly instantaneously, but this has yet to be realized. At the end of this chapter, new technologies that may be capable of delivering near-instantaneous results are discussed, but for now, we must operate within the constraints of the currently available systems. These have limits of detection of 10^2 to 10^5 CFU/g or ml and require enrichment for 6 to 24 h. Another factor in the selection of a rapid method is the number of samples that can be processed and analyzed at one time. Many tests have adopted the use of *microtiter plates* (small rectangular plates having 96 wells) or similar multiwell formats that can perform many tests for a single sample or that can be used to screen multiple samples simultaneously.

Although the method selected may be rapid and able to be completed within a time frame acceptable for a given operation, other factors, including the speed of sample processing, accuracy, and cost, must be considered. With respect to speed, single diagnostic tests may be acceptable if few tests are to be performed. If large numbers of samples must be processed, a high-throughput system, even though more costly, would be the best choice. The cost of the system selected must also be appropriate for the company using it. Many factors can influence cost, including the training of personnel, the purchase of specialized equipment to conduct the assay, service and maintenance contracts, disposable supplies, and reagents.

A significant factor to consider is whether specialized skills are required for laboratory personnel to perform the assay. Ideally, the assay should be technically easy to perform, the equipment easy to operate, and the results easy to interpret. If you haven't guessed by now, all aspects of the assay should be easy. Of course, the assay should be suitable for the food matrix that is being tested. In many cases, food constituents interfere with performance of the assay. Natural microflora of a sample and other debris may also interfere with the accuracy of the test. Finally, the method should be deemed acceptable by industry or government agencies.

There are several organizations, including the International Standards Organization, the International Dairy Federation, and the Association of Official Analytical Chemists (AOAC) International, that validate the effectiveness of testing methods for foods. The Food and Drug Administration (FDA) and U.S. Department of Agriculture, which do not validate methods, have manuals that outline standard methods used by each organization. In some instances, rapid methods are incorporated into the protocols;

however, this does not imply an official endorsement or approval of those tests. The AOAC publishes the *FDA Bacteriological Analytical Manual,* and the Food Safety Inspection Service of the U.S. Department of Agriculture publishes the *Microbiology Laboratory Guidebook.*

AOAC International is the most widely recognized and used service that provides third-party performance testing for manufacturers of test kits. There are two programs, the collaborative study program and the peer-verified program, used by AOAC to validate microbiological and chemical assays designed for the testing of foods. The collaborative study program is more rigorous. Methods that pass examination are reviewed by the AOAC Official Methods Board for approval as a first-action method. After 3 years of use by the scientific community, the methods are eligible for voting for final-action status. Rapid test kit manufacturers generally list on the product or in product information whether the test has undergone AOAC validation, i.e., AOAC first action or AOAC final action.

RAPID METHODS BASED ON TRADITIONAL METHODS

Traditionally, bacteria were differentiated and identified based on their biochemical profiles. To accomplish this, bacteria needed to be separated from a food matrix. This requires a number of labor-intensive steps to obtain a culture that can then be plated onto or inoculated into various differential media to determine a specific bacterial response. To facilitate the isolation of bacteria from food samples, the samples are mixed with diluent in sterile plastic bags and massaged using a stomacher. Previously, individual sterile blender cups were used, which meant that each cup required cleaning and sterilization after use. Additional dilution of samples can now be done using automated diluters. To further facilitate the process, automated plating systems have been developed that eliminate the need to make dilutions and that use a robotic arm to distribute liquid sample onto a rotating plate. This allows a >1,000-fold dilution range to be counted on a single plate. The process of counting colonies on a plate is time-consuming and open to human error, particularly if a large number of plates must be counted. Colony-counting systems use scanners and specialized software. An added advantage of these systems is that images of plates can be stored for future viewing and analysis.

A variety of products that replace the standard agar plate are available (Table 5.1). One of the most widely used is the Petrifilm system (3M). This system consists of rehydratable nutrients and a gelling agent embedded in disposable cardboard. Petrifilms have been developed for enumerating yeasts, molds, total bacteria, and specific bacteria. To rehydrate the film to support microbial growth, a 1-ml aliquot of sample is dispensed onto the center of the film. The film is incubated under the appropriate conditions, and colonies are counted directly. The Petrifilm system for determining coliforms in foods is comparable to the widely used violet red bile agar and has greater sensitivity.

The *hydrophobic grid membrane filter* (HGMF) (ISO-GRID system; QA Life Science) is more complex than the Petrifilm system and requires specialized equipment to complete. The HGMF requires filtration of the sample through a filter that contains a set of 1,600 grid cells. The HGMF is then placed onto an appropriate medium and incubated, and the colonies are counted. Grids have been designed for *Salmonella, E. coli* O157:H7, coliform-*E. coli,* yeast, and mold counts and aerobic plate counts.

Table 5.1   Representative manual and automated miniaturized biochemical tests

Organism	System	Manufacturer
Enterobacteriaceae	API	BioMerieux
	MICRO-ID	Remel
	IDS RapID	Remel
	BBL crystal	Becton Dickinson
	Vitek	BioMerieux
Coliforms	Bactometer	BioMerieux
Gram-negative bacteria	Vitek	BioMerieux
	Microlog	Biolog
	Omnilog ID	Biolog
Gram-positive bacteria	Vitek	BioMerieux
	Microlog	Biolog
	Omnilog ID	Biolog
Anaerobes	IDS RapID	Remel
	BBL crystal	Becton Dickinson
Listeria	MICRO-ID	Remel
	API	BioMerieux
Streptococci	RapID STR	Remel

Similar to the Petrifilm system and the HGMF system, the SimPlate system is based on the use of specialized dehydrated medium that is reconstituted with sterile water prior to use. The system is a modification of the most-probable number method. The petri dish device contains numerous wells, and the positive wells are counted and compared to a most-probable number chart. The system is well suited for testing a variety of products from fresh vegetables to ice cream.

The methods discussed thus far are basically kits and must be used as such. However, in recent years, a staggering number of media have been developed for the detection, enumeration, and identification of specific bacteria (Table 5.2), and media are now available for the detection and enumeration of yeasts and molds. For instance, media marketed by CHROMagar (Paris, France) contain specific chemicals that permit the differentiation and counting of specific bacteria. On CHROMagar ECC, coliform colonies appear pink and *E. coli* colonies appear blue. For the detection of *E. coli* O157:H7, the list of media just keeps growing. These media include MacConkey sorbitol agar, phenol red sorbitol agar containing 4-methylumbelliferyl-β-D-glucuronide, Levine eosin-methylene blue agar, Fluorcult *E. coli* O157:H7 agar (EM Sciences), BCM O157:H7(+) (Biosynth Biochemica and Synthetica), and MacConkey sorbitol agar containing either 5-bromo-4-chloro-3-indoxy,1-β-D-glucuronic acid or 4-methylumbelliferyl-β-D-glucuronide for the detection of β-D-glucuronidase activity.

IMMUNOLOGICALLY BASED METHODS

The most widely used rapid methods are based on immunoassay technology. The systems are acceptable for the screening and identification of specific bacteria associated with a variety of foods. The antibodies used in these systems may detect either many cellular targets (polyclonal antibodies) or specific targets (monoclonal antibodies). The use of monoclonal antibodies should eliminate some problems associated with cross-reaction

Table 5.2 Representative specialty medium rapid kits for detection of bacteria associated with food

Organism	Trade name	Assay format[a]	Manufacturer
E. coli	Petrifilm	Medium film	3M
	Redigel (Colichrome)	Medium	3M
	Coligel	MUG–X-Gal	Charm Science
	E-Colite	MUG–X-Gal	Charm Science
	Pathogel	MUG–X-Gal	Charm Science
Coliforms	Petrifilm	Medium film	3M
	Redigel (Colichrome)	Medium	3M
	Redigel (Violet Red Bile)	Medium	3M
	Coligel	MUG–X-Gal	Charm Science
	E-Colite	MUG–X-Gal	Charm Science
	Pathogel	MUG–X-Gal	Charm Science
Salmonella	SM ID	Medium	BioMerieux
Enterobacteriaceae	Petrifilm	Medium film	3M
	Pathogel	MUG–X-Gal	Charm Science
Staphylococci	Microdase Disk	Medium	Remel
	Bactistaph kit	Medium	Remel
	Novobiocin Disk	Medium	Remel
S. aureus	Petrifilm	Medium film	3M

[a]MUG, 4-methyl-umbelliferyl-β-D-glucuronide; X-Gal, 5-bromo-4-chloro-3-indolyl-β-D-galactopyranoside.

with bacteria other than the target organism. The systems range from simply mixing antibody-coated beads with a test sample to a multistep procedure that includes addition of an enzyme-labeled antibody to a well, followed by incubation, washing, and addition of a chromogen that is acted on and that can be detected either visually or using a spectrophotometer. The typical limits of detection for these systems range from 10^3 to 10^5 CFU/ml. Table 5.3 contains a list of some of the commercially available methods.

Coating of latex particles (small beads) or magnetic beads permits screening of samples for a specific bacterium or separation of a target bacterium from a sample, respectively. For the *latex agglutination test*, latex particles are coated with specific antibodies that upon contact with specific antigens of the target organism produce a visible clumping, or agglutination, reaction. The method is not that sensitive, since $>10^6$ CFU are required for a visible reaction. However, the method is rapid and provides a convenient means to screen for a target microbe. *Immunomagnetic separation* differs from latex agglutination in that the method is designed to separate the target organism from a sample. Basically, an aliquot of magnetic beads coated with a specific antibody is added to a sample (usually broth culture), the sample is incubated to facilitate binding of the target cells to the antibody, and then the complex is isolated from the sample using a magnet. The beads can then be used to inoculate broth, be plated onto selective agar medium, or be used directly in polymerase chain reaction (PCR) assays or in enzyme-linked immunosorbent assays (ELISA). Immunomagnetic separation systems are available for separation of *E. coli* O157:H7, *Listeria,* and *Salmonella* from a range of samples, including feces, raw milk, and ice cream.

The *Salmonella* 1-2 test (BioControl) is a popular and widely accepted method based on *immunodiffusion* (the movement of proteins through agar) and the formation of an antibody-antigen complex that forms a visible line

Table 5.3 Representative list of antibody-based assays

Organism	Trade name	Format[a]	Manufacturer
EHEC O157[b]	Reveal	Ab-ppt	Neogen
	Alert	ELISA	Neogen
	VIDAS	ELFA	BioMerieux
	Assurance EHEC EIA	ELISA	BioControl
	VIP	Ab-ppt	BioControl
	TECRA	ELISA	TECRA
E. coli O157:H7	DETEX	ELISA	Molecular Circuitry Inc.
Salmonella spp.	Reveal	Ab-ppt	Neogen
	Alert	ELISA	Neogen
	VIDAS	ELFA	BioMerieux
	1-2 Test	Diffusion	BioControl
	Assurance Salmonella EIA	ELISA	BioControl
	VIP	Ab-ppt	BioControl
	Assurance Gold Salmonella EIA	ELISA	BioControl
	TECRA	ELISA	TECRA
	UNIQUE	Capture EIA	TECRA
	DETX	ELISA	Molecular Circuitry Inc.
Listeria	Reveal	ELISA	
	VIP	Ab-ppt	BioControl
	Assurance	ELISA	BioControl
	TECRA	ELISA	TECRA
	UNIQUE	Capture EIA	TECRA
	DETEX	ELISA	Molecular Circuitry Inc.
L. monocytogenes	DETEX	ELISA	Molecular Circuitry Inc.
Campylobacter	Alert	ELISA	Neogen
	Uni-Lite XCEL	ELISA	Neogen
	VIDAS	ELFA	BioMerieux
	VIP	Ab-ppt	Biocontrol
	Campylobacter	ELISA	BioControl
	DETEX	ELISA	Molecular Circuitry Inc.
	TECRA	ELISA	TECRA
Bacillus thuringiensis	Agri-Screen	S-ELISA	Neogen
Pseudomonas	TECRA	ELISA	TECRA
S. aureus	TECRA	ELISA	TECRA

[a]EIA, enzyme immunoassay; S-ELISA, sandwich ELISA; Ab-ppt, antibody precipitation; ELFA, enzyme-linked fluorescent assay.
[b]EHEC, enterohemorrhagic E. coli.

of precipitation in a chamber. The antibody is specific for Salmonella flagellar antigen; therefore, nonmotile Salmonella organisms are not detected. Many immunoassays are self-contained units that are based on the migration of a sample along a chromatographic strip. Systems using this or similar technology include the VIP tests (BioControl), RevealTests (Neogen Corp.), SafePath (SafePath Laboratories LLC), and PATH-STIK (Lumac). Tests are available for Salmonella, E. coli O157:H7, Listeria, Campylobacter, and a variety of toxins. The systems have been used to screen fecal, beef, milk, and environmental samples. The specificity and sensitivity vary with the sample type and assay used.

There are several ELISA-based methods, such as TECRA for detection of Salmonella, Campylobacter, and E. coli O157:H7; Listeria-TEK (Organon Teknika); and the Assurance EIA system (BioControl). ELISA-based systems are also available for the detection of fungi at the genus or species level. Mycotoxins can also be detected by commercially available ELISA systems.

MOLECULAR METHODS

There has been an explosion in the past 10 years in the introduction of nucleic acid (RNA and DNA)-based assays for the differentiation and identification of food-borne pathogens (Table 5.4). DNA methods include, but are not limited to, PCR, pulsed-field gel electrophoresis, ribotyping, plasmid typing, randomly amplified polymorphic DNA, and restriction fragment length polymorphism. Some of these methods have been automated, and kits are available to facilitate the recovery of pure DNA. Perhaps the most widely recognized and used method is PCR.

PCR is a basic three-step process that is based on the amplification of a specific segment of cellular DNA (Fig. 5.1). In the first step, double-stranded DNA, also known as *template DNA,* is denatured into single strands, followed by annealing of *primers* (short segments of DNA complementary to a specific region on the template DNA strand) to the template DNA. The annealing temperature is critical, since it determines the *stringency* of the reaction, or how specific the attachment will be. The final step is extension of the primers, which is accomplished using a thermostable DNA polymerase. Within 2 h, PCR can amplify a single copy of DNA a millionfold. The PCR products can be visualized as a band on an agarose gel stained with ethidium bromide.

Currently, there are only a few commercially available PCR kits for the identification of foodborne pathogens. Although the test is extremely sensitive, equipment required to run the test is expensive, technical expertise is required, and PCR assays are affected by complex components of food. High levels of fats, protein, humic substances, and iron can interfere with the reaction. To overcome these problems, an enrichment step is often included in the assay, effectively diluting inhibitors and increasing cell numbers. The BAX system (Qualicon) is a commercially available PCR assay for the detection of *E. coli* O157:H7, *Salmonella,* and *Listeria.* The kit is unique in that all reagents required for the PCR (primers, enzyme, and deoxyribonucleosides) are *lyophilized* (freeze-dried) and included in a reaction tube. This eliminates potential problems with the introduction of contaminants or the use of incorrect amounts of a reagent.

There are a number of alternatives to conventional PCR, including reverse transcription-PCR (RT-PCR), real-time PCR, and the nucleic acid sequence-based amplification system. Commercial kits are available for nucleic acid sequence-based amplification (NucliSens Basic Kit; Organon Teknika) and for real-time PCR (Biotecon Diagnostics, Hamilton Square,

Table 5.4 Representative nucleic acid-based rapid methods

Organism	System	Format	Manufacturer
Salmonella	PCR-Light Cycler	PCR-ELISA	Biotecon Diagnostics
	Gene-Trak	Probe	Gene-Trak
	BAX	PCR	Qualicon
E. coli O157:H7	BAX	PCR	Qualicon
E. coli	Gene-Trak	Probe	Probe
L. monocytogenes	PCR-Light Cycler	PCR-ELISA	Biotecon Diagnostics
L. monocytogenes	Gene-Trak	Probe	Gene-Trak
Listeria spp.	BAX	PCR	Qualicon
Campylobacter	Gene-Trak	Probe	Gene-Trak
S. aureus	Gene-Trak	Probe	Gene-Trak

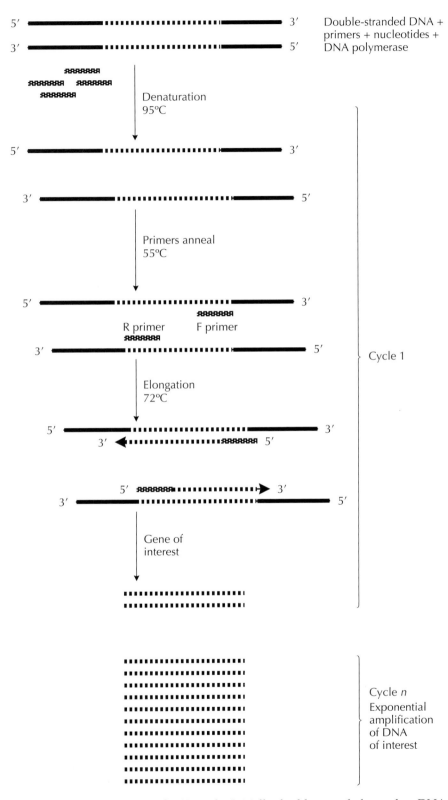

Figure 5.1   Representation of PCR cycle. Initially double-stranded template DNA (from a microbe) is denatured to form two single-stranded pieces of DNA. Primers designed to amplify a specific region on the template DNA are allowed to anneal to the single-stranded DNA. Elongation and extension of the primer make a complementary copy of the DNA template. These three steps make up a single cycle. The cycle is repeated a specified number of times.

N.J.). For assays for which kits are not available, the methods are well defined. Reagent packages that have been optimized for the isolation of RNA and DNA are available. RT-PCR uses the enzyme reverse transcriptase to convert RNA into DNA, which then serves as the template in PCR. RT-PCR is a useful method, since it is based on RNA, which has a very short half-life. When performed properly, RT-PCR detects only live cells, whereas in PCR, DNA from live and dead cells can potentially be detected.

The commercially available Riboprinter from Qualicon makes it easy to conduct ribotyping for the characterization and fingerprinting of strains and for epidemiological investigation. Results using the automated method can be obtained within 16 h, and riboprint patterns can be stored to create a unique database. Conducting ribotyping without the aid of an automated system is time-consuming and requires a high level of skill. Ribotyping has excellent reproducibility and discriminatory power.

POTPOURRI OF RAPID METHODS

Only a few of the many rapid methods that exist have been discussed. There are a few others that deserve mention, including luminescence-based assays, flow cytometry, and impedance. Of particular importance to the food industry are the bioluminescence-based methods, which can provide a rapid estimate of total microbial numbers. The presence of specific types of bacteria cannot be determined, but the production of light can be correlated to the number of microbes. The method makes use of the ubiquitous presence of adenosine triphosphate (ATP) in all living cells and the ability for it to react with the luciferase enzyme complex found in fireflies. The Uni-lite system (Biotrace, Inc., Plainsboro, N.J.), the Lighting bioluminescence system (IDEXX Laboratories, Inc., Westbrook, Maine), and the Lumac Hygiene Monitoring kit (Integrated BioSolutions, Inc., Monmouth Junction, N.J.) all make use of this technology. For example, the Uni-lite system consists of a transportable luminometer reading unit and swabs prepackaged in tubes containing reagents. The surface to be tested is swabbed, the reaction is activated by placing the swab into an enzyme solution, and the swab and tube are inserted into a chamber in the luminometer to obtain a reading. Results are obtained in <1 min.

The *direct epifluorescence filter technique* is another technique that is designed to illuminate bacteria. The method is simple and involves membrane filtration of a food sample to collect cells associated with the sample. The membrane is then processed by using fluorescent antibodies to stain target bacteria. Bacteria are visualized using an epifluorescence microscope. Problems can occur if food material trapped on the filter also fluoresces (i.e., it is difficult to differentiate bacteria from food).

Other methods that are available or in the developmental stages are based on nanotechnology, impedance, flow cytometry, and bacterial ice nucleation. The Bacterometer (BioMerieux) and the Malthus System V (Malthus Diagnostics) are commercially available automated systems based on impedance. As microbes grow, they metabolize substrates with low conductivity into products with high conductivity and therefore decrease impedance. Impedance systems are good for processing a large number of samples and can detect low levels of a given microbe.

The future development of rapid methods will likely exploit the use of nanotechnology. Nanotechnology will permit the development of real-time sensors capable of detecting several different target microbes of concern.

Although many different formats are available, silicon chips can be imprinted with specific DNA sequences for the detection and identification of target organisms. Chips could be imprinted with antibodies, RNA, or a host of other agents to target microorganisms. Microbial sensors are particularly applicable in fluid systems with limited organic substances. Presently, problems exist with sensor efficacy in food systems containing fats and proteins that coat the sensor and render it inoperable. These problems will be overcome as technology advances.

Concerns related to bioterrorism and the safety of the food supply have already inspired a massive effort for the development of rapid methods capable of detecting not only typical food-borne pathogens but also microbes of concern for human health that may be intentionally used to contaminate food. Handheld PCR units have been developed for the detection of *Bacillus anthracis* and *Yersinia pestis*. Research is ongoing to modify the unit for detection of *Salmonella*, *E. coli* O157:H7, and *L. monocytogenes*. These units have potential applications in the screening of food at ports of entry and border crossings and on farms.

Summary

- Specialized media that target specific biochemical reactions of the target microbe are available.
- The most widely used format for rapid methods is immunologically based assays.
- An enrichment step is required for most assays to increase the number of bacteria to a detectable level.
- Most assays require 10^2 to 10^5 CFU per g or ml.
- PCR-based assays are extremely sensitive but are subject to interference from the presence of inhibitors in food.
- Future rapid assays will likely be based on nanotechnology.

Suggested reading

Beumer, R. R., E. Brinkman, and F. M. Rombouts. 1991. Enzyme-linked immunoassays for the detection of *Salmonella* spp.: a comparison with other methods. *Int. J. Food Microbiol.* **12:**363–374.

Feng, P. 1995. Rapid methods for detecting foodborne pathogens, p. App.1.01–App.1.16. *In* R. I. Merker (ed.), *FDA Bacteriological Analytical Manual*, ed. 8A. AOAC International, Gaithersburg, Md.

Food and Drug Administration. 2001. *Bacteriological Analytical Manual Online.* http://vm.cfsan.fda.gov/~ebam/bam-mi.html.

Matthews, K. R. 2003. Rapid methods for microbial detection in minimally processed foods, p. 151–163. *In* J. S. Novak, G. M. Saper, and V. K. Juneja (ed.), *Microbial Safety of Minimally Processed Foods*. CRC Press, Boca Raton, Fla.

Questions for critical thought

1. The FDA has requested proposals for the development of PCR-based methods for the purpose of providing rapid confirmatory identification of microbial pathogens that are not usually associated with food and food-borne illness. Develop an outline for a proposal and include the pathogens that you would focus on and the type of PCR method you would use. Justify each of your responses.

2. Select a rapid assay listed in one of the tables. Gather literature from the company and find peer-reviewed journal articles related to the assay's efficacy. Based on your review of the literature, write a one-page abstract of the assay. Pretend the abstract is for a rapid-methods newsletter.

3. Using a table format, compare PCR, RT-PCR, and real-time PCR. Provide a brief narrative concerning when each method would be most appropriate to use.

4. Food constituents often interfere with the function of a rapid assay. For example, fat in a sample may coat a biosensor, inhibiting the binding of bacteria present in a sample to the sensor. Using immunologically based assays as an example, indicate from a scientific viewpoint why a given food constituent would interfere with the assay. Now, provide a protocol that would eliminate or prevent such interference.

5. You are working for a small development group that is interested in marketing a new rapid method for detection of food-borne pathogens. The only problem is, they do not have a new rapid method and have asked you to develop one. What type of method would you develop? It must be novel.

6. Many of the rapid bioluminescence assays are based on the detection of ATP. Since ATP is found in all living cells, what is a potential problem(s) associated with this assay in determining whether a surface is contaminated with bacteria?

6

Indicator Microorganisms and Microbiological Criteria*

LEARNING OBJECTIVES

The information in this chapter will enable the student to:
- differentiate among the various microbiological criteria
- identify national and international agencies involved in establishing microbiological criteria
- recognize how indicator organisms are used in microbiological criteria
- understand why some sampling plans are more stringent than others
- identify organisms and types of foods for which zero tolerance has been established
- identify and list steps required to manage microbiological hazards in foods

INTRODUCTION

The Purpose of Microbiological Criteria

Microorganisms cause foodborne illness and spoil food. Trained food microbiologists could probably isolate microbes from most raw and finished food products. Does that mean that the product is not safe to eat or will spoil rapidly? Not necessarily. Indeed, it is okay to have certain bacteria, even pathogenic bacteria, in food, depending on the product. Microbiological criteria ensure that a product has been produced under sanitary conditions and is microbiologically safe to consume.

Microbiological criteria are used to distinguish between an acceptable and an unacceptable product or between acceptable and unacceptable food-processing practices. The numbers and types of microorganisms associated with a food may be used to judge its microbiological safety and quality. Safety is determined by the absence, presence, or level of pathogenic microorganisms or their toxins and their expected control or destruction. The level of spoilage microorganisms reflects the microbiological quality, or wholesomeness, of a food, as well as the effectiveness of measures used to control or destroy such microorganisms. Indicator organisms may be used to assess either the microbiological quality or safety. Specifically, microbiological criteria are used to assess (i) the safety of food, (ii) adherence to good manufacturing practices (GMPs), (iii) the keeping quality (shelf life) of perishable foods, and (iv) the suitability of a food or ingredient for a particular purpose. Appropriately applied microbiological criteria ensure the safety and quality of foods. This in turn increases consumer confidence.

*This chapter was originally written by Merle D. Pierson and L. Michele Smoot for *Food Microbiology: Fundamentals and Frontiers*, 2nd ed., and has been adapted for use in an introductory text.

The Need To Establish Microbiological Criteria

A microbiological criterion should be established only in response to a need and when it is both effective and practical. There are many considerations to be taken into account when establishing microbiological criteria. Listed below are a few important factors for assessing the need for microbiological criteria.

- Evidence of a health hazard based on epidemiological data or a hazard analysis
- The nature of the food's normal microbial makeup and the ability of the food to support microbial growth
- The effect of processing on the microflora of the food
- The potential for microbial contamination and/or growth during processing, handling, storage, and distribution
- Spoilage potential, utility, and GMPs

Definitions

The National Research Council (NRC) of the U.S. National Academy of Sciences addressed the issue of microbiological criteria in their 1985 report entitled *An Evaluation of the Role of Microbiological Criteria for Foods and Food Ingredients*. The report indicates that a microbiological criterion should state what microorganism, group of microorganisms, or toxin produced by a microorganism is covered. The criterion should also indicate whether it can be present or present in only a limited number of samples or a given quantity of a food or food ingredient. In addition, a microbiological criterion should include the following information:

- Statement describing the identity of the food
- Statement identifying the contaminant
- Analytical method to be used for the detection, enumeration, or quantification of each contaminant
- Sampling plan (discussed below)
- Microbiological limits considered appropriate to the food and commensurate with the sampling plan

Criteria may be either mandatory or advisory. A *mandatory criterion* is a criterion that may not be exceeded. Food that does not meet the specified limit is required to be rejected, destroyed, reprocessed, or diverted. An *advisory criterion* permits acceptability judgments to be made. It serves as an alert to deficiencies in processing, distribution, storage, or marketing. For application purposes, there are three categories of criteria: standards, guidelines, and specifications. The following definitions were recommended by the NRC's Subcommittee on Microbiological Criteria for Foods and Food Ingredients.

Standard: A microbiological criterion that is part of a law, ordinance, or administrative regulation. A standard is a mandatory criterion. Failure to comply constitutes a violation of the law, ordinance, or regulation and will be subject to the enforcement policy of the regulatory agency having jurisdiction.

Guideline: A microbiological criterion often used by the food industry or regulatory agency to monitor a manufacturing process. Guidelines function as alert mechanisms to signal whether microbiological conditions prevailing at critical control points or in the finished product are within the normal range. Hence, they are used to assess processing efficiency at critical control points and conformity with good manufacturing practices. A microbiological guideline is advisory.

Specifications: Microbiological criteria that are used as a purchase requirement whereby conformance becomes a condition of purchase between buyer and vendor of a food ingredient. A microbiological specification may be advisory or mandatory.

Who Establishes Microbiological Criteria?

Different scientific organizations are involved in developing general principles for the application of microbiological criteria. The scientific organizations which have most influenced the U.S. food industry include the Joint Food and Agricultural Organization and World Health Organization Codex Alimentarius International Food Standards Program, The International Commission on Microbiological Specifications of Foods (ICMSF), the U.S. National Academy of Sciences, and the U.S. National Advisory Committee on Microbiological Criteria for Foods. The Codex Alimentarius Program first formulated *General Principles for the Establishment and Application of Microbiological Criteria* in 1981, and it has since been revised. In 1984, the NRC Subcommittee on Microbiological Criteria for Foods and Food Ingredients formulated general principles for the application of microbiological criteria to food and food ingredients as requested by four U.S. regulatory agencies. The *Codex Alimentarius* contains standards for all principal foods in the form (i.e., processed, semiprocessed, or raw) in which they are delivered to the consumer. Fresh perishable commodities not traded internationally are excluded from these standards. The World Trade Organization provides a framework for ensuring fair trade and harmonizing standards and import requirements for food traded through the Agreements on Sanitary and Phytosanitary Measures and Technical Barriers to Trade. Countries are required to base their standards on science, to base programs on risk analysis methodologies, and to develop ways of achieving equivalence among the different methods of inspection, analysis, and certification used by various trading countries. The World Trade Organization recommends the use of standards, guidelines, and recommendations developed by the Codex Alimentarius Program to make all standards similar.

Authors' note

The Codex Alimentarius *is a set of food standards and guidelines developed to protect the health of consumers and to ensure fair trade practices in the food trade.*

SAMPLING PLANS

A sampling plan includes both the sampling procedure and the decision criteria. To examine a food for the presence of microorganisms, a representative sample is examined by defined methods. A *lot* is a quantity of product produced, handled, and stored within a limited time under uniform conditions. Since it is impractical to examine the entire lot, statistical concepts of population probability and sampling are used to determine the number and size of sample units required from the lot and to provide conclusions drawn from the analytical results. The sampling plan is designed so that inferior lots will be rejected.

In a simplified example of a sampling plan, let us assume that 10 samples were taken and analyzed for the presence of a particular microorganism. Based on the decision criterion, only a certain number of the sample units could be positive for the presence of that microorganism for the lot to be considered acceptable. If in the criterion the maximum allowable number of positive units had been set at 2 ($c = 2$), then a positive result for >2 of the 10 sample units ($n = 10$) would result in rejection of the lot. Ideally, the decision criterion is set to accept lots that are of the desired quality and to reject lots that are not. However, since only part of the lot is examined, there is always the risk that an acceptable lot will be rejected or that an unacceptable lot will be accepted. The more samples examined, or the larger the size of n, the lower the risk of making an incorrect decision about the lot quality. However, as n increases, sampling becomes more time-consuming and costly. Generally, a compromise is made between the size of n and the level of risk that is acceptable.

Types of Sampling Plans

Sampling plans are divided into two main categories: variables and attributes. A variables plan depends on the frequency distribution of organisms in the food. For correct application of a variables plan, the organisms must have a log normal distribution (i.e., counts transformed to logarithms are normally distributed). When the food is from a common source and it is produced and/or processed under uniform conditions, log normal distribution of the organisms present is assumed. Attributes sampling is the preferred plan when microorganisms are not homogeneously distributed throughout the food or when the target microorganism is present at low levels. This is often the case with pathogenic microorganisms. Attributes plans are also widely used to determine the acceptance or rejection of products at ports or other points of entry, since there is little or no knowledge of how the food was processed and past performance records are not available. Attributes sampling plans may also be used to monitor performance relative to accepted GMPs. Attributes sampling, however, is not appropriate when there is no defined lot or when random sampling is not possible, as might occur when monitoring cleaning practices. Because attributes sampling plans are used more frequently than variables sampling plans, this text will not cover variables sampling in detail.

Attributes Sampling Plans

Two-class plans.   A two-class attributes sampling plan assigns the concentration of microorganisms of the sample units tested to a particular attribute class, depending on whether the microbiological counts are above or below some preset concentration, represented by the letter m. The decision criterion is based on (i) the number of sample units tested, n, and (ii) the maximum allowable number of sample units yielding unsatisfactory tests results, c. For example, when n is 5 and c is 2 in a two-class sampling plan designed to make a presence-absence decision about the lot (i.e., $m = 0$), the lot is rejected if more than two of the five sample units tested are positive. As n increases for the set number c, the stringency of the sampling plan also increases. Conversely, for a set sample size n, as c increases, the stringency of the sampling plan decreases, allowing a higher probability of accepting, P_a, food lots of a given quality. Two-class plans are applied most often in qualitative (semiquantitative) pathogen testing where the results are expressed as the presence or absence of the specific pathogen per sample weight analyzed.

Three-class plans.   Three-class sampling plans use the concentrations of microorganisms in the sample units to determine levels of quality and/or safety. Counts above a preset concentration M for any of the n sample units tested are considered unacceptable, and the lot is rejected. The level of the test organism acceptable in the food is represented by m. This concentration in a three-class attribute plan separates acceptable lots (i.e., those with counts less than m) from marginally acceptable lots (i.e., those with counts greater than m but not exceeding M). Counts above m and up to and including M are not desirable, but the lot can be accepted provided the number n of samples that exceed m is no greater than the preset number, c. Thus, in a three-class sampling plan, the food lot will be rejected if any one of the sample units exceeds M or if the number of sample units with contamination levels above m exceeds c. Similar to the two-class sampling plan, the stringency of the three-class sampling plan is also dependent on the two numbers represented by n and c. The larger the value of n for a given value of c, the better the food quality must be to have the same chance of passing, and vice versa. From n and c, it is then possible to find the probability of acceptance, P_a, for a food lot of a given microbiological quality.

ESTABLISHING LIMITS

Microbiological limits, as defined in a criterion, represent the level above which action is required. Levels should be realistic and should be determined based on knowledge of the raw materials and the effects of processing, product handling, storage, and the end use of the product. Limits should also take into account the likelihood of uneven distribution of microorganisms in the food, the inherent variability of the analytical procedure, the risk associated with the organisms, and the conditions under which the food is handled and consumed. Microbiological limits should include the sample weight to be analyzed, the method reference, and the confidence limits of the referenced method where applicable.

The shelf life of a perishable product is often determined by the number of microorganisms initially present. As a general rule, a food containing a large population of spoilage organisms will have a shorter shelf life than the same food containing fewer of the same spoilage organisms. However, the relationship between total counts and shelf life is not carved in stone. Some types of microorganisms have a greater impact on the sensory characteristics of a food than others.

Foods produced and stored under GMPs may be expected to have a better microbiological profile than those foods produced and stored under poor conditions. The use of poor-quality materials, improper handling, or unsanitary conditions may result in higher bacterial numbers in the finished product. However, low counts in the finished product do not necessarily mean that GMPs were adhered to. Processing steps, such as heat treatment, fermentation, freezing, or frozen storage, can reduce the counts of bacteria that have resulted from noncompliance with GMPs. Other products, such as ground beef, may normally contain high microbial counts even under the best conditions of manufacture due to the growth of psychrotrophic bacteria during refrigeration.

The use of quantitative risk assessment techniques to scientifically determine the probability of occurrence and the severity of known human exposure to food-borne hazards is rapidly gaining acceptance. The process consists of (i) hazard identification, (ii) hazard characterization, (iii) exposure assessment, and (iv) risk characterization. Though quantitative risk assessment techniques are well established for chemical agents, their application to food safety microbiology is relatively new.

INDICATORS OF MICROBIOLOGICAL QUALITY

Examination of a product for indicator organisms can provide simple, reliable, and rapid information about process failure, postprocessing contamination, contamination from the environment, and the general level of hygiene under which the food was processed and stored.

Ideal indicators of product quality or shelf life should meet the following criteria.

- They should be present and detectable in all foods whose quality is to be assessed.
- Their growth and numbers should have a direct negative correlation with product quality.
- They should be easily detected and enumerated and be clearly distinguishable from other organisms.
- They should be enumerable in a short period of time, ideally within a workday.

Indicator Microorganisms

Indicator microorganisms can be used in microbiological criteria. The presence of indicator microbes may suggest a microbial hazard. For example, the presence of generic *Escherichia coli* in a sample indicates possible fecal contamination. These criteria might be used to address existing product quality or to predict the shelf life of the food. Some examples of indicators and the products in which they are used are shown in Table 6.1. Those microorganisms listed in the table are the primary spoilage organisms of the specific products listed. Loss of quality in other products may not be limited to one organism but may involve a variety of organisms. In those types of products, it is often more practical to determine the counts of groups of microorganisms most likely to cause spoilage in each particular food.

The aerobic plate count (APC) is commonly used to determine the total number of microorganisms in a food product. By modifying the environment of incubation or the medium used, the APC can be used to preferentially screen for groups of microorganisms, such as those that are thermodurics, mesophiles, psychrophiles, thermophiles, proteolytic, or lipolytic. The APC may be a component of microbiological criteria assessing product

Table 6.1 Organisms highly correlated with product quality[a]

Organism	Product(s)
Acetobacter spp.	Fresh cider
Bacillus spp.	Bread dough
Byssochlamys spp.	Canned fruits
Clostridium spp.	Hard cheeses
Flat-sour spores	Canned vegetables
Lactic acid bacteria	Beers, wines
Lactococcus lactis	Raw milk (refrigerated)
Leuconostoc mesenteroides	Sugar (during refining)
Pectinatus cerevisiiphilus	Beers
"*Pseudomonas putrefaciens*"	Butter
Yeasts	Fruit juice concentrates
Zygosaccharomyces bailii	Mayonnaise, salad dressing

[a]Source: J. M. Jay, *Modern Food Microbiology*, 4th ed. (Chapman and Hall, New York, N.Y., 1992).

quality when those criteria are used to (i) monitor foods for compliance with standards or guidelines set by various regulatory agencies, (ii) monitor foods for compliance with purchase specifications, and (iii) monitor adherence to GMPs.

Microbiological criteria as specifications are used to determine the usefulness of a food or food ingredient for a particular purpose. For example, specifications are set for thermophilic spores in sugar and spices intended for use in the canning industry. Lots of sugar that fail to meet specifications may not be suitable for use in low-acid canning but could be diverted for other uses. The APC of refrigerated perishable foods, such as milk, meat, poultry, and fish, may be used to indicate the condition of the equipment and utensils used, as well as the time-temperature profile of storage and distribution of the food.

When evaluating the results of an APC for a particular food, it is important to remember that (i) APCs measure only live cells, and therefore, a grossly spoiled product may have a low APC if the spoilage organisms have died; (ii) APCs are of little value in assessing sensory quality, since high microbial counts are generally required for sensory-quality loss; and (iii) since the biochemical activities of different bacteria vary, quality loss may also occur at low total counts. With any food, specific causes of unexpectedly high counts can be identified by examination of samples at control points and by inspection of processing plants. Interpretation of the APC of a food requires knowledge of the expected microbial population at the point where the sample is collected. If counts are higher than expected, this indicates the need to determine why there has been a violation of the criterion.

The direct microscopic count (DMC) is used to give an estimate of the numbers of both viable and nonviable cells in samples containing a large number of microorganisms (i.e., $>10^5$ colony-forming units [CFU]/ml). Considering that the DMC does not differentiate between live and dead cells (unless a fluorescent dye, such as acridine orange, is employed) and that it requires that the total cell count exceed 10^5, the DMC has limited value as a part of the microbiological criteria for quality issues. The use of the DMC as a part of microbiological criteria for foods or ingredients is restricted to a few products, such as raw non-grade A milk, dried milk, liquid and frozen eggs, and dried eggs.

Other methods commonly used to indicate the quality of different food products include the Howard mold count, yeast and mold count, heat-resistant-mold count, and thermophilic-spore count. The Howard mold count (microscopic analysis) is used to detect the inclusion of moldy material in canned fruit and tomato products, as well as to evaluate the sanitary condition of processing machinery in vegetable canneries. Yeasts and molds grow on foods when conditions for bacterial growth are less favorable. Therefore, they can be a problem in fermented dairy products, fruits, fruit beverages, and soft drinks. Yeast and mold counts are used as parts of the microbiological standards for various dairy products, such as cottage cheese and frozen cream and sugar. Heat-resistant molds (e.g., *Byssochlamys fulva* and *Aspergillus fischerianus*) that may survive the thermal processes used for fruit and fruit products may need limits in purchase specifications for ingredients such as fruit concentrates. The canning industry is concerned about thermophilic spores in ingredients because they cause defects in foods held at elevated temperatures (i.e., due to inadequate cooling and/or storage at too high temperatures). Purchase specifications and verification criteria are often used for thermophilic-spore counts in ingredients intended for use in low-acid, heat-processed canned foods.

Metabolic Products

Bacterial levels in a food product can sometimes be estimated by testing for metabolic products made by the microorganisms in the food. When a correlation is established between the presence of a metabolic product and product quality loss, tests for the metabolite may be part of a microbiological criterion.

An example of the use of a test for metabolic products as part of a microbiological criterion is the sensory evaluation of imported shrimp. Trained personnel are able to classify the degree of decomposition (i.e., quality loss) into one of three classes through sensory examination. The shrimp are placed in one of the following quality classes: class 1, passable; class 2, decomposed (slight but definite); class 3, decomposed (advanced). The limits of acceptability of a lot are based on the number of shrimp in a sample that are placed in each of the three classes. Other commodities in which organoleptic examination (examination for changes in appearance, color, texture, and odor) is used to determine quality deterioration include raw milk, meat, poultry, and fish and other seafoods. The food industry also uses these examinations to classify certain foods into quality grades. Other examples of metabolic products used to assess product quality are listed in Table 6.2.

INDICATORS OF FOODBORNE PATHOGENS AND TOXINS

Microbiological criteria for product safety should be developed only when the application of a criterion can reduce or eliminate a foodborne hazard. Microbiological criteria verify that the process is adequate to eliminate the hazard. Each food type should be carefully evaluated through risk assessment to determine the potential hazards and their significance to consumers. When a food is repeatedly implicated as a vehicle in foodborne disease outbreaks, the application of microbiological criteria may be useful. Public health officials and the dairy industry responded to widespread outbreaks of milk-borne disease that occurred around the turn of the century in the United States. By imposing controls on milk production, developing safe and effective pasteurization procedures, and setting microbiological criteria, the safety of milk supplies was greatly improved.

Microbiological criteria may also be applied to food products, such as shellfish, which are frequently subject to contamination by harmful microorganisms. The National Shellfish Sanitation Program utilizes microbiolog-

Table 6.2 Some microbial metabolic products that correlate with food quality[a]

Metabolite(s)	Applicable food product(s)
Cadaverine and putrescine	Vacuum-packaged beef
Diacetyl	Frozen juice concentrate
Ethanol	Apple juice, fishery products
Histamine	Canned tuna
Lactic acid	Canned vegetables
Trimethylamine (TMA)	Fish
Total volatile bases (TVB), total volatile nitrogen (TVN)	Seafoods
Volatile fatty acids	Butter, cream

[a]Source: International Commission on Microbiological Specifications of Foods, *Microorganisms in Foods 2. Sampling for Microbiological Analysis: Principles and Applications,* 2nd ed. (University of Toronto Press, Toronto, Canada, 1986).

ical criteria in this manner to prevent the use of shellfish from polluted waters, which may contain various intestinal pathogens. Depending on the type and level of contamination anticipated, the imposition of microbiological criteria may or may not be justified. Contamination of food with pathogens that cannot grow to harmful levels does not warrant microbiological criteria. However, microbiological criteria would be warranted for a pathogen that has a low infectious dose. Though fresh vegetables are often contaminated with small numbers of *Clostridium botulinum*, *Clostridium perfringens*, and *Bacillus cereus* organisms, epidemiological evidence indicates that this contamination presents no health hazard. Thus, imposing microbiological criteria for these microorganisms would not be beneficial. However, criteria may be appropriate for enteric pathogens on produce, since there have been several outbreaks of food-borne illness resulting from fresh produce contaminated with enteric pathogens.

Often food processors alter the intrinsic or extrinsic parameters of a food (nutrients, pH, water activity, inhibitory chemicals, gaseous atmosphere, temperature of storage, and presence of competing organisms) to prevent the growth of undesirable microorganisms. If control over one or more of these parameters is lost, then there may be risk of a health hazard. For example, in the manufacture of cheese or fermented sausage, a lactic acid starter culture is used to produce acid quickly enough to inhibit the growth of *Staphylococcus aureus* to harmful levels. Process critical control points (CCPs), such as the rate of acid formation, are implemented to prevent the growth of harmful microorganisms or their contamination of the food and to ensure that process control is maintained.

Depending on the pathogen, low levels of the microorganism in the food product may or may not be of concern. Some microorganisms have such a low infectious dose that their mere presence in a food presents a significant public health risk. For such organisms, the concern is not the ability of the pathogen to grow in the food but its ability to survive in the food. Foods having intrinsic or extrinsic factors that prevent the survival of pathogens or toxigenic microorganisms may not be candidates for microbiological criteria related to safety. For example, the acidity of certain foods, such as fermented meat products, might be assumed to be sufficient for pathogen control. However, while the growth and toxin production of *S. aureus* might be prevented, enteric pathogens, such as *E. coli* O157:H7, could survive and produce a product that is unsafe for consumption.

An important and often overlooked variable in microbiological risk assessment is the consumer. More rigid microbiological requirements may be needed if a food is intended for use by infants, the elderly, or immunocompromised people, since they are more susceptible to infectious agents than are healthy adults. The ICMSF proposed a system for classification of foods according to risk into 15 hazard categories called cases, with suggested appropriate sampling plans, shown in Table 6.3.

The stringency of sampling plans is based on either the hazard to the consumer from pathogenic microorganisms and their toxins or toxic metabolites or the potential for quality deterioration. It should take into account the types of microorganisms present and their numbers. Food-borne pathogens are grouped into one of three categories based on the severity of the potential hazard (i.e., severe hazards, moderate hazards with potentially extensive spread, and moderate hazards with limited spread [Table 6.4]). Pathogens with the potential for extensive spread are often initially associated with specific foods. An example is fresh beef contaminated with

Table 6.3   Plan stringency (case) in relation to degree of health hazard and conditions of use[a]

Type of hazard	Conditions in which food is expected to be handled and consumed after sampling, in the usual course of events[b]		
No direct health hazard	**Increased shelf life**	**No change**	**Reduced shelf life**
Utility (e.g., general contamination, reduced shelf life, and spoilage)	Case 1 3-class $n = 5, c = 3$	Case 2 3-class $n = 5, c = 2$	Case 3 3-class $n = 5, c = 1$
Health hazard	**Reduced hazard**	**No change**	**Increased hazard**
Low, indirect (indicator)	Case 4 3-class $n = 5, c = 3$	Case 5 3-class $n = 5, c = 2$	Case 6 3-class $n = 5, c = 1$
Moderate, direct, limited spread[c]	Case 7 3-class $n = 5, c = 2$	Case 8 3-class $n = 5, c = 1$	Case 9 3-class $n = 5, c = 0$
Moderate, direct, potentially extensive spread	Case 10 2-class $n = 5, c = 0$	Case 11 2-class $n = 10, c = 0$	Case 12 2-class $n = 20, c = 0$
Severe, direct	Case 13 2-class $n = 15, c = 0$	Case 14 2-class $n = 30, c = 0$	Case 15 2-class $n = 60, c = 0$

[a]Reprinted with permission from International Commission on Microbiological Specifications of Foods, *Microorganisms in Foods 2. Sampling for Microbiological Analysis: Principles and Applications,* 2nd ed. (University of Toronto Press, Toronto, Canada, 1986).
[b]More stringent plans would generally be used for sensitive foods destined for susceptible populations.
[c]n, number of sample units drawn from lot; c, maximum allowable number of positive results.

E. coli O157:H7. One or a few contaminated pieces of meat can lead to widespread contamination of the product during processing, such as grinding to produce ground beef. There can also be cross-contamination if the fresh beef is improperly stored with ready-to-eat foods. Microbial pathogens in the lowest-risk group (moderate hazards within limited spread) are found in many foods, usually in small numbers. Generally, illness is caused only when ingested foods contain large numbers of a pathogen, e.g., *C. perfringens,* or have at some point contained large enough numbers of a pathogen, e.g., *S. aureus,* to produce sufficient toxin to cause illness. Outbreaks are usually restricted to consumers of a particular meal or a particular kind of food.

The effects of conditions to which the food will likely be exposed in relation to the growth or death of a relevant organism(s) must be considered before the appropriate case (Table 6.3) can be determined. In general, if a food will be subjected to treatment (e.g., *C. perfringens* in cooked meats held at inappropriate temperature) that would permit growth of the microorganism and thereby increase the hazard, then case 3, 6, 9, 12, or 15 would apply. Case 2, 5, 8, or 11 would apply if no change in the number of organisms present were expected, for example, in freezing of the product. Finally, case 1, 4, 7, or 10 would apply if the product was intended to be fully cooked, reducing the microbial hazard. Consider the case of foods derived from dried whole eggs which are cooked and eaten immediately (scrambled) or made into products (pasta). This would diminish the microbiological hazard. If the dried eggs were consumed dry or reconstituted and consumed immediately, the microbiological hazard would not change (no opportunity was provided for microorganisms present to multiply). A third scenario involves reconstituting the dried eggs and waiting for an extended period prior to consuming them, in which case any microorganisms present would have an opportunity to multiply and potentially cause illness.

Table 6.4 Hazardous microorganisms and parasites grouped on the basis of risk severity[a]

Category I (severe hazards)

Clostridium botulinum types A, B, E, and F

Shigella dysenteriae

Salmonella enterica serovars Paratyphi A and B

Enterohemorrhagic *Escherichia coli* (EHEC)

Hepatitis A and E viruses

Brucella abortus, Brucella suis

Vibrio cholerae O1

Vibrio vulnificus

Taenia solium

Category II (moderate hazards, potentially extensive spread[b])

Listeria monocytogenes

Salmonella spp.

Shigella spp.

Other enterovirulent *E. coli* (EEC)

Streptococcus pyogenes

Rotavirus

Norwalk virus group

Entamoeba histolytica

Diphyllobothrium latum

Ascaris lumbricoides

Cryptosporidium parvum

Category III (moderate hazards, limited spread)

Bacillus cereus

Campylobacter jejuni

Clostridium perfringens

Staphylococcus aureus

V. cholerae, non-O1

Vibrio parahaemolyticus

Yersinia enterocolitica

Giardia lamblia

Taenia saginata

[a]Source: A. W. Randell and A. J. Whitehead, *Rev. Sci. Tech.* **16:**313–321, 1997.
[b]Although these are classified as moderate hazards, complications and sequelae may be severe in certain susceptible populations.

Indicator Organisms

Microbiological criteria for food safety may use tests for indicator organisms that suggest the possibility of a microbial hazard. *E. coli* in drinking water, for example, indicates possible fecal contamination and therefore the potential presence of other enteric pathogens.

Microbial indicators should have the following characteristics:

- Easily and rapidly detectable
- Readily distinguishable from microbes commonly associated with the food
- A history of constant association with the pathogen whose presence it is to indicate

- Present when the pathogen of concern is present
- Numbers that ideally correlate with those of the pathogen of concern
- Growth requirements and a growth rate equaling those of the pathogen
- A die-off rate that at least parallels that of the pathogen and ideally has a slightly longer persistence than the pathogen of concern
- Absence from foods that are free of the pathogen except perhaps at certain minimum numbers

Additional criteria for fecal indicators used in food safety have been suggested and include the following. Ideally, the bacteria selected should demonstrate specificity, occurring only in intestinal environments. They should occur in very high numbers in feces so as to be encountered in high dilutions. They should possess a high resistance to the external environment, the contamination of which is to be assessed. They should permit relatively easy and fully reliable detection even when present in low numbers.

With the exception of those for *Listeria monocytogenes*, *E. coli* O157:H7, *Salmonella*, and *S. aureus*, most tests for ensuring safety use indicator organisms rather than direct tests for the specific hazard. An overview of some of the more common indicator microorganisms used for ensuring food safety is given below.

Fecal Coliforms and *E. coli*

Fecal coliforms, including *E. coli*, are easily destroyed by heat and may die during freezing and storage of frozen foods. Microbiological criteria involving *E. coli* are useful in those cases where it is desirable to determine if fecal contamination may have occurred. Contamination of a food with *E. coli* implies a risk that other enteric pathogens may also be present. Fecal coliform bacteria are used as a component of microbiological standards to monitor the wholesomeness of shellfish and the quality of shellfish-growing waters. The purpose is to reduce the risk of harvesting shellfish from waters polluted with fecal material. The fecal coliforms have a higher probability of containing organisms of fecal origin than do coliforms which are comprised of organisms of both fecal and nonfecal origins. Fecal coliforms can become established on equipment and utensils in the food-processing environment and contaminate processed foods. At present, *E. coli* is the most widely used indicator of fecal contamination.

The presence of *E. coli* in a heat-processed food means either process failure or, more commonly, postprocessing contamination. In the case of refrigerated ready-to-eat (RTE) products, such as shrimp and crabmeat, coliforms are indicators of process integrity with regard to the reintroduction of pathogens from environmental sources and maintenance of adequate refrigeration. The source of coliforms after thermal processing appears to be the processing environment, inadequate sanitation procedures, and/or poor temperature control. Coliforms are recommended over *E. coli* and APC, because coliforms are often present in higher numbers than *E. coli* and the levels of coliforms do not increase over time when the product is stored properly.

Metabolic Products

Certain microbiological criteria related to safety rely on tests for metabolites that indicate a potential hazard. Examples of metabolites as a component of microbiological criteria include (i) tests for themonuclease or thermostable deoxyribonuclease in foods containing or suspected of containing $\geq10^5$ *S. aureus* organisms per ml or g, (ii) illuminating grains under UV (ultraviolet) light to detect the presence of aflatoxin produced by *Aspergillus* spp., and (iii) assaying for the enzyme alkaline phosphatase, a natural constituent of milk that is inactivated during pasteurization, to determine whether milk has been pasteurized or whether pasteurized milk has been contaminated with raw milk.

APPLICATION AND SPECIFIC PROPOSALS FOR MICROBIOLOGICAL CRITERIA FOR FOOD AND FOOD INGREDIENTS

In this section, examples of the application of microbiological criteria to various foods are presented. There are no general criteria suitable for all foods. The relevant background literature that relates to the quality and safety of a specific food product should be consulted before implementation of microbiological criteria.

The utilization of microbiological criteria, as stated previously, may be either mandatory (standards) or advisory (guidelines or specifications). One obvious example of a mandatory criterion (standard) is the zero tolerance set for *Salmonella* in all RTE foods. The U.S. Department of Agriculture (USDA) Food Safety and Inspection Service has mandated zero tolerance for *E. coli* O157:H7 in fresh ground beef. Since the emergence of *L. monocytogenes* as a foodborne pathogen the Food and Drug Administration and USDA have also applied zero tolerance for the organism in all RTE foods. However, there is considerable debate as to whether zero tolerance is warranted for *L. monocytogenes*. The ICMSF has recommended that the international community allow tolerance of *L. monocytogenes* as outlined in Table 6.5. These recommendations are examples of advisory criteria. Other advisory criteria that have been established to address the safety of foods are shown in Table 6.6.

Mandatory criteria in the form of standards have also been employed for quality issues. Presented in Table 6.7 are examples of foods and food ingredients for which federal, state, and city, as well as international, microbiological standards have been developed.

Table 6.5 Summary of cases and sampling plans for *L. monocytogenes*[a]

Intended consumer	Health hazard	Conditions in which food is expected be handled and consumed after sampling in the usual course of events		
		Reduce degree of hazard	Cause no change in hazard	May increase hazard
Healthy individuals	Moderate, direct, potentially extensive spread	Case 10 $n = 5$ $\leq100/g$	Case 11 $n = 10$ $\leq100/g$	Case 12 $n = 20$ $\leq100/g$
Highly susceptible individuals	Severe, direct	Case 13 $n = 15$ $<1/375\,g$	Case 14 $n = 30$ $<1/750\,g$	Case 15 $n = 60$ $<1/1{,}500\,g$

[a]Reprinted with permission from International Commission on Microbiological Specifications of Foods, *Int. J. Food Microbiol.* **22:**89–96, 1994.

Table 6.6 Examples of various food products for which advisory microbiological criteria have been established

Product category	Test parameter(s)	Case	Plan class	Limit (per g)			
				n	c	m	M
Roast beef[a]	Salmonella	12	2	20	0	0	
Pâté	Salmonella	12	2	20	0	0	
Raw chicken	APC	1	3	5	3	5×10^5	10^7
Cooked poultry, frozen, RTE	S. aureus	8	3	5	1	10^3	10^4
Cooked poultry, frozen, to be reheated	S. aureus	8	3	5	1	10^3	10^4
	Salmonella	10	2	5	0	0	
Chocolate or confectionery	Salmonella	11	2	10^b	0	0	
Dried milk	APC	2	3	5	2	3×10^4	3×10^5
	Coliforms	5	3	5	1	10	10^2
	Salmonella[c]	10	2	5	0	0	
	(normal routine)	11	2	10	0	0	
		12	2	20	0	0	
	Salmonella[c]	10	2	15	0	0	
	(high-risk populations)	11	2	30	0	0	
		12	2	60	0	0	
Fresh cheese[d]	S. aureus			5	2	10^2	10^3
	Coliforms			5	2	10^2	10^3
Soft cheese[d]	S. aureus			5	2	10^2	10^3
	Coliforms			5	2	10^2	10^3
Pasteurized liquid; frozen and dried egg products	APC	2	3	5	2	5×10^4	10^6
	Coliforms	5	3	5	2	10^3	10^3
	Salmonella[b]	10	2	5	0	0	
	(normal routine)	11	2	10	0	0	
		12	2	20	0	0	
	Salmonella[b]	10	2	15	0	0	
	(high-risk populations)	11	2	30	0	0	
		12	2	60	0	0	
Fresh and frozen fish, to be cooked before being eaten	APC	1	3	5	3	5×10^3	10^7
	E. coli		4		3	11	500
	Salmonella[e]	10	2	5	0	0	
	Vibrio parahaemolyticus[e]	7	3	5	2	10^2	10^3
	S. aureus[e]	7	3	5	2	10^3	10^4
Coconut	Salmonella	1	3	5	3	5×10^5	10^7
	Growth not expected	11	2	10	0	0	
	Growth expected	12	2	20	0	0	

[a]Product is cooked, not a roast of beef.
[b]The 25-g analytical unit may be composited.
[c]The case is to be chosen based on whether the hazard is expected to be reduced, unchanged, or increased.
[d]Requirements only for fresh and soft cheese made from pasteurized milk.
[e]For fish known to derive from inshore or inland waters of doubtful bacteriological quality or where fish are to be eaten raw, additional tests may be desirable.

CURRENT STATUS

The ICMSF recommends that a series of steps be taken to manage microbiological hazards for foods intended for international trade. These steps include conducting a risk assessment and an assessment of risk management options, establishing a food safety objective (FSO), and confirming that the FSO is achievable by application of GMPs and Hazard Analysis CCPs (HACCPs).

An *FSO* is a statement of the frequency or maximum concentration of a microbiological hazard in a food that is considered acceptable for consumer protection. Examples of FSOs include staphylococcal enterotoxin levels in

Table 6.7 Examples of various food products for which mandatory microbiological criteria have been established

Product category	Test parameter	Comments
United States		
Dairy products		
Raw milk	Aerobic bacteria	Recommendations of U.S. Public Health Service
Grade A pasteurized milk	Aerobic bacteria Coliforms	Recommendations of U.S. Public Health Service
Grade A pasteurized (cultured) milk	Aerobic bacteria Coliforms	Recommendations of U.S. Public Health Service
Dry milk (whole)	SPC Coliforms	Recommendations of U.S. Public Health Service
Dry milk (nonfat)	SPC Coliforms	Standards of Agricultural Marketing Service (USDA)
Frozen desserts	SPC Coliforms	Recommendations of U.S. Public Health Service
Starch and sugars	Total thermophilic-spore count Flat-sour spores Thermophilic anaerobic spores Sulfide spoilage spores	National Canners Association (National Food Processors Association)
Breaded shrimp	APC *E. coli* *S. aureus*	Food and Drug Administration compliance policy guide
International		
Caseins and caseinates	Total bacterial count Thermophilic organisms Coliforms	Europe
Natural mineral waters	Aerobic mesophilic count Coliforms *E. coli* Fecal streptococci Sporulating sulfite-reducing anaerobes *Pseudomonas aeruginosa*, parasites, and pathogenic organisms	*Codex Alimentarius*
Hot meals served by airlines	*E. coli* *S. aureus* *B. cereus* *C. perfringens* *Salmonella*	Europe
Tomato juice	Mold count	Canada
Fish protein	Total plate count *E. coli*	Canada
Gelatin	Total plate count Coliforms *Salmonella*	Canada

cheese that must not exceed 1 μg/100 g or an aflatoxin concentration in peanuts that should not exceed 15 μg/kg. FSOs are broader in scope than microbiological criteria and are intended to communicate the acceptable level of hazard. *Control measures* are actions and activities that can be used to prevent or eliminate a food safety hazard or to reduce it to an acceptable level. The food industry is responsible for applying GMPs and establishing appropriate control measures, including CCPs, in their processes and HACCP plans. For foods in international commerce, quantitative FSOs provide an objective basis to identify what an importing country is willing to accept in relation to the microbiological safety of its food supply.

Summary

- Microbiological criteria are most effectively applied as part of quality assurance programs in which HACCP and other prerequisite programs are in place.
- Microbiological criteria for foods continue to evolve as new information becomes available.
- Quantitative risk assessment facilitates the establishment of microbiological criteria for foods in international trade and of guidelines for national standards and policies.
- Microbiological criteria differ based on the food and microorganism of concern.
- Microbiological criteria may be based on the presence or absence of specific metabolites.

Suggested reading

International Commission on Microbiological Specifications of Foods. 1994. Choice of sampling plan and criteria for *Listeria monocytogenes*. *Int. J. Food Microbiol.* **22:**89–96.

International Commission on Microbiological Specifications of Foods. 1986. *Microorganisms in Foods 2. Sampling for Microbiological Analysis: Principles and Applications,* 2nd ed. University of Toronto Press, Toronto, Canada.

Jay, J. M. 1992. *Modern Food Microbiology,* 4th ed. Chapman and Hall, New York, N.Y.

Pierson, M. D., and D. A. Corlett, Jr. (ed.). 1992. *HACCP: Principles and Applications.* Van Nostrand Reinhold, New York, N.Y.

Randell, A. W., and A. J. Whitehead. 1997. Codex Alimentarius: food quality and safety standards for international trade. *Rev. Sci. Tech.* **16:**313–321.

Questions for critical thought

1. Violation of which of the following could result in government action: a microbiological standard, guideline, or specification?

2. Fecal coliform counts are more informative under many circumstances than coliform counts. Briefly, describe the difference between the two indicator methods and provide a scenario for the use of each.

3. The microbiological criteria for *L. monocytogenes* in shrimp are as follows: $n = 5$, $c = 0$, and $m = M = 0$. Indicate what each term represents. Based on the information provided, if one sample had 10 CFU, would the product be accepted or rejected?

4. Explain why it is acceptable for some products to be positive for *S. aureus* but not for *Salmonella*.

5. How can the APC be used to screen specifically for a group of microorganisms? Explain your answer.

6. Explain why foods produced and stored under GMPs may be expected to have microbiological profiles different from those of foods produced and stored without the benefit of GMPs.

7. *C. botulinum* is a dangerous pathogen. Why is it acceptable for fresh vegetables to be contaminated with low numbers of this pathogen?

8. Why are microbiological criteria needed for products that will ultimately be cooked?

9. Zero tolerance is set for several pathogens. To what pathogens, and under what conditions, does zero tolerance apply? Speculate why zero tolerance was established for each of the pathogens.

10. How do microbiological criteria affect a food processor? Provide an example of a food and describe the microbiological criteria associated with the product and how those criteria could be met.

SECTION *II*

Gram-Negative Foodborne Pathogenic Bacteria

7

Salmonella Species*

LEARNING OBJECTIVES

The information in this chapter will enable the student to:
- use basic biochemical characteristics to identify *Salmonella*
- understand what conditions in foods favor *Salmonella* growth
- recognize, from symptoms and time of onset, a case of foodborne illness caused by *Salmonella*
- choose appropriate interventions (heat, preservatives, and formulation) to prevent the growth of *Salmonella*
- identify environmental sources of the organism
- understand the roles of *Salmonella* toxins and virulence factors in causing foodborne illness

Outbreak

On a cool fall evening, nearly 1,400 people gathered in a church meeting hall in Chaptico, Md., for the annual fall dinner. The meal included stuffed ham, turkey, fried oysters, mashed potatoes, and gravy. Many of the attendees commented that they had stuffed themselves and could not eat another bite. Others enjoyed the meal so much that they bought carryout platters. The gastric pleasure of the meal soon faded to gastric (stomach) upset for nearly half of the attendees. People who dined or bought carryout became sick, experiencing nausea, cramps, dehydration, and fever, classic symptoms of foodborne illness. Two elderly people died, 100 people sought care at the hospital emergency room, and 700 people reported becoming ill. Microbiological testing of each food demonstrated that the ham was positive for the strain of *Salmonella* associated with the outbreak. Improper cooking and holding of the stuffed ham likely resulted in growth of *Salmonella* to levels capable of causing illness.

INTRODUCTION

National epidemiologic registries continue to underscore the importance of *Salmonella* spp. as leading causes of foodborne bacterial illnesses in humans. Incidents of foodborne salmonellosis tend to dwarf those associated with most other foodborne pathogens (Table 7.1). It is noteworthy that the problem of human salmonellosis from the consumption of contaminated foods generally remains on the increase worldwide (Table 7.2). Notwithstanding

*This chapter was originally written by J. Stan Bailey and John J. Maurer for *Food Microbiology: Fundamentals and Frontiers,* 2nd ed., and has been adapted for use in an introductory text.

Table 7.1 Examples of major foodborne outbreaks of salmonellosis worldwide since 1973

Yr	Location(s)	Vehicle	Serovar(s)	No. of cases[a]	No. of deaths
1973	Canada, United States	Chocolate	Eastbourne	217	0
1973	Trinidad	Milk powder	Derby	3,000[b]	NS[c]
1974	United States	Potato salad	Newport	3,400[b]	0
1976	Spain	Egg salad	Typhimurium	702	6
1976	Australia	Raw milk	Typhimurium PT 9	>500	NS
1977	Sweden	Mustard dressing	Enteritidis PT 4	2,865	0
1981	The Netherlands	Salad base	Indiana	600[b]	0
1981	Scotland	Raw milk	Typhimurium PT 204	654	2
1984	Canada	Cheddar cheese	Typhimurium PT 10	2,700	0
1984	France, England	Liver pâté	Goldcoast	756	0
1984	International	Aspic glaze	Enteritidis PT 4	766	2
1984	United States	Salad bars	Typhimurium	751	0
1985	United States	Pasteurized milk	Typhimurium	16,284	7
1987	China (People's Republic)	Egg drink	Typhimurium	1,113	NS
1987	Norway	Chocolate	Typhimurium	361	0
1988	Japan	Cuttlefish	Champaign	330	0
1988	Japan	Cooked eggs	Salmonella spp.	10,476	NS
1991	United States, Canada	Cantaloupes	Poona	>400	NS
1991	Germany	Fruit soup	Enteritidis	600	NS
1993	France	Mayonnaise	Enteritidis	751	0
1993	United States	Tomatoes	Montevideo	100	0
1993	Germany	Paprika chips	Saint-Paul, Javiana, Rubislaw	>670	0
1994	Finland, Sweden	Alfalfa sprouts	Bovismorbificans	492	0
1994	United States	Ice cream	Enteritidis PT 8	740	0
1995	United States	Orange juice	Hartford, Gaminara	62	0
1995	United States	Restaurant foods	Newport	850	0
1996	United States	Alfalfa sprouts	Montevideo, Meleagrides	481	1
1997	United States	Stuffed ham	Heidelberg	746	1
1998	United States	Toasted oat cereal	Agona	209	0
1998	Canada	Cheddar cheese	Enteritidis PT 8	700	0
1999	Japan	Dried squid	Salmonella spp.	>453	0
1999	Australia	Orange juice	Typhimurium	427	NS
1999	Japan	Cuttlefish chips	Oranienburg	≥1,500	NS
1999	Canada	Alfalfa sprouts	Paratyphi B (Java)	>53	NS
1999	Japan	Peanut sauce	Enteritidis PT 1	644	NS
1999	United States, Canada	Orange juice	Muenchen	>220	0
2000	United States	Orange juice	Enteritidis	>74	0
2000	United States	Mung bean sprouts	Enteritidis	>45	0

[a]Confirmed cases unless stated otherwise.
[b]Estimated number of cases.
[c]Not specified.

recent improvements in procedures for the epidemiologic investigation of foodborne outbreaks in many countries, the global increases in foodborne salmonellosis are considered real. They are not a result of enhanced surveillance programs and/or greater resourcefulness of medical and public health officials. Events such as the ongoing pandemic of egg-borne *Salmonella enterica* serovar Enteritidis infection clearly have an impact on current disease statistics. Major outbreaks of foodborne salmonellosis in the last

Table 7.2 Trends in foodborne salmonellosis

| Country | Annual no. of reported incidents | | | | | No. of cases[a] |
	1985	1987	1989	1991	1993	
Austria	124	151	440	963	—[b]	NA[c]
Bulgaria	15	10	12	13	—	2.1–13.9
Canada	59	53	51	28	43	0.5–3.8
Czechoslovakia	94	120	258	—	—	94.2–258.0
Denmark	12	5	5	11	—	27.0–69.0
England and Wales	372	421	935	936	—	22.0–45.0
Finland	11	3	2	4	—	89.1–153.2
France	7	178	462	477	626	NA
Germany	ND[d]	ND[d]	23	62	—	103.5–242.7
Hungary	116	122	131	110	—	86.7
Japan	—	90	146	159	—	2.9–8.2
Poland	380	690	709	625[e]	—	44.1–91.8
Scotland	133	180	151	115	—	35.0–46.0
United States	79	52	117	122	—	1.1–2.5

[a]Range of cases per 100,000 population within the review period (1985 to 1993).
[b]—, data not published.
[c]NA, not available.
[d]ND, no data; German reunification did not occur until 1989.
[e]Estimated from graphic representation of data.

few decades are of interest because they underline the multiplicity of foods and *Salmonella* serovars that have been implicated in human illness. In 1974, temperature abuse of egg-containing potato salad served at an outdoor barbecue led to an estimated 3,400 human cases of *S. enterica* serovar Newport infection in which cross-contamination of the salad by an infected food handler was suspected. The large Swedish outbreak of serovar Enteritidis (PT) 4 phage type infection in 1977 was attributed to the consumption of a mayonnaise dressing in a school cafeteria. In 1984, Canada experienced its largest outbreak of foodborne salmonellosis. It was attributed to the consumption of Cheddar cheese manufactured from heat-treated and pasteurized milk, and it resulted in >2,700 confirmed cases of serovar Typhimurium PT 10 infection. Manual override of a flow diversion valve reportedly led to the entry of raw milk into vats of thermized and pasteurized cheese milk. The following year, the largest outbreak of foodborne salmonellosis in the United States occurred, involving 16,284 confirmed cases of illness. Although the cause of this outbreak was never determined, a cross-connection between raw and pasteurized milk lines was likely the cause. A large outbreak of salmonellosis affecting >10,000 Japanese consumers was attributed to a cooked egg dish. In 1993, paprika imported from South America was incriminated as the contaminated ingredient in the manufacture of potato chips distributed in Germany. A major outbreak of foodborne salmonellosis that occurred in the United States involved ice cream contaminated with serovar Enteritidis. The transportation of ice cream mix in an unsanitized truck that had previously carried raw eggs was the source of contamination. Despite the general perception about the involvement of meat and egg products as the primary sources of human *Salmonella* infections, many outbreaks in recent years have been associated with tomatoes, orange juice, and vegetable sprouts.

CHARACTERISTICS OF THE ORGANISM

In the early 19th century, clinical pathologists in France first documented the association of human intestinal ulceration with a contagious agent. The disease was later identified as typhoid fever. During the first quarter of the 20th century, great advances occurred in the serological detection of somatic (O) and flagellar (H) antigens within the *Salmonella* group. An antigenic scheme for the classification of salmonellae was later developed and now includes >2,400 serovars.

Biochemical Identification

Salmonella spp. are facultatively anaerobic gram-negative rod-shaped bacteria belonging to the family *Enterobacteriaceae*. Although members of the genus are motile via peritrichous flagella (flagella distributed uniformly over the cell surface), nonflagellated variants, such as *S. enterica* serovars Pullorum and Gallinarum, and nonmotile strains resulting from dysfunctional flagella do occur. Salmonellae are chemoorganotrophic (able to utilize a wide range of organic substrates), with an ability to metabolize nutrients by both respiratory and fermentative pathways. The bacteria grow optimally at 37°C and catabolize D-glucose and other carbohydrates with the production of acid and gas. Salmonellae are oxidase negative and catalase negative and grow on citrate as a sole carbon source. They generally produce hydrogen sulfide, decarboxylate lysine, and ornithine and do not hydrolyze urea. Many of these traits have formed the basis for the presumptive biochemical identification of *Salmonella* isolates. Accordingly, a typical *Salmonella* isolate would produce acid and gas from glucose in triple sugar iron (TSI) agar medium and would not utilize lactose or sucrose in TSI or in differential plating media, such as brilliant green, xylose lysine desoxycholate, and Hektoen enteric agars. Additionally, typical salmonellae readily produce an alkaline reaction from the decarboxylation of lysine to cadaverine in lysine ion agar, generate hydrogen sulfide gas in TSI and lysine ion media, and fail to hydrolyze urea. The dynamics of genetic variability (the result of bacterial mutations and conjugative intra- and intergeneric exchange of plasmids encoding determinant biochemical traits) continue to reduce the proportion of "typical" *Salmonella* biotypes. *Salmonella* utilization of lactose and sucrose is plasmid mediated, and the occurrence of lac^+ and/or suc^+ biotypes in clinical isolates and food materials is of public health concern. These isolates could escape detection on the disaccharide-dependent plating media that are commonly used in hospital and food industry laboratories. Bismuth sulfite agar remains a medium of choice for isolating salmonellae because, in addition to its high level of selectivity, it responds effectively to the production of extremely low levels of hydrogen sulfide gas. The diagnostic hurdles created by the changing patterns of disaccharide utilization by *Salmonella* are being further complicated by the increasing occurrence of biotypes that cannot decarboxylate lysine, that possess urease activity, that produce indole, and that readily grow in the presence of KCN. The variability in biochemical traits is leading to the replacement of traditional testing protocols with molecular technologies targeted at the identification of stable genes and/or their products that are unique to the genus *Salmonella*.

Taxonomy and Nomenclature

According to the World Health Organization (WHO) Collaborating Centre for Reference and Research on *Salmonella* (Institut Pasteur, Paris, France), *S. enterica* and *Salmonella bongori* currently include 2,443 and 20 serovars,

Table 7.3 The genus *Salmonella*[a]

Salmonella species and subspecies (no.)	No. of serovars
S. enterica subsp. *enterica* (I)	1,454
S. enterica subsp. *salamae* (II)	489
S. enterica subsp. *arizonae* (IIIa)	94
S. enterica subsp. *diarizonae* (IIIb)	324
S. enterica subsp. *houtenae* (IV)	70
S. enterica subsp. *indica* (VI)	12
S. bongori (V)	20
Total	2,463

[a]Source: J.-Y. D'Aoust, *In* B. M. Lund, A. C. Baird-Parker, and G. W. Gould (ed.), *The Microbiological Safety and Quality of Food* (Aspen Publishers Inc., Gaithersburg, Md., 2000).

respectively (Table 7.3). The Centers for Disease Control and Prevention have adopted as their official nomenclature the following scheme. The genus *Salmonella* contains two species, each of which contains multiple serovars (Table 7.3). The two species are *S. enterica*, the type species, and *S. bongori*, which was formerly subspecies V. *S. enterica* is divided into six subspecies, which are referred to by a roman numeral and a name (I, *S. enterica* subsp. *enterica*; II, *S. enterica* subsp. *salamae*; IIIa, *S. enterica* subsp. *arizonae*; IIIb, *S. enterica* subsp. *diarizonae*; IV, *S. enterica* subsp. *houtenae*; and VI, *S. enterica* subsp. *indica*). *S. enterica* subspecies are differentiated biochemically and by genomic relatedness. *Salmonella* serovars that have been linked to recent cases of foodborne illness include *S. enterica* serovars Enteritidis, Typhimurium, Newport, and Stanley.

Serological Identification

The aim of serological testing procedures is to determine the complete antigenic formulas of individual *Salmonella* isolates. Antigenic determinants include somatic (O) lipopolysaccharides (LPS) on the external surface of the bacterial outer membrane, flagellar (H) antigens associated with the peritrichous flagella, and the capsular (Vi [virulence]) antigen, which occurs only in serovars Typhi, Paratyphi C, and Dublin. The heat-stable somatic antigens are classified as major or minor antigens. The former category consists of antigens, such as the somatic factors O:4 and O:3, which are specific determinants for the somatic groups B and E, respectively. In contrast, minor somatic antigenic components, such as O:12, are nondiscriminatory, as evidenced by their presence in different somatic groups. Smooth (S) variants are strains with well-developed serotypic LPS that readily agglutinate with specific antibodies, whereas rough (R) variants exhibit incomplete LPS antigens, resulting in weak or no agglutination with *Salmonella* somatic antibodies. Flagellar (H) antigens are heat-labile proteins, and individual *Salmonella* strains may produce one (monophasic) or two (diphasic) sets of flagellar antigens. Although serovars such as Dublin produce a single set of flagellar antigens, most serovars can elaborate either of two sets of antigens, i.e., phase 1 and phase 2 antigens. Capsular (K) antigens, commonly encountered in members of the family *Enterobacteriaceae*, are limited to the Vi antigen in the genus *Salmonella*.

We will use serovar Infantis (6,7:r:1,5 [see below]) as an example. Commercially available polyvalent somatic antisera each consist of a mixture of antibodies specific for a limited number of major antigens, e.g., polyvalent B (poly-B) antiserum (Difco Laboratories, Detroit, Mich.) recognizes somatic groups C_1, C_2, F, G, and H. Following a positive agglutination with poly-B antiserum, single-group antisera representing the five somatic groups included in the poly-B reagent would be used to define the serogroup of the isolate. The test isolate would react with the C_1 group antiserum, indicating that antigens 6 and 7 are present. Flagellar antigens would then be determined by broth agglutination reactions using poly-H antisera or the Spicer-Edwards series of antisera. In the former assay, a positive agglutination reaction with one of the five polyvalent antisera (poly-A to -E; Difco) would lead to testing with single-factor antisera to specifically identify the phase 1 and/or phase 2 flagellar antigens present. Agglutination in poly-C flagellar antiserum and subsequent reaction of the isolate with single-group H antisera would confirm the presence of the r antigen (phase 1). The empirical antigenic formula of the isolate would then be (6,7:r). Phase reversal in semisolid agar supplemented with r antiserum would immobilize phase 1 salmonellae at or near the point of inoculation, thereby facilitating the recovery of phase 2 cells from the edge of the zone of migration. Serological testing of phase 2 cells with poly-E and 1-complex antisera would confirm the presence of the flagellar 1 factor. Confirmation of the flagellar 5 antigen with single-factor antiserum would yield the final antigenic formula 6,7:r:1,5, which corresponds to serovar Infantis. A similar analytical approach would be used with the Spicer-Edwards poly-H antisera in which the identification of flagellar antigens would arise from the pattern of agglutination reactions among the four Spicer-Edwards antisera and with three additional polyvalent antisera, including the L, 1, and e,n complexes.

Physiology

Growth and Survival

Temperature. *Salmonella* spp. are resilient and can adapt to extreme environmental conditions. Some *Salmonella* strains can grow at elevated temperatures (54°C), and others exhibit psychrotrophic properties (they are able to grow in foods stored at 2 to 4°C) (Table 7.4). Moreover, preconditioning of cells to low temperatures can greatly increase the growth and

Table 7.4 Physiological limits for the growth of *Salmonella* spp. in foods and bacteriological media

Parameter	Limits (time to double in number)		Product	Serovar(s)
	Minimum	Maximum		
Temp (°C)	2 (24 h)		Minced beef[a]	Typhimurium
	2 (2 days)		Minced chicken[b]	Typhimurium
	4 (≤10 days)		Shell eggs[b]	Enteritidis
		54.0[c]	Agar medium	Typhimurium
pH	3.99[d]		Tomatoes	Infantis
	4.05[e]		Liquid medium	Anatum, Tennessee, Senftenberg
		9.5	Egg washwater[b]	Typhimurium
a_w	0.93[f]		Rehydrated dried soup[b]	Oranienburg

[a]Naturally contaminated.
[b]Artificially contaminated.
[c]Mutants selected to grow at elevated temperature.
[d]Growth within 24 h at 22°C.
[e]Acidified with HCl or citric acid; growth within 24 h at 30°C.
[f]Growth within 3 days at 30°C.

survival of salmonellae in refrigerated food products. Studies of the maximum temperatures for growth of *Salmonella* spp. in foods are generally lacking. Notwithstanding an early report of growth of salmonellae in inoculated custard and chicken à la king at 45.6°C, more recent evidence indicates that prolonged exposure of mesophilic strains to thermal stress conditions results in mutants of serovar Typhimurium capable of growth at 54°C.

Salmonella can survive for extended periods in foods stored at freezing and room temperatures. Several factors can influence the survival of salmonellae during frozen storage of foods: (i) the composition of the freezing matrix, (ii) the kinetics of the freezing process, (iii) the physiological state of the foodborne salmonellae, and (iv) serovar-specific responses. The viability of salmonellae in dry foods stored at 25°C decreases with increasing storage temperature and with increasing moisture content.

Many factors may enhance the heat resistance of *Salmonella* and other foodborne bacterial pathogens in food ingredients and finished products. The heat resistance of *Salmonella* spp. increases as the water activity (a_w) of the heating matrix (food) decreases. The type of solutes used to alter the a_w of the heating menstruum can influence the level of acquired heat resistance. Other important features associated with this adaptive response include the greater heat resistance of salmonellae grown in nutritionally rich versus minimal media, of stationary-phase versus logarithmic-phase cells, and of salmonellae previously stored in a dry environment. The ability of *Salmonella* to acquire greater heat resistance following exposure to sublethal temperatures is equally important. This adaptive response has potentially serious implications for the safety of thermal processes that expose or maintain food products at marginally lethal temperatures. Changes in the fatty acid composition of cell membranes in heat-stressed *Salmonella* provide a greater proportion of saturated membrane phospholipids. This results in reduced fluidity of the bacterial cell membrane and an associated increase in membrane resistance to heat damage. The likelihood that other protective cellular functions are triggered by heat shock stimuli cannot be ruled out.

The following scenario involving *Salmonella* contamination of a chocolate confectionery product demonstrates the complex interactions of food, temperature, and organism. The survival of salmonellae in dry-roasted cocoa beans can lead to contamination of in-line and finished products. Thermal inactivation of salmonellae in molten chocolate is difficult. This is because the time-temperature conditions that are required to eliminate the pathogen in this sucrose-containing product with low a_w would likely result in an organoleptically (off taste and off odor) unacceptable product. The problem is further compounded by the ability of salmonellae to survive for many years in the finished product when it is stored at room temperature. Clearly, effective decontamination of raw cocoa beans and stringent in-plant control measures to prevent cross-contamination of in-line products are of great importance in this food industry.

pH. The physiological adaptability of *Salmonella* spp. is further demonstrated by their ability to grow at pH values ranging from 4.5 to 9.5, with an optimum pH for growth of 6.5 to 7.5 (Table 7.4). Of the various organic and inorganic acids used in product acidification, propionic and acetic acids are more bactericidal than the common food-associated lactic and citric acids. Interestingly, the antibacterial action of organic acids decreases with increasing length of the fatty acid chain. Early research on the ability of *Salmonella* to grow in acidic environments revealed that wild-type

strains preconditioned on pH gradient plates could grow in liquid and solid media at considerably lower pH values than the starting wild-type strains. This raises concerns regarding the safety of fermented foods, such as cured sausages and fermented raw milk products. The starter culture-dependent acidification of fermented foods could provide a favorable environment for the growth of salmonellae in the product to a state of increased acid tolerance. The growth and/or enhanced survival of salmonellae during the fermentative process would result in a contaminated ready-to-eat product. Ultimately, this may promote survival of *Salmonella* in the host.

Brief exposure of serovar Typhimurium to mild acid environments of pH 5.5 to 6.0 (preshock) followed by exposure of the adapted cells to a pH of ≤4.5 (acid shock) triggers a complex acid tolerance response (ATR). This permits the survival of the microorganism under extreme acid environments (pH 3.0 to 4.0). The response translates into an induced synthesis of 43 acid shock and outer membrane proteins, a reduced growth rate, and pH homeostasis. It is demonstrated by the bacterial maintenance of internal pH values of 7.0 to 7.1 and 5.0 to 5.5 upon sequential exposure of cells to an external pH of 5.0 and 3.3, respectively. Aside from the log-phase ATR, *Salmonella* also possesses a stationary-phase ATR that provides greater acid resistance than the log-phase ATR. This is induced at pHs of <5.5 and functions maximally at pH 4.3. This stationary-phase ATR induces the synthesis of only 15 shock proteins. The induction of the remaining acid-protective mechanism associated with stationary-phase *Salmonella* is independent of the external pH and dependent on the alternative sigma factor (σ^s) encoded by the *rpoS* locus. The mechanism enhances the ability of stationary-phase cells to survive in hostile environmental conditions. Thus, three possibly overlapping cellular systems confer acid tolerance on *Salmonella* spp. These are (i) the pH-dependent, *rpoS*-independent log-phase ATR; (ii) the pH-dependent and *rpoS*-independent stationary-phase ATR; and (iii) the pH-independent, *rpoS*-dependent stationary-phase acid resistance. These systems likely operate in the acidic environments that prevail in fermented and in acidified foods and in phagocytic cells of the infected host.

Acid stress can also enhance bacterial resistance to other adverse environmental conditions. The growth of serovar Typhimurium at pH 5.8 increases thermal resistance at 50°C, enhances tolerance to high osmotic stress (2.5 M NaCl), and produces greater surface hydrophobicity and an increased resistance to the antibacterial lactoperoxidase system and surface-active agents, such as crystal violet and polymyxin B.

Osmolarity. High salt concentrations have long been recognized for their ability to extend the shelf life of foods by inhibiting microbial growth. Foods with a_w values of <0.93 do not support the growth of salmonellae. Although *Salmonella* is generally inhibited in the presence of 3 to 4% NaCl, bacterial salt tolerance increases with increasing temperature in the range of 10 to 30°C.

The pH, salt concentration, and temperature of the microenvironment can exert profound effects on the growth kinetics of *Salmonella* spp. It is now accepted that *Salmonella* spp. have the ability to grow under acidic (pH < 5.0) conditions or in environments of high salinity (>2% NaCl) with increasing temperature. Interestingly, the presence of salt in acidified foods can reduce the antibacterial action of organic acids in which low concentrations of NaCl or KCl stimulate the growth of serovar Enteritidis in broth medium acidified to pH 5.19 with acetic acid.

Modified atmosphere. Concerns about the ability of *Salmonella* to survive under extremes of pH, temperature, and salinity are further heightened by the widespread refrigerated storage of foods packaged under vacuum or modified atmosphere to prolong shelf life. Gaseous mixtures consisting of 60 to 80% (vol/vol) CO_2 with varying proportions of N_2 and/or O_2 can inhibit the growth of aerobic spoilage microorganisms, such as *Pseudomonas* spp., without promoting the growth of *Salmonella* spp. The safety of modified-atmosphere- and vacuum-packaged foods that contain high levels of salt is coming under question, since anaerobic conditions may enhance *Salmonella* salt tolerance.

RESERVOIRS

Salmonella spp. will continue to be significant human pathogens in the global food supply for several reasons, including their presence in the environment; intensive farming practices used in the meat, fish, and shellfish industries (which promotes spread among animals); and the recycling of slaughterhouse by-products into animal feeds. Poultry meat and eggs are predominant reservoirs of *Salmonella* spp. in many countries. This overshadows the importance of other animal meats, such as pork, beef, and mutton, as potential vehicles of infection. To address the problem of *Salmonella* in meat products, the U.S. Department of Agriculture Food Safety Inspection Service requires the meat and poultry industries to implement Hazard Analysis Critical Control Point (HACCP) plans in all of their plants. The Food Safety Inspection Service conducts *Salmonella* testing to verify that the implemented HACCP is helping to control *Salmonella* on finished products. The rate of *Salmonella* contamination has been reduced for all meat animals since the implementation of HACCP.

The continuing pandemic of human serovar Enteritidis infections associated with the eating of raw or lightly cooked shell eggs and egg-containing products further emphasizes the importance of poultry as vehicles of human salmonellosis. The need for sustained and stringent bacteriological control of poultry husbandry practices is urgent. This egg-related pandemic is of particular concern, because the problem arises from transovarian transmission of serovar Enteritidis into the interior of the egg prior to shell deposition. The viability of these internalized serovar Enteritidis organisms remains unaffected by egg surface sanitizing practices. Increased efforts to intervene and reduce the prevalence of serovar Enteritidis have led to significant reductions in egg-borne outbreaks in the United Kingdom and the United States.

Rapid depletion of wild stocks of fish and shellfish in recent years have greatly increased the importance of the international aquaculture industry. The feeding of raw meat scraps and offal, feces potentially contaminated with typhoid and paratyphoid salmonellae, and animal feeds that may harbor *Salmonella* to reared species is common in developing countries. Accordingly, aquaculture farmers are relying heavily on antibiotics applied at subtherapeutic levels to safeguard the vigor of farmed fish and shellfish. The use of antibiotics creates a serious public health concern, since these antimicrobial agents are the mainstay treatment for systemic salmonellosis in humans.

Fruits and vegetables have gained notoriety in recent years as vehicles of human salmonellosis. The situation has developed from the increased global export of fresh and dehydrated fruits and vegetables from countries that enjoy tropical and subtropical climates. The prevailing hygienic conditions during the production, harvesting, and distribution of products in these countries do not always meet minimum standards and may facilitate product contamination. Operational changes in favor of field irrigation with treated effluents, washing of fruits and vegetables with disinfected water, education of local workers on the hygienic handling of fresh produce, and greater protection of products from environmental contamination during all phases of handling would greatly enhance the safety of fresh fruits and vegetables.

CHARACTERISTICS OF DISEASE

Symptoms and Treatment

Human *Salmonella* infections can lead to several clinical conditions, including enteric (typhoid) fever, uncomplicated enterocolitis, and systemic infections by nontyphoid microorganisms. Enteric fever is a serious human disease associated with the typhoid and paratyphoid strains. Symptoms of enteric fever appear after a period of incubation ranging from 7 to 28 days and may include diarrhea, prolonged and spiking fever, abdominal pain, headache, and prostration. Diagnosis of the disease relies on the isolation of the infective agent from blood or urine samples in the early stages of the disease or from stools after the onset of clinical symptoms. An asymptomatic chronic carrier state commonly follows the acute phase of enteric fever. The treatment of enteric fever is based on supportive therapy and/or the use of chloramphenicol, ampicillin, or trimethoprim-sulfamethoxazole to eliminate the systemic infection. Marked global increases in the resistance of typhoid and paratyphoid organisms to these antibacterial drugs in the last decade has limited their effectiveness in human therapy.

Human infections with nontyphoid *Salmonella* strains commonly result in symptoms 8 to 72 h after ingestion of the invasive pathogen. The illness is usually self-limiting, and remission of the characteristic nonbloody diarrheal stools and abdominal pain usually occurs within 5 days of the onset of symptoms. The successful treatment of uncomplicated cases may require only supportive therapy, such as fluid and electrolyte replacement. Antibiotics are not used in such episodes, because they prolong the carrier state. Asymptomatic persistence of salmonellae in the gut is probably due to antibiotic-dependent repression of native gut microflora. Human infections with nontyphoid strains can also degenerate into systemic infections and precipitate various chronic conditions.

Salmonella can induce chronic conditions, such as aseptic reactive arthritis, Reiter's syndrome, and ankylosing spondylitis. Bacterial prerequisites for the onset of these chronic diseases include the ability of the bacterial strain to infect mucosal surfaces, the presence of outer membrane LPS, and the ability to invade host cells.

Preventative Measures

The global impact of typhoid and paratyphoid salmonellae on human health led to the early development of parenteral vaccines consisting of heat-, alcohol-, or acetone-killed cells. The phenol- and heat-killed typhoid vaccine is widely used. However, this vaccine can cause adverse reactions, including fever, headache, pain, and swelling at the site of injection. The pandemic of poultry- and egg-borne serovar Enteritidis infections that continues to afflict consumers underscores the potential benefit of vaccines for human and veterinary applications. Live attenuated vaccines continue to generate great research interest because such preparations induce strong and durable immune responses in vaccines.

Antibiotic Resistance

Current global increases in the number of antibiotic-resistant *Salmonella* serovars in humans and farm animals are alarming. The major public health concern is that *Salmonella* will become resistant to antibiotics used in human medicine, thereby greatly reducing therapeutic options and threatening the lives of infected individuals. The liberal administration of antimicrobial agents in hospitals and other treatment centers contributes significantly to the emergence of many resistant strains. The emergence of antimicrobial resistance in bacterial pathogens, many of which can be traced to food-producing animals, has generated much controversy and public media attention.

Because evolutionary processes offer genetic and phenotypic variability for resistance genes, there will always be some level of background resistance in bacterial populations. However, without selective pressures, resistance levels are low, and so are the chances for therapy failures. Resistance in *Salmonella* to a single antibiotic was first noted in the early 1960s. The appearance of serovar Typhimurium definitive type 104 (DT104) in the late 1980s raised major concerns because of its multiple resistances against ampicillin, chloramphenicol, streptomycin, sulfonamides, and tetracycline. Some strains are also resistant to gentamicin, trimethoprim, fluoroquinolones, and other antibiotics. Fluoroquinolones belong to a class of newer antimicrobial agents with a broad spectrum of activity, high efficiency, and widespread application in human and veterinary medicine. An increase in the resistance of DT104 strains to fluoroquinolones has led many European countries to ban the nonhuman use of fluoroquinolones.

The genetic determinants for *Salmonella* virulence and antibiotic resistance can occur on the same plasmid. This increases the disease potential of *Salmonella* DT104. In the next few decades (or perhaps sooner), we may well witness the international scientific community feverishly trying to develop new drugs to replace those currently in use, since they will no longer be effective against the multiple-antibiotic-resistant *Salmonella*.

No studies have conclusively linked the subtherapeutic use of antibiotics in agriculture with the development of resistance. The obvious need for data led to the joint development in 1996 of the National Antimicrobial Resistance Monitoring System by the Food and Drug Administration's Center for Veterinary Medicine, the Centers for Disease Control and Prevention, and the U.S. Department of Agriculture. The purpose of this program is to provide baseline data so that changes in the antimicrobial resistance and susceptibility of bacteria can be identified. Additionally, in June 2000 the World Health Organization released a set of global principles aimed at reducing the risks related to the use of antimicrobials in food animals. The new recommendations were designed for use by governments, veterinary and other professional societies, industry, and academia.

INFECTIOUS DOSE

Newborns, infants, the elderly, and immunocompromised individuals are more susceptible to *Salmonella* infections than healthy adults. The incompletely developed immune system in newborns and infants, the frequently

Table 7.5 Human infectious doses of *Salmonella*[a]

Food	Serovar(s)	Infectious dose (CFU)[b]
Eggnog	Meleagridis	10^6–10^7
	Anatum	10^5–10^7
Goat cheese	Zanzibar	10^5–10^{11}
Carmine dye	Cubana	10^4
Imitation ice cream	Typhimurium	10^4
Chocolate	Eastbourne	10^2
Hamburger	Newport	10^1–10^2
Cheddar cheese	Heidelberg	10^2
Chocolate	Napoli	10^1–10^2
Cheddar cheese	Typhimurium	10^0–10^1
Chocolate	Typhimurium	$\leq 10^1$
Paprika potato chips	Saint-Paul, Javiana, Rubislaw	$\leq 4.5 \times 10^1$
Alfalfa sprouts	Newport	$\leq 4.6 \times 10^2$
Ice cream	Enteritidis	$\leq 2.8 \times 10^1$

[a]Adapted from J.-Y. D'Aoust, *Int. J. Food Microbiol.* **24:**11–31, 1994.
[b]CFU, colony-forming units.

weak and/or delayed immunological responses in the elderly and debilitated persons, and the generally low gastric acid production in infants and seniors facilitate the intestinal colonization and systemic spread of salmonellae in these populations.

The ingestion of only a few *Salmonella* cells can be infectious (Table 7.5). Determinant factors in salmonellosis are not limited to the immunological heterogeneity within human populations and the virulence of infecting strains but may also include the chemical composition of contaminated food. A common denominator of foods associated with low infectious doses is a high fat content (e.g., cocoa butter in chocolate, milk fat in cheese, and animal fat in meat). *Salmonella* organisms may become entrapped within hydrophobic lipid micelles and be protected against the bactericidal action of gastric acidity. Food producers, processors, and distributors need to be reminded that low levels of salmonellae in a finished food product could lead to serious public health consequences. This could undermine the reputation and economic viability of the incriminated food manufacturer.

PATHOGENICITY AND VIRULENCE FACTORS

Specific and Nonspecific Human Responses

The presence of viable salmonellae in the human intestinal tract confirms the successful evasion of ingested organisms from nonspecific host defenses. Antibacterial lactoperoxidase in saliva, gastric acidity, mucoid secretions from intestinal cells, intestinal peristalsis, and sloughing of luminal epithelial cells synergistically inhibit bacterial colonization of the intestine. In addition to these hurdles to bacterial infection, the antibacterial action of phagocytic cells coupled with the immune responses mounts a formidable defense against the systemic spread of *Salmonella*. The human diarrheagenic response to foodborne salmonellosis results from the migration of the pathogen in the oral cavity to intestinal tissues and mesenteric lymph follicles (enterocolitis). The failure of host defense systems to hold the invasive *Salmonella* in check can degenerate into septicemia and other chronic clinical conditions.

Attachment and Invasion

The establishment of a human *Salmonella* infection depends on the ability of the bacterium to attach (colonization) and enter (invasion) intestinal cells. Salmonellae must successfully compete with indigenous gut microflora for suitable attachment sites on the intestinal wall. Upon contact with epithelial cells, *Salmonella* produces proteinaceous appendages on its surface. Following bacterial attachment, signalling between the pathogen and the host cell results in *Salmonella* invasion of intestinal cells. Histologically, *Salmonella* invasion is characterized by membrane ruffling of epithelial cells and programmed cell death (apoptosis) of epithelial and phagocytic cells. Diarrhea associated with salmonellae now appears to be in response to bacterial invasion of intestinal cells rather than the action of a putative enterotoxin.

Several sets of genes are involved in the invasion process. The *inv* pathogenicity island is a multigenic locus consisting of 30 genes. Many of these genes encode the synthesis of enzymes and transcriptional activators responsible for the regulation, expression, and translocation of important effectors to the surfaces of host cells. Genes in the pathogenicity island are useful targets for polymerase chain reaction (PCR)-based detection of *Salmonella* in foods.

Growth and Survival within Host Cells

In contrast to several bacterial pathogens, such as *Yersinia, Shigella,* and enteroinvasive *Escherichia coli,* which replicate within the cytoplasm of host cells, salmonellae are confined to endocytotic vacuoles in which bacterial replication begins within hours following internalization. The infected vacuoles move from the apical to the basal pole of the host cell, where *Salmonella* organisms are released into the tissue. A bacterial surface mechanism that facilitates the migration of salmonellae deeper into layers of tissue upon their release has been tentatively identified.

The systemic migration of *Salmonella* exposes it to phagocytosis and to the antibacterial conditions in the cytoplasm of the host defense cells. Survival of the bacterial cell within the hostile confines of the phagocytic cells determines the host's fate. Whether the host develops enteric fever from *Salmonella* infection is determined partly by the genetics of the host and the salmonellae. Intracellular pathogens have developed different strategies to survive within phagocytic cells. These include (i) escaping the phagosomes, (ii) inhibiting acidification of the phagosomes, (iii) preventing phagosome-lysosome fusion, and (iv) withstanding the toxic environment of the phagolysosome. Although there is evidence to support the ability of *Salmonella* to prevent acidification of the phagosomes, fusion with the lysosome, or maturation of the phagolysosome, *Salmonella* is a hardy bacterium that can survive and replicate within the bactericidal milieu of the acidified phagolysosome.

Virulence Plasmids

Virulence plasmids are large DNA structures that replicate in synchrony with the bacterial chromosome. These plasmids contain many virulence loci ranging from 30 to 60 megadaltons in size and occur with a frequency of one or two copies per chromosome. The presence of virulence plasmids within the genus *Salmonella* is limited and has been confirmed in serovars Typhimurium, Dublin, Gallinarum-Pullorum, Enteritidis, Choleraesuis, and Abortusovis. The highly infectious serovar Typhi does not carry a virulence plasmid. Although limited in its distribution in nature, the plasmid is self-transmissible. Gene products from the transcription of this plasmid aid the systemic spread and infection of tissues other than the intestine but not *Salmonella* adhesion to and invasion of epithelial cells. More specifically, virulence plasmids may enable carrier strains to rapidly multiply within host cells and overwhelm host defense mechanisms. These plasmids also confer on the salmonellae the ability to induce lysis of macrophages, illicit inflammatory response, and enteritis in their animal hosts. The *Salmonella* virulence plasmids contain highly conserved nucleotide sequences and exhibit functional homology in which plasmid transfer from a wild-type serovar to another, plasmid-cured serovar restores the virulence of the recipient strain. The *Salmonella* virulence plasmid also contains a fimbrial operon that encodes an adhesin involved in colonization of the small intestine.

Other Virulence Factors

Siderophores are yet another part of the *Salmonella* virulence weaponry. These elements retrieve essential iron from host tissues to drive key cellular functions, such as the electron transport chain and enzymes associated with iron cofactors. To this end, *Salmonella* must compete with host transferrin, lactoferrin, and ferritin ligands for available iron. For example, transferrin scavenges tissue fluids for Fe^{3+} ions to form Fe^{3+}-transferrin complexes that bind to surface host cell receptors. Upon internalization, the complexes dissociate and the released Fe^{3+} is complexed with ferritin for intracellular storage. Siderophores do not appear to be the only mechanism by which *Salmonella* can acquire iron from its host.

Diarrheagenic enterotoxin may be a *Salmonella* virulence factor. The release of toxin into the cytoplasm of infected host cells precipitates an activation of adenyl cyclase localized in the epithelial cell membrane and a marked increase in the cytoplasmic concentration of cyclic adenosine monophosphate (AMP) in host cells. The simultaneous fluid exsorption into the intestine results from a net secretion of Cl^- ions and depressed Na^+ absorption at the level of the intestinal villi. Enterotoxigenicity is a virulence phenotype of *Salmonella*, including serovar Typhi, which is expressed within hours following bacterial contact with the targeted host cells.

In addition to enterotoxin, *Salmonella* strains generally elaborate a thermolabile cytotoxic protein, which is localized in the bacterial outer membrane. Hostile environments, such as acidic pH and elevated (42°C) temperature, cause release of the toxin, possibly as a result of induced bacterial lysis. The virulence attribute of cytotoxin stems from its inhibition of protein synthesis and its lysis of host cells, thereby promoting the dissemination of viable salmonellae into host tissues.

Three additional virulence determinants located within or on the external surface of the *Salmonella* outer membrane are worth mentioning. The capsular polysaccharide Vi antigen occurs in most strains of serovar Typhi, in a few strains of serovar Paratyphi C, and rarely in serovar Dublin. The length of LPS that protrudes from the bacterial outer membrane not only defines the rough (short-LPS) and smooth (LPS) phenotypes, but also plays an important role in preventing attack by the host immune system. Porins are outer membrane proteins that function as transmembrane (outer membrane) channels in regulating the influx of nutrients, antibiotics, and other small molecular species. Low osmolarity, low nutrient availability, and low temperature can regulate the expression of these genes.

Summary

- Intense animal husbandry practices in meat and poultry production and processing industries continue to make raw poultry and meats principal vehicles of human foodborne salmonellosis.
- Unless changes in agricultural and aquacultural practices are implemented, the prevalence of human foodborne salmonellosis will continue.
- Salmonellae are classified as serovars of two species.
- The emergence of multiple-antibiotic-resistant strains may be linked to the use of antibiotics in agriculture.
- Salmonellae quickly adapt to environments of high salinity and low pH, creating a problem with respect to control in food matrices.
- Virulence genes are located on the chromosome in pathogenicity islands and in large virulence plasmids.
- Epidemiologic evidence suggests a low infective dose.
- *Salmonella* invasion proteins may activate apoptosis in epithelial and phagocytic cell types.

Suggested reading

Brenner, F. W., R. G. Villar, F. J. Angulo, R. Tauxe, and B. Swaminathan. 2000. *Salmonella* nomenclature. *J. Clin. Microbiol.* **38:**2465–2467.

D'Aoust, J.-Y. 1991. Pathogenicity of foodborne *Salmonella*. *Int. J. Food Microbiol.* **12:**17–40.

D'Aoust, J.-Y. 1994. *Salmonella* and the international food trade. *Int. J. Food Microbiol.* **24:**11–31.

D'Aoust, J.-Y. 2000. *Salmonella*, p. 1233–1299. *In* B. M. Lund, A. C. Baird-Parker, and G. W. Gould (ed.), *The Microbiological Safety and Quality of Food.* Aspen Publishers Inc., Gaithersburg, Md.

Falkow, S., R. R. Isberg, and D. A. Portnoy. 1992. The interaction of bacteria with mammalian cells. *Annu. Rev. Cell Biol.* **8:**333–363.

Glynn, J. R., and D. J. Bradley. 1992. The relationship of infecting dose and severity of disease in reported outbreaks of *Salmonella* infections. *Epidemiol. Infect.* **109:**371–388.

Questions for critical thought

1. The typing scheme for *Salmonella* is based on two antigens. What is meant by the nomenclature (6,7:r:1,5)?
2. Serovar Enteritidis has been difficult to control in poultry. What unique characteristic(s) of this pathogen contributes to the control problem?
3. *Salmonella* DT104 is resistant to multiple antibiotics. Within the bacterial cell, where are the genes located and why is that important?
4. How might the ATR contribute to bacterial survival in food and subsequently in the host?
5. How could motility and antibodies against the flagella be used for detection of *Salmonella?*
6. To prevent foodborne illness associated with large meals, such as holiday dinners, what food-handling practices should be employed?
7. Individuals on antibiotic therapy are more prone to infection with *Salmonella*. Explain why.
8. Based on the impacts of pH, temperature, salinity, and a_w on the growth and survival of *Salmonella*, develop a hurdle technology to prevent the growth of *Salmonella* in the food of your choice.

8

Campylobacter Species*

LEARNING OBJECTIVES

The information in this chapter will enable the student to:

- use basic biochemical characteristics to identify *Campylobacter* spp.
- understand what conditions in foods favor *Campylobacter* sp. growth
- recognize, from symptoms and time of onset, a case of foodborne illness caused by a *Campylobacter* sp.
- choose appropriate interventions (heat, preservatives, and formulation) to prevent the growth of *Campylobacter* spp.
- identify environmental sources of the organism
- understand the roles of *Campylobacter* sp. toxins and virulence factors in causing foodborne illness

Outbreak

In many states, the sale of unpasteurized milk is illegal, since it is an important vehicle for the transmission of a number of foodborne pathogens, including *Campylobacter jejuni*. In Wisconsin, to circumvent the law against the sale of unpasteurized milk, an organic dairy operation initiated a cow-leasing program. Customers paid a fee to lease "part" of a cow, which farm operators milked; the milk from all leased cows was stored in a bulk tank (a large stainless steel tank). Customers could either pick up the milk at the farm or have the milk delivered to their homes. The farm operation was well run, meeting grade A standards for a farm shipping milk to a pasteurization plant. Customers participating in the cow lease program found out the hard way about campylobacteriosis. A total of 75 persons developed diarrhea, abdominal cramps, fever, and nausea after drinking unpasteurized milk from the organic dairy farm. The genetic fingerprints of *C. jejuni* isolates obtained from stool samples of patients and from milk samples collected on the farm were identical. *C. jejuni* can survive in unpasteurized milk held at 4°C. Finding *C. jejuni* in a farm environment is not unusual, and there are many routes by which milk can be contaminated. Pasteurization effectively kills *C. jejuni*.

INTRODUCTION

Campylobacter jejuni subsp. *jejuni* (hereafter referred to as *C. jejuni*) is one of many species and subspecies within the genus *Campylobacter*, family *Campylobacteraceae*. During the 1980s, there was an explosion of information published on these bacteria, as is evident from the 1984 edition of *Bergey's*

*This chapter was originally written by Irving Nachamkin for *Food Microbiology: Fundamentals and Frontiers,* 2nd ed., and has been adapted for use in an introductory text.

101

Manual of Systematic Bacteriology, published at a time when there were only eight species and subspecies within the genus *Campylobacter.* Since then, many investigators have become interested in studying the taxonomy and clinical importance of campylobacters. This has greatly expanded the number of genera and species associated with this group of bacteria.

CHARACTERISTICS OF THE ORGANISM

The genera *Campylobacter* and *Arcobacter* are included in the family *Campylobacteraceae.* The family *Campylobacteraceae* includes at present 18 species and subspecies within the genus *Campylobacter* and 4 species in the genus *Arcobacter. Campylobacter* is the type genus within the family *Campylobacteraceae.* The organisms are curved, S-shaped, or spiral rods that are 0.2 to 0.9 μm wide and 0.5 to 5 μm long. They are gram-negative, nonsporeforming rods that may form spherical or coccoid bodies in old cultures or cultures exposed to air for prolonged periods. The organisms are motile by means of a single polar unsheathed flagellum at one or both ends. The various species are microaerobic (5% oxygen and 10% carbon dioxide), with a respiratory type of metabolism. Some strains grow aerobically or anaerobically. An atmosphere containing increased hydrogen may be required by some species for microaerobic growth. *C. jejuni* and *Campylobacter coli* have genomes ~1.7 megabases in size, as determined by pulsed-field gel electrophoresis (PFGE), which is about one-third the size of the *Escherichia coli* genome.

ENVIRONMENTAL SUSCEPTIBILITY

C. jejuni is susceptible to a variety of environmental conditions that make it unlikely to survive for long periods of time outside the host. The organism does not grow at temperatures below 30°C, is microaerobic, and is sensitive to drying, high-oxygen conditions, and low pH. The decimal reduction time for campylobacters varies and depends upon the food source and temperature. Thus, the organism should not survive in food products brought to adequate cooking temperatures. The organisms are susceptible to gamma irradiation (1 kilogray), but the rate of killing depends on the type of product being processed. Irradiation is less effective for frozen materials than for refrigerated or room temperature meats. Early-log-phase cells are more susceptible than cells grown to log or stationary phase. *Campylobacter* spp. are more radiation sensitive than other foodborne pathogens, such as salmonellae and *Listeria monocytogenes.* Irradiation treatment that is effective against salmonellae and *L. monocytogenes* should be sufficient to kill *Campylobacter* spp.

   Campylobacter spp. are susceptible to low pH and are killed at pH 2.3. Campylobacters remain viable and grow in bile at 37°C and survive better in feces, milk, water, and urine held at 4°C than in material held at 25°C. The maximum periods of viability of *Campylobacter* spp. at 4°C were 3 weeks in feces, 4 weeks in water, and 5 weeks in urine. Freezing reduces the number of *Campylobacter* organisms in contaminated poultry, but even after being frozen to −20°C, small numbers of *Campylobacter* organisms can be recovered.

RESERVOIRS AND FOODBORNE OUTBREAKS

Many animals serve as reservoirs of *C. jejuni* for human disease. Reservoirs for infection include rabbits, rodents, wild birds, sheep, horses, cows, pigs, poultry, and domestic pets (Table 8.1). Contaminated vegetables and shell-

Table 8.1 Reservoirs and disease-associated species in the family *Campylobacteraceae*[a]

Organism	Reservoir(s)	Disease, sequelae, or comments	
		Human	**Animals**
C. jejuni subsp. *jejuni*	Human, other mammals, birds	Diarrhea, systemic illness, GBS	Diarrhea in primates
C. jejuni subsp. *doylei*	Unknown	Diarrhea	
C. fetus subsp. *fetus*	Cattle, sheep	Systemic illness, diarrhea	Abortion
C. fetus subsp. *venerealis*	Cattle		Infertility
C. coli	Pigs, birds	Diarrhea	
C. lari	Birds, dogs	Diarrhea	
C. upsaliensis	Domestic pets	Diarrhea	Diarrhea
C. hyointestinalis subsp. *hyointestinalis*	Cattle, pigs, hamsters, deer	Rare; proctitis, diarrhea	Proliferative enteritis
C. hyointestinalis subsp. *lawsonii*	Pigs		
C. mucosalis	Pigs	Rare; diarrhea	Proliferative enteritis
C. hyoilei	Pigs		
C. sputorum biovar sputorum	Humans	Oral cavity abscesses	Genital tract of bulls; abortion in sheep
C. sputorum biovar paraureolyticus	Cattle	Diarrhea	
C. sputorum biovar faecalis	Cattle, sheep		Enteritis
C. concisus	Humans	Periodontal disease	
C. curvus	Humans	Periodontal disease	
C. rectus	Humans	Periodontal disease, pulmonary infections	
C. showae	Humans	Periodontal disease	
C. helveticus	Domestic pets		Diarrhea
C. gracilis	Humans	Infections of head, neck, other sites	
Arcobacter butzleri	Cattle, pigs	Diarrhea	Diarrhea, abortion
Arcobacter cryaerophilus	Cattle, sheep, pigs	Diarrhea, bacteremia	Abortion
Arcobacter skirrowii	Cattle, sheep, pigs		Abortion, diarrhea; isolated from genital tract of bulls
Arcobacter nitrofigilis	Plants		

[a]Source: M. B. Skirrow, *J. Comp. Pathol.* **111:**113–149, 1994.

fish may also be vehicles of infection. Campylobacters are frequently isolated from water and water supplies, and water has been the source of infection in some outbreaks. *C. jejuni* can remain dormant in water in a state that has been called *viable but nonculturable* (VBNC). Under unfavorable conditions, the organism essentially remains dormant and cannot be easily recovered on growth media. The role of these forms as a source of infection for humans is not clear.

From 1978 to 1996, there were 111 outbreaks of *Campylobacter* enteritis reported to the Centers for Disease Control and Prevention, affecting 9,913 individuals (Table 8.2). The vehicles of *Campylobacter* outbreaks have changed over the past 2 decades. Water and unpasteurized milk were responsible for over half of the outbreaks between 1978 and 1987, whereas other foods accounted for >80% of outbreaks between 1988 and 1996. In a 10-year review of outbreaks from 1981 to 1990, 20 outbreaks occurred that affected 1,013 individuals who drank raw milk. The attack rate was 45%. At least one outbreak occurred each year, and most of the outbreaks involved children who had gone on field trips to dairy farms.

Table 8.2   Foodborne and waterborne outbreaks of *Campylobacter* infections reported in the United States, by vehicle, 1978 to 1996[a]

Origin	No. of outbreaks	No. of outbreak-associated cases
Foodborne		
Milk	30	1,212
Chicken	2	16
Turkey	1	11
Beef	1	24
Other meat	2	30
Eggs	1	26
Fruits	4	227
Other foods	4	251
Multiple foods	10	411
Unknown food	42	2,775
Waterborne		
Community water supply	8	5,068
Other water supply	4	104
Total	109	10,155

[a]Reprinted from C. R. Friedman, J. Neimann, H. C. Wegener, and R. V. Tauxe, p. 121–138, *in* I. Nachamkin and M. J. Blaser (ed.), *Campylobacter,* 2nd ed. (ASM Press, Washington, D.C., 2000).

The seasonal distribution of outbreaks is somewhat different from that of sporadic cases. Milkborne and waterborne outbreaks tend to occur in the spring and fall but do not occur frequently in the summer months. Sporadic cases are usually most frequent during the summertime. In contrast to the relatively low occurrence of outbreaks caused by campylobacters, the bacteria have been identified in many studies as being among the most common causes of foodborne illness in the United States. In 1995, the Centers for Disease Control and Prevention, the U.S. Department of Agriculture, and the Food and Drug Administration developed an active surveillance system for foodborne diseases, including *Campylobacter* infections, called the Foodborne Diseases Active Surveillance Network, also known as FoodNet. Using a variety of data, including FoodNet incidence data, U.S. Census Bureau data, and data from other sources, it is estimated that there are 2.4 million *Campylobacter* infections in the United States each year. Since campylobacteriosis is self-limiting, the true number of cases that occur each year can only be estimated. This is similar to the incidence of infection in the United Kingdom and other developed nations. Cases occur more often during the summer months and usually follow the ingestion of improperly handled or cooked food, primarily poultry products. Other exposures include drinking raw milk, contaminated surface water, overseas travel, and contact with domestic pets (Table 8.3).

Table 8.3   Isolation of *Campylobacter* from different food sources[a]

Product	% Positive samples
Chicken	14–98
Turkey	3–25
Duck	48
Goose	38
Cow milk	0–12.3
Goat milk	0
Ewe's milk	0
Beef	0–23.6
Pork	1–23.5
Lamb	0–15.5
Sheep	3
Offal	47
Mussels	47–69
Oysters	6–27
Vegetables	<5

[a]Source: W. Jacobs-Reitsma, p. 467–481, *in* I. Nachamkin and M. J. Blaser (ed.), *Campylobacter,* 2nd ed. (ASM Press, Washington, D.C., 2000).

CHARACTERISTICS OF DISEASE

C. jejuni and *C. coli*

Most *Campylobacter* species are associated with intestinal tract infection. Infections in other areas of the body are common with some species, such as *Campylobacter fetus* subsp. *fetus* (hereafter referred to as *C. fetus*). *C. jejuni* and *C. coli* have been recognized since the 1970s as agents of gastrointestinal tract infection.

C. jejuni and *C. coli* are the most common *Campylobacter* species associated with diarrheal illness, and they are indistinguishable. The ratio of *C. jejuni* to *C. coli* is not known, because most laboratories do not routinely distinguish between the organisms. In the United States, an estimated ~5 to 10% of cases attributed to *C. jejuni* are actually due to *C. coli;* this figure may be higher in other parts of the world.

A spectrum of illness can occur during *C. jejuni* or *C. coli* infection, and patients may be *asymptomatic* (show no sign of illness) to severely ill. Symptoms and signs usually include fever, abdominal cramping, and diarrhea (with or without blood) that lasts several days to >1 week. Symptomatic infections are usually self-limited, but relapses may occur in 5 to 10% of untreated patients. Reoccurrence of abdominal pain is common. Other infections and diseases include bacteremia, bursitis, urinary tract infection, meningitis, endocarditis, peritonitis, erythema nodosum, pancreatitis, abortion and neonatal sepsis, reactive arthritis, and Guillain-Barré syndrome (GBS). Deaths directly attributable to *C. jejuni* infection are rarely reported.

C. jejuni and *C. coli* are susceptible to many antibiotics, including macrolides, fluoroquinolones, aminoglycosides, chloramphenicol, and tetracycline. Erythromycin is the drug of choice for treating *C. jejuni* gastrointestinal tract infections, but ciprofloxacin is a good alternative drug. Early therapy of *Campylobacter* infection with erythromycin or ciprofloxacin is effective in eliminating the campylobacters from stool and may also reduce the duration of symptoms associated with infection.

C. jejuni is generally susceptible to erythromycin, with resistance rates of <5%. Rates of erythromycin resistance in *C. coli* vary, with up to 80% of strains having resistance in some studies. Although ciprofloxacin is effective in treating *Campylobacter* infections, resistant strains have emerged. The resistance may be linked to the use of similar antibiotics (fluoroquinolones) in poultry. The Food and Drug Administration Center for Veterinary Medicine has proposed banning the use of fluoroquinolones in poultry.

Other *Campylobacter* Species

In contrast to *C. jejuni*, *C. fetus* is primarily associated with bacteremia and non-intestinal tract infections. *C. fetus* is also associated with abortions, septic arthritis, abscesses, meningitis, endocarditis, mycotic aneurysm, thrombophlebitis, peritonitis, and salpingitis. Although intestinal tract infections occur with this species, the number of cases is low, because the organism does not grow well at 42°C. *C. fetus* is susceptible to the antibiotic cephalothin. *C. fetus* subsp. *venerealis* is not associated with human infection.

EPIDEMIOLOGIC SUBTYPING SYSTEMS USEFUL FOR INVESTIGATING FOODBORNE ILLNESSES

Many typing systems have been devised to study the epidemiology of *Campylobacter* infections. These methods include biotyping, serotyping, bacteriocin sensitivity, detection of preformed enzymes, lectin binding, phage typing, multilocus enzyme electrophoresis, and molecularly based methods, such as PFGE, ribotyping, and restriction fragment length polymorphism combined with polymerase chain reaction (PCR) methods (see chapter 5).

The most frequently used phenotypic systems are biotyping and serotyping. The two major serotyping schemes used worldwide detect heat-labile and O antigens. Serotyping systems are simple to perform and have good ability to discriminate between strains.

During the past 5 years, molecularly based methods have come to the forefront for studying the epidemiology of *Campylobacter* infections. PFGE is one of the most powerful methods for molecular analysis of *Campylobacter* species. It has excellent discriminatory power. A variety of PCR-based methods, including ribotyping, random amplified polymorphic DNA, amplified fragment length polymorphism, multiplex PCR-restriction fragment length polymorphism, and real-time PCR have all been used in various applications to study *Campylobacter* spp. (see chapter 5 for explanations of these methods).

INFECTIVE DOSE AND SUSCEPTIBLE POPULATIONS

C. jejuni is susceptible to low pH, and hence stomach acidity kills most campylobacters. The infective dose of *C. jejuni* is not high, with <1,000 organisms causing illness. The ability to cause infection may vary greatly between strains. Human volunteers (18%) became ill when infected with 10^8 colony-forming units of a strain designated A3249, but 46% of the volunteers became ill when infected with another strain, 81-176. In an outbreak of *Campylobacter* infection after the ingestion of raw milk, the number of individuals who became ill and the severity of illness seemed to be linked to the amount of milk ingested (the more milk consumed, the greater the number of *Campylobacter* organisms consumed).

The greatest number of infections in the United States occur in young children and young adults 20 to 40 years of age. The incidence of *Campylobacter* infection in developing countries, such as Mexico and Thailand, may be 10 to 100 times higher than in the United States. In developing countries, in contrast to developed countries, campylobacters are frequently isolated from individuals who may or may not have diarrheal disease. Most infections occur in infancy and early childhood, and the incidence decreases with age. As a child matures, its immune system can fight off *Campylobacter* infections, thereby decreasing the number of illnesses. Travelers to developing countries may become infected with *Campylobacter.*

The elderly are more likely to get *bacteremia* (presence of bacteria in the blood) than any other group (infants, young adults, etc.). Continual diarrheal illness and bacteremia may occur in immunocompromised individuals, such as in patients with human immunodeficiency virus infection.

VIRULENCE FACTORS AND MECHANISMS OF PATHOGENICITY

Little is known about the mechanism by which *C. jejuni* causes human disease. *C. jejuni* can cause an enterotoxigenic-like illness with loose or watery diarrhea or an inflammatory intestinal disease with fever. The mechanism of disease may be *invasive* (entering into intestinal cells), since some individuals have blood in their feces and may develop bacteremia. A major problem in understanding the pathogenesis of *Campylobacter* infection is the lack of suitable animal models.

Cell Association and Invasion

The uptake of *C. jejuni* by host cells occurs by intimate binding of the bacterium with the host cell surface by using many types of adhesins. Following contact, *C. jejuni* produces at least 14 new proteins. Production corresponds to a rapid increase in uptake and changes in the host cell membrane. At a later point, additional proteins are made and secreted into the cytoplasm of host target cells. Following attachment and internalization, *Campylobacter*-infected cells release molecules that promote the recruitment of white blood cells to the site of the infection. Induction of host cell death by *C. jejuni* may enhance the survival and spread of the pathogen.

Flagella and Motility

Campylobacter species are motile and have a single polar, unsheathed flagellum at one or both ends. Motility and flagella are important determinants for the entry process. *Campylobacter* colonization and/or infection in a variety of animal models is dependent on intact motility and full-length flagella; however, other factors may be involved. Two genes are involved in the expression of the flagellar filament, and motility is essential for colonization. Components of intestinal mucin, particularly L-fucose, attract *C. jejuni.* Movement of *C. jejuni* toward such compounds may be important in the pathogenesis of infection.

Toxins

C. jejuni may produce an enterotoxin similar to cholera toxin. Evidence for production of this enterotoxin is provided by damage to the intestines of rabbits in the rabbit ileal loop model. The strains tested produced intestinal bleeding, swelling, and accumulation damage. No genetic evidence is available to support the presence of this toxin. The role of cytolethal distending toxin (CDT) in pathogenesis is unknown. Most strains of *C. jejuni* have *cdt* genes, but the amounts of toxin produced by the strains vary.

Other Factors

The regulation of genes in response to environmental changes is becoming an important area of *Campylobacter* research. Campylobacters are microaerobic, and *C. jejuni* has a higher temperature for optimal growth, 42°C, than other intestinal bacterial pathogens. Understanding the effects of environmental signals on the growth, metabolism, and pathogenicity of *Campylobacter* strains will have a major impact on the ability to control campylobacters in the environment and food chain. Such pathways include response to iron; oxidative stress; temperature regulation, including cold and heat shock responses; and starvation.

Autoimmune Diseases

Campylobacter infection is a major trigger of GBS, an acute immune-mediated paralytic disorder affecting the peripheral nervous system. The pathogenesis of GBS induced by *C. jejuni* is not clear. The bacterial lipopolysaccharide causes an immune response. The antibodies produced recognize not only the lipopolysaccharide but also peripheral nerve tissue. This is likely a major mechanism of *Campylobacter*-induced GBS.

IMMUNITY

Infection with *Campylobacter* species results in protective immunity and is likely antibody mediated. Rechallenge with the same strains 28 days after the initial challenge resulted in protection against illness but not necessarily against colonization by *C. jejuni*. People who are regularly exposed to the pathogen through drinking raw milk are less likely to become ill. In developing countries, where *Campylobacter* infections are endemic, immunity to *Campylobacter* infection appears to be age dependent. As a child matures, he or she is less likely to become ill.

Flagellin is an important immunogen during *Campylobacter* infection. Antibodies against this protein correlate to some degree with protective immunity. Breast-feeding provides protection against *Campylobacter* infection in developing countries. Antibodies against flagellin present in breast milk appear to be associated with protection of infants against infection.

Other strategies to prevent the spread of infection to humans include improved hygiene practices during broiler chicken production, such as decontamination of water supplies; use of competitive exclusion flora, which may prevent *C. jejuni* colonization of young chicks; and immunological approaches through the use of animal vaccines.

Summary _____

- Campylobacteriosis linked to consumption of contaminated food is the most prevalent form of foodborne illness in the United States.
- *C. jejuni* is commonly isolated from poultry.
- Thermoprocessing easily inactivates campylobacters.
- *Campylobacter* infection is a major trigger of GBS.
- Motility is essential for colonization.
- Five to 10% of cases attributed to *C. jejuni* are actually due to *C. coli*.
- The infective dose of *C. jejuni* may be <1,000 organisms.

Suggested reading

Blaser, M. J. 2000. *Campylobacter jejuni* and related species, p. 2276–2285. *In* G. L. Mandell, J. E. Bennett, and R. Dolin (ed.), *Principles and Practice of Infectious Diseases.* Churchill Livingstone, Philadelphia, Pa.

Friedman, C. R., J. Neimann, H. C. Wegener, and R. V. Tauxe. 2000. Epidemiology of *Campylobacter jejuni* infections in the United States and other industrialized nations, p. 121–138. *In* I. Nachamkin and M. J. Blaser (ed.), *Campylobacter,* 2nd ed. ASM Press, Washington, D.C.

Jacobs-Reitsma, W. 2000. *Campylobacter* in the food supply, p. 467–481. *In* I. Nachamkin and M. J. Blaser (ed.), *Campylobacter,* 2nd ed. ASM Press, Washington, D.C.

Konkel, M. E., L. A. Joens, and P. F. Mixter. 2000. Molecular characterization of *Campylobacter jejuni* virulence determinants, p. 217–240. *In* I. Nachamkin and M. J. Blaser (ed.), *Campylobacter,* 2nd ed. ASM Press, Washington, D.C.

Pickett, C. L. 2000. *Campylobacter* toxins and their role in pathogenesis, p. 179–190. *In* I. Nachamkin and M. J. Blaser (ed.), *Campylobacter,* 2nd ed. ASM Press, Washington, D.C.

Ransom, G. M., B. Kaplan, A. M. McNamara, and I. K. Wachsmuth. 2000. *Campylobacter* prevention and control: the USDA-Food Safety and Inspection Service role and new food safety approaches, p. 511–528. *In* I. Nachamkin and M. J. Blaser (ed.), *Campylobacter,* 2nd ed. ASM Press, Washington, D.C.

Skirrow, M. B. 1994. Diseases due to *Campylobacter, Helicobacter* and related bacteria. *J. Comp. Pathol.* **111:**113–149.

Questions for critical thought

1. How does oxygen concentration influence the growth of *C. jejuni*?
2. How is *C. jejuni* infection linked to GBS?
3. What is the impact of antibiotic-resistant *C. jejuni* on human health? How might the use of antibiotics in poultry production practices influence resistance?
4. Campylobacters can exist in a VBNC state. In what environments do they enter this state? Can the organism recover from this state? Why is entry into the VBNC state important for survival of the organism?
5. What practices can be implemented to control campylobacters in food?
6. Describe how you would differentiate *C. jejuni* from *E. coli* O157:H7 microbiologically and based on signs and symptoms of illness.
7. The incidence of *Campylobacter* infection in developing countries, such as Mexico and Thailand, is severalfold higher than in the United States. What factors may account for such a great difference in the infection rate?
8. *C. jejuni* is zoonotic. What does this statement mean?
9. You are a new employee at a start-up biotechnology firm and have been asked to develop a vaccine against *C. jejuni*. Outline your plan of attack. Consider what you know about the pathogen and the illness it causes and what others have done in the past with respect to vaccine development.

9

Enterohemorrhagic
*Escherichia coli**

LEARNING OBJECTIVES

The information in this chapter will enable the student to:

- use basic biochemical characteristics to identify *Escherichia coli* O157:H7
- understand what conditions in foods favor *E. coli* O157:H7 growth
- recognize, from symptoms and time of onset, a case of foodborne illness caused by *E. coli* O157:H7
- choose appropriate interventions (heat, preservatives, and formulation) to prevent the growth of *E. coli* O157:H7
- identify environmental sources of *E. coli* O157:H7
- understand the role of *E. coli* O157:H7 toxins and virulence factors in causing foodborne illness

Outbreak

In June 1998, physicians in a small Wyoming town noted an increase in bloody diarrhea among town residents. Further investigation revealed that visitors attending a family reunion and tourists from surrounding states had also become ill. A total of 157 people became ill, but no deaths occurred. Epidemiological evidence demonstrated that illness was significantly associated with nonchlorinated water from the municipal water system. This outbreak underscores the ability of *E. coli* O157:H7 to survive in water and the need to use only protected waters that have been sufficiently processed.

Attending the local county fair is often the highlight of the summer for people living in rural Lane County, Oregon. Walking through the barns to observe the sheep, goats, and other farm animals is entertaining and fascinating. However, for 82 people who attended the fair during the summer of 2002, the visit left them sick from *E. coli* O157:H7 infection. The culprit was not food, water, or even direct contact with animals but rather *E. coli* O157:H7 organisms that were in the air, based on microbiological analysis of dust and grime samples taken inside the goat and sheep barn. There were 74 confirmed cases (based on positive stool culture) and 8 presumed cases; approximately two-thirds of the cases involved children <6 years old. Other outbreaks linked to airborne *E. coli* O157:H7 in barns have occurred.

Consuming a fresh, cool, crisp salad on a hot summer day pleases the palate and cools the body. For teenagers attending a cheerleading camp in Washington, consuming a healthy salad nearly turned into a deadly experience. Twenty-nine teenagers became ill, with one requiring dialysis as a result of extensive kidney

---

*This chapter was originally written by Jianghong Meng, Michael P. Doyle, Tong Zhao, and Shaohua Zhao for *Food Microbiology: Fundamentals and Frontiers*, 2nd ed., and has been adapted for use in an introductory text.

damage. Investigators determined that the lettuce used in the salad was contaminated with *E. coli* O157:H7. How, when, and where the lettuce became contaminated could not be determined. During investigation of the outbreak, the U.S. Food and Drug Administration issued a warning that manure- or feces-contaminated irrigation water sometimes taints fresh produce.

INTRODUCTION

Large outbreaks of *E. coli* O157:H7 infection involving hundreds of cases have occurred in the United States, Canada, Japan, and the United Kingdom. The largest outbreak worldwide occurred from May to December 1996 in Japan, involving >11,000 reported cases. In the same year, 21 elderly people died in a large outbreak involving 501 cases in central Scotland. Although *E. coli* O157:H7 is still the predominant serotype of enterohemorrhagic *E. coli* (EHEC) in the United States, Canada, the United Kingdom, and Japan, an increasing number of outbreaks and sporadic cases related to other EHEC serotypes have occurred. A large epidemic involving several thousand cases of *E. coli* O157:NM infection occurred in Swaziland and South Africa following consumption of contaminated surface water. In continental Europe, Australia, and Latin America, non-O157 EHEC infections are more common than *E. coli* O157:H7 infections. Details of many reported foodborne and waterborne outbreaks of EHEC infection are provided in Table 9.1. There are no distinguishing biochemical phenotypes for non-O157 EHEC serotypes, making screening for these bacteria difficult and labor-intensive. The prevalence of non-O157 EHEC infections may be greatly underestimated.

Categories of *E. coli*

Diarrheagenic *E. coli* is categorized into specific groups based on virulence properties, mechanisms of pathogenicity, clinical syndromes, and distinct O:H serotypes. *E. coli* isolates are serologically differentiated based on three major surface antigens that allow serotyping: the O (somatic), H (flagellar), and K (capsule) antigens. The categories of diarrheagenic *E. coli* are enteropathogenic *E. coli* (EPEC), enterotoxigenic *E. coli* (ETEC), enteroinvasive *E. coli* (EIEC), diffusely adhering *E. coli* (DAEC), enteroaggregative *E. coli* (EAEC), and EHEC. This chapter focuses primarily on EHEC, which causes the most severe illness.

EPEC

EPEC can cause severe diarrhea in infants, especially in developing countries. EPEC was previously associated with outbreaks of diarrhea in nurseries in developed countries. The major O serogroups associated with illness include O55, O86, O111ab, O119, O125ac, O126, O127, O128ab, and O142. Humans are an important reservoir. EPEC organisms have been shown to induce lesions in cells to which they adhere, and they can invade epithelial cells.

ETEC

ETEC is a major cause of infantile diarrhea in developing countries. It is also the bacterium most frequently responsible for traveler's diarrhea. ETEC colonizes the small intestine via fimbrial colonization factors and produces a heat-labile or heat-stable enterotoxin that causes fluid accumulation and diarrhea. The most frequent ETEC serogroups include O6, O8,

Table 9.1 Representative foodborne and waterborne outbreaks of *E. coli* O157:H7 and other EHEC infections[a]

Yr	Mo	Location	No. of cases (deaths)	Setting	Vehicle
1982	2	Oregon	26	Community	Ground beef
1982	5	Michigan	21	Community	Ground beef
1985		Canada	73 (17)	Nursing home	Sandwiches
1987	6	Utah	51	Custodial institution	Ground beef; person to person
1988	10	Minnesota	54	School	Precooked ground beef
1989	12	Missouri	243	Community	Water
1990	7	North Dakota	65	Community	Roast beef
1991	11	Massachusetts	23	Community	Apple cider
1991	7	Oregon	21	Community	Swimming water
1992	12	Oregon	9	Community	Raw milk
1993	1	California, Idaho, Nevada, Washington	732 (4)	Restaurant	Ground beef
1993	3	Oregon	47	Restaurant	Mayonnaise?
1993	7	Washington	16	Church picnic	Pea salad
1993	8	Oregon	27	Restaurant	Cantaloupe
1994	11	Washington, California	19	Home	Salami
1995	10	Kansas	21	Wedding	Punch, fruit salad
1995	11	Oregon	11	Home	Venison jerky
1995	7	Montana	74	Community	Leaf lettuce
1995	9	Maine	37	Camp	Lettuce
1996	5, 6	Connecticut, Illinois	47	Community	Mesclun lettuce
1996	7	Osaka, Japan	7,966 (3)	Community	White radish, sprouts
1996	10	California, Washington, Colorado	71 (1)	Community	Apple juice
1996	11	Central Scotland	501 (21)	Community	Cooked meat
1997	5	Illinois	3	School	Ice cream bars
1997	7	Michigan	60	Community	Alfalfa sprouts
1997	11	Wisconsin	13	Church banquet	Meatballs, coleslaw
1998	6	Wisconsin	63	Community	Cheese curds
1998	6	Wyoming	114	Community	Water
1998	7	North Carolina	142	Restaurant	Coleslaw
1998	7	California	28	Prison	Milk
1998	8	New York	11	Delicatessen	Macaroni salad
1998	9	California	20	Church	Cake
1999	8	New York	900 (2)	Fair	Well water

[a] *E. coli* O157:H7 unless otherwise noted.

O15, O20, O25, O27, O63, O78, O85, O115, O128ac, O148, O159, and O167. Humans are the principal reservoir of ETEC strains that cause human illness.

EIEC

EIEC causes nonbloody diarrhea and dysentery similar to that caused by *Shigella* spp. by invading and multiplying within intestinal epithelial cells. Like that of *Shigella,* the invasive capacity of EIEC is associated with the presence of a large plasmid (ca. 140 megadaltons [MDa]) that encodes several outer membrane proteins involved in invasiveness. The antigenicities of these outer membrane proteins and the O antigens of EIEC are closely related. The principal site of bacterial localization is the colon, where EIEC organisms invade and grow within epithelial cells, causing cell death. Humans are a major reservoir, and the serogroups most frequently associated with illness include O28ac, O29, O112, O124, O136, O143, O144, O152, O164, and O167. Among these serogroups, O124 is the serogroup most commonly encountered.

DAEC

DAEC has been associated with diarrhea primarily in young children who are older than infants. The relative risk of DAEC-associated diarrhea increases with age from 1 to 5 years. The reason for this age-related infection is unknown. DAEC is most commonly of serogroup O1, O2, O21, or O75. Typical symptoms of DAEC infection are mild diarrhea without blood. DAEC strains adhere in a random fashion to HEp-2 or HeLa cell lines. DAEC does not usually produce heat-labile or heat-stable enterotoxin or elevated levels of Shiga toxin (Stx).

EAEC

EAEC is associated with persistent diarrhea in infants and children in several countries worldwide. These *E. coli* strains are different from the other types of pathogenic *E. coli* because of their ability to produce a characteristic pattern of aggregative adherence on HEp-2 cells. EAEC adheres with the appearance of stacked bricks to the surfaces of HEp-2 cells. Serogroups associated with EAEC include O3, O15, O44, O77, O86, O92, O111, and O127. A gene probe derived from a plasmid associated with EAEC strains has been developed to identify *E. coli* strains of this type. More epidemiological information is needed to determine the significance of EAEC as a cause of diarrheal disease.

EHEC

EHEC was first recognized as a human pathogen in 1982, when *E. coli* O157:H7 was identified as the cause of two outbreaks of bloody diarrhea. Since then, other serotypes of *E. coli*, such as O26, O111, and sorbitol-fermenting O157:NM, also have been associated with cases of bloody diarrhea and are classified as EHEC. Serotype O157:H7 is the main cause of EHEC-associated disease in the United States and many other countries. All EHEC strains produce factors *cytotoxic* (deadly) to African green monkey kidney (Vero) cells. Thus, they are named verotoxins or (Stxs) because of their similarity to the Stx produced by *Shigella dysenteriae*. Stx-producing *E. coli* infections are linked to a severe and sometimes fatal condition, hemolytic-uremic syndrome (HUS). *E. coli* of many different serotypes produce Stxs; hence, they have been named Stx-producing *E. coli* (or EHEC). More than 200 serotypes of EHEC have been isolated from humans, but only those strains that cause bloody diarrhea are considered to be EHEC. Major non-O157:H7 EHEC serotypes include O26:H11, O111:H8, and O157:NM. Several outbreaks of infection with the EHEC *E. coli* O111 have occurred worldwide. In June 1999, an outbreak of *E. coli* O111:H8 infection involving 58 cases occurred at a teenage cheerleading camp in Texas. Contaminated ice was the implicated vehicle. Since *E. coli* O157:H7 is the most common serotype of EHEC and because more is known about this serotype than about other serotypes of EHEC, this chapter focuses on *E. coli* O157:H7 (Box 9.1).

BOX 9.1

E. coli **O157:H7**

E. coli O157:H7 emerged as a pathogen in 1982. The initial cases of illness were predominantly linked to the consumption of beef products, but recently, illnesses have been linked to contaminated fruits and vegetables, recreational water (lakes, swimming pools, and water parks), and person-to-person spread. The organism has acquired many tools (e.g., production of Stx and the ability to produce AE lesions) that facilitate its survival in a host but that can cause debilitating illness. As with *Shigella*, illness may be caused by <100 cells. In the United States, *E. coli* O157:H7 is the predominant EHEC strain. In other parts of the world, this is not necessarily the case.

CHARACTERISTICS OF *E. COLI* O157:H7 AND NON-O157 EHEC

E. coli is a common part of the normal microbial population in the intestinal tracts of humans and other warm-blooded animals. Most *E. coli* strains are harmless; however, some strains are pathogenic and cause diarrheal disease. At present, a total of 167 O antigens, 53 H antigens, and 74 K antigens have been identified. It is considered necessary to determine only the O and H antigens in order to serotype strains of *E. coli* associated with diarrheal disease. The O antigen identifies the serogroup of a strain, and the H antigen identifies its serotype. The application of serotyping to isolates associated with diarrheal disease has shown that particular serogroups often fall into one category of diarrheagenic *E. coli*. However, some serogroups, such as O55, O111, O126, and O128, appear in more than one category.

E. coli O157:H7 was first identified as a foodborne pathogen in 1982, although there had been prior isolation of the organism, which was identified later among isolates at the Centers for Disease Control and Prevention. The earlier isolate was from a Californian woman who had bloody diarrhea in 1975. In addition to production of Stx(s), most strains of *E. coli* O157:H7 also have several characteristics uncommon in most other *E. coli* strains, i.e., the inability to grow well, if at all, at temperatures of ≥44.5°C in *E. coli* broth; inability to ferment sorbitol within 24 h; inability to produce β-glucuronidase (i.e., inability to hydrolyze 4-methylumbelliferyl-D-glucuronide), possession of an attaching-and-effacing (AE) gene (*eae*), and carriage of a 60-MDa plasmid. Non-O157 EHEC strains do not share these growth and metabolic characteristics, although they all produce Stx(s) and most carry the *eae* gene and large plasmid.

Acid Tolerance

Unlike most foodborne pathogens, many strains of *E. coli* O157:H7 are unusually tolerant of acidic environments. The minimum pH for *E. coli* O157:H7 growth is 4.0 to 4.5, but this is dependent upon the interaction of the pH with other factors. For instance, organic acid sprays containing acetic, citric, or lactic acid do not affect the level of *E. coli* O157:H7 on beef. *E. coli* O157:H7, when inoculated at high levels, has survived fermentation, drying, and storage of fermented sausage (pH 4.5) for up to 2 months at 4°C and has survived in mayonnaise (pH 3.6 to 3.9) for 5 to 7 weeks at 5°C and for 1 to 3 weeks at 20°C and in apple cider (pH 3.6 to 4.0) for 10 to 31 days or for 2 to 3 days at 8 or 25°C, respectively.

Three systems in *E. coli* O157:H7 are involved in acid tolerance: an acid-induced oxidative system, an acid-induced arginine-dependent system, and a glutamate-dependent system. The oxidative system is less effective in protecting the organism from acid stress than the arginine-dependent and glutamate-dependent systems. The alternate sigma factor RpoS is required for oxidative acid tolerance but is only partially involved with the other two systems. Once induced, the acid-tolerant state can persist for a prolonged period (≥28 days) at refrigeration temperature. More importantly, induction of acid tolerance in *E. coli* O157:H7 can also increase tolerance of other environmental stresses, such as heating, radiation, and antimicrobials.

Antibiotic Resistance

When *E. coli* O157:H7 was first associated with human illness, the pathogen was susceptible to most antibiotics affecting gram-negative bacteria. However, it is becoming increasingly resistant to antibiotics. *E. coli* O157:H7 strains isolated from humans, animals, and food have developed resistance to multiple antibiotics, with streptomycin-sulfisoxazole-tetracycline being the most common resistance profile. Non-O157 EHEC strains isolated from humans and animals also have acquired antibiotic resistance, and some are resistant to multiple antimicrobials commonly used in human and animal medicine. Antibiotic-resistant EHEC strains possess a selective advantage over other bacteria colonizing the intestines of animals that are treated with antibiotics (therapeutically or subtherapeutically). Therefore, antibiotic-resistant EHEC strains may become the primary *E. coli* strains present under antibiotic selective pressure and thus more prevalent in feces. Perhaps fortunately, antibiotic therapy is not recommended in cases of human illness, since the disease may become worse through the release of endotoxin and Stxs from the dead bacteria.

Inactivation by Heat and Irradiation

E. coli O157:H7 is not more heat resistant than other pathogens. The presence of certain compounds in food can protect the organism. The presence of fat protects *E. coli* O157:H7 in ground beef, with *D* values (see chapter 2) for lean (2.0% fat) and fatty (30.5% fat) ground beef of 4.1 and 5.3 min at 57.2°C and 0.3 and 0.5 min at 62.8°C, respectively. Pasteurization of milk (72°C; 16.2 s) is an effective treatment that kills $>10^4$ *E. coli* O157:H7 organisms per ml. Proper heating of foods of animal origin, e.g., heating foods to an internal temperature of at least 68.3°C for several seconds, is an important critical control point to ensure inactivation of *E. coli* O157:H7.

Many countries have approved the use of irradiation to eliminate foodborne pathogens in food. Unlike other processing technologies, irradiation eliminates foodborne pathogens yet maintains the raw character of foods. In the United States, 4.5 kilograys (kGy) is approved for refrigerated raw ground beef and 7.5 kGy is approved for frozen raw ground beef. Foodborne bacterial pathogens are relatively sensitive to irradiation. D_{10} values for *E. coli* O157:H7 in raw ground beef patties range from 0.241 to 0.307 kGy, depending on the temperature, with D_{10} values significantly higher for patties irradiated at -16 than at 4°C. Hence, an irradiation dose of 1.5 kGy should be sufficient to eliminate *E. coli* O157:H7 at the levels that may occur in ground beef. Irradiated ground beef is now available in supermarkets across the United States; however, the product has received limited acceptance (due to general consumer fear associated with irradiation).

RESERVOIRS OF *E. COLI* O157:H7

Detection of *E. coli* O157:H7 and EHEC on Farms

Undercooked ground beef, and less frequently unpasteurized milk, have caused many outbreaks of *E. coli* O157:H7 infection, and hence, cattle have been the focus of many studies of their role as a reservoir of *E. coli* O157:H7. The first link between *E. coli* O157:H7 and cattle was in Argentina in 1977 and was associated with a <3-week-old calf. Rates of EHEC carriage as high as 60% have been found in bovine herds in many countries, but in most cases, the rates range from 10 to 25%. The isolation rates of *E. coli* O157:H7 are much lower than those of non-O157 EHEC. Rates of EHEC carriage by dairy cattle on farms in Canada range from 36% of cows and 57% of calves in 80 herds tested. In the United States, 31 (3.2%) of 965 dairy calves and 191 (1.6%) of 11,881 feedlot cattle tested were positive for *E. coli* O157:H7. Young, weaned animals more frequently carry *E. coli* O157:H7 than adult cattle. Cattle are more likely to test positive for *E. coli* O157:H7 during warmer months of the year. This agrees with the seasonal variation in human disease.

Factors Associated with Bovine Carriage of *E. coli* O157:H7

There may be an association between fecal shedding of *E. coli* O157:H7 and feed or environmental factors. For example, some calf starter feed regimens or environmental factors and feed components, such as whole cottonseed, are linked with a reduction of *E. coli* O157:H7. In contrast, keeping calves in groups before weaning, sharing of calf-feeding utensils without sanitation, and early feeding of grain are associated with increased carriage of *E. coli* O157:H7. Grain and hay feeding programs may influence *E. coli* O157:H7 carriage by cattle.

Environmental factors, such as water and feed sources, or farm management practices, such as manure handling, may play important roles in influencing the prevalence of *E. coli* O157:H7 on dairy farms. The pathogen is frequently found in water troughs on farms. *E. coli* O157:H7 can survive for weeks or months in bovine feces and water.

The susceptibility of cattle to intestinal colonization with *E. coli* O157:H7 is largely a function of age. Young animals are more likely to be positive than older animals in the same herd. For many cattle, *E. coli* O157:H7 is transiently carried in the gastrointestinal tract and is intermittently excreted for a few weeks to months by young calves and heifers. Cows can carry more than one strain of *E. coli* O157:H7.

Cattle Model for Infection by *E. coli* O157:H7

E. coli O157:H7 is not a pathogen of weaned calves and adult cattle; hence, animals that carry the pathogen are not ill. However, there is evidence that *E. coli* O157:H7 can cause diarrhea and lesions in newborn calves. The initial sites of localization of *E. coli* O157:H7 in cattle are the forestomachs (rumen, omasum, and reticulum).

Domestic Animals and Wildlife

Although cattle are thought to be the main source of EHEC in the food chain, EHEC strains have also been isolated from other domestic animals and wildlife, such as sheep, goat, deer, dogs, horses, swine, and cats. *E. coli* O157:H7 was also isolated from seagulls and rats. The prevalence of *E. coli* O157:H7 and EHEC in sheep is generally higher than in other animals. In a survey of seven animal species in Germany, EHEC strains were isolated most frequently from sheep (66.6%), goats (56.1%), and cattle (21.1%), with lower prevalence rates in chickens (0.1%), pigs (7.5%), cats (13.8%), and dogs (4.8%).

Humans

Fecal shedding of *E. coli* O157:H7 by patients with hemorrhagic colitis or HUS usually lasts 13 to 21 days following the onset of symptoms. However, in some instances, the pathogen can be excreted in feces for weeks. A child infected during a day care center outbreak excreted the pathogen for 62 days. People living on dairy farms have elevated antibody titers against *E. coli* O157:H7; however, the pathogen was not isolated from feces. An asymptomatic long-term carrier state has not been identified. Fecal carriage of *E. coli* O157:H7 by humans is significant because of the potential for person-to-person spread of the pathogen. A factor contributing to person-to-person spread is the bacterium's extraordinarily low infectious dose. Fewer than 100 cells, and possibly as few as 10 cells, can cause illness. Inadequate attention to personal hygiene, especially after using the bathroom, can transfer the pathogen through contaminated hands, resulting in secondary transmission.

DISEASE OUTBREAKS

Geographic Distribution

E. coli O157:H7 is a cause of many major outbreaks worldwide. At least 30 countries on six continents have reported *E. coli* O157:H7 infection in humans. In the United States, 196 outbreaks or clusters of *E. coli* O157:H7 infection were documented through 1998. The number of outbreaks has increased from an average of 2 per year between 1982 and 1992 to 29 annually between 1993 and 1998. This dramatic increase may be due in part to improved recognition of *E. coli* O157:H7 infection following the publicity of a large multistate outbreak in the western United States in 1993. Since 1 January 1994, individual cases of *E. coli* O157:H7 infection have been reportable to the National Notifiable Diseases Surveillance System. The numbers of reported outbreaks, clusters, and cases of infection in the United States between 1982 and 1998 are presented in Table 9.2. The exact number of *E. coli* O157:H7 illnesses in the United States is not known because infected persons with mild or no symptoms and persons with nonbloody diarrhea often do not seek medical attention, and these cases would not be reported. The Foodborne Diseases Active Surveillance Network (FoodNet [http://www.cdc.gov/ncidod/dbmd/foodnet/]) reports that the annual rate of *E. coli* O157:H7 infection at several surveillance sites in the United States ranged from 2.1 to 2.8 cases per 100,000 population for 1996 to 1999. The Centers for Disease Control and Prevention estimate that *E. coli* O157:H7 causes 73,480 illnesses and 61 deaths annually in the United States

Table 9.2   Number of reported outbreaks, clusters, and cases of *E. coli* O157:H7 infection in the United States 1982 to 1998[a]

Yr	No. of outbreaks	No. of cases
1982	2	47
1984	2	70
1986	2	52
1987	1	52
1988	3	153
1989	2	246
1990	2	75
1991	4	54
1992	4	75
1993	17	1,000
1994	32	543
1995	32	455
1996	29	488
1997	22	298
1998	42	477
Total	196	4,085

[a]Source: P. M. Griffin, personal communication, 2000.

and that non-O157 EHEC strains account for an additional 37,740 cases with 30 deaths. Eighty-five percent of these cases are due to foodborne transmission.

Seasonality of *E. coli* O157:H7 Infection

Outbreaks and clusters of *E. coli* O157:H7 infection peak during the warmest months of the year. Approximately 86% of outbreaks and clusters reported in the United States occur from May to October. The reasons for this seasonal pattern are unknown but may include (i) an increased prevalence of the pathogen in cattle during the summer, (ii) greater human exposure to ground beef or other *E. coli* O157:H7-contaminated foods during the cookout months, and/or (iii) more improper handling (temperature abuse and cross-contamination) or incomplete cooking of products such as ground beef during warm months than during other months.

Age of Patients

All age groups can be infected by *E. coli* O157:H7, but the very young and the elderly most frequently experience severe illness with complications. HUS usually occurs in children, whereas thrombotic thrombocytopenic purpura (TTP) occurs in adults. Children 2 to 10 years of age are more likely to become infected with *E. coli* O157:H7. The high rate of infection in this age group is likely the result of increased exposure to contaminated foods, contaminated environments, and infected animals, as well as more opportunities for person-to-person spread between infected children with relatively undeveloped hygiene skills and undeveloped immune systems.

Transmission of *E. coli* O157:H7

Many foods are identified as vehicles of *E. coli* O157:H7 infection. Examples include ground beef, roast beef, cooked meats, venison jerky, salami, raw milk, pasteurized milk, yogurt, cheese, ice cream bars and cake, lettuce, unpasteurized apple cider and juice, cantaloupe, potatoes, radish sprouts, alfalfa sprouts, fruit or vegetable salad, and coleslaw. Among the 196 outbreaks reported in the United States for which a vehicle has been identified, 48 (33.1%) were associated with ground beef, 4 were associated (2.8%) with raw milk, and 3 (2.1%) were associated with roast beef (Table 9.3). Contact of foods with meat or feces (human or bovine) contaminated with *E. coli*

Table 9.3 Leading vehicles or modes of transmission associated with *E. coli* O157:H7 outbreaks in the United States, 1982 to 1998[a]

Rank	Vehicle	No. of outbreaks (% of known modes of transmission)
1	Ground beef	48 (33.1)
2	Person to person	37 (25.5)
3	Vegetables, salad bars	18 (12.4)
3	Water, swimming water	18 (12.4)
5	Raw milk, milk	4 (2.8)
5	Apple cider, juice	4 (2.8)
7	Roast beef	3 (2.1)
	Others	13 (9.0)
	Unknown	51

[a]Source: P. M. Griffin, personal communication, 2000.

O157:H7 is a likely source of cross-contamination. Outbreaks attributed to person-to-person (including secondary) transmission (25.5%) and waterborne (particularly recreational water) transmission (12.4%) have occurred.

Examples of Foodborne and Waterborne Outbreaks

The Original Outbreaks

The first documented outbreak of *E. coli* O157:H7 infection occurred in Oregon in 1982, with 26 cases and 19 persons hospitalized. All of the patients had bloody diarrhea and severe abdominal pain. This outbreak was associated with eating undercooked hamburgers from fast-food restaurants of a specific chain. *E. coli* O157:H7 was recovered from the stools of patients. A second outbreak occurred 3 months later and was associated with the same fast-food restaurant chain in Michigan, with 21 cases and 14 persons hospitalized, with an age range of 4 to 58 years. Contaminated hamburgers again were the cause. *E. coli* O157:H7 was isolated both from patients and from a frozen ground beef patty. *E. coli* O157:H7 was identified as the cause by its association with the food and by recovery of bacteria with identical microbiological characteristics from both the patients and the meat from the implicated supplier.

Waterborne Outbreaks

Reported waterborne outbreaks of *E. coli* O157:H7 infection have increased alarmingly in recent years. Swimming water, drinking water, well water, and ice have all been implicated. Investigations of lake-associated outbreaks revealed that in some instances the water was likely contaminated with *E. coli* O157:H7 by toddlers defecating while swimming. Swallowing lake water was subsequently identified as the cause. A large waterborne outbreak of *E. coli* O157:H7 among attendees of a county fair in New York occurred in August 1999. More than 900 persons were infected, 65 of whom were hospitalized. Two persons, including a 3-year-old girl, died from HUS (kidney failure), and a 79-year-old man died from HUS-TTP. Unchlorinated well water used to make beverages and ice was identified as the vehicle, and *E. coli* O157:H7 was isolated from samples of well water. Waterborne outbreaks of *E. coli* O157 infections have also been reported in Scotland, southern Africa, and Japan.

Outbreaks from Apple Cider and Juice

The first confirmed outbreak of *E. coli* O157:H7 infection associated with apple cider occurred in Massachusetts in 1991, involving 23 cases. In 1996, three outbreaks of *E. coli* O157:H7 infection associated with unpasteurized apple juice or cider were reported in the United States. The largest of the three occurred in three western states (California, Colorado, and Washington) and British Columbia, Canada, with 71 confirmed cases and one death. *E. coli* O157:H7 was isolated from the apple juice involved. An outbreak also occurred in Connecticut, with 14 cases. Manure contamination of apples was the suspected source of *E. coli* O157:H7 in several of the outbreaks. Using apple drops (i.e., apples picked up from the ground) for making apple cider is a common practice. Apples can become contaminated by resting on soil contaminated with manure. Apples also can become contaminated if transported or stored in areas that contain manure or if they are treated with contaminated water. Investigation of the 1991 outbreak in Massachusetts revealed that the cider press processor also raised cattle that grazed in a field adjacent to the cider mill. These outbreaks led to the U.S. Food and Drug Administration Hazard Analysis Critical Control Point regulations for juice. These require processors to achieve a 5-log-unit reduction in the numbers of the most resistant pathogen in their finished products.

A Large Multistate Outbreak

A large multistate outbreak of *E. coli* O157:H7 infection in the United States occurred in Washington, Idaho, California, and Nevada in early 1993. Approximately 90% of the primary cases were associated with eating at a single fast-food restaurant chain (chain A), at which *E. coli* O157:H7 was isolated from hamburger patties. The number of people who became ill increased because of secondary spread (48 patients in Washington alone) from person to person. One hundred seventy-eight people were hospitalized, 56 developed HUS, and four children died. The outbreak resulted from insufficient cooking of hamburgers by chain A restaurants. Hamburgers cooked according to chain A's cooking procedures in Washington State had internal temperatures below 60°C, which is substantially less than the minimum internal temperature of 68.3°C required by the state of Washington. Cooking the patties to an internal temperature of 68.3°C would have killed the low numbers of *E. coli* O157:H7 organisms in the ground beef.

Outbreaks Associated with Vegetables

Raw vegetables, particularly lettuce and alfalfa sprouts, have been implicated in outbreaks of *E. coli* O157:H7 infection in North America, Europe, and Japan. In the United States, outbreaks associated with lettuce have continued to increase during the past decade. In May 1996, mesclun lettuce was associated with a multistate outbreak of 47 cases in Illinois and Connecticut. Traceback studies implicated one grower as the likely source of the contaminated lettuce and revealed that cattle were present in the lettuce-growing and -processing areas.

Between May and December 1996, multiple outbreaks of *E. coli* O157:H7 infection occurred in Japan, involving 11,826 cases and 12 deaths. The largest outbreak affected 7,892 schoolchildren and 74 teachers and staff in Osaka in July 1996, among whom 606 individuals were hospitalized, 106 had HUS, and 3 died. Contaminated white radish sprouts were the source of *E. coli* O157:H7.

CHARACTERISTICS OF DISEASE

The spectrum of human illness caused by *E. coli* O157:H7 infection includes nonbloody diarrhea, bloody diarrhea (hemorrhagic colitis), kidney disease (HUS), and TTP (the adult form of HUS). Some persons may be infected but exhibit no signs or symptoms of illness (this is known as *asymptomatic infection*). Ingestion of the organism is followed by a 3- to 4-day incubation period (range, 2 to 12 days), during which colonization of the large intestine occurs. Illness begins with nonbloody diarrhea and severe abdominal cramps for 1 to 2 days. This progresses in the second or third day of illness to bloody diarrhea that lasts for 4 to 10 days. Many outbreak investigations revealed that >90% of microbiologically documented cases of diarrhea caused by *E. coli* O157:H7 were bloody, but in some outbreaks 30% of cases have involved nonbloody diarrhea. Symptoms usually end after a week, but ~6% of patients progress to HUS. The case-fatality rate from *E. coli* O157:H7 infection is ~1%.

HUS largely affects children, for whom it is the leading cause of acute kidney failure. The syndrome is characterized by three features: acute renal insufficiency, microangiopathic hemolytic anemia, and thrombocytopenia. TTP mainly affects adults and resembles HUS histologically. It is accompanied by distinct neurological abnormalities resulting from blood clots in the brain.

INFECTIOUS DOSE

The infectious dose of *E. coli* O157:H7 is thought to be extremely low. For example, between 0.3 and 15 colony-forming units of *E. coli* O157:H7 per g were enumerated in lots of frozen ground beef patties associated with a 1993 multistate outbreak in the western United States. Similarly, 0.3 to 0.4 colony-forming units of *E. coli* O157:H7 per g were detected in several intact packages of salami that were associated with a foodborne outbreak. These data suggest that the infectious dose of *E. coli* O157:H7 may be <100 cells. Additional evidence for a low infectious dose is the capability for person-to-person and waterborne transmission of EHEC infection. Age, immune status, and preexisting debilitating health conditions can individually or collectively influence whether an individual becomes ill.

MECHANISMS OF PATHOGENICITY

The exact mechanism of pathogenicity of EHEC is not known. General knowledge about the pathogenicity of EHEC indicates that the bacteria cause disease by their ability to adhere to the host cell membrane and colonize the large intestine. They then produce one or more Stxs. However, the mechanisms of intestinal colonization are not well understood. The roles of various other potential virulence factors and host factors remain to be determined. The virulence factors involved in the pathogenesis of EHEC are summarized in Table 9.4.

Attaching and Effacing

Adherence factors and their genes have been characterized, but the mechanisms of adherence and colonization by *E. coli* O157:H7 and other EHEC strains have not been characterized completely. Studies with animal models revealed that the pathogen can colonize the colons of orally infected an-

Table 9.4 Proteins and genes involved in pathogenesis of EHEC[a]

Genetic locus	Protein description	Gene(s)	Function
Chromosome (locus of enterocyte effacement)	Intimin	*eae*	Adherence
	Tir	*tir*	Intimin receptor
	Secretion proteins	*espA, espB, espD*	Induces signal transduction
	Type III secretion system	*escC, escD, escF, escJ, escN, escR, escS, escT, escU, escV, sepQ, sepZ*	Apparatus for extracellular protein secretion
Phage	Stx	*stx1, stx2, stx2c, stx2d*	Inhibits protein synthesis
Plasmid	EHEC hemolysin	EHEC-*hylA*	Disrupts cell membrane permeability
	Catalase-peroxidase	*katP*	

[a]Sources: A. D. O'Brien et al., *Curr. Top. Microbiol. Immunol.* **180**:65–94, 1992; N. T. Perna et al., *Infect. Immun.* **66**:3810–3817, 1998.

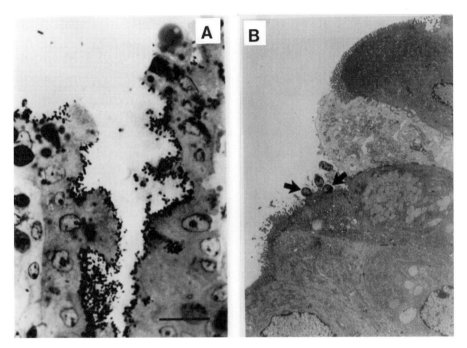

Figure 9.1   AE lesion caused by *E. coli* O157:H7. **(A)** Irregularly shaped cuboidal to low-columnar crypt neck cells with bacteria adherent to the luminal plasma membrane of a gnotobiotic pig (toluidine blue stain). Reproduced with permission from D. H. Francis, J. E. Collins, and J. R. Duimstra, *Infect. Immun.* **51:**953–956, 1986. **(B)** Transmission electron photomicrograph of rabbit ileum with adherent bacteria (arrow). Reproduced with permission from P. Sherman, R. Soni, and M. Karmali, *Infect. Immun.* **56:**756–761, 1988.

imals by an AE mechanism (Fig. 9.1). The AE lesion is characterized by intimate attachment of the bacteria to intestinal cells. This results in loss of microvilli on the epithelial cells and accumulation of actin (F actin) in the cytoplasm. It is believed that similar mechanisms regulate the processes of AE formation by *E. coli* O157:H7.

Locus of Enterocyte Effacement

Genes on the locus of enterocyte effacement of the EPEC chromosome are involved in producing the AE lesion. The locus of enterocyte effacement is not present in *E. coli* strains that are part of the normal intestinal flora, *E. coli* K-12, or ETEC strains. The locus of enterocyte effacement is found in those EPEC and EHEC strains that produce the AE lesion. The locus of enterocyte effacement consists of three segments with genes of known function. Genes associated with the locus of enterocyte effacement encode proteins involved in attachment and recognition that are known to interact directly with the host.

Intimin is a outer membrane protein encoded by the *eae* (from *E. coli* AE) gene. The protein is produced by EPEC, EHEC, *Hafnia alvei,* and *Citrobacter rodentium.* Intimin is the only described adherence factor of *E. coli* O157:H7 and is important for intestinal colonization in animal models.

The Tir protein is produced by the bacterial cell and moves to the host cell, where it serves as a receptor for intimin, via the type III secretion system. Most bacterial pathogens use existing host cell receptors. The ability of EPEC and EHEC to supply their own receptors may provide additional advantages for initiating the process of adhesion, motility, and signal transduction. The type III secretion system, present in many gram-negative bacterial pathogens, is induced upon contact with the host cells and only exports virulence factors to eukaryotic cells.

The 60-MDa Plasmid (pO157)

E. coli O157:H7 isolates carry a plasmid (pO157) of ~60 MDa (unrelated to the 60-MDa plasmid present in EPEC). Because of the association of the 60-MDa plasmid with EHEC, the plasmid is linked to the pathogenesis of EHEC infections. However, its exact role in virulence has not been determined. The plasmid carries potential virulence genes, including those encoding EHEC hemolysin and catalase-peroxidase. The EHEC catalase-peroxidase is a enzyme that protects the bacterium against oxidative stress, a defense mechanism of mammalian cells during bacterial infection.

Stxs

EHEC produces one or two Stxs. The nomenclature of the Stx family and their important characteristics are listed in Table 9.5. Molecular studies of Stx1 proteins from different *E. coli* strains revealed that Stx1 is either completely identical to the Stx of *S. dysenteriae* type 1 or differs by only one amino acid. Unlike Stx1, toxins of the Stx2 group are not neutralized by serum raised against Stx and do not cross-react with Stx1-specific DNA probes. There is sequence and antigenic variation within toxins of the Stx2 family produced by *E. coli* O157:H7 and other EHEC strains. The subgroups of Stx2 are Stx2, Stx2c, Stx2d, and Stx2e.

Structure of the Stx Family

Stxs are composed of a single enzymatic 32-kDa A subunit in association with a pentamer of 7.7-kDa receptor-binding B subunits. The Stx A subunit can be split by trypsin into an enzymatic A1 fragment (~27 kDa) and a carboxyl-terminal A2 fragment (~4 kDa) that links A1 to the B subunits.

Genetics of Stxs

While all *stx1* operons appear identical and are located on the genomes of lysogenic lambdoid bacteriophages, there is considerable heterogeneity in the *stx2* family. Unlike the genes for other Stx2 proteins, which are located on bacteriophages that integrate into the chromosome, Stx of *S. dysenteriae*

Table 9.5 Nomenclature and biological characteristics of Stx[a]

Nomenclature	Genetic locus	% Amino acid homology to Stx2 subunit:		Receptor	Activated by intestinal mucus	Disease
		A	B			
Stx	Chromosome	55	57	Gb_3	No	Human diarrhea, HC[b], HUS
Stx1	Phage	55	57	Gb_3	No	Human diarrhea, HC, HUS
Stx2	Phage	100	100	Gb_3	No	Human diarrhea, HC, HUS
Stx2c	Phage	100	97	Gb_3	No	Human diarrhea, HC, HUS
Stx2d	Phage	99	97	Gb_3	Yes	Human diarrhea, HC, HUS
Stx2e	Chromosome	93	84	Gb_4[c]	No	Pig edema disease

[a]Source: A. R. Melton-Celsa and A. D. O'Brien, p. 121–128, *in* J. B. Kaper and A. D. O'Brien (ed.), Escherichia coli *O157:H7 and Other Shiga Toxin-Producing* E. coli *Strains* (ASM Press, Washington, D.C., 1998).
[b]HC, hemorrhagic colitis.
[c]Gb4, glycolipid globotetraosylceramide.

type 1 and Stx2e are encoded by chromosomal genes. The genes of additional variants of Stx2 have been isolated from EHEC. Finally, Stx and Stx1 production are negatively regulated at the transcriptional level by an iron-Fur protein complex that binds at the *stx1* promoter. This not affected by temperature, whereas Stx2 production is neither iron nor temperature regulated.

Receptors

All members of the Stx family bind to glycolipids on the eukaryotic cell surface. The alteration of binding specificity between Stx2e and the rest of the Stx family is related to the carbohydrate specificities of receptors. Stx1 binds preferentially to Gb_3 (glycolipid globotriaosylceramide) containing C20:1 fatty acid, whereas Stx2c prefers Gb_3 containing C18:1 fatty acid.

Mode of Action of the Stxs

Stxs act by inhibiting protein synthesis. While it appears that transfer of the toxin to the Golgi apparatus is essential for intoxication, the mechanism of entry of the A subunit, and particularly the role of the B subunit, remains unclear. Although the entire toxin is necessary for its toxic effect on whole cells, the A1 subunit can cleave the *N*-glycoside bond in one adenosine position of the 28S ribosomal RNA (rRNA) that comprises 60S ribosomal subunits. This elimination of a single adenine nucleotide inhibits the elongation factor-dependent binding to ribosomes of aminoacyl-bound transfer RNA (tRNA) molecules. Peptide chain elongation is stopped, and overall protein synthesis is suppressed, resulting in cell death.

The Roles of Stxs in Disease

The precise roles of Stxs in mediating colonic disease, HUS, and neurological disorders are not understood. There is no satisfactory animal model for bloody diarrhea (hemorrhagic colitis) or HUS. The severity of the diseases prevents the study of experimental infections in humans. Therefore, our understanding of the role of Stxs in causing disease is obtained from histopathology of diseased human tissues, animal models, and endothelial tissue culture cells. It appears that Stxs contribute to pathogenesis by directly damaging vascular endothelial cells in certain organs, thereby disrupting the homeostatic properties of the cells.

There is a correlation between infection with *E. coli* O157:H7 and development of HUS in humans. Histopathologic examination of kidney tissue from HUS patients revealed extensive structural damage in the glomeruli, the basic filtration unit of the kidney. Endotoxin in the presence of Stxs also can activate phagocytic cells to synthesize and release cytokines, superoxide radicals, or proteinases and can amplify endothelial cell damage.

E. coli O157:H7 strains isolated from patients with bloody diarrhea usually produce both Stx1 and Stx2 or Stx2 only; isolates producing only Stx1 are uncommon. Patients infected with *E. coli* O157:H7 producing only Stx2 or Stx2 in combination with Stx1 are likely to develop serious kidney or circulatory complications compared with patients infected with EHEC strains producing Stx1 only.

CONCLUSION

The serious nature of hemorrhagic colitis (bloody diarrhea) and HUS caused by *E. coli* O157:H7 places this pathogen in a category apart from other foodborne pathogens that typically cause only mild symptoms. The severity of the illness, combined with its low infectious dose (<100 cells), qualifies *E. coli* O157:H7 to be among the most dangerous of foodborne pathogens. Illness caused by this pathogen has been linked to the consumption of a variety of foods. Recreational and drinking water are also vehicles of transmission of *E. coli* O157:H7 infection. The pathogenic mechanisms of *E. coli* O157:H7 are not fully understood; however, production of one or more Stxs and AE adherence are important virulence factors. Although other EHEC strains cause disease, *E. coli* O157:H7 is still by far the most important serotype of EHEC in North America. Isolation of non-O157:H7 EHEC strains requires techniques not generally used in clinical laboratories; hence, these bacteria are rarely detected in routine practice. Recognition of non-O157 EHEC strains in foodborne illness requires the identification of serotypes of EHEC other than O157:H7 in persons with bloody diarrhea and/or HUS and preferably in the implicated food. The increased availability in clinical laboratories of techniques such as testing for Stxs or their genes and identification of other virulence markers unique to EHEC may enhance the detection of disease attributable to non-O157 EHEC.

Summary

- Cattle are a major reservoir of *E. coli* O157:H7, with undercooked ground beef being the single most frequently implicated vehicle of transmission.
- Acid tolerance is an important feature of this pathogen.
- EHEC strains other than O157:H7 have been increasingly associated with cases of HUS.
- Production of Stx is an important virulence feature of *E. coli* O157:H7.
- *E. coli* O157:H7 can be readily distinguished from other *E. coli* strains based on biochemical characteristic, such as inability to ferment sorbitol and lack of production of β-glucuronidase.

Suggested reading

Besser, R. E., P. M. Griffin, and L. Slutsker. 1999. *Escherichia coli* O157:H7 gastroenteritis and the hemolytic uremic syndrome: an emerging infectious disease. *Annu. Rev. Med.* **50:**355–367.

Centers for Disease Control and Prevention. 2000. Preliminary FoodNet data on the incidence of foodborne illnesses—selected sites, United States, 1999. *Morb. Mortal. Wkly. Rep.* **49:**201–205.

Melton-Celsa, A. R, and A. D. O'Brien. 1998. Structure, biology, and relative toxicity of Shiga toxin family members for cells and animals, p. 121–128. *In* J. B. Kaper and A. D. O'Brien (ed.), Escherichia coli *O157:H7 and Other Shiga Toxin-Producing* E. coli *Strains.* ASM Press, Washington, D.C.

Meng, J., and M. P. Doyle. 1998. Microbiology of Shiga toxin-producing *Escherichia coli* in foods, p. 92–111. *In* J. Kaper and A. O'Brien (ed.), Escherichia coli *O157:H7 and Other Shiga Toxin-Producing* E. coli *Strains.* ASM Press, Washington, D.C.

O'Brien, A. D., V. L. Tesh, A. Donohue-Rolfe, M. P. Jackson, S. Olsnes, K. Sandvig, A. A. Lindberg, and G. T. Keusch. 1992. Shiga toxin: biochemistry, genetics, mode of action, and role in pathogenesis. *Curr. Top. Microbiol. Immunol.* **180:**65–94.

Perna, N. T., G. F. Mayhew, G. Posfai, S. Elliott, M. S. Donnenberg, J. B. Kaper, and F. R. Blattner. 1998. Molecular evolution of a pathogenicity island from enterohemorrhagic *Escherichia coli* O157:H7. *Infect. Immun.* **66:**3810–3817.

Philpott, D., and F. Ebel (ed.). 2003. E. coli *Shiga Toxin Methods and Protocols.* Humana Press, Totowa, N.J.

Questions for critical thought

1. The infectious dose of *E. coli* O157:H7 may be as low as 10 cells. What characteristics of the pathogen contribute to the low infectious dose?

2. *E. coli* O157:H7 produces Stx. Explain how Stxs affect host cells and the role they play in disease.

3. How might acid tolerance contribute to foodborne illness associated with *E. coli* O157:H7?

4. On-farm feeding practices can influence shedding of the pathogen by cattle. How could this information be used to reduce contamination of beef products?

5. Based on the outbreaks described, what measures could be implemented to reduce cases of foodborne illness linked to *E. coli* O157:H7?

6. Provide a scenario that may result in cross-contamination of lettuce (or other types of produce) with *E. coli* O157:H7.

7. Why are young children (10 years of age and younger) more likely than adults to develop complications (HUS) when infected with *E. coli* O157:H7?

8. Based on characteristics of *E. coli* O157:H7, what tests could be performed to distinguish the pathogen from other *E. coli* strains?

9. Speculate as to why produce-linked *E. coli* O157:H7-related outbreaks have increased during the past decade.

10

Yersinia enterocolitica*

LEARNING OBJECTIVES

The information in this chapter will enable the student to:

- use basic biochemical characteristics to identify *Yersinia*
- understand what conditions in foods favor *Yersinia* growth
- recognize, from symptoms and time of onset, a case of foodborne illness caused by *Yersinia*
- choose appropriate interventions (heat, preservatives, and formulation) to prevent the growth of *Yersinia*
- identify environmental sources of the organism
- understand the role of *Yersinia* toxins and virulence factors in causing foodborne illness

Outbreak

One of the largest outbreaks of foodborne disease caused by *Yersinia enterocolitica* occurred between June and August of 1982. At least 172 *Y. enterocolitica* culture-positive cases were identified, and 41% occurred among children <5 years of age. Consumption of pasteurized milk was epidemiologically implicated as the vehicle of transmission of *Y. enterocolitica*. Although *Y. enterocolitica* was not isolated from the milk, the same serotype found in the outbreak was isolated from a milk crate on a hog farm where the outdated milk from the implicated processing plant was fed to hogs. The milk crates, if reused without proper sanitizing, might have resulted in cross-contamination of the outside of milk cartons destined for consumers. The symptoms of *Yersinia* enteritis, severe abdominal pain and fever, mimic those of appendicitis. Indeed, 17 patients underwent unnecessary *appendectomies* (removal of the appendix).

INTRODUCTION

Y. enterocolitica first emerged as a human pathogen during the 1930s. *Y. enterocolitica* shows between 10 and 30% DNA homology with other genera in the family *Enterobacteriaceae* and is ~50% related to *Yersinia pseudotuberculosis* and *Yersinia pestis* (the cause of plague). The last two species show >90% DNA homology, and genetic analysis has revealed that *Y. pestis* is a clone of *Y. pseudotuberculosis* that evolved some 1,500 to 20,000 years ago, shortly before the first known pandemics of human plague.

---

*This chapter was originally written by Roy M. Robins-Browne for *Food Microbiology: Fundamentals and Frontiers*, 2nd ed., and has been adapted for use in an introductory text.

Characteristics of the Organism

The genus *Yersinia* has 11 species and is classified within the family *Enterobacteriaceae*. As with other members of this family, yersiniae are gram-negative, oxidase-negative, rod-shaped, facultative anaerobes that ferment glucose. The genus includes three well-characterized pathogens of mammals, one pathogen of fish, and several other species whose etiologic role in disease is uncertain. The four known pathogenic species are *Y. pestis*, the causative agent of bubonic and pneumonic plague (the black death); *Y. pseudotuberculosis*, a rodent pathogen that occasionally causes disease in humans; *Yersinia ruckeri*, a cause of enteric disease in salmonids and other freshwater fish; and *Y. enterocolitica*, a versatile intestinal pathogen, which is the most prevalent *Yersinia* species among humans.

Y. pestis is transmitted to its host by the bites of fleas or respiratory aerosols, whereas *Y. pseudotuberculosis* and *Y. enterocolitica* are foodborne pathogens. Nevertheless, these three species share a number of essential virulence determinants that enable them to overcome host defenses. Analogs of these virulence determinants occur in several other enterobacteria, such as enteropathogenic and enterohemorrhagic *Escherichia coli* and *Salmonella* and *Shigella* species, as well as in various pathogens of animals (e.g., *Pseudomonas aeruginosa* and *Bordetella* species) and plants (e.g., *Erwinia amylovora*, *Xanthomonas campestris*, and *Pseudomonas syringae*). This provides evidence for the horizontal transfer of virulence genes among diverse bacterial pathogens. In addition, yersiniae may have acquired a number of human genes that enable them to undermine key aspects of the physiological response to infection.

Classification

Y. enterocolitica is a heterogeneous species that is divisible into a large number of subgroups, largely according to biochemical activity and lipopolysaccharide (LPS) O antigens (Table 10.1). Biotyping is based on the ability of *Y. enterocolitica* to metabolize selected organic substrates and provides a convenient means to subdivide the species into subtypes of clinical and epidemiologic significance (Table 1). Most pathogenic strains of humans and domestic animals occur within biovars 1B, 2, 3, 4, and 5. By contrast, *Y. enterocolitica* strains of biovar 1A are commonly obtained from ter-

Table 10.1 Biotyping scheme of *Y. enterocolitica*

Test	Reaction of biovar[a]:					
	1A	1B	2	3	4	5
Lipase (Tween hydrolysis)	+	+	−	−	−	−
Esculin hydrolysis	D	−	−	−	−	−
Indole production	+	+	(+)	−	−	−
D-Xylose fermentation	+	+	+	+	−	D
Voges-Proskauer reaction	+	+	+	+	+	(+)
Trehalose fermentation	+	+	+	+	+	−
Nitrate reduction	+	+	+	+	+	−
Pyrazinamidase	+	−	−	−	−	−
β-D-Glucosidase	+	−	−	−	−	−
Proline peptidase	D	−	−	−	−	−

[a]+, positive; (+) delayed positive; −, negative; D, different reactions.

restrial and freshwater ecosystems. For this reason, they are often referred to as environmental strains, although some of them may be responsible for intestinal infections. Not all isolates of *Y. enterocolitica* obtained from soil, water, or unprocessed foods can be assigned to a biovar. These strains lack the characteristic virulence determinants of biovars 1B though 5 (see below) and may represent novel nonpathogenic subtypes or even new *Yersinia* species.

Serotyping of *Y. enterocolitica*, based on LPS surface O antigens, coincides to some extent with biovar typing and provides a useful additional tool to subdivide this species in a way that relates to its pathologic significance. Serogroup O:3 is the variety most frequently isolated from humans. Almost all of these isolates belong to biovar 4. Other serogroups commonly obtained from humans include O:9 (biovar 2) and O:5,27 (biovar 2 or 3), particularly in northern Europe. The most frequent *Y. enterocolitica* biovar obtained from human clinical samples worldwide is biovar 4. Biovar 1B bacteria are usually isolated from patients in the United States and are referred to as American strains. They have caused several foodborne outbreaks of yersiniosis in the United States.

At least 18 flagellar (H) antigens of *Y. enterocolitica*, designated by lowercase letters (a,b; b,c; b,c,e,f,k; m; etc.), have also been identified. There is some overlap between the H antigens of *Y. enterocolitica* and those of related species, but complete O and H serotyping is seldom done.

Other schemes for subtyping *Yersinia* species include bacteriophage typing, multienzyme electrophoresis, and the demonstration of restriction fragment length polymorphisms of chromosomal and plasmid DNA. These techniques can be used to facilitate epidemiologic investigations of outbreaks or to trace the sources of sporadic infections.

Susceptibility and Tolerance

Y. enterocolitica is unusual because it can grow at temperatures below 4°C. The doubling time at the optimum growth temperature (ca. 28 to 30°C) is ~34 min, which increases to 1 h at 22°C, 5 h at 7°C, and ~40 h at 1°C. *Y. enterocolitica* readily withstands freezing and can survive in frozen foods for extended periods even after repeated freezing and thawing. *Y. enterocolitica* generally survives better at room temperature and refrigeration temperature than at intermediate temperatures. *Y. enterocolitica* persists longer in cooked foods than in raw foods, probably due to increased availability of nutrients in cooked foods. Also, the presence of other psychrotrophic (growing between 5 and 35°C) bacteria, including nonpathogenic strains of *Y. enterocolitica*, in unprocessed food may restrict bacterial growth. The number of viable *Y. enterocolitica* organisms may increase more than a millionfold on cooked beef or pork within 24 h at 25°C or within 10 days at 7°C. Growth rates are slower on raw beef and pork. *Y. enterocolitica* can grow at refrigeration temperature in vacuum-packed meat, boiled eggs, boiled fish, pasteurized liquid eggs, pasteurized whole milk, cottage cheese, and tofu (soybean curd). Growth also occurs in refrigerated seafoods, such as oysters, raw shrimp, and cooked crabmeat, but at a lower rate than in pork or beef. Yersinae can also persist for extended periods in refrigerated vegetables and cottage cheese.

Y. enterocolitica and *Y. pseudotuberculosis* can grow over a pH range of approximately pH 4 to 10, with an optimum pH of ca. 7.6. They tolerate alkaline conditions extremely well, but their acid tolerance is less apparent and depends on the acidulent used, the environmental temperature, the composition of the medium, and the growth phase of the bacteria. The acid tolerance of *Y. enterocolitica* is enhanced by the production of urease, which hydrolyzes urea to release ammonia and elevates the cytoplasmic pH.

Y. enterocolitica and *Y. pseudotuberculosis* are susceptible to heat and are easily killed by pasteurization at 71.8°C for 18 s or 62.8°C for 30 min. Exposure of surface-contaminated meat to hot water (80°C) for 10 to 20 s reduced bacterial viability by at least 99.9%. *Y. enterocolitica* is also easily killed by ionizing and ultraviolet (UV) irradiation and by sodium nitrate and nitrite added to food; although it is relatively resistant to these salts in solution. The pathogen can also tolerate NaCl at concentrations of up to 5%. *Y. enterocolitica* is susceptible to organic acids, such as lactic and acetic acids, and to chlorine. However, some resistance to chlorine occurs among yersiniae grown under conditions that are similar to natural aquatic environments.

CHARACTERISTICS OF INFECTION

Infections with *Y. enterocolitica* typically manifest as nonspecific, self-limiting diarrhea, but they may lead to a variety of autoimmune diseases (Table 10.2). The risk of these diseases is determined partly by host factors, in particular, age and immune status. *Y. enterocolitica* enters the gastrointestinal tract after ingestion of contaminated food or water. The median infectious dose for humans is not known, but it likely exceeds 10^4 colony-forming units. Stomach acid is a significant barrier to infection with *Y. enterocolitica*, and in individuals with decreased levels of stomach acid, the infectious dose may be lower.

Table 10.2 Clinical symptoms and diseases associated with *Y. enterocolitica* infections

Common symptoms and diseases

Diarrhea (gastroenteritis), especially in young children

Enterocolitis

Pseudoappendicular syndrome due to terminal ileitis; acute mesenteric lymphadenitis

Pharyngitis

Postinfection autoimmune sequelae
 Arthritis, especially associated with HLA-B27
 Erythema nodosum
 Uveitis, associated with HLA-B27
 Glomerulonephritis (uncommon)
 Myocarditis (uncommon)
 Thyroiditis (uncommon)

Less common symptoms and diseases

Septicemia

Visceral abscesses, e.g., in liver, spleen, or lung

Skin infection (pustules, wound infection, pyomyositis, etc.)

Pneumonia

Endocarditis

Osteomyelitis

Peritonitis

Meningitis

Intussusception

Eye infection (conjunctivitis, panophthalmitis)

Most symptomatic infections with *Y. enterocolitica* occur in children, especially in those <5 years of age. In these patients, yersiniosis causes diarrhea, often accompanied by low fever and abdominal pain. The diarrhea varies from watery to mucoid. The illness typically lasts from a few days to 3 weeks, although some patients develop diarrhea that may persist for several months.

In children older than 5 years of age and adolescents, acute yersiniosis often causes pain in the abdomen that is mistaken for appendicitis. This is referred to as pseudoappendicular syndrome. Some children who experience this also have fever but little or no diarrhea. The pseudoappendicular syndrome happens more frequently with the more virulent strains of *Y. enterocolitica*, notably strains of biovar 1B. *Y. enterocolitica* is rarely found in patients with true appendicitis.

Although *Y. enterocolitica* is seldom isolated outside of the intestine, there appears to be no tissue in which it cannot grow. Factors that may make an individual more susceptible to yersiniosis bacteremia include decreased immune function, malnutrition, chronic kidney disease, liver disease, alcoholism, diabetes, and acute and chronic iron overload states. Spread of the organism in the body can lead to various diseases, including abscesses, catheter-associated infections, heart disease, and meningitis. *Yersinia* bacteremia has a fatality rate between 30 and 60%.

Although most individuals with yersiniosis recover without long-term complications, infections with *Y. enterocolitica* are noteworthy for the large variety of immunological complications, such as reactive arthritis, carditis, and thyroiditis, which follow acute infection. Of these, reactive arthritis is the most widely recognized.

RESERVOIRS

Infections with *Yersinia* species are *zoonoses* (diseases transmitted from animals to humans). The subgroups of *Y. enterocolitica* that commonly occur in humans also occur in domestic animals, whereas those which are infrequent in humans generally reside in wild rodents. *Y. enterocolitica* can survive in many environments and has been isolated from the intestinal tracts of many different mammalian species, as well as from birds, frogs, fish, flies, fleas, crabs, and oysters.

Foods that are frequently positive for *Y. enterocolitica* include pork, beef, lamb, poultry, and dairy products (notably milk, cream, and ice cream). *Y. enterocolitica* is also commonly found in terrestrial and freshwater systems, including soil, vegetation, lakes, rivers, wells, and streams. It can survive for extended periods in soil, vegetation, streams, lakes, wells, and spring water, particularly at low environmental temperatures. Many environmental isolates of *Y. enterocolitica* lack markers of bacterial virulence and may not be a risk to human or animal health.

Although *Y. enterocolitica* has been recovered from a variety of wild and domesticated animals, pigs are the only animal species from which *Y. enterocolitica* of biovar 4, serogroup O:3 (the variety most commonly associated with human disease), has been isolated with any degree of frequency. Pigs may also carry *Y. enterocolitica* of serogroups O:9 and O:5,27, particularly in regions where human infections with these varieties are common. In countries with a high incidence of human yersiniosis, *Y. enterocolitica* is commonly isolated from pigs at slaughterhouses. The tissue most frequently culture positive at slaughter is the tonsils. This appears to be the preferred site of *Y. enterocolitica* infection in pigs. Yersiniae have also been isolated from tongue, cecum, rectum, feces, and gut tissue. *Y. enterocolitica* is seldom isolated from meat offered for retail sale. However, standard methods of bacterial isolation and detection may underestimate the true incidence of contamination. Further evidence that pigs are a significant reservoir of human infections is based on epidemiologic studies linking consumption of raw or undercooked pork with yersiniosis. Infection also occurs after handling contaminated pig intestines while preparing chitterlings.

Food animals are seldom infected with biovar 1B strains of *Y. enterocolitica*, the reservoir of which remains unknown. The relatively low incidence of human yersiniosis caused by these strains, despite their high virulence, suggests limited contact between their reservoir and humans. Yersiniae of this biovar are pathogens of rodents; therefore, rats or mice are likely the natural reservoir of these strains.

FOODBORNE OUTBREAKS

Considering the widespread occurrence of *Y. enterocolitica* in nature and its ability to colonize food animals, to persist within animals and the environment, and to grow at refrigeration temperature, outbreaks of yersiniosis are surprisingly uncommon. Most foodborne outbreaks in which a source was identified have been traced to milk (Table 10.3). *Y. enterocolitica* is easily destroyed by pasteurization. Therefore, infection results from the consumption of raw milk or milk that is contaminated after pasteurization. During the mid 1970s, two outbreaks of yersiniosis caused by *Y. enterocolitica* O:5,27 occurred among 138 Canadian schoolchildren who had consumed raw milk, but the organism was not recovered from the suspected source. In 1976, serogroup O:8 *Y. enterocolitica* was responsible for an outbreak in New York State which affected 217 people, 38 of whom were culture positive. The source of infection was chocolate milk, which evidently became contaminated after pasteurization.

In 1981, an outbreak of infection with *Y. enterocolitica* O:8 affected 35% of 455 individuals at a diet camp in New York state. Seven patients were hospitalized as a result of infection, five of whom underwent appendectomies. The source of the infection was reconstituted powdered milk and/or chow mein. An infected food handler probably contaminated the food during preparation.

Table 10.3   Selected foodborne outbreaks of infection with *Y. enterocolitica*

Location	Yr	Mo	No. of cases	Serogroup	Source
Canada	1976	April	138	O:5,27	Raw milk?[a]
New York	1976	September	38	O:8	Chocolate-flavored milk
Japan	1980	April	1,051	O:3	Milk
New York	1981	July	159	O:8	Powdered milk; chow mein
Washington	1981	December	50	O:8	Tofu and spring water
Pennsylvania	1982	February	16	O:8	Bean sprouts and well water
Southern United States	1982	June	172	O:13a,13b	Milk?
Hungary	1983	December	8	O:3	Pork cheese (sausage)
Georgia (United States)	1989	November	15	O:3	Pork chitterlings
Northeastern United States	1995	October	10	O:8	Pasteurized milk?

[a]?, bacteria were not isolated from the suspected source.

More recently, an outbreak of infection with *Y. enterocolitica* O:3 affected 15 infants and children in metropolitan Atlanta. In this instance, bacteria were transmitted from raw chitterlings to the affected children on the hands of food handlers. Other foods that have been responsible for outbreaks of yersiniosis include pork cheese (a type of sausage prepared from chitterlings), bean sprouts, and tofu. In the outbreaks associated with bean sprouts and tofu, contaminated well or spring water was the probable source of yersiniae. Water was also the likely source of infection in a case of *Y. enterocolitica* bacteremia in a 75-year-old man in New York state and a small family outbreak in Ontario, Canada. Several outbreaks of presumed foodborne infection with *Y. enterocolitica* O:3 have occurred in the United Kingdom and Japan, but in most cases, the source of these outbreaks was not identified.

MECHANISMS OF PATHOGENICITY

Y. enterocolitica is an invasive *enteric* (intestinal) pathogen whose virulence determinants have been the subject of intensive investigation. Not all strains of *Y. enterocolitica* are equally virulent. *Y. enterocolitica* strains of biovars 1B, 2, 3, 4, and 5 possess many interactive virulence determinants.

Pathological Changes

Examination of surgical specimens from patients with yersiniosis shows that *Y. enterocolitica* is an invasive pathogen. The distal ileum (part of the intestine) is the main infection site. Human volunteers cannot be used to study yersiniosis, since they may develop other diseases. Thus, most information regarding the pathogenesis of yersiniosis has been obtained from animal models, in particular mice and rabbits. Studies of experimentally infected rabbits and pigs show that after penetrating the epithelium of the intestine, the pathogens can spread throughout the body. If the bacteria reach the lymph nodes, they can enter the blood stream and can disseminate to any organ. However, they preferentially localize in the liver and spleen.

VIRULENCE DETERMINANTS

Chromosomal Determinants of Virulence

The ability of *Y. enterocolitica* to enter mammalian cells is associated with invasin, an outer membrane protein encoded by the chromosomal *inv* gene. Invasins are related to intimin, an essential virulence determinant of enteropathogenic and enterohemorrhagic strains of *E. coli,* which require the protein in order to produce the distinctive attaching-effacing lesions that characterize infection with these bacteria.

The amino terminus of invasin is inserted in the bacterial outer membrane, while the carboxyl terminus is exposed on the surface. The carboxyl terminus binds the host cell integrins. The internalization process is controlled entirely by the host cell, because dead bacteria and even latex particles coated with invasin are internalized. Although DNA sequences the same as that of *inv* occur in all *Yersinia* species (except *Y. ruckeri*), the gene is functional only in *Y. pseudotuberculosis* and the classical pathogenic biovars (1B through 5) of *Y. enterocolitica*. This suggests that invasin plays a key role in virulence.

Most strains of *Y. enterocolitica* secrete a heat-stable enterotoxin, known as Yst (or Yst-a), which is active in infant mice. The contribution of Yst to diarrhea associated with yersiniosis is uncertain. Some strains of *Y. entero-colitica* can produce Yst or the other enterotoxins over a wide range of temperatures, from 4 to 37°C. Since these toxins are relatively acid stable, they could resist inactivation by stomach acid. If they were ingested preformed in food, they could cause foodborne illness. In artificially inoculated foods, however, these toxins are produced mainly at 25°C during the stationary phase of bacterial growth. The storage conditions required for their production in food generally results in severe spoilage, making the ingestion of preformed Yst unlikely.

Under laboratory conditions, *Y. enterocolitica* is motile when grown at 25°C but not when grown at 37°C. The expression of the genes encoding both flagellin and invasin appears to be regulated at the level of transcription. Mutants of *Y. enterocolitica* that are defective in the expression of invasin are extremely motile. Flagellar proteins are made in individuals with yersiniosis, based on assays for the detection of flagellar proteins. Motility does not appear to contribute to the virulence of *Y. enterocolitica* in mice.

Other Virulence Determinants

The observation that patients suffering from iron overload have increased susceptibility to severe infections with *Y. enterocolitica* suggests that the availability of iron in tissues may determine the outcome of yersiniosis. Some isolates of *Y. enterocolitica* are *hemolytic* (they lyse red blood cells) due to the production of phospholipase A. A strain of *Y. enterocolitica* in which the *yplA* gene encoding this enzyme was deleted had decreased virulence for inoculated mice. YplA is secreted by *Y. enterocolitica* via the same export system used for flagellar proteins.

In *Y. enterocolitica*, acid tolerance is associated with the production of urease. This enzyme catalyzes the release of ammonia from urea and enables the bacteria to resist pHs as low as 2.5. Urease also contributes to the survival of *Y. enterocolitica* in host tissues, but the mechanism by which this occurs is not known. The urease of *Y. enterocolitica* is also unusual in that it displays optimal activity at pH 3.5 to 4.5, suggesting a physiological role in protecting the bacteria from acid.

All fully virulent, highly invasive strains of *Y. enterocolitica* carry a ca. 70-kb plasmid, called pYV (for plasmid for *Yersinia* virulence), which is found in *Y. enterocolitica*, *Y. pestis*, and *Y. pseudotuberculosis*. pYV permits the bacteria to resist phagocytosis (ingestion by specific host cells) and lysis. Thus, the pathogen can grow within tissues. When pYV-negative strains of *Y. enterocolitica* are incubated with host epithelial cells or phagocytes, they penetrate the cells in large numbers without causing damage. By contrast, pYV-positive bacteria generally remain outside of host cells (i.e., resist phagocytosis).

Enteropathogenic yersinae use a special secretory pathway (the type III secretory pathway) to inject proteins into the cytosol of eukaryotic cells. This facilitates pathogenesis. The transport of *Yersinia* outer membrane proteins (Yops) from the bacterial cytoplasm into the host cell cytosol may occur in one step from bacteria that are closely bound to the host cell.

Pathogenesis of *Yersinia*-Induced Autoimmunity

Arthritis

Following an infection with pYV-positive *Y. enterocolitica*, a small number of patients develop *autoimmune* (the person's own immune system starts to attack them) arthritis. A similar condition can also occur after infections with *Campylobacter, Salmonella, Shigella,* or *Chlamydia* species. The pathogenesis of reactive arthritis is poorly understood. Men and women are affected equally. Arthritis typically follows the onset of diarrhea or the pseudoappendicular syndrome by 1 to 2 weeks, with a range of 1 to 38 days. The joints most commonly involved are the knees, ankles, toes, tarsal joints, fingers, wrists, and elbows. The duration of arthritis is typically <3 months.

Thyroid Diseases

Y. enterocolitica has been implicated in various thyroid disorders, including autoimmune thyroiditis and Graves' disease hyperthyroidism. The latter is an immune system disorder caused by a person producing antibodies to the thyrotropin (TSH) receptor. The main link between *Y. enterocolitica* and thyroid diseases is that patients with these disorders frequently have elevated levels of antibodies to *Y. enterocolitica* O:3.

Other autoimmune complications of yersiniosis, including Reiter's syndrome, uveitis, acute proliferative glomerulonephritis, collagenous colitis, and rheumatic-like carditis, have been reported, mostly from Scandinavian countries. Yersiniosis has also been linked to various thyroid disorders, including nontoxic goiter and Hashimoto's thyroiditis, although the causative role of yersiniae in these conditions is uncertain. In Japan, *Y. pseudotuberculosis* has been linked to Kawasaki's disease.

Summary

- *Y. enterocolitica* expresses factors, such as urease, flagella, and smooth LPS, which facilitate its passage through the stomach and the mucus layer of the small intestine.

- Once *Y. enterocolitica* begins to replicate in the intestine at 37°C, LPS becomes rough, exposing Ail and YadA on the bacterial surface.

- The higher infectivity of *Y. enterocolitica* when grown at ambient temperature compared with its infectivity when grown at 37°C may account for the small number of reports of human-to-human transmission of yersiniosis.

- The well-defined life cycle of *Y. enterocolitica* with its distinctive temperature-induced phases is reminiscent of the flea-rat-flea cycle of *Y. pestis.*

- Outbreaks of yersiniosis are surprisingly uncommon, considering the ubiquity of *Y. enterocolitica* in nature and its abilities to colonize food animals, to persist within animals and the environment, and to proliferate at refrigeration temperature.

Suggested reading

Andersen, J. K., R. Sorensen, and M. Glensbjerg. 1991. Aspects of the epidemiology of *Yersinia enterocolitica:* a review. *Int. J. Food Microbiol.* **13:**231–237.

Burnens, A. P., A. Frey, and J. Nicolet. 1996. Association between clinical presentation, biogroups and virulence attributes of *Yersinia enterocolitica* strains in human diarrhoeal disease. *Epidemiol. Infect.* **116:**27–34.

Cover, T. L., and R. C. Aber. 1989. *Yersinia enterocolitica. N. Engl. J. Med.* **321:**16–24.

Schiemann, D. A. 1989. *Yersinia enterocolitica* and *Yersinia pseudotuberculosis,* p. 601–672. *In* M. P. Doyle (ed.), *Foodborne Bacterial Pathogens.* Marcel Dekker, New York, N.Y.

Wauters, G., K. Kandolo, and M. Janssens. 1987. Revised biogrouping scheme of *Yersinia enterocolitica. Contrib. Microbiol. Immunol.* **9:**14–21.

Questions for critical thought

1. Why is *Y. enterocolitica* considered unusual among pathogenic enterobacteria? Explain.

2. What domestic animal is considered a reservoir for *Y. enterocolitica*? How was this conclusion drawn?

3. Which group is most likely to develop pseudoappendicular syndrome? What is pseudoappendicular syndrome?

4. What four key factors are associated with the virulence of *Y. enterocolitica*?

5. Infection with *Y. enterocolitica* can result in other diseases. What diseases can develop, and how does infection with *Y. enterocolitica* play a role in those diseases?

6. Identify unique characteristics of *Y. enterocolitica* that would facilitate identification of the pathogen.

7. In reference to the large *Y. enterocolitica* outbreak linked to consumption of pasteurized milk, what measure(s) should have been implemented to prevent such an outbreak?

8. *Y. enterocolitica* is motile when grown at 25°C but not when grown at 37°C. This seems counter to properties required for an enteric pathogen. Provide several reasons why the organism would prefer motility at 25°C to motility at 37°C.

9. Select two of the autoimmune diseases from Table 10.2 and use your investigative skills to determine the characteristics of each disease.

10. Virulent strains of *Yersinia* carry the plasmid for *Yersinia* virulence (pYV). What properties that enhance virulence does the plasmid impart to strains? Outline a set of experiments that could be done to prove the role of pYV in virulence.

11

Shigella Species*

LEARNING OBJECTIVES

The information in this chapter will enable the student to:

- use basic biochemical characteristics to identify *Shigella*
- understand what conditions in foods favor *Shigella* growth
- recognize, from symptoms and time of onset, a case of foodborne illness caused by *Shigella*
- choose appropriate interventions (heat, preservatives, and formulation) to prevent the growth of *Shigella*
- identify environmental sources of the organism
- understand the role of *Shigella* toxins and virulence factors in causing foodborne illness

Outbreak

In a 2-month period during the summer of 1998, seven outbreaks of *Shigella sonnei* infection associated with eating fresh parsley occurred. Fresh parsley is often used as a garnish or sprinkled on top of a meal prior to serving to add taste and color. Hundreds of people became ill, with most experiencing severe short-term diarrhea. Most of the patients ate at restaurants that served chopped, uncooked parsley. The causative agent was identified as *S. sonnei*, and based on pulsed-field gel electrophoresis, epidemiologic traceback, and data from other investigations, one farm in Mexico was implicated as the source of the contaminated parsley. As a result of these outbreaks, changes in food-handling practices were proposed, eliminating practices such as chopping and holding large quantities of parsley and chopping of parsley at room temperature. *Shigella* is spread through contaminated water used for irrigation and postharvest processing of fresh produce. Transmission occurs through the fecal-oral route, with as few as 10 to 100 organisms capable of causing infection. Humans and other primates are the only reservoirs for *S. sonnei*.

INTRODUCTION

Bacillary dysentery, or shigellosis, is caused by *Shigella* species. Dysentery was the term used by Hippocrates to describe an illness characterized by frequent passage of stools containing blood and mucus accompanied by painful abdominal cramps. Perhaps one of the greatest historical impacts of

*This chapter was originally written by Keith A. Lampel and Anthony T. Maurelli for *Food Microbiology: Fundamentals and Frontiers,* 2nd ed., and has been adapted for use in an introductory text.

139

this disease has been its powerful influence on military operations. Protracted military campaigns and sieges have almost always spawned epidemics of dysentery, causing large numbers of military and civilian casualties. With a low infectious dose required to cause disease coupled with oral transmission by fecally contaminated food and water, it is not surprising that dysentery caused by *Shigella* spp. follows in the wake of many natural (earthquakes, floods, and famine) and man-made (war) disasters. Apart from these special circumstances, shigellosis remains an important disease in developed and developing countries.

During the past 2 decades, several large outbreaks of shigellosis (listed below) have been linked to the consumption of contaminated food. Disease is caused by ingestion of these contaminated foods, and in some instances it subsequently leads to rapid dissemination through contaminated feces from infected individuals.

1987: Rainbow Family gathering. As many as half of the 12,700 people in attendance at an annual gathering of the Rainbow Family (a naturalist organization) may have had shigellosis. *S. sonnei* was isolated from stool cultures of tested attendees. Spread of the organism most likely occurred through the fecal-oral route in a crowded environment by contamination of food or water or both. Subsequent outbreaks reported in three other states were due to attendees returning home and infecting others.

1989 and 1994: Shigellosis aboard cruise ships. In October 1989, 14% of passengers and 3% of crew members aboard a cruise ship reported having gastrointestinal symptoms. A multiple-antibiotic-resistant strain of *Shigella flexneri* 2a was isolated from several ill passengers and crew. The vehicle of the outbreak was German potato salad. Contamination was introduced by infected food handlers, initially in the country where the food was originally prepared and subsequently by a member of the galley crew on the cruise ship. Another outbreak of shigellosis occurred in August 1994 on the cruise ship S.S. Viking Serenade. Thirty-seven percent (586) of the passengers and 4% (24) of the crew reported having diarrhea, and one death occurred. *S. flexneri* 2a was isolated from patients, and the suspected vehicle was spring onions.

1990: Operation Desert Shield. Diarrheal diseases during a military operation can be a major factor in reducing troop readiness. Enteric pathogens were isolated from 214 U.S. soldiers in Operation Desert Shield, and of those, 113 cases were diagnosed as shigellosis; *S. sonnei* was the most prevalent species isolated. Shigellosis accounted for more time lost from military duties and was responsible for more severe morbidity than enterotoxigenic *Escherichia coli*, the most common enteric pathogen isolated from U.S. troops in Saudi Arabia. The suspected vehicle was contaminated fresh vegetables, specifically, lettuce. Twelve heads of lettuce were tested, and enteric pathogens were isolated from all of them.

1991: Moose soup in Alaska. In September 1991 in Galena, Alaska, 25 people who participated in a gathering of local residents contracted shigellosis associated with eating homemade moose soup. One of five women who made the soup reported having gastroenteritis while preparing it. *S. sonnei* was isolated from a hospitalized patient.

1994: Contaminated produce. Lettuce and green onions have been implicated in illness. An outbreak in Norway of 110 culture-confirmed cases of shigellosis caused by *S. sonnei* was reported in 1994. Iceberg lettuce from Spain, served in a salad bar, was suspected as the source of the outbreak in Norway and was likely responsible for increases in shigellosis in other European countries, including the United Kingdom and Sweden. *S. sonnei* was isolated from patients from several northwest European countries but was not isolated from any foods. Epidemiologic evidence indicated that imported lettuce was the vehicle of these outbreaks. An outbreak of *S. flexneri* serotype 6 (mannitol-negative) infection occurred in the Midwest in 1994. Although not confirmed, the suspected vehicle was Mexican green onions (scallions, or spring onions). Seventeen cases of shigellosis occurred at a church potluck meal in Indiana, 29 cases occurred at an anniversary reception in Indiana, and 26 culture-confirmed mannitol-negative *S. flexneri* or *Shigella* sp. cases were reported to the Illinois State Department of Health. *S. flexneri* serotype 6 was also isolated from patients in Missouri, Minnesota, Wisconsin, Michigan, and Kentucky. Ingestion of green onions was implicated as the vehicle of infection. An infected worker most likely contaminated the onions at the time of harvest or packing.

2000: Five-layer bean dip. An outbreak of shigellosis associated with contaminated five-layer (bean, salsa, guacamole, nacho cheese, and sour cream) party dip occurred in three West Coast states. The causative agent, *S. sonnei*, was isolated from at least 30 patients. The pathogen was isolated from only one layer (cheese) of the dip and was initially detected by a polymerase chain reaction (PCR) assay targeting shigellae. *Shigella* was isolated subsequently by enrichment, followed by plating on selective agar.

One of the striking features regarding foodborne outbreaks of shigellosis is that contamination of foods usually does not occur at the processing plant but rather through an infected food handler. As is evident from the examples above and in Table 11.1, these incidents can be anything from contamination of foods by infected food handlers at small-town gatherings and picnics to large-scale outbreaks, such as those on cruise ships and at institutions.

Table 11.1 Examples of foodborne outbreaks caused by *Shigella* spp.

Yr	Location	Source of contamination[a]	Isolate
1986	Texas	Shredded lettuce	*S. sonnei*
1987	Rainbow Family gathering	Food handlers	*S. sonnei*
1988–1989	Monroe, N.Y.	Multiple sources	*S. sonnei*
1988	Outdoor music festival, Michigan	Food handlers	*S. sonnei*
1988	Commercial airline	Cold sandwiches	*S. sonnei*
1989	Cruise ship	Potato salad	*S. flexneri*
1990	Operation Desert Shield (U.S. troops)	Fresh produce	*Shigella* spp.
1991	Alaska	Moose soup	*S. sonnei*
1992–1993	Operation Restore Hope, Somalia (U.S. troops)		*Shigella* spp.
1994	Europe	Shredded lettuce from Spain	*S. sonnei*
1994	Midwest	Green onions	*S. flexneri*
1994	Cruise ship		*S. flexneri*
1998	Various U.S. locations	Fresh parsley	*S. sonnei*
2000	West Coast	Bean dip	*S. sonnei*

[a]The source of contamination is listed when known.

Classification and Biochemical Characteristics

There are four species in the genus *Shigella,* serologically grouped (41 serotypes) based on their somatic (O) antigens: *Shigella dysenteriae* (group A), *S. flexneri* (group B), *Shigella boydii* (group C), and *S. sonnei* (group D). As members of the family *Enterobacteriaceae,* they are genetically almost identical to the escherichiae and closely related to the salmonellae. *Shigella* spp. are nonmotile, oxidase-negative, gram-negative rods. An important biochemical characteristic that distinguishes these bacteria from other enterics is their inability to ferment lactose; however, some strains of *S. sonnei* may ferment lactose slowly or utilize citric acid as a sole carbon source. They do not produce H$_2$S, except for *S. flexneri* 6 and *S. boydii* serotypes 13 and 14, and do not produce gas from glucose. *Shigella* spp. are inhibited by potassium cyanide and do not synthesize lysine decarboxylase. Enteroinvasive *E. coli* (EIEC) has pathogenic and biochemical properties similar to those of *Shigella* spp. This similarity poses a problem in distinguishing these pathogens. For example, EIEC is nonmotile and is unable to ferment lactose. Some serotypes of EIEC also have O antigens identical to those of *Shigella.*

Shigella spp. are not particular in their growth requirements and are routinely cultivated in the laboratory on artificial medium. Cultures of *Shigella* are easily isolated and grown from analytical samples, including water and clinical specimens. In the latter case, *Shigella* spp. are present in fecal specimens in large numbers (10^3 to 10^9 per g of stool) during the acute phase of infection, and therefore, identification is readily accomplished using culture media, biochemical analysis, and serological typing. Shigellae are shed and continue to be detected from convalescent patients (10^2 to 10^3 per g of stool) for weeks or longer after the initial infection. Isolation of *Shigella* at this stage of infection is more difficult because a selective enrichment broth for shigellae is not available, and therefore, shigellae can be outgrown by resident bacterial fecal flora.

Isolation of *Shigella* spp. from foods is not as easy as from other sources. Foods have many different physical attributes that may affect the recovery of shigellae. These factors include composition, such as the fat content of the food; physical parameters, such as pH and salt; and the natural microbial flora of the food. In the last case, other microbes in a sample may overgrow shigellae during culture in broth media. The amount of time from the clinical report of a suspected outbreak to the analysis of the food samples can be considerable, thus lessening the chances of identifying the causative agent. The physiological state of shigellae present in the food is a contributing factor in the successful recovery of the pathogen. *Shigella* spp. may be present in low numbers or in a poor physiological state in suspected food samples. Under these conditions, special enrichment procedures are required for successful isolation and detection of shigellae.

Shigella in Foods

Shigella spp. are not associated with any specific foods. Common foods that have been implicated in outbreaks caused by shigellae include potato salad, chicken, tossed salad, and shellfish. Establishments where contaminated foods have been served include the home, restaurants, camps, picnics, schools, airlines, sorority houses, and military mess halls. In many cases, the source (food) was not identified. From 1983 to 1987, 2,397 foodborne outbreaks representing 54,453 cases were reported to the Centers for Disease Control and Prevention (CDC). In only 38% of the cases was the source of the etiological agent identified. Whereas epidemiologic methods may strongly imply a common food source, *Shigella* spp. are not often recovered from foods and identified by using standard bacteriological methods. Also, since shigellae are not commonly associated with any particular food, routine testing of foods to identify these pathogens is not usually performed.

The traditional approach to address the problem of microbially contaminated foods in the processing plant is to inspect the final product. There are several drawbacks to this approach. Current bacteriological methods are often time-consuming and laborious. An alternative to end-product testing is the Hazard Analysis Critical Control Point (HACCP) system. The HACCP system identifies certain points of the processing system that may be most vulnerable to microbial contamination and chemical and physical hazards.

In contrast, establishing specific critical control points for preventing *Shigella* contamination of foods is not always suitable for the HACCP concept. The pathogen is usually introduced into the food supply by an infected person, such as a food handler with poor personal hygiene. In some cases, this may occur at the manufacturing site, but more likely it happens at a point between the processing plant and the consumer. Another factor is the fact that foods, such as vegetables (lettuce is a good example), can be contaminated at the site of collection and shipped directly to market. Although the HACCP system is a method for controlling food safety and preventing foodborne outbreaks, pathogens such as *Shigella* that are not indigenous to but rather introduced into foods are most likely to be undetected.

Survival and Growth in Foods

Depending upon growth conditions, *Shigella* spp. can survive in media with a pH range of 2 to 3 for several hours. However, shigellae do not usually survive well in low-pH foods or in stool samples. Studies using citrus juices (orange and lemon), carbonated beverages, and wine revealed that shigellae were recovered after 1 to 6 days. In neutral-pH foods, such as butter or margarine, shigellae were recovered after 100 days when stored frozen or at 6°C.

Shigella can survive a temperature range of −20°C to room temperature. However, shigellae survive longer in foods stored frozen or at refrigeration temperature than in those stored at room temperature. In foods such as salads containing mayonnaise and some cheese products, *Shigella* has survived for 13 to 92 days. *Shigella* can survive for an extended period on dry surfaces and in foods such as frozen shrimp, ice cream, and minced pork. Growth of *Shigella* is impeded in the presence of 3.8 to 5.2% NaCl at pH 4.8 to 5.0, in 300 to 700 mg of $NaNO_2$/liter, and in 0.5 to 1.5 mg of sodium hypochlorite (NaClO)/liter of water at 4°C. *Shigella* is sensitive to ionizing radiation, with a reduction of 10^7 colony-forming units/g at 3 kilogray.

CHARACTERISTICS OF DISEASE

Shigellosis is differentiated from diseases caused by most other foodborne pathogens described in this book by at least two important characteristics: (i) the production of bloody diarrhea or dysentery and (ii) the low infectious dose. Dysentery involves bloody diarrhea, but the passage of bloody mucoid stools is accompanied by severe abdominal and rectal pain, cramps, and fever. While abdominal pain and diarrhea are experienced by

Shigella **is easily transmitted**

Shigella is easily passed from person to person by food. Outbreaks are often associated with poor hygiene, especially improper hand washing after use of the bathroom. The pathogen multiplies rapidly in food at room temperature. *Shigella* produces a powerful toxin, Shiga toxin, once in the host. Shigellosis is an infection, not intoxication. That is, vegetative cells must be ingested, and the cells multiply and cause illness. As few as 100 cells may cause illness.

nearly all patients with shigellosis, fever occurs in about one-third and gross blood in the stools occurs in ~40% of cases. The clinical features of shigellosis range from a mild watery diarrhea to severe dysentery. The dysentery stage caused by *Shigella* spp. may or may not be preceded by watery diarrhea. During the dysentery stage, there is extensive bacterial colonization of the colon and invasion of the cells of the colon. As the infection progresses, dead cells of the mucosal surface slough off. This leads to the presence of blood, pus, and mucus in the stools.

Shigella is a serious pathogen that causes disease in otherwise healthy individuals (Box 11.1). The greatest frequency of illness is among children <6 years of age. The incubation period for shigellosis is 1 to 7 days, but the symptoms usually begin within 3 days. The severity of illness will differ depending on the strain involved; however, regardless of the severity of the illness, shigellosis is self-limiting. If left untreated, clinical illness usually persists for 1 to 2 weeks (although it may last as long as a month), and the patient recovers. Since the infection is self-limited in normally healthy patients and full recovery occurs without the use of antibiotics, drug therapy is usually not indicated. However, the antibiotic of choice for treatment of shigellosis is trimethoprim-sulfamethoxazole. Complications arising from the disease include severe dehydration, intestinal perforation, septicemia, seizures, hemolytic-uremic syndrome, and Reiter's syndrome.

FOODBORNE OUTBREAKS

Although the number of reported foodborne outbreaks of shigellosis in the United States has declined recently, shigellosis continues to be a major public health concern. Data from 1982 to 1997 estimated that there were approximately 448,000 cases of shigellosis in the United States during that period. This made it the third leading cause of foodborne outbreaks by bacterial pathogens. Worldwide, the World Health Organization estimates that *Shigella* spp. are responsible for 164.7 million cases of shigellosis annually in developing countries and 1.5 million cases in developed countries.

Humans are the natural reservoir of *Shigella*. Human-to-human transmission of *Shigella* is through the fecal-oral route. Most cases of shigellosis result from the ingestion of fecally contaminated food or water. With foods, the major cause of contamination is poor personal hygiene of food handlers. From infected carriers, shigellae are spread by several routes, including food, fingers, feces, and flies. The highest incidence of shigellosis occurs during the warmer months of the year. Improper storage of contaminated foods is the second most common factor contributing to foodborne outbreaks of shigellosis. Other contributing factors are inadequate cooking, contaminated equipment, and food obtained from unsafe sources. To reduce the spread of shigellosis, infected patients should be monitored until stool samples are negative for *Shigella*. The low infectious dose of *Shigella* underlies the high rate of transmission. As few as 100 cells of *Shigella* can cause illness, facilitating person-to-person spread, as well as foodborne and waterborne outbreaks of diarrhea.

Shigellosis can be widespread in institutional settings, such as prisons, mental hospitals, and nursing homes, where crowding and/or insufficiently hygienic conditions create an environment for direct fecal-oral contamination. The occurrence of disasters that destroy the sanitary waste treatment and water purification infrastructure is often associated with large outbreaks of shigellosis.

VIRULENCE FACTORS

The clinical symptoms of shigellosis can be directly attributed to the hallmarks of *Shigella* virulence: the ability to induce diarrhea, invade epithelial cells of the intestine, multiply intracellularly, and spread from cell to cell. The production of enterotoxins by the bacteria while they are in the small bowel probably causes the diarrhea that precedes dysentery. The ability of *Shigella* to invade epithelial cells and move from cell to cell is regulated by an array of genes. A bacterium's mechanism for regulating expression of the genes involved in virulence is important for pathogenicity. *S. dysenteriae* produces a thermolabile toxin, designated Shiga toxin, that is involved in the pathogenesis of *Shigella* diarrhea. Strains that are invasive and produce the toxin cause the most severe infection.

The growth temperature is an important factor in controlling virulence. Virulent strains of *Shigella* spp. are invasive when grown at 37°C but noninvasive when grown at 30°C. This strategy ensures that the organism conserves energy by synthesizing virulence products only when the bacterium is in the host.

Genetic Regulation

Given the complexity of the interactions between host and pathogen, it is not surprising that *Shigella* virulence requires several genes. These include both chromosomal and plasmid-encoded genes. The large plasmid has an indispensable role in invasion by *S. sonnei* and *S. flexneri*. Other *Shigella* spp., as well as strains of EIEC, contain similar plasmids, which are functionally interchangeable and show significant degrees of DNA relatedness. The plasmids of *Shigella* and EIEC are probably derived from a common ancestor.

The *ipa* group of genes encode invasion plasmid antigens, the main antigens detected with sera from convalescent patients and experimentally challenged monkeys. These genes are required for the invasion of mammalian cells. Ultimately, the proteins form a complex on the bacterial cell surface and are responsible for transducing the signal leading to entry of *Shigella* into the host cells via bacterium-directed phagocytosis. The products of the *ipa* genes have also been postulated to be the contact hemolysin responsible for lysis of the phagocytic vacuole minutes after entry of the bacterium into the host cell. The ability of *S. flexneri* to induce programmed cell death in infected macrophages is an additional property assigned to IpaB.

In contrast to the genes of the virulence plasmid that are responsible for the invasion of mammalian tissues, most of the chromosomal loci associated with *Shigella* virulence are involved in regulation or survival within the host. Although *Shigella* and *E. coli* are very closely related at the genetic level, there are significant differences beyond the presence of the virulence plasmid in *Shigella*. In addition to extra genes in the *Shigella* chromosome, there are genes present in the closely related *E. coli* that are missing from the chromosome of *Shigella*; genetic divergence accounts for differences in the virulence and pathogenicity of *Shigella* and *E. coli*.

CONCLUSION

While foodborne infections due to *Shigella* spp. may not be as frequent as those caused by other foodborne pathogens, they have the potential for explosive spread due to the extremely low infectious dose that can cause overt clinical disease. In addition, cases of bacillary dysentery frequently require medical attention (even hospitalization), resulting in time lost from work, as the severity and duration of symptoms can be incapacitating. There is no effective vaccine against dysentery caused by *Shigella*. These features, coupled with the wide geographical distribution of the strains and the sensitivity of the human population to *Shigella* infection, make *Shigella* a formidable public health threat.

Summary

- The infective dose may be as low as 100 cells.
- Shigellosis is self-limiting.
- Humans are the natural reservoir of *Shigella.*
- *Shigella* is spread through the fecal-oral route; therefore, hand washing is one of the most effective control measures.
- Expression of virulence genes is regulated by temperature.
- *Shigella* shares many virulence genes with *E. coli.*
- Genes responsible for virulence are located on a virulence plasmid.

Suggested reading

Parsot, C., and P. J. Sansonetti. 1996. Invasion and the pathogenesis of *Shigella* infections. *Curr. Top. Microbiol. Immunol.* **209:**25–42.

Smith, J. L. 1987. *Shigella* as a foodborne pathogen. *J. Food Prot.* **50:**788–801.

Questions for critical thought

1. Shigellosis is distinguished from diseases caused by other foodborne pathogens by at least two characteristics. What are the two key characteristics?
2. Why is the use of antibiotics to treat shigellosis controversial?
3. What is meant by the term infectious dose? Compared to other foodborne pathogens, does *Shigella* have a low infectious dose?
4. How does temperature play a role in *Shigella* pathogenesis?
5. Once *Shigella* has invaded epithelial cells of the host, what unique function can the pathogen perform and how does that influence pathogenesis?
6. What is the most likely route by which food becomes contaminated with *Shigella?*
7. What characteristics of the pathogen with respect to reservoir and shedding pattern may exacerbate its spread and the subsequent outbreak of shigellosis?
8. Many cases of shigellosis are linked to consumption of fresh produce. What measures could be implemented to reduce the risk of shigellosis from consumption of fresh vegetables?

12

Vibrio Species*

LEARNING OBJECTIVES

The information in this chapter will enable the student to:
- use basic biochemical characteristics to identify *Vibrio*
- understand what conditions in foods favor *Vibrio* growth
- recognize, from symptoms and time of onset, a case of foodborne illness caused by *Vibrio*
- choose appropriate interventions (heat, preservatives, and formulation) to prevent the growth of *Vibrio*
- identify environmental sources of the organism
- understand the role of *Vibrio* toxins and virulence factors in causing foodborne illness

Outbreak

Raw oysters are consumed for their taste and because of the old wives' tale that they act as an aphrodisiac. People rarely consider the hazards associated with the consumption of raw shellfish. During a 2-month period (July and August) in 1997, one of the largest outbreaks of culture-confirmed *Vibrio parahaemolyticus* infection occurred in North America. At least 209 persons became ill with gastroenteritis from eating raw oysters. Symptoms included diarrhea, abdominal cramps, nausea, and vomiting. Although symptoms of gastroenteritis may attract the attention of others, they would certainly not be considered attractive. Oysters sampled from the beds implicated in the outbreak yielded <200 *V. parahaemolyticus* colony-forming units (CFU)/g of oyster meat, suggesting that human illness may occur at levels of bacteria lower than the current action level. The United States and Canada permit the sale of oysters if there are $<10^4$ CFU of *V. parahaemolyticus* per g of oyster meat.

INTRODUCTION

Over 20 *Vibrio* species have now been described. At least 12 are capable of causing infection in humans, although with the exception of *Vibrio cholerae* and *V. parahaemolyticus*, little is known about the virulence mechanisms they employ. Of the 12 pathogens, 8 are directly food associated, and these are the subject of this chapter. (*Vibrio carchariae* and *Vibrio damsela* infections appear to result solely from wound infections, and the routes of infection for *Vibrio metschnikovii* and *Vibrio cincinnatiensis* are unclear.)

*This chapter was originally written by James D. Oliver and James B. Kaper for *Food Microbiology: Fundamentals and Frontiers,* 2nd ed., and has been adapted for use in an introductory text.

One of the most consistent aspects of vibrio infections is a recent history of seafood consumption. Vibrios, the predominant bacterial genus in estuarine waters, are associated with a great variety of seafoods. Approximately 40 to 60% of finfish and shellfish at supermarkets may contain *Vibrio* spp., with *V. parahaemolyticus* and *Vibrio alginolyticus* most commonly isolated. Vibrios are most frequently isolated from molluscan shellfish during the summer months.

CHARACTERISTICS OF THE ORGANISM

Many different enrichment broths have been described for the isolation of vibrios, although alkaline peptone water remains the most commonly used. These are frequently coupled with thiosulfate-citrate-bile salts-sucrose (TCBS) agar or other plating media. In order to distinguish vibrios from the *Enterobacteriaceae*, sucrose-positive colonies are subjected to the oxidase test (used to detect the presence of cytochrome *c*); however, erroneous results may arise when colonies are obtained directly from TCBS agar. Therefore, sucrose-positive colonies on TCBS should be subcultured by heavy inoculation onto a nonselective medium, such as blood agar, and allowed to grow for 5 to 8 h before being tested for oxidase activity.

Epidemiology

The numbers of vibrios in both surface waters and shellfish correlate with seasonality, generally being greater during the warm-weather months between April and October (Fig. 12.1). Similarly, vibrios are more commonly isolated from the warmer waters of the Gulf and East Coasts than from those of the West and Pacific Northwest. Seasonality is most notable for *Vibrio vulnificus* and *V. parahaemolyticus* infections, whereas some vibrios, such as *Vibrio fluvialis*, occur throughout the year.

Figure 12.1 Numbers of CFU of *V. vulnificus* (± standard deviations) isolated from northeastern United States coastal waters as a function of time of year and water temperature. Reprinted from R. C. Tilton and R. W. Ryan, *Diagn. Microbiol. Infect. Dis.* **6:**109–117, 1987, with permission of the publisher.

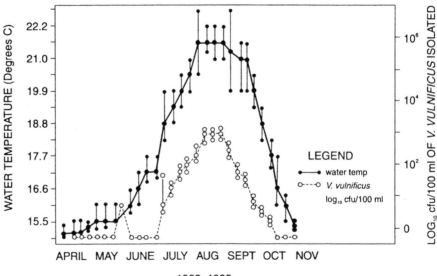

CHARACTERISTICS OF DISEASE

There is considerable variation in the severities of the various *Vibrio*-associated diseases. The outcomes of infections depend on an individual's underlying health (e.g., patients with chronic liver disease are more likely to succumb to severe illness). An exception to this generalization is *V. cholerae* O1/O139, which can readily make noncompromised individuals sick. Most cases are associated with a recent history of seafood consumption, especially eating raw oysters, according to an article by J. C. Desenclos, K. C. Klontz, L. E. Wolfe, and S. Hoecherl published in 1991. Because vibrios are part of the normal estuarine microflora caused by fecal contamination, *Vibrio* infections will not be controlled through shellfish sanitation programs. It is therefore essential that raw seafood be kept cold to prevent significant bacterial growth and that it be properly cooked before consumption.

Symptoms of foodborne illness for the two most common *Vibrio* species differ and are worth noting. Symptoms of *V. vulnificus* foodborne illness generally include fever, chills, and nausea. Diarrhea is generally not experienced. Mortality occurs in 40 to 60% of cases. The most common symptoms of *V. parahaemolyticus* infection are diarrhea, abdominal cramps, nausea, and vomiting. Mortality is extremely low.

SUSCEPTIBILITY TO PHYSICAL AND CHEMICAL TREATMENTS

With the exception of *V. cholerae* and *V. parahaemolyticus*, relatively little is known about the effects of preservation methods on inactivation of vibrios. Generally, vibrios are sensitive to cold, and seafoods can be protective for vibrios at refrigeration temperatures. Vibrios have survived for extended periods in products stored at refrigeration temperatures and after prolonged frozen storage. Thermal processing is a very effective means of reducing populations of *Vibrio* in foods. The U.S. Food and Drug Administration recommends steaming shellstock oysters, clams, and mussels for 4 to 9 min; frying shucked oysters for 10 min at 375°C; or baking oysters for 10 min at 450°C. Thorough heating of shellfish to an internal temperature of at least 60°C for several minutes should kill pathogenic vibrios. Irradiation and high hydrostatic pressure also reduce the numbers of *Vibrio* organisms. High hydrostatic pressure has an additional advantage in that the process perfectly shucks the oyster. A large variety of dried spices, the oils of several herbs, tomato sauce, and several organic acids exhibit bactericidal activity. The process of depuration, in which filter-feeding bivalves are purified of certain bacteria by pumping bacterium-free water through the bivalve tissues, removes *Salmonella* and *Escherichia coli*, but not *Vibrio*.

V. cholerae

V. cholerae O1 causes cholera, one of the few foodborne illnesses with *epidemic* (many people in one area become sick at the same time) and *pandemic* (similar to an epidemic but over a larger area) potential. Important distinctions within the species are made on the basis of production of cholera enterotoxin (cholera toxin [CT]), serogroup, and potential for epidemic spread. Not all *V. cholerae* strains of the O1 serogroup produce CT; however, *V. cholerae* O1 CT-producing strains have long been associated with epidemic and pandemic cholera. Nearly 200 serogroups of *V. cholerae* have been described. Serotyping of *V. cholerae* strains is based on the lipopolysaccharide (LPS), also known as the somatic (O) antigen; H antigens are not useful in serotyping.

Isolation and Identification

Suspected *V. cholerae* isolates can be cultured on a standard series of biochemical media used for identification of *Enterobacteriaceae* and *Vibrionaceae*. Both conventional tube tests and commercially available enteric identification systems work well.

Nucleic acid probes are not routinely used for the identification of *V. cholerae* due to the ease of identifying this species by conventional methods. DNA probes are useful in distinguishing CT-producing strains of *V. cholerae* through the detection of CT (*ctx*) genes. Restriction fragment length polymorphism analysis has been useful in epidemiological studies of *V. cholerae*.

Reservoirs

V. cholerae is part of the normal, free-living bacterial flora in estuarine areas. Non-O1/non-O139 strains are much more commonly isolated from the environment than are O1 strains, even in epidemic settings in which fecal contamination of the environment might be expected. Periodic introduction of such environmental isolates into the human population through the ingestion of uncooked or undercooked shellfish appears to be responsible for localized outbreaks along the U.S. Gulf Coast and in Australia. The persistence of *V. cholerae* within the environment may be facilitated by its ability to assume survival forms, including a viable but nonculturable (VBNC) state. In this dormant state, the cells are reduced in size and become ovoid. Although VBNC vibrios are not culturable with nonselective enrichment broth or plates, studies have demonstrated that VBNC *V. cholerae* O1 injected into ligated rabbit ileal loops or ingested by volunteers have yielded culturable *V. cholerae* O1 in the intestinal contents or stool specimens, respectively.

Long-term carriage of *V. cholerae* in humans is extremely rare and is not important in the transmission of disease. However, even after the cessation of symptoms, patients who have not been treated with antibiotics may continue to excrete vibrios for 1 to 2 weeks. Asymptomatic carriers are most commonly identified among household members of persons with acute illness: in various studies, the rate of asymptomatic carriage in this group has ranged from 4 to almost 22%.

Foodborne Outbreaks

The critical role of water in the transmission of cholera has been recognized for more than a century. In 1854, the London physician and epidemiologist John Snow determined that illness was associated with consumption of water derived from the Thames River at a point below major sewage inflows. The range of food items implicated in the transmission of cholera includes crabs, shrimp, raw fish, mussels, cockles, squid, oysters, clams, rice, raw pork, millet gruel, cooked rice, food bought from street vendors, frozen coconut milk, and raw vegetables and fruit. One shared characteristic of the foods is their neutral or nearly neutral pH. When foodborne outbreaks occur and a number of different foods are suspected, foods with an acid pH can be eliminated. Food acts to buffer *V. cholerae* O1 against killing by gastric acid. In the United States, crabs, shrimp, and oysters have been the most frequently implicated vehicles of *V. cholerae* illness.

Characteristics of Disease

The explosive, dehydrating diarrhea characteristic of cholera actually occurs in only a minority of persons infected with CT-producing *V. cholerae* O1/O139. Most infections with *V. cholerae* O1 are mild or even asymptomatic. The incubation period of cholera can range from several hours to

5 days and is dependent in part on the *inoculum size* (the number of cells ingested). The onset of illness may be sudden, with watery diarrhea, or there can be loss of appetite, abdominal pain, and simple diarrhea. Initially, the stool is brown with fecal matter, but once diarrhea starts, it becomes a pale gray color with a slightly fishy odor. Mucus in the stool gives the characteristic "rice water" appearance. Vomiting can occur a few hours after the onset of diarrhea. In healthy North American volunteers, doses of 10^{11} CFU of *V. cholerae* were required to cause diarrhea when the inoculum was given in buffered saline (pH 7.2). However, a lower number of bacteria caused illness in volunteers when 10^6 vibrios were given, with food (fish and rice) acting as a buffer.

Virulence Mechanisms

Infection due to *V. cholerae* O1/O139 begins with the ingestion of food or water contaminated with the pathogen. After passage through the acid barrier of the stomach, vibrios colonize the small intestine using one or more adherence factors. Production of CT (and possibly other toxins) disrupts ion transport by intestinal epithelial cells. This leads to the severe diarrhea characteristic of cholera with excessive loss of water and electrolytes. Cholera halotoxin is composed of five identical B subunits and a single A subunit; the individual subunits are not sufficient to cause secretogenic activity (an increase in secretion of cellular components) in animals or intact cell culture systems.

 V. cholerae produces a variety of extracellular products that are harmful to eukaryotic cells. In addition to the well-characterized toxins (soluble hemagglutinin and hemolysins), *V. cholerae* can produce a number of other toxic factors, but the responsible proteins and genes have not yet been purified or cloned. A number of other virulence factors are involved in illness, including colonization factors, flagella, LPS, and polysaccharide capsule.

V. mimicus

Prior to 1981, *Vibrio mimicus* was known as sucrose-negative *V. cholerae* non-O1. This species was determined to be a distinctly different species on the basis of biochemical reactions and DNA hybridization studies, and the name *mimicus* was given because of its similarity to *V. cholerae*. This organism is mainly isolated from cases of gastroenteritis but can also cause ear infections. The reservoir of *V. mimicus* is the aquatic environment. Besides being present free in the water, *V. mimicus* was also isolated from the roots of aquatic plants, from sediments, and from plankton at levels up to 6×10^4 CFU per 100 g of plankton.

Foodborne Outbreaks

Gastroenteritis due to *V. mimicus* has been associated only with consumption of seafood. In the United States, consumption of raw oysters is the main cause of illness due to this species. In Japan, consumption of raw fish has resulted in at least two outbreaks involving *V. mimicus* of serogroup O41.

Characteristics of Disease

Disease due to *V. mimicus* is characterized by diarrhea, nausea, vomiting, and abdominal cramps in most patients. In some individuals, fever, headache, and bloody diarrhea also occur. There are no volunteer or epidemiologic data to enable estimation of an infectious dose for *V. mimicus*. There is no particularly susceptible population, other than people who eat raw oysters.

Virulence Factors

V. mimicus does not produce unique enterotoxins, but many strains produce toxins that were first described in other *Vibrio* species, including CT, thermostable direct hemolysin (TDH), Zot, and a heat-stable enterotoxin apparently identical to the NAG-ST produced by *V. cholerae* non-O1/non-O139 strains. There is little information about potential intestinal colonization factors of *V. mimicus*.

V. parahaemolyticus

Along with *V. cholerae*, *V. parahaemolyticus* is the best described of the pathogenic vibrios, with numerous studies reported since the first description of its involvement in a major outbreak of food poisoning in 1950. Between 1973 and 1998, a total of 40 outbreaks of *V. parahaemolyticus* infection in the United States were reported to the Centers for Disease Control and Prevention, with >1,000 persons involved.

Classification

V. parahaemolyticus is serotyped according to both its somatic (O) and capsular polysaccharide (K) antigens. There are presently 12 O (LPS) antigens and 59 K (capsular polysaccharide) antigens recognized. Although many environmental and some clinical isolates are untypeable by the K antigen, most clinical strains can be classified based on their O types. A special consideration in the taxonomy of *V. parahaemolyticus* is the ability of certain strains to produce a hemolysin, called the TDH, or Kanagawa, hemolysin, which is linked to virulence in the species.

Reservoirs

V. parahaemolyticus occurs naturally in estuarine waters throughout the world and is easily isolated from coastal waters of the United States, as well as from sediment, suspended particles, plankton, and a variety of fish and shellfish. The last source includes at least 30 different species, among them clams, oysters, lobster, scallops, shrimp, and crab. A high percentage of seafood samples tested positive for the species. *V. parahaemolyticus* counts are season dependent, so that samples analyzed in January and February are often free of *V. parahaemolyticus*.

Foodborne Outbreaks

Gastroenteritis from *V. parahaemolyticus* is almost exclusively associated with seafood that is consumed raw, inadequately cooked, or cooked but recontaminated. In Japan, *V. parahaemolyticus* is a major cause of foodborne illness. Approximately 70% of all bacterial foodborne illnesses in the 1960s were the result of this pathogen. Seafood, including fish, crab, shrimp, lobster, and oysters, are primarily associated with U.S. outbreaks. The first major outbreak (with 320 persons ill) in the United States occurred in Maryland in 1971, a result of eating improperly steamed crabs. Subsequent outbreaks occurred throughout U.S. coastal regions and Hawaii. The largest U.S. outbreak of *V. parahaemolyticus* infection occurred during the summer of 1978 and affected 1,133 of 1,700 persons attending a dinner in Port Allen, La.

Characteristics of Disease

V. parahaemolyticus has generation times of 8 to 9 min at 37°C and 12 to 18 min in seafood. Hence, *V. parahaemolyticus* has the ability to grow rapidly, both in vitro and in vivo, contributing to the infectious dose required for illness. Symptoms may begin 4 to >30 h after the ingestion of contaminated food, with a mean onset time of 23.6 h. Primary symptoms include diarrhea and abdominal cramps, along with nausea, vomiting, and fever. The symptoms subside in 3 to 5 days in most individuals. Approximately 10^5 to 10^7 CFU is required for illness. The numbers of *V. parahaemolyticus* organisms present in fish and shellfish are usually no greater than 10^4 Kanagawa phenotype-positive cells, suggesting that temperature abuse of contaminated food occurs prior to consumption. The temperature abuse permits the growth of the pathogen to levels that result in illness.

Virulence Mechanisms
Although the epidemiologic linkage between virulence for humans and the ability of *V. parahaemolyticus* isolates to produce the Kanagawa hemolysin has long been established, the molecular mechanisms by which this factor can cause diarrhea have only recently been elucidated. *V. parahaemolyticus* possesses at least three hemolytic components; a thermolabile hemolysin gene (*tlh*), a TDH gene (*tdh*), and a TDH-related gene (*trh*), linked to disease.

While much is understood regarding the toxins of *V. parahaemolyticus*, little is known of the adherence process. This is an essential step in the pathogenesis of most enteropathogens (pathogens that infect the intestines). Several adhesive factors have been proposed, including the outer membrane, lateral flagella, and pili and a mannose-resistant, cell-associated hemagglutinin; however, the importance of any of these factors in human disease is unknown.

V. vulnificus
V. vulnificus is the most serious of the pathogenic vibrios in the United States. It is responsible for 95% of all seafood-borne deaths in this country. In Florida, *V. vulnificus* is the leading cause of reported deaths due to foodborne illness. Among the susceptible population at risk for infection by this bacterium, primary septicemia cases resulting from raw oyster consumption typically have fatality rates of 60%. This is the highest death rate for any foodborne disease agent in the United States. The Centers for Disease Control and Prevention estimate that there are ~50 cases of foodborne *V. vulnificus* infection annually. The bacterium can produce wound infections in addition to gastroenteritis and primary septicemias. Wound infections carry a 20 to 25% fatality rate and are also seawater and/or shellfish associated. Surgery is usually required to clean the infected tissue. In some cases, amputation of the infected limb is required.

Classification
The phenotypic traits of this species have been fully described in several studies. The isolation of *V. vulnificus* from blood samples is straightforward, as the bacterium grows readily on TCBS, MacConkey, and blood agars. Isolation from the environment is much more difficult. Vibrios comprise 50% or more of estuarine bacterial populations, and most have not been characterized. Considerable variation exists in the phenotypic traits of *V. vulnificus*, including lactose and sucrose fermentation, considered among the most important characteristics in identifying the species.

Susceptibility to Control Methods
V. vulnificus is susceptible to freezing, low-temperature pasteurization, high hydrostatic pressure, and ionizing radiation. The pathogen can be killed through exposure to horseradish-based sauces. Application of these

sauces to raw oysters will not make the oysters safe to eat, since they do not kill bacteria within the oyster.

Reservoirs

V. vulnificus is widespread in estuarine environments, from the Gulf, Atlantic, and Pacific coasts of the United States to locations around the world. The presence of *V. vulnificus* in water is not associated with the presence of fecal coliforms. Water temperature does have an impact on the presence and levels of the pathogen in water. *V. vulnificus* is seldom isolated from water or oysters when water temperatures are low. This was thought to result from cold-induced death during cold-weather months. However, it may be related to a cold-induced VBNC state, in which the cells remain viable but are no longer culturable on the routine media normally employed for their isolation.

Foodborne Outbreaks

There is no report of more than one person developing *V. vulnificus* infection following consumption of the same lot of oysters. Raw oysters from the same lot, or even the same serving, may have very different levels of the pathogen. In fact, two oysters taken from the same estuarine location may have vastly different populations of *V. vulnificus*. In most cases *V. vulnificus* illness occurs during the months of April through October, when the water is warmer. Nearly all cases of *V. vulnificus* infection result from consumption of raw oysters, and most of these infections result in primary septicemias.

Characteristics of Disease

Discussion here is limited to the primary foodborne form of infection caused by *V. vulnificus*. Symptoms vary considerably among cases, with onset times ranging from 7 h to several days, with a median of 26 h. Symptoms include fever, chills, nausea, and hypotension. Surprisingly, symptoms typical of gastroenteritis—abdominal pain, vomiting, and diarrhea—are not common. In severe cases, survival depends to a great extent directly on prompt antibiotic administration. The infectious dose of *V. vulnificus* is not known. *V. vulnificus* is susceptible to most antibiotics.

Virulence Mechanisms

The polysaccharide capsule, LPS, and a large number of extracellular compounds contribute to virulence. The polysaccharide capsule, which is produced by nearly all strains of *V. vulnificus*, is essential to the bacterium's ability to initiate infection. Elevated levels of serum iron appear to be essential for *V. vulnificus* to multiply in the human host. Avirulent strains produce "translucent" acapsular colonies (Fig. 12.2). Symptoms which occur during *V. vulnificus* septicemia, including fever, tissue edema, hemorrhage, and especially the significant hypotension, are those classically associated with endotoxic shock caused by gram-negative bacteria.

V. fluvialis, V. furnissii, V. hollisae, and *V. alginolyticus*

V. fluvialis has been isolated frequently from brackish and marine waters and sediments in the United States, as well as other countries. It also has been isolated from fish and shellfish from the Pacific Northwest and Gulf Coast. *Vibrio furnissii* has been isolated from river and estuarine waters, marine mollusks, and crustacea throughout the world. The distribution of

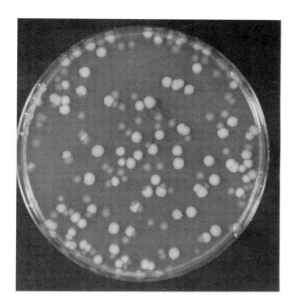

Figure 12.2 Opaque (encapsulated) and translucent (acapsular) colonies of *V. vulnificus*. Virulent strains are generally encapsulated, whereas nonencapsulated cells are usually avirulent. Reprinted from L. M. Simpson, V. K. White, S. F. Zane, and J. D. Oliver, *Infect. Immun.* **55:**269–272, 1987, with permission.

Vibrio hollisae is not well documented, although it is likely a marine species. It appears that the bacterium prefers warm waters. *V. alginolyticus* is often found in high numbers in seawater and seafood obtained throughout the world. It is easily isolated from fish, clams, crabs, oysters, mussels, and shrimp, as well as water. Human infections caused by vibrios such as *V. fluvialis*, *V. furnissii*, *V. hollisae*, and *V. alginolyticus* are less common and usually less severe than those caused by the pathogens discussed above, although deaths have been reported. Consumption of contaminated seafood is usually the source of infection.

Summary

- Illness associated with *V. cholerae* is not common in the United States.
- *V. vulnificus* and *V. parahaemolyticus* are most commonly associated with foodborne illness.
- *V. vulnificus* is solely responsible for 95% of all deaths due to seafood-borne infections in the United States.
- Numbers of vibrios in water are usually greater during the warm-weather months between April and October.
- Proper thermal processing of seafood will kill *Vibrio* species.

Suggested reading

Daniels, N. A., L. MacKinnon, R. Bishop, S. Altekruse, B. Ray, R. M. Hammond, S. Thompson, S. Wilson, N. H. Bean, P. M. Griffin, and L. Slutsker. 2000. *Vibrio parahaemolyticus* infections in the United States, 1973–1998. *J. Infect. Dis.* **181:**1661–1666.

Farmer, J. J., III, F. W. Hickman-Brenner, and M. T. Kelly. 1985. *Vibrio*, p. 282–301. *In Manual of Clinical Microbiology*, 4th ed. American Society for Microbiology, Washington, D.C.

Janda, J. M., C. Powers, R. G. Bryant, and S. L. Abbot. 1988. Current perspectives on the epidemiology and pathogenesis of clinically significant *Vibrio* spp. *Clin. Microbiol. Rev.* **1:**245–267.

Karaolis, D. K. R., J. A. Johnson, C. C. Bailey, E. C. Boedeker, J. B. Kaper, and P. R. Reeves. 1998. A *Vibrio cholerae* pathogenicity island associated with epidemic and pandemic strains. *Proc. Natl. Acad. Sci. USA* **95:**3134–3139.

Linkous, D. A., and J. D. Oliver. 1999. Pathogenesis of *Vibrio vulnificus. FEMS Microbiol. Lett.* **174:**207–214.

Strom, M. S., and R. N. Paranjpye. 2000. Epidemiology and pathogenesis of *Vibrio vulnificus. Microbes Infect.* **2:**177–188.

Tilton, R. C., and R. W. Ryan. 1987. Clinical and ecological characteristics of *Vibrio vulnificus* in the Northeastern United States. *Diagn. Microbiol. Infect. Dis.* **6:**109–117.

Questions for critical thought

1. Vibrio may enter a VBNC state in water. What are key characteristics of this state, and how is this state associated with survival of the organism?

2. Why would treatment of oysters with Tabasco sauce or other similar agents be ineffective in eliminating all *Vibrio* spp. from an oyster?

3. Sanitation programs are often of limited value in the control of vibrios associated with seafood. Why?

4. Which *Vibrio* species is associated with cutaneous infection? What group of people is most at risk? Are cutaneous infections a significant health risk?

5. What do the common virulence characteristics of *V. parahaemolyticus* and *V. vulnificus* include? How do these contribute to pathogenesis?

6. Upon review of a patient's symptoms, how could you presumptively state whether the individual was ill from *V. vulnificus* or *V. parahaemolyticus*?

SECTION III

Gram-Positive Foodborne Pathogenic Bacteria

13

Listeria monocytogenes*

LEARNING OBJECTIVES

The information in this chapter should enable the student to:
- use basic biochemical characteristics to identify *Listeria monocytogenes*
- understand what conditions in foods favor *L. monocytogenes* growth
- recognize, from symptoms and time of onset, a case of foodborne illness caused by *L. monocytogenes*
- choose appropriate interventions (heat, preservatives, or formulation) to prevent *L. monocytogenes* growth
- identify environmental sources of *L. monocytogenes*
- understand the role of *L. monocytogenes* toxins and virulence factors in causing foodborne illness

Outbreak

How long does it take to detect an outbreak of a sporadic disease? How much evidence does it take to launch a recall? What do you do with 35 million pounds of recalled hot dogs?

These are just a few of the questions asked in a listeriosis outbreak that killed 12 people and made 79 people in 17 states ill in 1998 and 1999. The sequence of events is outlined below.

Spring and summer, 1998: A hot dog-processing plant had an unusually high rate of listeria organisms isolated from the environment.

July 4 weekend, 1998: An overhead air-conditioning unit thought to be the source of listeriae was cut apart and removed from the plant.

October, 1998: The Centers for Disease Control and Prevention (CDC) began investigating four listeriosis cases from Tennessee. A 74-year-old mother of four who loved hot dogs was dead. Her flu-like illness had made her drowsy, and headachy, and then killed her. Soon, more cases from Ohio, New York, and Connecticut were detected.

November, 1998: A 31-year-old camp counselor died in Ohio. The genetic fingerprints of bacteria isolated from the listeriosis victims in the different states were identical. They must have eaten a common food.

Early December, 1998: The food was statistically identified as hot dogs.

Mid-December, 1998: Hot dog brands packed at Sara Lee's Bil Mar plant were implicated. With four people dead, the plant stopped shipping.

*This chapter was originally written by Bala Swaminathan for *Food Microbiology: Fundamentals and Frontiers,* 2nd ed., and has been revised for use in an introductory text.

December 22, 1998: Sara Lee announced the recall of all meat made at the Bil Mar plant.

December 23, 1998: The outbreak strain of *L. monocytogenes* was isolated from an unopened package of Bil Mar hot dogs.

December 25, 1998: Another victim died in upstate in New York.

January, 1999: The U.S. Department of Agriculture (USDA) issued its first press release on the recall at the end of January. It was of no help to a pregnant 27-year-old video store manager. She survived a severe "case of the flu" only to see her premature twin babies be born dead. Listeria organisms were isolated from her placenta.

May, 1999: The USDA advised consumers with immunity problems not to eat hot dogs or deli meats unless the foods were thoroughly heated.

This is but one listeria recall. The USDA issued ~30 other recalls for listeria contamination in 1999, 2000, 2001, and 2002. It was not until 2001 that listeriosis became a nationally notifiable disease.

INTRODUCTION

Over the last 25 years, listeriosis has become a major foodborne disease. The first outbreak was traced to coleslaw in 1981. Outbreaks were soon linked to cheeses, luncheon meats, milk, chicken nuggets, and fish. There are now ~2,000 cases of listeriosis reported every year, with 500 fatalities. Many factors contribute to the emergence of listeriosis and other "new" foodborne diseases. These include the following.

- Eliminating spoilage bacteria that prevent listeria from growing. *L. monocytogenes* is a relatively poor competitor with other bacteria but thrives under refrigerated conditions when the competitors have been eliminated.
- Changing demographics, with more people at risk due to old age, immunosuppressants, organ transplants, etc. Some pathogens, such as *Clostridium botulinum,* will make anyone sick. Others, like *L. monocytogenes,* are *opportunistic* pathogens that attack only people whose defenses are weak.
- Changing food production practices, particularly centralization and consolidation. Small institutional kitchens are increasingly being consolidated into larger centralized facilities, which makes good hygienic practices more challenging.
- Increasing use of refrigeration to preserve foods. *L. monocytogenes* grows better than other organisms in the cold.
- Changing eating habits, with increased consumer demand for "fresh," "minimally processed," and "natural" foods that require little cooking or preparation and do not contain preservatives that would prevent listerial growth.
- Changing awareness (have you ever noticed that once you buy a silver Nissan Altima, the road is full of silver Nissan Altimas?) and ability to detect sporadic outbreaks. Computerized databases and the Internet have played key roles in enabling us to detect previously unrecognized sporadic outbreaks.

Listeriosis is not a typical foodborne illness. The disease is sporadic and rare, but severe. It can cause meningitis, septicemia, and abortion

Table 13.1 Symptoms caused by *L. monocytogenes*

Low-grade flu-like infection—not serious, except in pregnant women (who abort)
Listeric meningitis—headache, drowsiness, coma
Perinatal infection
Encephalitis
Psychosis
Infectious mononucleosis
Septicemia

(Table 13.1). Twenty to 30% of people who get listeriosis die from it (i.e., the fatality/case ratio is high), and there is a long time between when food is eaten and when people get sick from it. *L. monocytogenes* differs from most other foodborne pathogens. The organism

- is widely distributed
- is resistant to adverse environmental conditions
- is microaerobic
- is psychrotrophic
- grows in human phagocytes

In addition, listeriae survive in or on food for long periods. They also survive in soil, plants, and water. These traits make listeriae a major concern for the food industry. It is unclear how *L. monocytogenes* should be viewed. The severity and case fatality ratio of the disease require that it be controlled, but the organism's characteristics make it unrealistic to expect all food to be *Listeria* free.

Authors' note
Once the Listeria *cell is phagocytized, it can continue growing and move from cell to cell within the human without reentering the bloodstream.*

CHARACTERISTICS OF THE ORGANISM

Classification

The Genus *Listeria*

The genus *Listeria* belongs to the *Clostridium* subbranch, together with *Staphylococcus, Streptococcus, Lactobacillus,* and *Brochothrix. L. monocytogenes* is one of six species in the genus *Listeria*. The other species are *Listeria ivanovii, Listeria innocua, Listeria seeligeri, Listeria welshimeri,* and *Listeria grayi*. Within the genus *Listeria,* only *L. monocytogenes* and *L. ivanovii* are pathogens. *L. monocytogenes* is a human pathogen. *L. ivanovii* is primarily an animal pathogen.

Listeria species are identified by a few biochemical traits. The biochemical tests that are used to characterize the species are acid production from D-xylose, L-rhamnose, α-methyl-D-mannoside, and D-mannitol. The ability to lyse red blood cells differentiates *L. monocytogenes* from nonpathogenic *Listeria* species. "Rapid" tests based on genetic or immunological traits identify listeria to the genus or species level.

Further Characterization of *L. monocytogenes*

L. monocytogenes isolates are characterized below the species level to help public health officials investigate outbreaks. This can be done by genetic fingerprinting or serotyping. There are 13 serotypes of pathogenic *L. monocytogenes*. Ninety-five percent of the organisms isolated from listeriosis victims belong to three serotypes: 1/2a, 1/2b, and 4b. In the United States, the

Table 13.2 Temperature influence on doubling time

Temp (°C)	$t_d{}^a$ (h)
4	43
10	6.6
37	1.1

$^a t_d$, doubling time.

Table 13.3 Some foods that have been associated with listeriosis

Cheese
Fish
Raw milk
Turkey franks
Smoked fish
Deli meat
Vegetable rennet
Ice cream
Homemade sausage
Alfalfa tablets
Chicken nuggets
Cod roe
Cook-and-chill chicken
Deli salads
Hot dogs
Human breast milk
Pork sausage
Salted mushrooms

CDC has established a network (PulseNet) of public health laboratories to subtype foodborne pathogenic bacteria. PulseNet laboratories use standardized methods for subtyping bacteria by pulsed-field gel electrophoresis (PFGE). They can quickly compare PFGE patterns of foodborne pathogens from different locations via the Internet to identify a common-source outbreak.

Susceptibility to Physical and Chemical Agents

L. monocytogenes grows from 0 to 45°C but grows slowly at colder temperatures (Table 13.2). Freezing does not significantly reduce the size of the bacterial population. Survival and injury during frozen storage depend on the food and the freezing rate. *L. monocytogenes* is killed at temperatures of >50°C.

L. monocytogenes grows in laboratory media at pH values as low as 4.4. At pH values below 4.3, cells may survive, but they do not grow. Organic acids, such as acetic, citric, and lactic acids, at 0.1% can inhibit *L. monocytogenes* growth. The antilisterial activities of these acids are related to their degrees of dissociation. Not all acids are created equal; citric and lactic acids are less harmful at an equivalent pH than acetic acid. HCl is least effective of all.

L. monocytogenes grows best at a water activity (a_w) of ≥0.97. For most strains, the minimum a_w for growth is 0.93, but some strains may grow at a_w values as low as 0.90. The bacterium may survive for long periods at a_w values as low as 0.83. *L. monocytogenes* heat resistance increases as the a_w of the food in which it is heated decreases. This is problematic for food manufacturers who combine low a_w and heat treatments to maintain safety.

L. monocytogenes grows to high levels at moderate salt concentrations (6.5%). It can even grow in the presence of 10 to 12% sodium chloride. The bacterium survives for long periods at higher salt concentrations. Lowering the temperature increases survival in high-salt environments. Thus, cured meats (like hot dogs, bologna, and ham) are very hospitable environments for listeria growth.

Listeriosis and Ready-to-Eat Foods

Some ready-to-eat foods (such as soft cheeses, frankfurters, delicatessen meats, and poultry products) pose a high listeriosis risk for susceptible populations. Some foods that have caused listeriosis are shown in Table 13.3. Because 20% of refrigerators have temperatures of >50°F, refrigeration cannot ensure the safety of ready-to-eat foods. Refrigeration offers a good environment for *L. monocytogenes* growth.

Milk Products

Raw milk is a source of *L. monocytogenes* cells. A 1985 outbreak in California was associated with Mexican-style cheese that was made with unpasteurized milk. Heat inactivation studies of *L. monocytogenes* in milk have given conflicting results about its heat resistance and ability to survive pasteurization. This could be due to the use of different testing methods, the use of listeria levels much higher than those normally found in milk, possible protection of cells located within phagocytes, and other factors. However, pasteurization is still considered an adequate safety intervention. The World Health Organization states, "Pasteurization is a safe process which reduces the number of *L. monocytogenes* in raw milk to levels that do not pose an appreciable risk to human health." U.S. regulatory authorities share this view. However, since *L. monocytogenes* grows well in pasteurized milk, *postprocess contamination* (the contamination of food by environmental bacteria after the food has undergone antimicrobial processing) is a major concern. The number of bacteria can increase 10-fold in 7 days at 4°C. Listeriae grow more rapidly in pasteurized milk than in raw milk at 7°C. Bacterial populations in milk contaminated after pasteurization and then refrigerated can reach very high levels after 1 week. Temperature abuse can then enhance listeria growth. In a recent outbreak caused by chocolate milk in Illinois, listeria populations reached the very high level of 10^9 colony-forming units (CFU)/ml. This gave the milk an off flavor, but people still drank it.

Authors' note

10^9 CFU/ml is the level of bacteria that can be reached in a laboratory on a good day.

Cheeses

L. monocytogenes survives cheese manufacturing and ripening because of its temperature hardiness, ability to grow in the cold, and salt tolerance. During manufacturing, *L. monocytogenes* is concentrated in the cheese curd. The behavior of listeriae in the curd is influenced by the type of cheese, ranging from growth in feta cheese to significant death in cottage cheese. For example, during cheese ripening, *L. monocytogenes* can grow in Camembert cheese, die gradually in Cheddar or Colby cheeses, or decrease rapidly during early ripening and then stabilize, as in blue cheese. Consumption of soft cheeses by susceptible persons is a risk factor for listeriosis. The CDC recommends that pregnant women and other immunocompromised people avoid eating soft cheeses.

Meat and Poultry Products

L. monocytogenes growth in meat and poultry depends on the type of meat, the pH, and the presence of other bacteria. *L. monocytogenes* grows better in poultry than in other meats. Roast beef and summer sausage support the least growth. Contamination of animal muscle tissue can be caused by *L. monocytogenes* in or on the animal before slaughter or by contamination of the carcass after slaughter. Because *L. monocytogenes* concentrates and multiplies in the kidney, lymph nodes, liver, and spleen, eating organ meat may be more hazardous than eating muscle tissue. *L. monocytogenes* attaches to the surface of raw meats and is difficult to remove or kill. *L. monocytogenes* grows readily in meat products, including vacuum-packaged beef, at pH values near 6.0. There is little or no growth around pH 5.0. Ready-to-eat meats that are heated and then cooled in brine before being packaged may be especially good for *L. monocytogenes* growth because competitive bacteria are reduced and listeriae have a high salt tolerance.

A 1993–1996 USDA monitoring program for *L. monocytogenes* in cooked ready-to-eat meats found it in 0.2 to 5.0% of beef jerky, cooked sausages, salads, and spreads. In sliced ham and sliced luncheon meats, the incidence was 5.1 to 81%. This has decreased dramatically over the last 10 years. In Canada, the incidence of *L. monocytogenes* in domestic ready-to-eat cooked meat products was 24% in 1989 and 1990 but declined to ≤3% during 1991 and 1992. However, the environmental prevalence of listeria in these establishments remained constant from 1989 to 1992. This suggests that once established in a food plant environment, *L. monocytogenes* may persist for years.

Cooked, ready-to-eat meat and poultry cause sporadic and epidemic listeriosis. Eating unheated frankfurters and undercooked chicken is a risk factor for sporadic listeriosis. The CDC identified contaminated turkey hot dogs as the source of sporadic *L. monocytogenes* infection in a cancer patient in 1989. Multistate outbreaks of listeriosis were linked to contaminated frankfurters and turkey deli meat.

Seafoods

L. monocytogenes has been isolated from fresh, frozen, and processed seafood products, including crustaceans, molluscan shellfish, and finfish. Its presence in shrimp, smoked mussels, and imitation crab meat has caused illness. Rainbow trout were implicated in two outbreaks. A U.S. Food and Drug Administration (FDA) survey of refrigerated or frozen cooked crabmeat showed contamination levels of 4.1% for domestic products and 8.3% for imported products. Crab and smoked fish were contaminated with *L. monocytogenes* at 7.5 and 13.6%, respectively.

Seafoods that may be high-risk foods for listeriosis include

- molluscs, including mussels, clams, and oysters
- raw fish
- lightly preserved fish products, including salted, marinated, fermented, and cold-smoked fish
- mildly heat-processed fish products and crustaceans

Other Methods of Food Preservation

A survey of vacuum-packaged processed meat revealed that 53% was contaminated with *L. monocytogenes* and that 4% contained >1,000 CFU/g (this is a high level for food that is ready to eat). This corroborates experimental evidence that the growth of *L. monocytogenes* (a facultative anaerobe) is not significantly affected by vacuum packaging. Studies of meat juice, raw chicken, and precooked chicken nuggets revealed that modified atmospheres do not protect against *L. monocytogenes* growth.

Many surviving cells are injured by heating, freezing, or various other treatments. Heat-stressed *L. monocytogenes* may be less pathogenic than nonstressed cells. On nonselective agar, injured cells can repair the damage induced by stress and grow. When injured cells are subjected to additional stress in selective agar, they may not form colonies.

Sources of *L. monocytogenes* in the Environment

L. monocytogenes is ubiquitous in the environment and has many routes to infect humans (Fig. 13.1). It survives and grows in soil and water. *Listeria* species are found in many water environments: surface water of canals and lakes, ditches of polders in The Netherlands, freshwater tributaries draining into a California bay, and sewage. *Listeria* species have been isolated from alfalfa plants and crops grown on soil treated with sewage sludge. Half the radish samples grown in soil inoculated with *L. monocytogenes* still harbored the organism 3 months later. *L. monocytogenes* is also present in pasture grasses and grass silages. Decaying plant and fecal material probably contribute to *L. monocytogenes* presence in soil. The soil provides a cool, moist environment, and the decaying material provides the nutrients.

L. monocytogenes has been isolated from the feces of healthy animals. Many animal species can get listeriosis. Humans having listeriosis and humans who carry the organism without becoming sick both shed the organism in their feces. Figure 13.1 illustrates how *L. monocytogenes* is spread from the environment to animals and humans and back to the environment.

Food-Processing Plants

L. monocytogenes enters food-processing plants through soil on workers shoes and clothing and on vehicles. It can also enter the plant with contaminated raw plant and animal tissue and human carriers. The high humidity

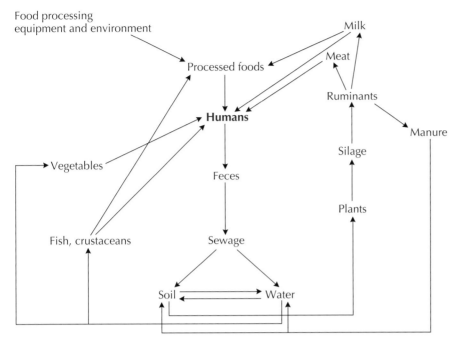

Figure 13.1 Potential routes of transmission of *L. monocytogenes.* (Based on B. Swaminathan, p. 383–409, *in* M. P. Doyle, L. R. Beuchat, and T. J. Montville [ed.], *Food Microbiology: Fundamentals and Frontiers,* 2nd ed. [ASM Press, Washington, D.C., 2001].)

and nutrient levels of food plants promote listeria growth. *L. monocytogenes* is often detected in moist areas, such as floor drains, condensed and stagnant water, floors, residues, and processing equipment. *L. monocytogenes* attaches to surfaces, including stainless steel, glass, and rubber.

L. monocytogenes gets on carcasses by fecal contamination during slaughter. A high percentage (11 to 52%) of healthy animals are fecal carriers. *L. monocytogenes* is present in both unclean and clean zones in slaughterhouses, especially on workers' hands. The most heavily contaminated working areas are those where cow dehiding, pig stunning, and hoisting take place. Studies in turkey and poultry slaughterhouses found *L. monocytogenes* in feather plucker drip water, chill water overflow, and recycled cleaning water.

L. monocytogenes does not survive the heat processing used to make foods safe. It gets into processed foods primarily by postprocessing contamination. *L. monocytogenes* is particularly hard to eliminate from food-processing plants because it adheres to food contact surfaces and forms biofilms in hard-to-reach areas, such as drains. This makes proper sanitation difficult. *L. monocytogenes* is often in raw materials used in food-processing plants, so there are ample opportunities for reintroduction of listeriae into processing facilities that have previously eradicated it.

Prevalence and Regulatory Status of *L. monocytogenes*

L. monocytogenes contamination of food is widespread. It is a major cause of USDA recalls (Fig. 13.2). Contamination levels range from 0% in bakery goods to 16% in some ready-to-eat foods. The contamination of raw foods is much higher: up to 60% in raw chicken. Preserved, but not heat-treated, fish and meat products are more frequently contaminated than heat-treated

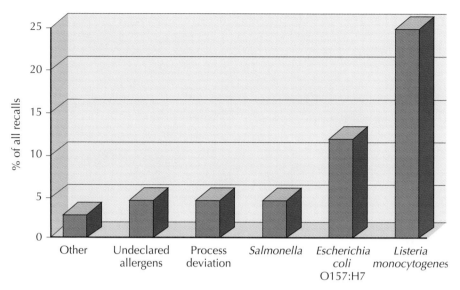

Figure 13.2 Causes of USDA recalls, 1997.

meat foods. *L. monocytogenes* can be isolated from many vegetables, including bean sprouts, cabbage, cucumbers, leafy vegetables, potatoes, prepackaged salads, radishes, salad vegetables, and tomatoes.

There is an international debate about how *L. monocytogenes* should be regulated (Box 13.1). Because *L. monocytogenes* is so common, regulatory agencies in many countries assert that it is impossible to produce *L. monocytogenes*-free foods. They have established "tolerance levels" for *L. monocytogenes*. Products that have caused human listeriosis are placed in a special category and regulated more strictly than foods that have never caused listeriosis. Most European Union countries feel that foods should be *L. mono-*

BOX 13.1

Group exercise: industrial and regulatory responses to *Listeria monocytogenes*

Objective
To understand some of the real-world factors involved in setting industrial policies through use of a simulation.

Scenario
You work for MSF Foods, a large cookie company that added a premium pie line 8 months ago. This line has been a huge success, causing a 4% increase in corporate profits. Marketing has determined that adding cheesecake and cream pie products to the line could fill a new market niche and increase next year's profits by a similar extent.

The product line manager questioned if these products might pose a *Listeria* risk and what kind of end-product testing would be required to ensure product safety. (In the cookie line, the company tests for *Salmonella* and holds the product until negative results are received. Should a similar program be instituted for *Listeria* in the pies?) The production side of the company says that a test-and-hold policy would be the kiss of death for this short-shelf-life, refrigerated project. The microbiology department insists that end-product testing is the only way to ensure the safety of the product.

A meeting has been called to resolve this issue. Your job at the end of the meeting is to write a two-page recommendation to the Vice President of Cookies, Creams, and Cakes.

Your assignment

1. THE PRE-MEETING: Split into groups and discuss the issues. Decide what additional information is required for you to make a decision. Decide which side of the "testing controversy" you stand on and will fight for during the meeting.
2. THE MEETING: The group tries to reach consensus. (Self-explanatory.)
3. THE WORK PRODUCT: Write a two-page memo to the Vice President of Cookies, Creams, and Cakes (who knows a lot about cookies but nothing about microbiology) outlining your recommendation as to what the company should do and why.

Table 13.4 Relative risk of listeriosis for different populations

Population	Relative risk
Total	1[a]
Over 70 years old	3
Pregnant	17
Human immunodeficiency virus positive	200

[a]Absolute risk, seven cases per million people.

cytogenes free, if possible, or have the lowest level possible. Thus, foods intended for susceptible populations must be listeria free. Other foods may contain listeriae at <100 CFU/g. In contrast, the United Kingdom and the United States have "zero tolerance" (in a 25-g sample) for *L. monocytogenes*. Both countries argue that the infectious dose must be known before an "acceptable" level for *L. monocytogenes* can be set. The infectious dose is unknown and may differ for different types of people.

There are two arguments against the U.S. zero-tolerance policy. The first is that the FDA has legal authority only to issue regulations that are based on sound science to protect the health of the American people. However, different parts of the population require different degrees of protection (Table 13.4). The incidence of listeriosis (~0.7 per 100,000 people) is the same in European countries with a <100-CFU/g tolerance as it is in the United States, where there is a zero tolerance. Thus, the zero-tolerance regulation appears to offer no additional benefit. Second, microbial specifications require harmonization for international trade. When the United States bars the importation of foods meeting the European Union's <100-CFU/g tolerance, it can be charged with inhibition of free trade through "nontariff trade barriers." The debate over zero tolerance continues. The current dilemma is discussed in Box 13.2.

Authors' note
There is zero tolerance for L. monocytogenes *in ready-to-eat foods, but no 100% accurate test for detecting it.*

BOX 13.2

The present dilemma

The United States is spending hundreds of millions of dollars to decrease the incidence of listeriosis. But is the money being spent effectively? Is the obsession with listeriosis cost-effective? It is hard to say. The organism is widespread in the environment and difficult to eradicate, yet relatively few foods have been implicated in listeriosis outbreaks. Should all ready-to-eat foods be tested for *Listeria*, or only those with a historical problem? There are relatively few cases of listeriosis, and when outbreaks occur, only a tiny percentage of people who eat the product get sick. However, although the attack rate is low, the fatality rate is high. Might it be more cost-effective to educate at-risk populations

to avoid certain foods rather than test for *Listeria*? The methods for isolating and identifying *Listeria* leave much to be desired; they have a 10 to 15% false-negative rate, i.e., if 100 samples containing *Listeria* were tested, it would not be detected in 10 to 15 of the samples. Is it fair to demand zero tolerance when there is no 100% reliable test? Does it make sense to test products in which the presence of *Listeria* is unlikely or that have received listericidal treatments? The International Commission for Microbial Specifications for Foods has suggested that these foods need not be tested. The current regulatory emphasis *has* dramatically decreased the isolation of listeria from food, but the number of cases of listeriosis has not had a corresponding decrease. Regulatory agen-

cies, food industry groups, and consumer advocates all agree that the present regulations need to be reexamined, but they disagree on how or why. For the moment, there is emphasis on developing better surveillance for listeriosis so that progress (or lack of progress) can be monitored. Foods that have been associated with listeriosis are under increased scrutiny, but this has not prevented additional recalls. There are efforts to educate at-risk groups, especially pregnant women, who can be reached through their obstetricians, but these need to be more effective. The FDA is considering a consumer petition to allow <100 CFU of *L. monocytogenes*/g in foods that do not support its growth. The debate is sure to continue for years.

Human Carriers

L. monocytogenes can be carried asymptomatically in the feces of many groups of people. These include healthy people, pregnant women, patients with gastroenteritis, slaughterhouse workers, laboratory workers handling *Listeria*, food handlers, and patients undergoing hemodialysis. *L. monocytogenes* has been isolated from 2 to 6% of fecal samples from healthy people. Patients with listeriosis often excrete high numbers of *L. monocytogenes* organisms. Specimens from 21% of patients had $>10^4$ *L. monocytogenes* organisms/g of feces. The same strain of *L. monocytogenes* was fecally shed by 18% of their housemates. Fecal carriers amplify outbreaks through secondary transmission. (The primary transmission is through the food to the first victim. The secondary transmission is from the feces of the first victim to another person.) This secondary fecal-oral transmission of bacteria that originate in foodborne outbreaks also occurs with other pathogens.

FOODBORNE OUTBREAKS

Transmission of listeriosis through food was first documented in 1981 during an outbreak in Canada. The use of a case-control study and strain typing proved that food was the source. The 1981 Canadian outbreak was in Nova Scotia. There were 34 cases in pregnant woman and seven cases in nonpregnant adults over a 6-month period. A case control study implicated coleslaw as the source of the outbreak. The epidemic strain was later isolated from an unopened package of coleslaw. Cabbage fertilized with manure from sheep suspected to have had *Listeria* meningitis was the probable source. Harvested cabbage was stored over the winter and spring in an unheated shed. The cold provided a growth advantage for the psychrotrophic *L. monocytogenes*.

Contaminated Mexican-style cheese caused a 1985 listeriosis outbreak in California. There were 142 cases over 8 months. Pregnant women accounted for 93 cases. The remaining 49 were nonpregnant adults. Most of the victims had a predisposing condition for listeriosis. Almost one third of the people infected died. However, ~10,000 people consumed the cheese, so the overall attack rate was very low. Inadequate pasteurization of milk and mixing of raw milk with pasteurized milk caused the outbreak.

L. monocytogenes in soft cheese was responsible for a 4-year outbreak of 122 cases in Switzerland. A contaminated pâté caused a 300-case outbreak in the United Kingdom. Recalling the food, advising people not to eat the contaminated products, and taking action to prevent *L. monocytogenes* contamination at processing facilities controlled these large outbreaks.

There was a 10-state outbreak of listeriosis between May and November 2000. The *L. monocytogenes* isolates were all serotype 1/2a and had identical PFGE patterns. Eight perinatal (the perinatal period includes the time shortly before birth to the age of about 1 month) and 21 nonperinatal cases were reported. Among the 21 nonperinatal-case patients, the median age was 65 years (range, 29 to 92 years); 62% were female. This outbreak resulted in four deaths and three miscarriages or stillbirths. A case-control study implicated deli turkey meat.

CHARACTERISTICS OF DISEASE

L. monocytogenes causes disease in well-defined high-risk groups. These include pregnant women, neonates, and immunocompromised adults. Listeriosis occasionally occurs in otherwise-healthy people. In nonpregnant adults, *L. monocytogenes* causes septicemia, meningitis, and meningoencephalitis. The mortality rate is 20 to 25%. Clinical conditions that predispose people to listeriosis include malignancy, organ transplants, immunosuppressive therapy, infection with human immunodeficiency virus, and advanced age. Pregnant women, particularly in the third trimester, may experience only mild flu-like symptoms as a result of *L. monocytogenes* infection. The infection, however, has disastrous consequences for the fetus, leading to stillbirth or abortion. While listeriosis outbreaks attract the most attention, most cases of human listeriosis are sporadic. Some sporadic cases may be unrecognized common-source outbreaks. Although foods are implicated in some cases, the mode of infection in most cases remains unknown. Many cases are not associated with food because it is so hard to investigate sporadic cases. Incubation times of up to 5 weeks make it hard to get accurate food histories and to examine suspect foods. Understanding the epidemiology of sporadic cases is critical for developing effective control strategies.

Sporadic listeriosis is a rare disease. In Europe and Canada, passive surveillance suggests that there are two to seven cases per million people. An active surveillance study in the United States gave a more precise estimate of 7.4 cases per million people. The almost 50% decrease in sporadic listeriosis cases and deaths in the United States since 1990 suggests that current preventive measures have made a difference.

Although exposure to *L. monocytogenes* is common, listeriosis is rare. It is unclear whether this is due to human resistance or to most strains being only weak pathogens. Human listeriosis is poorly understood. Some healthy individuals are asymptomatic fecal carriers of *L. monocytogenes*. The risk of clinical disease in carriers is unknown. Although listeriae seem to target pregnant women, some pregnant women have asymptomatic fecal listeria carriage and have normal pregnancy outcomes. Women infected in one pregnancy do not necessarily suffer the same problem in later pregnancies.

L. monocytogenes can also cause a second, different illness, feverish gastroenteritis. The gastroenteritis outbreaks differ from the invasive outbreaks described above. They affect people with no predisposing risk factors. The infectious dose is higher (1.9×10^5 to 1×10^9 colony-forming units [CFU]/g or ml of food) than that for invasive listeriosis. Finally, the symptoms appear within several hours (18 to 27 h) of exposure in contrast to the weeks observed for invasive listeriosis. The first case of listeric gastroenteritis was reported in 1994. It was apparently caused by the postprocess contamination of chocolate milk held in an unrefrigerated processing vat. The contaminated milk was then stored unrefrigerated (i.e., incubated) by the buyers, who served it at a large summer gathering where ~75% of those who drank the milk got sick. The investigation was relatively easy, since most of the victims contacted suggested testing the milk, which "tasted bad."

INFECTIOUS DOSE AND SUSCEPTIBLE POPULATIONS

The infectious dose of *L. monocytogenes* depends on many factors, including the immunological status of the host, the virulence of the microbe, and the food. Studies with human volunteers are impossible (i.e., they would be unethical), but in animals, reducing exposure levels reduces clinical disease. Populations of *L. monocytogenes* in foods responsible for foodborne cases are usually >100 CFU/g of food. However, the frankfurters implicated in the 1998 listeriosis outbreak had <0.3 CFU/g. More epidemiologic data are needed to accurately assess the infectious dose.

VIRULENCE FACTORS AND MECHANISMS OF PATHOGENICITY

L. monocytogenes is unique among foodborne pathogens. Other pathogens excrete toxins or multiply in the blood. In contrast, *L. monocytogenes* enters the host's cells, grows inside the cell, and passes directly into nearby cells (Fig. 13.3). Cell-to-cell transmission reduces the bacterium's exposure to antibiotics and circulating antibodies. This membrane-penetrating ability allows *L. monocytogenes* to cross into the brain and placenta.

Pathogenicity of *L. monocytogenes*

There are many tests for studying *L. monocytogenes* pathogenicity, including a fertilized hen's egg test, tissue culture assays, and tests using laboratory animals, particularly mice. In these studies, mice are infected by injection of the bacteria into the stomach cavity, intravenously, or by mouth. Virulence

Figure 13.3 Schematic representation of *L. monocytogenes* cell-to-cell spread. (Reprinted from B. Swaminathan, p. 383–409, *in* M. P. Doyle, L. R. Beuchat, and T. J. Montville [ed.], *Food Microbiology: Fundamentals and Frontiers,* 2nd ed. [ASM Press, Washington, D.C., 2001].)

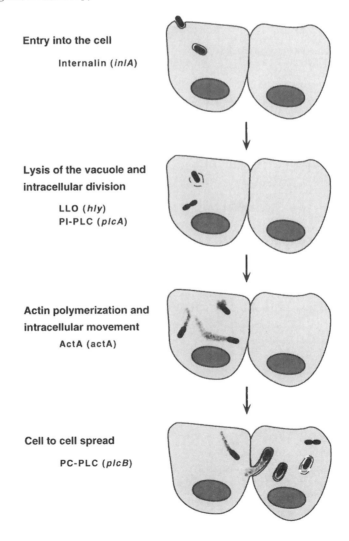

Entry into the cell

Internalin (*inlA*)

Lysis of the vacuole and intracellular division

LLO (*hly*)
PI-PLC (*plcA*)

Actin polymerization and intracellular movement
ActA (actA)

Cell to cell spread

PC-PLC (*plcB*)

is then evaluated by comparing the 50% lethal dose or by counting bacteria in the spleen or liver. The 50% lethal doses range from 10^3 to 10^7 CFU.

When mice are exposed to *L. monocytogenes,* the infection follows a well-defined course, lasting for ~1 week. Within 10 min after intravenous injection, 90% of the bacteria are taken up by the liver and 5 to 10% are taken up by the spleen. During the first 6 h, the number of viable listeriae in the liver decreases 10-fold, indicating their rapid destruction. Surviving listeriae then infect susceptible macrophages and multiply. They grow exponentially in the spleen and liver for the next 48 h. If inactivation ensues during the next 3 to 4 days, the host recovers.

In humans, *L. monocytogenes* crosses the intestinal barrier after entering by the oral route. The bacteria are then internalized by macrophages, in which they replicate. The blood then transports them to the lymph nodes. When they reach the liver and the spleen, most listeriae are rapidly killed. The survivors travel via the blood to the brain or the placenta.

Specific Genes Mediate Pathogenicity

L. monocytogenes cells can fight their way into nonphagocytic human cells. Our understanding of how gene regulation influences pathogenicity is in its infancy. A series of *L. monocytogenes* genes gives the instructions for each step of this infectious process. The protein internalin is needed to start the process. Internalin is coded for by the gene *inl*A. The gene is thermoregulated and is expressed more readily at 37 than at 20°C. Mutants lacking the gene cannot form the phagosome that introduces the bacteria to the inside of the mammalian cell. To do real damage, the listeriae must be released from the phagosome and liberated into the cytoplasm of the host cell. The *hly* gene is responsible for this. It codes for listeriolysin, an enzyme that breaks open the phagosome. Listeriolysin breaks down red blood cells; its activity is measured as hemolytic activity. *L. monocytogenes* mutants lacking the *hly* gene are not pathogenic. These cells can be entrapped by phagosomes but are not freed into the cytoplasm.

Once the listeria cells are free in the host cytoplasm, the genes that regulate growth and multiplication work in the normal fashion. The listeria cells have the unusual ability to use the host's actin molecules to propel them across human cells. Actin, together with myosin, is the major component of human muscles. It occurs in nonmuscle cells at lower concentrations. The *act*A gene in listeriae codes for a protein, ActA, which is made at one end of the bacterium. ActA makes the actin propel the cell forward. The force generated by the actin pushes the listeria cell through the infected host's membrane and through the membrane of the adjacent host cell. The final gene needed for pathogenicity is *plc*B. This gene codes for a membrane-hydrolyzing enzyme that helps listeriolysin O liberate the bacterium into the cytoplasm of the adjacent human cell.

Summary

- *L. monocytogenes* is a ubiquitous organism that can grow at refrigeration temperature.
- There are relatively few (5,000/year) cases of listeriosis, but there is a high (~25%) fatality rate among people who get it.
- *L. monocytogenes* is a hearty organism that is relatively resistant to dehydration, low pH, and low a_w.

- Ready-to-eat foods that are preserved by refrigeration pose a special challenge with regard to *L. monocytogenes* infection.
- Pasteurization is adequate to control *L. monocytogenes.*
- Symptomatic and asymptomatic people can shed *L. monocytogenes* in their feces.
- The United States has zero tolerance for *L. monocytogenes* in ready-to-eat foods, but in Europe, there is some tolerance for it in certain foods.
- Outbreaks of listeriosis generally extend over a long period, are caused by refrigerated foods, have high case fatality rates, and involve a disproportionate number of pregnant women.
- *L. monocytogenes* enters the host's cells, grows inside the cells, and passes directly into nearby cells.
- Internalin, listeriolysin, ActA, and a membrane-hydrolyzing enzyme are the four major proteins that govern the pathogenic process.

Suggested reading

National Advisory Committee on Microbiological Criteria for Foods. 1991. *Listeria monocytogenes. Int. J. Food Microbiol.* **14:**185–246.

Ryser, E. T., and E. H. Marth. 1991. *Listeria, Listeriosis, and Food Safety.* Marcel Dekker, New York, N.Y.

Swaminathan, B. 2001. *Listeria monocytogenes*, p. 383–409. *In* M. P. Doyle, L. R. Beuchat, and T. J. Montville (ed.), *Food Microbiology: Fundamentals and Frontiers*, 2nd ed. ASM Press, Washington, D.C.

Tompkin, R. B. 2002. Control of *Listeria monocytogenes* in the food-processing environment. *J. Food Prot.* **65:**709–725.

Questions for critical thought

1. What reasons might explain our lack of awareness of *L. monocytogenes* prior to 1980?
2. What are the pros and cons of a tolerance versus zero tolerance for *L. monocytogenes* in ready-to-eat food?
3. "Make up" an outbreak of listeriosis. Include food, setting, population, symptoms, the chain of events that caused the outbreak, and how it might have been prevented.
4. Diagram the life of *L. monocytogenes* as an intracellular pathogen, noting important gene products.
5. What characteristics make *L. monocytogenes* a "successful" foodborne pathogen?
6. Why do listeriae grow more rapidly in pasteurized milk than in raw milk?
7. Write down one thing you do not understand about *L. monocytogenes.* Discuss it with a classmate.
8. Choose one of the issues in "More questions than answers" below and write a page on it.
9. Use the CDC, USDA, or FDA website to find a recent listeria outbreak or recall. How do its characteristics fit with those described in the text?
10. Write a really good question about *Listeria* that might be used on your exam.

11. There is evidence that a <100-CFU/g tolerance for *L. monocytogenes* in food where it cannot grow may afford a greater level of safety than a zero tolerance. Research this issue and discuss the pros and cons of the proposed tolerance.

More questions than answers

1. Are all *L. monocytogenes* strains equally pathogenic?
2. Are all of them capable of causing outbreaks?
3. What are the infectious doses of *L. monocytogenes* for nonpregnant adults, persons with immune system deficiencies, and pregnant women?
4. Do the infectious doses vary significantly between strains?
5. Is the expression of virulence genes influenced by the food the listeriae are in?
6. Do all strains cause feverish gastroenteritis in humans?
7. Are some *L. monocytogenes* strains better able to form biofilms and persist in the environment?
8. Is it realistic to expect all ready-to-eat foods to be *Listeria* free?
9. How can we ensure that ready-to-eat salads are not contaminated with *L. monocytogenes*?
10. How can we prevent *Listeria* contamination of other ready-to-eat foods, such as frankfurters, soft cheeses, and smoked fish?

14

*Staphylococcus aureus**

LEARNING OBJECTIVES

The information in this chapter should enable the student to:

- use basic biochemical characteristics to identify *Staphylococcus aureus*
- understand what conditions in foods favor *S. aureus* growth
- recognize, from symptoms and time of onset, a case of *S. aureus* food poisoning
- choose appropriate interventions (heat, preservatives, or formulation) to prevent the growth of *S. aureus*
- identify environmental sources of *S. aureus*
- understand the role of staphylococcal enterotoxins in causing foodborne illness

Outbreak

My fear of flying is related to bad food, not hijackings, crashes, or boring in-flight movies. As a young college student, I boarded a flight in Tokyo, Japan. It stopped in Anchorage, Alaska, for a fresh crew and (not so fresh) food. The first leg of the trip was uneventful. An hour out of Anchorage, we ate peanuts and soda. Breakfast would break the boredom of the flight—in a way I could not imagine. Breakfast smelled good. It was an omelet with ham and cheese, served nice and hot. But before the service cart reached the back of the plane, there were rumblings in the front. About 30 min after eating, a passenger rushed to the rear of the plane, ignoring the attendant's admonition that the seat belt sign was on, and vomited on a blue-haired lady before finally reaching the toilet. Although the air was calm, most of the passengers started to turn green. The trickle of passengers trying to reach the toilet turned into a stream. The flow crested 2.5 h after breakfast. The first man to become ill was lucky. He got a toilet. There were only four bathrooms for the remaining 250 people with vomiting and diarrhea! A burly first-class passenger blocked the way to the front of the cabin. In first class, the passenger-to-toilet ratio was 20 to 2, and first-class food hadn't even made them sick! About 80% of the passengers had diarrhea, 70% had vomiting, and 85% had nausea. I'll spare you what happened when they ran out of "discomfort bags." Fortunately, the crew was served a "microbiologically insensitive" meal. Only one of them got sick. She'll never pass up steak in favor of an omelet again.

After we landed, two men in lab coats told me I would be exempt from the customs line if I would "do them a favor." My rectal swabs tested positive for *S. aureus* toxin and the organism. They found *S. aureus* toxin and the organism in the cheese omelets and the ham, too. Other investigators visited the kitchen. They

*This chapter was originally written by Lynn M. Jablonski and Gregory A. Bohach for *Food Microbiology: Fundamentals and Frontiers*, 2nd ed., and has been adapted for use in an introductory text.

discovered that the meals had been prepared 14 h before the flight and held at room temperature. They were held under inadequate refrigeration during the flight and were microwaved to heat them up just before they were served. When the public health investigators went to shake hands with the head chef, they saw a pus-filled boil on his hand. They cultured it, and guess what they found—*S. aureus.*

CHARACTERISTICS OF THE ORGANISM

Historical Aspects and General Considerations

Staphylococcal food poisoning is a common cause of gastroenteritis. *S. aureus* is the main agent of staphylococcal food poisoning. Unlike many other forms of gastroenteritis, one doesn't get staphylococcal food poisoning by eating live bacteria. One gets it by eating staphylococcal toxins that have already been made in the contaminated food. This form of food poisoning is *intoxication,* or poisoning, because it does not require bacterial growth in the victim. Indeed, outbreaks have been caused by foods in which the organism has been killed but the toxin remains. Staphylococcal toxin is unique because it is not destroyed by heating or even by canning.

The link between staphylococci and foodborne illness was made in 1914 when researchers found that drinking contaminated milk caused vomiting and diarrhea. In 1930, G. M. Dack and coworkers voluntarily consumed fluids from cultures of "a yellow hemolytic *Staphylococcus*" isolated from contaminated cake. They became ill with "vomiting, abdominal cramps, and diarrhea."

S. aureus is well characterized. It excretes a variety of compounds. Many of these, including the staphylococcal enterotoxins, are virulence factors. The staphylococcal toxins cause at least two human diseases, toxic shock syndrome and staphylococcal food poisoning.

Nomenclature, Characteristics, and Distribution of Staphylococcal-Enterotoxin-Producing Staphylococci

The term "staphylococci" informally describes a group of small, spherical, gram-positive bacteria. Their cells have diameters ranging from ~0.5 to 1.5 μm (Fig. 14.1). They are *catalase positive* (i.e., they have enzymes that break down hydrogen peroxide). Staphylococci have typical gram-positive cell walls containing peptidoglycan and teichoic acids. Some characteristics that differentiate *S. aureus* from other staphylococcal species are summarized in Table 14.1.

Bergey's Manual of Determinative Bacteriology (the "dictionary" for the classification of bacteria) puts staphylococci in the family *Micrococcaceae.* This family includes the genera *Micrococcus, Staphylococcus,* and *Planococcus.* The genus *Staphylococcus* is further subdivided into >23 species and subspecies. Many of these contaminate food. Several species of *Staphylococcus,* including both coagulase-negative and coagulase-positive isolates, can produce staphylococcal enterotoxins. (Coagulase is an enzyme closely associated with toxin production.) Although several species can cause gastroenteritis, nearly all staphylococcal food poisoning is attributed to *S. aureus.*

Introduction and Nomenclature of the Staphylococcal Enterotoxins

The current classification scheme is based on antigenicity. M. S. Bergdoll and colleagues at the University of Wisconsin were the first to produce purified staphylococcal enterotoxin and to develop specific antisera. They used purified toxins to show that protective antibodies were made in ani-

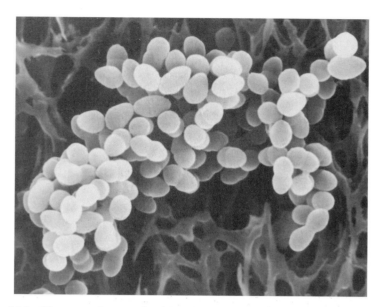

Figure 14.1 Electron micrograph of *S. aureus* cells.

mals. Immunity did not provide protection against other strains, however. It soon became clear that *S. aureus* makes multiple toxins with different immunological properties.

Staphylococcal enterotoxins are named by letter in the order of their discovery. Staphylococcal enterotoxins A, B, C, D, and E are the major types. Protein sequencing and recombinant-DNA methods have yielded the primary sequences of the enterotoxins (Fig. 14.2). They have a high degree of homology. Staphylococcal enterotoxins G, H, J, and I are more recent discoveries. An exotoxin produced by the *S. aureus* strain associated with toxic shock syndrome was initially called staphylococcal enterotoxin F. When it was discovered that this exotoxin did not cause vomiting in animals as the other enterotoxins do, its name was changed to toxic shock syndrome toxin 1 (Box 14.1).

Table 14.1 General characteristics of selected species of *Staphylococcus*[a]

Characteristic	Presence of phenotype[b]					
	S. aureus	*S. chromogenes*	*S. hyicus*	*S. intermedius*	*S. epidermidis*	*S. saprophyticus*
Coagulase	+	−	+	+	−	−
Thermostable nuclease	+	−	+	+	±	−
Clumping factor	+	−	−	+	−	−
Yellow pigment	+	+	−	−	−	±
Hemolytic activity	+	−	−	+	±	−
Phosphatase	+	+	+	+	±	−
Lysostaphin	Sensitive	Sensitive	Sensitive	Sensitive	Slightly sensitive	ND
Hyaluronidase	+	−	+	−	±	ND
Mannitol fermentation	+	±	−	±	−	+/−
Novobiocin resistance	−	−	−	−	−	+

[a]Reprinted from L. M. Jablonski and G. A. Bohach, p. 411–434, *in* M. P. Doyle, L. R. Beuchat, and T. J. Montville (ed.), *Food Microbiology: Fundamentals and Frontiers*, 2nd ed. (ASM Press, Washington, D.C., 2001).
[b]ND, not determined; +, present; −, absent; ±, variable.

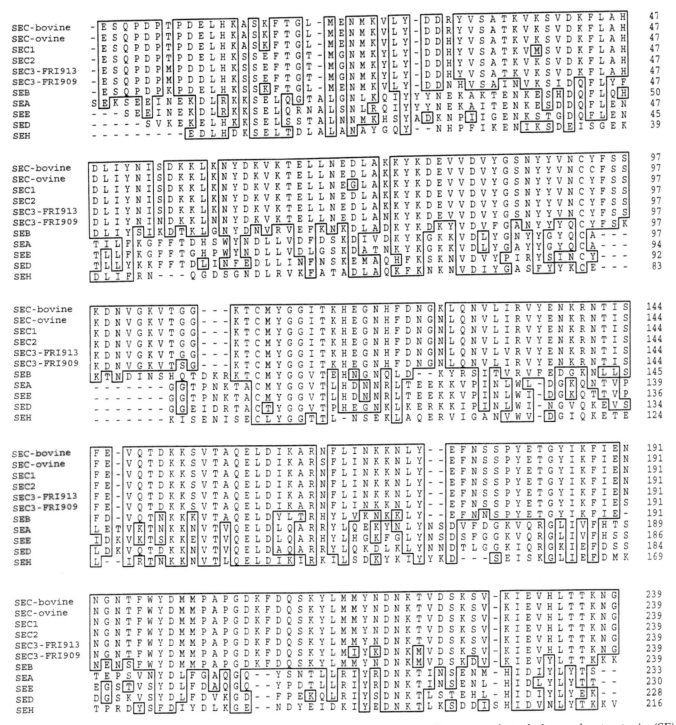

Figure 14.2 Amino acid sequence alignment of staphylococcal enterotoxin (SE) types A, E, J, D, C, B, G, H, and I. Identical residues are boxed. Reprinted from L. M. Jablonski and G. A. Bohach, p. 411–434, *in* M. P. Doyle, L. R. Beuchat, and T. J. Montville (ed.), *Food Microbiology: Fundamentals and Frontiers,* 2nd ed. (ASM Press, Washington, D.C., 2001).

BOX 14.1

Toxic shock syndrome: changes in microbial ecology give rise to a new threat from an old pathogen

Toxic shock syndrome is not caused by food, but it is caused by *S. aureus*. In women, toxic shock syndrome causes vomiting, diarrhea, renal prob-

lems, central nervous system symptoms, and the peeling of skin from the hands and feet. It can be fatal. Toxic shock syndrome emerged in the mid-1970s when superabsorbent tampons were introduced into the highly competitive feminine hygiene market. The superabsorbency led to extended periods of usage. This led to extended time for the growth of, and toxin produc-

tion by, *S. aureus* in a moist, warm, nutrient-rich environment. Toxic shock syndrome toxin 1 causes 75% of the cases, but staphylococcal enterotoxin B (the same toxin that causes food poisoning) is responsible for 25% of the cases. The federal government now regulates tampon absorbency and advises menstruating women to change their tampons frequently.

Staphylococcal Regulation of Enterotoxin Expression

General Considerations

Staphylococcal enterotoxins are produced in small quantities throughout most of the exponential growth phase. The amount made is strain dependent. Staphylococcal enterotoxins B and C are produced in the greatest quantities, up to 350 µg/ml of culture. Staphylococcal enterotoxins A, D, and E are easily detectable by gel diffusion assays, which detect as little as 100 ng of enterotoxin per ml of culture. Some strains produce very low levels of toxins that require more sensitive methods for detection. Of these, staphylococcal enterotoxins D and J are most likely to be undetected.

Molecular Regulation of Staphylococcal-Enterotoxin Production

Three genes that regulate the expression of *S. aureus* virulence factors are *agr* (for accessory gene regulator), *sar* (for staphylococcal accessory regulator), and *sae* (for *S. aureus* exoprotein expression). The best characterized is *agr*. Mutations in *agr* decrease toxin expression. Gene regulation by *agr* can be transcriptional or translational. Not all staphylococcal toxins are regulated by *agr*. Staphylococcal enterotoxin A expression is not affected by *agr* mutations. Since at least 15 genes are under the control of *agr*, it is considered a *global regulator*.

Regulation of Staphylococcal Enterotoxin Gene Expression

The properties of *agr* explain several aspects of enterotoxin production and, in many cases, explain what were once known as "glucose effects." For example, *agr* expression coincides with the expression of enterotoxins B and C during growth. All have their highest expression during late exponential and postexponential growth. Staphylococcal enterotoxin A, which is not regulated by *agr*, is produced earlier. Furthermore, the production of several staphylococcal enterotoxins is inhibited by growth in media containing glucose, which affects *agr*. Enterotoxin C expression is affected by glucose in two different ways. First, glucose metabolism indirectly influences enterotoxin C production by reducing the pH: *agr* is most highly expressed at neutral pH, but growth in media containing glucose lowers the pH. This directly reduces *agr* expression. The expression of *sec* and other *agr* target genes is also affected. Glucose also reduces *sec* expression in *agr* mutant strains. This suggests a second glucose-dependent mechanism for stopping toxin production, independent of *agr* and not influenced by pH.

In many organisms, a process known as *autoinduction* coordinates the expression of proteins. In autoinduction, the production of one compound

Authors' note_____
Quorum sensing is the way bacteria determine the size of their population and, if it is big enough, act in a certain fashion. This is analogous to legislative bodies that require a quorum (a certain number of people) in order to do business.

Authors' note_____
Signal transduction is the mechanism by which bacteria sense some external factor and transduce it to an internal response.

induces the production of others in the same organism. The *agr* locus is at the center of the *S. aureus* autoinduction response. Components of the *agr* operon comprise a *quorum-sensing* apparatus analogous to the *signal transduction* pathway of bacterial two-component (receiver and response) systems.

Other Relevant Molecular Aspects of Staphylococcal-Enterotoxin Expression

Many factors selectively inhibit enterotoxin expression. Their effects on regulation and signal transduction are only beginning to be defined. *agr* is not the only signal transduction mechanism for *S. aureus*. The negative effect of glucose is not entirely due to higher acid levels. Cultures containing glucose produce less enterotoxin even when the acid is neutralized.

   S. aureus is osmotolerant. It grows at relatively low water activity and is commonly associated with foods having high salt (e.g., ham) or sugar (e.g., cream fillings) concentrations. When it grows in foods with low water activities, less enterotoxin is produced. In experiments with enterotoxin C-producing strains, levels of *sec* mRNA and enterotoxin C protein are both reduced in response to high NaCl concentrations.

SOURCES OF *S. AUREUS*

Sources of Staphylococcal Food Contamination

People are the main reservoir of *S. aureus*. Most staphylococcal species inhabit the body's exterior. Humans are natural carriers and spread staphylococci to other people and to food. In humans, the interior of the nose is the main colonization site. *S. aureus* also occurs on the skin. *S. aureus* spreads by direct contact, through skin fragments, or through respiratory droplets produced when people cough or sneeze.

   The many human and animal reservoirs make controlling staphylococcal food poisoning especially challenging. *S. aureus* persists in sites such as the mucosal surface due to its ability for intracellular survival. *S. aureus* can bind to and be internalized by many different cell types. Once bound, *S. aureus* triggers a series of host-specific changes that are similar to those caused by other facultative intracellular pathogens (e.g., listeriae): phagosome formation, protein tyrosine kinase activation, and changes in cell morphology.

   Most staphylococcal food poisoning is traced to food contaminated by humans during preparation. In addition to contamination by food handlers, meat grinders, knives, storage containers, cutting blocks, and saw blades may also introduce *S. aureus* into food. Conditions often associated with outbreaks of staphylococcal illness are

- inadequate refrigeration
- preparing foods too far in advance
- poor personal hygiene
- inadequate cooking or heating of food
- prolonged use of warming plates when serving foods

   Animals are also *S. aureus* sources. For example, bovine mastitis (the infection of cow teats) is a serious problem for the dairy industry. It is often caused by *S. aureus*. Bovine mastitis is the single most costly agricultural disease in the United States. Mastitis is also a public health concern because

Table 14.2 Prevalence of *S. aureus* in several common food products[a]

Product	No. of samples tested	% Positive for *S. aureus*	*S. aureus* content (CFU/g)
Ground beef	74	57	$\geq$100
	1,830	8	$\geq$1,000
	1,090	9	>100
Big game	112	46	$\geq$10
Pork sausage	67	25	100
Ground turkey	50	6	>10
	75	80	>3.4
Salmon steaks	86	2	>3.6
Oysters	59	10	>3.6
Blue crab meat	896	52	$\geq$3
Peeled shrimp	1,468	27	$\geq$3
Lobster tail	1,315	24	$\geq$3
Assorted cream pies	465	1	$\geq$25
Tuna pot pies	1,290	2	$\geq$10
Delicatessen salads	517	12	$\geq$3

[a]Adapted from L. M. Jablonski and G. A. Bohach, p. 411–434, *in* M. P. Doyle, L. R. Beuchat, and T. J. Montville (ed.), *Food Microbiology: Fundamentals and Frontiers*, 2nd ed. (ASM Press, Washington, D.C., 2001).
[b]Determined by either direct plate count or most-probable-number technique.

the bacteria can contaminate milk and dairy products. This can be controlled through strict hygiene of automated milkers, milk handlers, and facilities.

The source of staphylococcal food contamination for a specific outbreak is not always known. Regardless of its source, *S. aureus* is present in many foods (Table 14.2). The staphylococcus levels are usually low initially, <100 colony-forming units (CFU)/g. However, they can grow to high levels, >10[6] CFU/g, and cause staphylococcal food poisoning under favorable conditions.

Resistance to Adverse Environmental Conditions

S. aureus's unique resistance traits help it grow in food. It is one of the most resistant non-spore-forming pathogens. It can survive in a dry state and is isolated from air, dust, sewage, and water relatively easily. Environmental sources cause many staphylococcal-food-poisoning outbreaks.

S. aureus can acquire genetic resistance to heavy metals and antibiotics. However, its resistance to food preservation methods is only average. One exception is its osmotolerance. *S. aureus* grows in medium containing ~20% NaCl and survives at water activities of <0.86. This is important, because these conditions inhibit other microbes that normally outcompete *S. aureus.*

Staphylococci have an efficient osmoprotectant system. Several compounds accumulate in the cell or enhance its growth under osmotic stress. Glycine betaine is the most important osmoprotectant. To varying degrees, other compounds, including L-proline, proline betaine, choline, and taurine, also act as compatible solutes. Proline and glycine betaine accumulate to very high levels in *S. aureus* in response to low water activity. This lowers the intracellular water activity to match the external water activity. However, the transport of these compounds requires energy, diverting it from other cellular purposes. Under extreme conditions of growth, the organism may not produce toxin.

FOODBORNE OUTBREAKS

Incidence of Staphylococcal Food Poisoning

There is little reason to report staphylococcal food poisoning, because people usually recover in 24 to 48 h and have no reason to go to the doctor. Although there is national surveillance for staphylococcal food poisoning, it is not an officially reportable disease. Only 1 to 5% of all staphylococcal food-poisoning cases in the United States are reported. Most of these are highly publicized outbreaks. Sporadic cases in the home are usually unreported. Staphylococcal food poisoning accounts for ~14% of the total outbreaks of foodborne illness within the United States. It is the third most common confirmed bacterial foodborne illness. There are ~25 major outbreaks of staphylococcal food poisoning annually in the United States. Their occurrence is seasonal. Most cases are in the late summer, when temperatures are warm and food is often stored improperly. A second peak occurs in November and December, presumably due to the incorrect handling of leftover holiday food. Although international reporting is also poor, staphylococcal food poisoning is a leading cause of foodborne illness worldwide. In one study, 40% of foodborne gastroenteritis outbreaks in Hungary were found to be due to staphylococcal food poisoning. The percentage is slightly lower in Japan at 20 to 25%. In Great Britain, meat or poultry products caused 75% of the 359 staphylococcal food-poisoning cases reported between 1969 and 1990.

A Typical Large Staphylococcal Food-Poisoning Outbreak

The FDA reported an outbreak that had many elements typical of staphylococcal food poisoning. The type of food involved, means of contamination, inadequate food-handling measures, and symptoms all pointed to *S. aureus*. The outbreak was traced to one meal fed to 5,824 school children at 16 sites in Texas. A total of 1,364 children developed typical staphylococcal food poisoning. Investigations revealed that 95% of the ill children had eaten chicken salad. *S. aureus* was found at high levels in the chicken salad.

The meal was prepared in a central kitchen the day before. Frozen chickens were boiled for 3 h. After being cooked, the chickens were deboned, cooled to room temperature with a fan, ground into small pieces, placed into 30.5-cm-deep pans, and stored overnight in a walk-in refrigerator at 5.5 to 7°C. The following morning, the other salad ingredients were added and the mixture was blended. The food was placed in containers and trucked to the schools between 9:30 and 10:30 a.m. It was kept at room temperature until it was served between 11:30 a.m. and noon.

The chicken was probably contaminated after being cooked, when it was deboned. The storage of the warm chicken in the deep pans prevented rapid cooling, which provided a good environment for staphylococcal growth and toxin production. Holding the food in warm classrooms provided an additional opportunity for growth. The screening of food handlers to identify *S. aureus* carriers, cooling the chicken more rapidly, and refrigerating the salad after preparation could have prevented the incident.

Another outbreak is described in Box 14.2.

CHARACTERISTICS OF DISEASE

Staphylococcal food poisoning is a self-limiting illness causing *emesis* (vomiting) after an unusually short time before onset (as little as 30 min after eating). However, vomiting is not the only symptom, and con-

*Authors' note*_____

The period between the consumption of tainted food and the onset of illness can be an important clue in foodborne illness.

BOX 14.2

Everything affects everything else

By now, the reader should realize that microbes grow in food ecosystems and that the foods' intrinsic and extrinsic factors determine their safety. Sometimes, these factors can reach far beyond the realm of food microbiology, even into politics and economic systems. Consider the case of the Chinese canned mushrooms.

Late in 1989, the Centers for Disease Control and Prevention received four reports of staphylococcal food-poisoning outbreaks from Mississippi, New York, and Pennsylvania. Almost 100 people were ill with nausea, vomiting, diarrhea, and cramps. All of the victims had eaten omelets, salads, pizza, or tomato sauce containing mushrooms. All of the mushrooms were traced to large institutional cans of mushrooms that had been imported from the People's Republic of China. The U.S. Food and Drug Administration then banned the importation of institutional-size cans of mushrooms from China, creating an international incident by disrupting this 50 million-pound-per-year mushroom business. A team of U.S. investi-

gators was quickly sent to the China to determine the cause of the tainted mushrooms and to help correct the situation.

Processing records indicated that the mushrooms had been properly canned and not contaminated after processing. This meant that the heat-resistant staphylococcus enterotoxin had been formed in the mushrooms before they were canned. The Chinese had a long history of safe canning for export. What had changed?

The chain "from farm to fork" was short and fast in the era when the communist state centralized growing and canning (see the figure). The mushrooms were grown in fields around the cannery, harvested, and canned within a few hours. There was total vertical integration and control. When free-market agriculture was introduced to the region, brokers who bought and sold mushrooms entered the scene. The brokers would pack mushrooms in polyvinyl chloride bags and buy and sell them like any other commodity. The cannery usually ended up getting mushrooms that had been in storage rather than premium fresh mushrooms. The *S. aureus* organisms, which did not compete with the aerobic spoilage organisms in

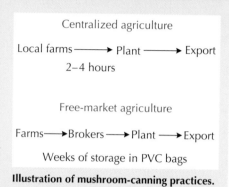

Illustration of mushroom-canning practices.

fresh mushrooms, thrived in the anaerobic conditions of the plastic bags. They made toxin without causing obvious spoilage to the mushrooms. The mystery was solved. The *S. aureus* cells grew and made toxin in the mushrooms before they were canned. When the mushrooms were heated during the canning process, the cells were killed but the toxin survived.

Although the case of the Chinese canned mushrooms may seem unique, food safety always depends on every link of the food chain that reaches from farm to fork. Whenever any part of the chain is changed, the effect on the rest of the chain should be considered.

versely, many patients with staphylococcal food poisoning do not vomit. Nausea, cramps, diarrhea, headaches, and/or prostration are other common symptoms. In a summary of clinical symptoms involving 2,992 patients diagnosed with staphylococcal food poisoning, 82% complained of vomiting, 74% felt nauseated, 68% had diarrhea, and 64% exhibited abdominal pain. In all cases of diarrhea, vomiting was also present. The lack of fever is consistent with the illness being caused by a toxin, not an infection.

Symptoms usually develop within 6 h after eating. In one report, 75% of the victims had symptoms of staphylococcal food poisoning within 6 to 10 h after eating. The mean incubation period is 4.4 h, although illness can appear in ~30 min. Death due to staphylococcal food poisoning is rare. The fatality rate ranges from 0.03% for the general public to 4.4% for more susceptible populations, such as children and the elderly. Approximately 10% of patients with confirmed staphylococcal food poisoning seek medical help. Treatment is usually minimal, although fluids are given when diarrhea and vomiting are severe.

INFECTIVE DOSE AND SUSCEPTIBLE POPULATIONS

Numbers of Staphylococci Required

According to the Food and Drug Administration, when levels of *S. aureus* are $>10^5$ cells/g of food, there may be enough enterotoxin to cause illness. Other studies suggest that 10^5 to 10^8 cells is the typical range, although smaller cell populations are sometimes implicated.

Toxin Dose Required

The best data on the minimum toxic dose come from analysis of food recovered from outbreaks. Although staphylococcal enterotoxins are quite potent, and 1 ng (10^{-9} g) per g of food can cause illness, the staphylococcal-enterotoxin level in outbreak foods is relatively large compared to those of other exotoxins. Many outbreaks are caused by 1 to 5 μg of ingested toxin per person.

Many factors contribute to the likelihood of getting staphylococcal food poisoning and its severity. These include individual susceptibility to the toxin, how much food was eaten, and a person's overall health. The toxin type may also be important. Although enterotoxin A causes more outbreaks, staphylococcal enterotoxin B produces more severe symptoms. Forty-six percent of 2,291 individuals exposed to staphylococcal enterotoxin B had symptoms severe enough for them to be hospitalized. Only 5% of 1,813 individuals exposed to enterotoxin A required hospitalization. This may reflect different levels of toxin expression, as enterotoxin B is generally produced at higher levels than enterotoxin A.

The amount of oral enterotoxins required to make human volunteers and monkeys vomit varies. Generally, monkeys are less susceptible than humans. In a study in which human volunteers ingested partially purified toxin, 20 to 25 μg of staphylococcal enterotoxin B (0.4 μg/kg of body weight) caused vomiting. In rhesus monkeys, the 50% emetic dose is ~1 μg/kg.

The staphylococcal enterotoxin acts on the viscera (gut). The emetic response results from a stimulation of the neural receptors. These transmit impulses through the nerves, ultimately stimulating the brain's vomiting center.

VIRULENCE FACTORS AND MECHANISMS OF PATHOGENICITY: STAPHYLOCOCCAL ENTEROTOXIN STRUCTURE-FUNCTION ASSOCIATIONS

Basic Structural and Biophysical Features

Staphylococcal enterotoxins are single 25- to 28-kilodalton polypeptides. Most are neutral or basic proteins that have no net charge at pH values ranging from 7 to 8.6. All staphylococcal enterotoxins are *monomeric* (single-unit) proteins. They are made as larger precursors with a signal peptide that is cut off during export from the cell.

Enterotoxin sequence analysis, relatedness, and diversity have been discussed above. The three-dimensional shapes of all staphylococcal enterotoxins are similar. Staphylococcal enterotoxins A, B, C, and E have a low α-helix content ($<10\%$) compared to β-pleated sheets or β-turn structures (~60 to 85%).

Staphylococcal enterotoxins are very stable. Their toxicity and antigenicity are not completely destroyed by boiling or even canning. The temperatures needed to inactivate staphylococcal toxins are much higher than those needed to kill *S. aureus* cells. In many cases of staphylococcal food poisoning, no live bacteria are found in the food.

Antigenic Properties of Staphylococcal Enterotoxin

Each enterotoxin type has enough antigenic distinctness to be differentiated from other toxins by using antibodies. However, some cross-reactivity can occur. The level of cross-reactivity generally correlates with the extent of similar amino acid sequences. The type C enterotoxin subtypes and their molecular variants are cross-reactive, as are enterotoxins A and E, the two major serological types with the greatest sequence similarity. The two most distantly related enterotoxins recognized by a common antibody are staphylococcal enterotoxins A and D.

Summary

- *S. aureus* is commonly associated with humans.
- Poor sanitation causes many outbreaks of staphylococcal food poisoning.
- Symptoms of vomiting and diarrhea develop 0.5 to 6 h after eating.
- *S. aureus* makes several antigenically distinct enterotoxins.
- The cells are heat sensitive, but the toxins are heat resistant.
- *S. aureus* is the most osmotolerant foodborne pathogen.
- There is considerable knowledge of enterotoxin genetics and mechanism of action.

Suggested reading

Bergdoll, M. S. 1989. *Staphylococcus aureus*, p. 247–254. *In* M. P. Doyle (ed.), *Foodborne Bacterial Pathogens.* Marcel Dekker, Inc., New York, N.Y.

Dinges, M. M., P. M. Orwin, and P. M. Schievery. 2000. Exotoxins of *Staphylococcus aureus. Clin. Microbiol. Rev.* **13:**16–34.

Jablonski, L. M., and G. A. Bohach. 2001. *Staphylococcus aureus*, p. 411–434. *In* M. P. Doyle, L. R. Beuchat and T. J. Montville (ed.), *Food Microbiology: Fundamentals and Frontiers,* 2nd ed. ASM Press, Washington, D.C.

Questions for critical thought

1. What issues contributed to the outbreak described at the beginning of the chapter? How could it have been prevented?
2. Why did the passengers get sick even though the food was properly heated?
3. What would *you* do if you were served undercooked chicken on a transatlantic flight?
4. What is the difference between enterotoxins and exotoxins? Name an exotoxin (from another chapter) that is not an enterotoxin.
5. Why do food microbiologists sometimes refer to staphylococcal food poisoning as a "double-bucket" syndrome?
6. What is signal transduction? Why do bacteria need it?
7. Briefly discuss genetic aspects of staphylococcal enterotoxin production.

8. Both botulinal toxin and staphylococcal toxin are proteins, but the former is very heat sensitive and the latter is very heat resistant. Why?

9. You work for Chubby Chuck's Chickenry (CCC), which produces (among other chicken specialties) about 1,000 pounds of chicken salad per day. It is packed in 10-pound tubs and distributed under refrigeration to restaurants and delicatessens. Chubby Chuck heard about the terrible Texas outbreak and tells you it is your job to make sure this never happens at CCC. What do you do?

10. As the diligent plant manager of CCC, you test the product for staphylococcal enterotoxins. On Friday, you get the results of Thursday's testing. Two batches of chicken salad test positive for *Staphylococcus* enterotoxin A. Batch 200498-6 was packaged and shipped out. Batch 200491-6 was batched into 10-gallon tubs, which are still in the cold-storage room of the plant. What do you do? What (and when) do you tell Chubby Chuck?

15

*Clostridium botulinum**

> **LEARNING OBJECTIVES**
>
> The information in this chapter should enable the student to:
> - use basic biochemical characteristics to identify *Clostridium botulinum*
> - understand what conditions in foods favor the growth of *C. botulinum*
> - recognize, from symptoms and time of onset, a case of botulism
> - choose appropriate interventions (heat, preservatives, and formulation) to prevent the growth of *C. botulinum*
> - identify environmental sources of *C. botulinum*
> - understand the roles of spores, anaerobic conditions, and heat sensitivity of the toxin in causing or preventing botulism

INTRODUCTION

Botulism is a rare but sometimes deadly disease. There are an average of 24 cases of foodborne botulism, three cases of wound botulism, and 71 cases of intestinal botulism reported annually to the Centers for Disease Control and Prevention. Although botulism from commercial foods is rare, many countries report relatively frequent outbreaks (Table 15.1).

Four Faces of Botulism

Botulism is traditionally associated with canned foods, usually home canned. The last commercial cases occurred in New York State in the early 1970s. An elderly couple ate some vichyssoise soup on a hot summer night. By the next morning, they had started to see double, had difficulty swallowing, and experienced paralysis in their arms and legs. If untreated, the toxin would have paralyzed their diaphragms and suffocated them. Fortunately, they were quickly diagnosed with botulism and put on ventilators. They lived, but they never fully recovered. The investigation revealed that the workers responsible for ensuring the commercial sterilization of the cans had misinterpreted a new process schedule. They underprocessed the soup, allowing *C. botulinum* to grow and make toxin. In response to this incident, the Food and Drug Administration (FDA) instituted good manufacturing practices for low-acid foods. The good manufacturing practices require that all retort operators attend a better processing control school and be certified. Processing schedules and changes to them can be made only by recognized processing authorities.

One of the biggest outbreaks of botulism occurred in Clovis, N.Mex. Over 40 people who ate at a local salad bar started seeing double, became

---

*This chapter was originally written by John W. Austin for *Food Microbiology: Fundamentals and Frontiers,* 2nd ed., and has been adapted for use in an introductory text.

Table 15.1 Reported foodborne botulism cases

Country	Period	No. of cases	Usual type	Usual food
Argentina	1979–1997	277	A	Preserved vegetables
Belgium	1988–1998	10	B	Meats
Canada	1985–1999	183	E	Traditional Inuit fermented marine mammal meat
China	1958–1989	2,861	A, B	Fermented bean products
France	1988–1998	72	B	Home-cured ham
Germany	1988–1998	177	B	Meats
Iran	1972–1974	314	E	Fish
Italy	1988–1998	412	B	Vegetables preserved in oil or water
Japan	1951–1987	479	E	Fish or fish products
Norway	1975–1997	26	E	*Rakfisk* (traditional fermented fish)
Poland	1988–1998	1,995	B	Home-preserved meats
Russia	1988–1992	2,300	B	Home-preserved mushrooms, fish
Spain	1988–1998	92	B	Vegetables
United States	1950–1996	1,087	A	Vegetables

short of breath and weak, had slurred speech, and showed various signs of paralysis. The local doctor, to his credit, quickly diagnosed botulism. Remembering its association with canned foods, he (prematurely) proclaimed the three-bean salad to be the culprit. Public health authorities from the Centers for Disease Control and Prevention and the FDA did an exhaustive investigation that cleared the beans and implicated the potato salad. They discovered that, at the end of the night in restaurants, leftover potatoes are often put in a box and stored at room temperature. When enough potatoes accumulate, they are used for potato salad with no further heating. This turns out to be a perfect scenario for botulism poisoning. Botulinal spores are normally found on potatoes. Baking kills any other competing bacteria and drives off oxygen, creating the anaerobic conditions under which *C. botulinum* thrives. The surviving spores turn into cells that grow in the potatoes and make botulinal toxin. Since the potato salad is not cooked, the toxin is not destroyed. It is fully lethal. Victims of this outbreak required years to recover.

Commercial products generally have good safety records. However, commercially bottled garlic in oil has caused many outbreaks. As a result, garlic in oil can be sold in North America only if a second barrier, such as acidification, is present and the product is refrigerated. In northern Canada, Alaska, Scandinavia, and northern Japan, most botulism outbreaks involve fish. *C. botulinum* type E is implicated in most outbreaks involving northern native foods. Many of these foods are fermented products. However, the level of fermentable carbohydrates is too low to ensure sufficiently rapid acidification to inhibit botulinal growth.

Scientists at the California Department of Public Health in the 1970s discovered the newest form of botulism, infant botulism. They noticed infants who were not thriving, could not lift themselves up, and had poor muscle tone. Their deaths resembled "sudden infant death." Early investigation of hospitalized infants revealed an association with breast-feeding. It was soon realized, however, that the protective effect of breast-feeding allowed the diagnosis and hospitalization of these infants. Infants who were not breast-fed were much sicker and usually did not make it to the hospital. Eventually, an association with eating honey and other raw agricultural products led to a new scenario for botulism. We all consume botulinal spores, but we are protected by the bacteria colonizing our guts. When infants, who have no protective intestinal microbiota, eat spores, these spores can germinate, become cells, colonize the gut, and produce toxin in place. Adults who are immunocompromised or who have undergone antibiotic therapy that kills their intestinal bacteria can also get "infant" botulism, so it has been renamed intestinal botulism. To prevent botulism in infants, the American Academy of Pediatrics recommends that children under the age of 2 should not be fed any raw agricultural products.

Wound botulism is not foodborne. It is caused when spores are introduced into body tissue below the skin. In one recent year, the number of botulism cases caused by drug users injecting contaminated heroin was greater than the number of foodborne cases. A fourth face of botulism is the intentional use of *C. botulinum* as an agent of bioterrorism (Box 15.1).

BOX 15.1

Biosecurity and biosafety in the age of terrorism

C. botulinum produces the most potent toxin known, yet there was a time when food microbiologists traded botulinum cultures like baseball cards. If they had nothing to trade, they could buy cultures at minimal cost with minimal red tape. That time ended in October 2001, when a terrorist sent spore-laden letters to politicians and celebrities. *Bacillus anthracis* became a household word—a word associated with terror. Microbiology was suddenly not such a noble field. Microbiologists who worked with spores were "visited" by Federal Bureau of Investigation agents.

By the end of January 2002, Congress passed and President George W. Bush signed the Providing Appropriate Tools Required to Intercept and Obstruct Terrorism Act (i.e., the "Patriot Act") and the Public Health Security and Bioterrorism Preparedness and Response Act. Together, these laws give the federal government strong oversight of microbiologists who work with *select agents*, microbes that could be used for bioterrorism. Select agents that food microbiologists might normally work with include *C. botulinum*, botulinal toxin, *Clostridium perfringens* epsilon toxin, staphylococcal enterotoxin, and T-2 toxin (a trichothecene mycotoxin produced by the fungus *Fusarium sporotrichioides*). It is now a federal offense to possess any of these unless the laboratory has been registered, inspected, and approved by the Centers for Disease Control and Prevention. The Department of Justice must run background checks and approve scientists before they can work with select agents. "Restricted persons," i.e., anyone convicted of a felony or a crime that could carry a penalty of more than a year in jail, citizens of countries hostile to the United States, and people who are "mentally defective" or were committed to a mental institution, are prohibited from having access to select agents.

The select agent act requires that the agents be inventoried, kept locked in a room accessible only to people cleared by the Department of Justice, and handled under a biosafety cabinet. In addition to adhering to the normal biological safety level (BSL) guidelines, the laboratory must conduct a risk assessment, have a security plan in place, and maintain detailed records of laboratory entry, culture maintenance, and experimental procedures.

Biosafety levels are designated according to how dangerous the organism is and how easy it is to treat. Microbes that do not consistently cause sickness in healthy adults can be used in a BSL-1 laboratory. BSL-1 labs must use standard microbiological practices, such as no smoking, eating, or drinking in the lab and no mouth pipetting. No other special equipment is required. Most foodborne pathogens need a BSL-2 laboratory. These organisms can make people sick if ingested or if they enter the body through cuts or membranes. These sicknesses are relatively easy to cure. In addition to the BSL-1 requirements, access to BSL-2 labs is restricted to lab personnel, there is a biohazard warning sign, procedures that produce aerosols are done under a biosafety cabinet, and laboratory personnel must wear lab coats and gloves. BSL-3 labs are much more specialized and are required for microbes that can be transmitted through aerosols. Organisms causing lethal illnesses require BSL-3. In addition to BSL-2 requirements, BSL-3 requires controlled access, decontamination of laboratory waste and laboratory clothing, and engineering safeguards. The laboratory cannot be off of a common hallway, must have negative airflow into the laboratory, and must have an exhausted air supply that is not recirculated to other parts of the building. Biosafety cabinets are used for all culture manipulations. There are also BSL-4 laboratories for organisms like Ebola virus that are highly infectious and have no cure. Fortunately, no foodborne organisms require BSL-4 precautions.

These security and safety precautions will undoubtedly tighten over the next decade. They will make it more difficult, costly, and restrictive to conduct research on select agents. The lengthy time required for Department of Justice clearance and immunizations may prevent graduate students and postdoctoral associates from working with these organisms. This will reduce the number of scientists trained to respond to bioterroristic attacks. Hopefully, future legislation will make it easier to conduct legitimate research while still protecting us from the misuse of microbes.

CHARACTERISTICS OF DISEASE

Foodborne botulism can range from a mild illness, which may be disregarded or misdiagnosed, to a serious disease that can kill within a day. Symptoms typically appear 12 to 36 h after ingestion of the neurotoxin but may appear within a few hours or not for up to 14 days. The earlier symptoms appear, the more serious the illness. The first symptoms are generally nausea and vomiting. These are followed by neurological signs and symptoms, including visual impairments (blurred or double vision, ptosis, or fixed and dilated pupils), loss of normal mouth and throat functions (difficulty in speaking and swallowing; dry mouth, throat, and tongue; or sore throat), general fatigue and lack of muscle coordination, and respiratory impairment. Other symptoms may include abdominal pain, diarrhea, or constipation. Respiratory failure and airway obstruction are the main causes of death. Fatality rates in the early 1900s were ≥50%. The availability of antisera and modern respiratory support systems have decreased the fatality rate to ~10%. However, it may take 3 or 4 months on a ventilator and then years of therapy to fully recover.

Botulism is often incorrectly diagnosed as other illnesses, including other forms of food poisoning, stroke, poliomyelitis, organophosphate poisoning, tick paralysis, myasthenia gravis, and carbon monoxide poisoning, but is most commonly misdiagnosed as Guillain-Barré syndrome. The initial symptom of intestinal botulism is usually constipation.

Initial treatment of botulism involves removing or inactivating the neurotoxin by (i) neutralizing the circulating neurotoxin with antiserum, (ii) using enemas to remove residual neurotoxin from the bowel, and (iii) gastric lavage or treatment with emetics if the food might still be in the stomach. Antiserum is most effective in the early stages of illness. Subsequent treatment is mainly mechanical ventilation to counteract paralysis of the respiratory muscles. The optimal treatment for intestinal botulism consists primarily of high-quality supportive care.

TOXIC AND INFECTIOUS DOSES AND SUSCEPTIBLE POPULATIONS

Since toxicity experiments with human subjects are not ethical, little is known about the minimum toxic (or infectious) doses of *C. botulinum* and its neurotoxins. From a food safety perspective, the presence of neurotoxin or conditions permitting *C. botulinum* growth are simply not tolerated. Botulinal toxin is the most toxic natural substance known. The 50% lethal dose (LD_{50}) (the amount required to kill 50% of subjects) for botulinal toxin injected under the skin in monkeys is ~0.4 ng/kg of body weight. This suggests that the LD_{50} for a 150-pound human would be 0.000000001 (10^{-9}) ounce! This toxicity, and the relatively low technology level required to grow *C. botulinum*, creates concern about its use in biological warfare. Indeed, in 2003, the United Nations Special Commission reported that Iraq had produced ~50 gallons of concentrated botulinal toxin. This was allegedly weaponized in 16 missiles and 100 bombs.

The only way to prevent foodborne botulism is by preventing neurotoxin production in foods. Immunization of high-risk populations has been considered, but it is not cost-effective. Currently, only laboratory workers who work with the organism and members of the U.S. military serving in the Middle East are immunized. The preservation of high-moisture foods is usually designed to prevent *C. botulinum* growth. The thermal processing of shelf-stable canned foods kills *C. botulinum* spores, while control of *C. botulinum* in minimally processed foods is achieved by growth inhibition, typically by using a combination of factors. Controlling *C. botulinum* growth usually controls other foodborne pathogens and spoilage microorganisms.

CHARACTERISTICS OF *C. BOTULINUM*

Classification

C. botulinum is a gram-positive, anaerobic, rod-shaped bacterium. Oval endospores are formed in stationary-phase cultures (Fig. 15.1). There are seven types of *C. botulinum*, A through G, based on the serological specificity of the neurotoxin produced. Human botulism, including foodborne, wound, and intestinal botulism, is associated with types A, B, E, and very rarely, F. Types C and D cause botulism in animals. To date, there is no direct evidence linking type G to disease.

The species is also divided into four groups based on physiological differences (Table 15.2). Group I contains all type A strains and *proteolytic strains* (i.e., those that produce enzymes that break down proteins) of types B and F. Group II contains all type E strains and nonproteolytic strains of types B and F. Group III contains type C and D strains, and group IV contains *C. botulinum* type G. This grouping agrees with the results of DNA homology studies. Sequence studies using 16S and 23S rRNAs show a high degree of relatedness among strains within each group but little relatedness between groups.

Group I strains are proteolytic and are typified by strains that produce neurotoxin type A. The optimal temperature for growth is 37°C, with

Figure 15.1 Scanning electron micrograph of *C. botulinum* vegetative cells (rods) and spores (irregular ovals). Originally taken by B. Maleeff and kindly provided by P. Cooke, both of the U.S. Department of Agriculture Eastern Regional Research Center.

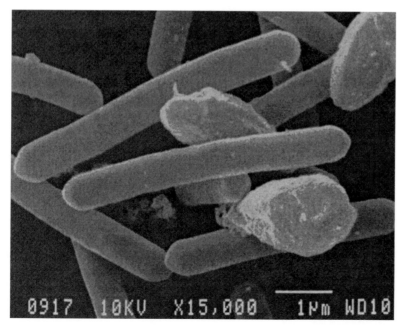

Table 15.2 Grouping and characteristics of *C. botulinum*[a]

Characteristic	Value for group:			
	I	II	III	IV
Neurotoxin type(s)	A, B, F	B, E, F	C, D	G
Minimum temp for growth (°C)	10	3	15	ND
Optimum temp for growth (°C)	35–40	18–25	40	37
Minimum pH for growth	4.6	ca. 5	ND	ND
Inhibitory NaCl concentration (%)	10	5	ND	ND
Minimum a_w for growth	0.94	0.97	ND	ND
$D_{100°C}$ of spores (min)	25	<0.1	0.1–0.9	0.8–1.12
$D_{121°C}$ of spores (min)	0.1–0.2	<0.001	ND	ND

[a]ND, not determined. Reprinted from J. W. Austin, p. 329–349, *in* M. P. Doyle, L. R. Beuchat, and T. J. Montville (ed.), *Food Microbiology: Fundamentals and Frontiers* (ASM Press, Washington, D.C., 2001).

growth occurring between 10 and 48°C. High levels of neurotoxin (10^6 mouse LD_{50}/ml [1 mouse LD_{50} is the amount of neurotoxin required to kill 50% of injected mice within 4 days]) are typically produced in cultures. Spores have high heat resistance, with $D_{100°C}$ values of ~25 min (the $D_{n°C}$ value is the time required to inactivate 90% of the population at a temperature of n°C). To inhibit growth, the pH must be <4.6, the salt concentration must be >10%, or the water activity (a_w) must be <0.94.

Group II strains are nonproteolytic, have a lower optimum growth temperature (30°C), and grow at temperatures as low as 3°C. The spores have much lower heat resistance, with $D_{100°C}$ values of <0.1 min. Group II strains are inhibited by a pH of <5.0, salt concentrations of >5%, or an a_w of <0.97. The focus in this chapter is on groups I and II, because they include the strains involved in human illness.

Tolerance of Preservation Methods

Temperature, pH, a_w, redox potential (E_h), added preservatives, and the presence of competing microorganisms are the main factors controlling *C. botulinum* growth in foods. Historically, studies have established maximum and/or minimum limits for these parameters that control the growth of *C. botulinum* (Table 15.2). These factors seldom function by themselves. Usually, they act in concert with other inhibitory factors, often having synergistic effects. This is the conceptual basis of the "(multiple) hurdle theory" for food preservation.

Low Temperature

Refrigerated storage prevents *C. botulinum* growth. The lower limits are 10°C for group I and 3.0°C for group II. However, these limits depend on otherwise-optimal growth conditions. Irrespective of the actual minimum growth temperature, the production of neurotoxin generally requires weeks at the lower temperature limits. The optimum growth temperatures are between 35 and 40°C for group I and between 25 and 30°C for group II.

Thermal Inactivation

Thermal processing inactivates *C. botulinum* spores and is the most common method of producing shelf-stable foods. *C. botulinum* spores of group I are very heat resistant and are the target organisms for most thermal processes. *D* values vary considerably among *C. botulinum* strains. The *D* values depend on how the spores are produced and treated, the heating environment, and the recovery system. Spores of group I strains are the most heat resistant, having $D_{121°C}$ values between 0.1 and 0.2 min (121°C may seem to be an arbitrary reference temperature, but it is the temperature of industrial steam at 15 pounds/square inch, and thus it has great relevance in canning factories). Group I spores are of particular concern in the commercial sterilization of canned low-acid foods. The canning industry has adopted a *D* value of 0.2 min at 121°C as a standard for calculating thermal processes. The *z* values (the temperature change necessary to cause a 10-fold change in the *D* value) for the most resistant strains are ~10°C. Despite variations in *D* and *z* values, the adoption of a 12-*D* process as the minimum thermal process for commercially canned low-acid foods has resulted in a remarkable record of safety for such foods.

Strains of group II are considerably less heat resistant ($D_{100°C}$ < 0.1 min) than those of group I. Spores of *C. botulinum* group II can be inactivated at moderate temperatures (40 to 50°C) when heating is combined with high pressures of up to 827 megapascals. However, the survival of spores of *C. botulinum* group II in pasteurized refrigerated products could be dangerous, because they can grow at refrigeration temperatures. These products require special attention. $D_{82°C}$ values for *C. botulinum* type E in neutral phosphate buffer are in the range of 0.2 to 1.0 min. Values ranging from 0.15 to >4.90 min have been reported for type E strains, depending on the heating environment, strain, and plating agar and the presence or absence of *lysozyme* (an enzyme that aids the growth of spores that are injured but not killed).

pH

The minimum pH for growth of *C. botulinum* group I is 4.6. For group II, it is ~5. Many fruits and vegetables are sufficiently acidic to inhibit *C. botulinum*, whereas acidulants are used to preserve other products. The substrate, temperature, nature of the acidulant, presence of preservatives, a_w, and E_h all influence the acid tolerance of *C. botulinum*. Acid-tolerant microorganisms, such as yeasts and molds, may grow in acidic products and raise the pH in their immediate vicinity to a level that allows *C. botulinum* growth.

Salt and a_w

Sodium chloride is one of the most important factors used to control *C. botulinum* in foods. It acts primarily by decreasing the a_w. Consequently, its concentration in the water phase, called the brine concentration (percent brine = percent NaCl × 100/percent H_2O + percent NaCl), is critical. The growth-limiting brine concentrations are ~10% for group I and 5% for group II under otherwise-optimal conditions. These concentrations correspond well to the limiting a_w values of 0.94 for group I and 0.97 for group II in foods where NaCl is the main a_w depressant. The solute used to control the a_w may influence these limits. Generally, NaCl, KCl, glucose, and sucrose show similar effects, whereas growth occurs in glycerol at lower a_w. The limiting a_w may be increased substantially by other factors, such as increased acidity or the use of preservatives.

Oxygen and Redox Potential

While it is assumed that *C. botulinum* cannot grow in foods exposed to O_2, the E_h of most foods is usually low enough to permit growth. There have been many outbreaks of botulism caused by foods exposed to oxygen when the foods are heated enough to drive out the dissolved oxygen. Oxygen diffusion back into these foods is so slow that they can remain anaerobic for many hours. *C. botulinum* grows optimally at an E_h of −350 millivolts, but growth initiation may occur in the E_h range of $+30$ to $+250$ millivolts. Once growth is initiated, the E_h declines rapidly.

Initial atmospheres containing 20% O_2 did not delay neurotoxin production by *C. botulinum* in inoculated pork compared to that in samples packaged with 100% N_2. CO_2 is used in modified-atmosphere packaging to inhibit spoilage and pathogenic microorganisms but can stimulate *C. botulinum* growth. Pressurized CO_2 is lethal to *C. botulinum*, with lethality increasing with the pressure of the CO_2. The safety of specific atmospheres with respect to *C. botulinum* in foods should be carefully investigated before use.

Preservatives

Nitrite in cured meat products inhibits *C. botulinum*. It also contributes to color and flavor. Nitrite's effectiveness depends upon complex interactions among pH, sodium chloride, heat treatment, time and temperature of storage, and the composition of food. Nitrite reacts with many cellular constituents and appears to inhibit *C. botulinum* by more than one mechanism, including reaction with essential iron-sulfur proteins to inhibit energy-yielding systems in the cell. The reaction of nitrite, or nitric oxide, with secondary amines in meats to produce nitrosamines, some of which are carcinogenic, has led to regulations limiting the amount of nitrite used.

Sorbates, parabens, nisin, phenolic antioxidants, polyphosphates, ascorbates, EDTA, metabisulfite, *n*-monoalkyl maleates and fumarates, and lactate salts are also active against *C. botulinum*. The use of natural or liquid smoke inhibits *C. botulinum* in fish but not in meats.

Competitive and Growth-Enhancing Microorganisms

The growth of competitive and growth-promoting microorganisms in foods has a very significant effect on the fate of *C. botulinum*. Acid-tolerant molds, such as *Cladosporium* spp. or *Penicillium* spp., can provide an environment that enhances the growth of *C. botulinum*. Other microorganisms may inhibit *C. botulinum* by changing the environment, by producing specific in-

Pett © 2003 *Lexington Herald-Leader*. Distributed by Universal Press Syndicate. Reprinted with permission. All rights reserved.

hibitory substances, or both. Lactic acid bacteria, including *Lactobacillus, Pediococcus,* and *Streptococcus,* can inhibit *C. botulinum* growth in foods, largely by reducing the pH but also by the production of bacteriocins. The use of lactic acid bacteria and a fermentable carbohydrate, the "Wisconsin process," is permitted in the United States for producing bacon with a decreased level of nitrite. If the bacon is temperature abused, the lactic acid ferments the sugar and drops the pH to <4.5 before toxin can be produced.

Inactivation by Irradiation

C. botulinum spores are probably the most radiation-resistant spores of public health concern. The D values (the irradiation dose required to inactivate 90% of the population [this is different from the D value discussed above, used in thermal processing]) of group I strains at −50 to −10°C are between 2.0 and 4.5 kilograys in neutral buffers and in foods. Spores of type E are more sensitive, having D values between 1 and 2 kilograys. *Radappertization* ("canning" using radiation as the energy source) is designed to reduce the number of viable *C. botulinum* spores by 12 log cycles. Different environmental conditions, such as the presence of O_2, changes in irradiation temperature, and irradiation and recovery environments, can affect the D values of spores. Generally, spores in the presence of O_2 or preservatives and at temperatures above 20°C have greater sensitivity to irradiation.

SOURCES OF *C. BOTULINUM*

Occurrence of *C. botulinum* in the Environment

Contamination of food depends on the environmental incidence of *C. botulinum.* Many surveys of different environments for *C. botulinum* spores have revealed that they are common in soils and sediments, but their numbers and types vary depending on the location (Table 15.3). Type A spores predominate in soils in the western United States, China, Brazil, and

Table 15.3 Incidence of *C. botulinum* in soils and sediments[a]

Location	% Positive samples	MPN/kg	% Type[b]:				
			A	B	C/D	E	F
Eastern United States, soil	19	21	12	64	12	12	0
Western United States, soil	29	33	62	16	14	8	0
Green Bay, Wis., sediment	77	1,280	0	0	0	100	0
Alaska, soil	41	660	0	0	0	100	0
Britain, soil	6	2	0	100	0	0	0
Scandinavian coast, sediment	100	>780	0	0	0	100	0
Finland, Baltic Sea, offshore	88	1,020	0	0	0	100	0
Netherlands, soil	94	2,500	0	22	46	32	0
Switzerland, soil	44	48	28	83	6	0	27
Rome, Italy, soil	1	2	86	14	0	0	0
Iran, Caspian Sea, sediment	17	93	0	8	0	92	0
Sinkiang, China, soil	70	25,000	47	32	19	2	0
Ishikawa, Japan, soil	56	16	0	0	100	0	0
Brazil, soil	35	86	57	7	29	0	7
South Africa, soil	3	1	0	100	0	0	0

[a]Adapted from J. W. Austin, p. 329–349, *in* M. P. Doyle, L. R. Beuchat, and T. J. Montville (ed.), *Food Microbiology: Fundamentals and Frontiers* (ASM Press, Washington, D.C., 2001).
[b]Percentage for each type represents percentage out of all types identified.

Table 15.4 Prevalence of *C. botulinum* spores in food[a]

Product	Origin	% Positive samples	MPN/kg	Type(s) identified
Eviscerated whitefish chubs	Great Lakes	12	14	E, C
Vacuum-packed frozen flounder	Atlantic Ocean	10	70	E
Dressed rockfish	California	100	2,400	A, E
Salmon	Alaska	100	190	A
Smoked salmon	Denmark	2	<1	B
Salted carp	Caspian Sea	63	490	E
Fish and seafood	Osaka, Japan	8	3	C, D
Raw meat	North America	<1	0.1	C
Cured meat	Canada	2	0.2	A
Raw pork	United Kingdom	0–14	<0.1–5	A, B, C
Random honey samples	United States	1	0.4	A, B
Honey samples associated with infant botulism	United States	100	8×10^4	A, B

[a]Reprinted from J. W. Austin, p. 329–349, *in* M. P. Doyle, L. R. Beuchat, and T. J. Montville (ed.), *Food Microbiology: Fundamentals and Frontiers* (ASM Press, Washington, D.C., 2001).

Argentina. Type B spores predominate in soils in the eastern United States, the United Kingdom, and much of continental Europe. Most American type B strains are proteolytic, whereas most European strains are nonproteolytic. Type E predominates in northern regions and in most temperate aquatic regions and their surroundings, whereas types C and D are present more frequently in warmer environments.

Occurrence of *C. botulinum* in Foods

Many surveys have determined the incidence of *C. botulinum* spores in foods (Table 15.4). Food surveys have focused largely on fish, meats, and infant foods, especially honey. *C. botulinum* type E spores are common in fish and aquatic animals. Meat and meat products generally have low levels of contamination. These products are less likely than fish to be contaminated with spores because there is considerably less contamination of the farm environment than of the aquatic environment. In North America, the average most probable number (MPN) is ca. 0.1 spore per kg of meat products, whereas in Europe, the average MPN is ca. 2.5 spores per kg. The spore types most often associated with meats are A and B.

C. botulinum spores, usually type A or B, can contaminate fruits and vegetables, particularly those in close contact with the soil. Products in which contamination is often detected include asparagus, beans, cabbage, carrots, celery, corn, onions, potatoes, turnips, olives, apricots, cherries, peaches, and tomatoes. The overall incidence of *C. botulinum* spores in commercially available precut modified-atmosphere-packaged vegetables is low, ~0.36%; however, spores of both *C. botulinum* types A and B have been detected in these products. Cultivated mushrooms are of special concern, with up to 2.1×10^3 type B spores per kg.

VIRULENCE FACTORS AND MECHANISMS OF PATHOGENICITY

Botulinal toxin has been called the most poisonous poison known. *C. botulinum* produces eight antigenically distinct toxins, designated types A, B, C_1, C_2, D, E, F, and G. All of the toxins except C_2 are neurotoxins. C_2 and exoenzyme C_3 are adenosine diphosphate-ribosylating enzymes.

All seven of the neurotoxins are similar in structure and mode of action. *C. botulinum* neurotoxins are high-molecular-mass (150 kilodalton) two-chain proteins that are among the most toxic substances known. The neurotoxins block neurotransmission at nerve endings by preventing acetylcholine release. Fortunately, botulinal toxins are very heat sensitive. Heating a sample to 80°C for 10 min or bringing it to a boil completely inactivates the toxin.

Structure of the Neurotoxins

Botulinal toxins are water-soluble proteins produced as single polypeptides with an approximate relative molecular weight (M_r) of 150,000. They are cleaved by a protease approximately one-third of the distance from the N terminus. This produces an active neurotoxin composed of one heavy (M_r = 100,000) and one light (M_r = 50,000) chain linked by a single disulfide bond.

Bacterial proteases or proteases such as trypsin can cause proteolytic cleavage. The light chain remains bound to the N-terminal half of the heavy chain by a disulfide bond between Cys-429 and Cys-453 and noncovalent bonds. The two chains individually are nontoxic.

Genetic Regulation of the Neurotoxins

Complete gene sequences have been determined for the neurotoxins produced by *C. botulinum* types A, B (proteolytic and nonproteolytic), C, D, E, F, and G; *Clostridium barati* type F; and *Clostridium butyricum* type E. (Note that "botulinal" toxins can be produced by other clostridial species, although this is rare.) The degrees of relatedness of the various neurotoxins have been determined on the basis of sequence homologies. Different-serotype neurotoxins display less sequence homology than neurotoxins of the same serotype, even if the same serotypes are from different species.

The locations of the genes coding for botulinal toxin and the associated nontoxic proteins vary depending upon the serotype. The genes coding for botulinal toxins A, B, E, and F and the associated nontoxic proteins are located on the bacterial chromosome.

Mode of Action of the Neurotoxins

Botulinal toxin blocks the release of acetylcholine from synaptic vesicles at nerve terminals. This results in the flaccid paralysis observed in botulism poisoning and has been utilized in the therapy of several neurologic disorders. Botox (a trade name for type A botulinal toxin) is also being used as an alternative to cosmetic surgery to "erase" facial wrinkle lines (Box 15.2).

BOX 15.2

The dose makes the poison—or the cure

Botulinal toxin is the most deadly poison known, but when it is diluted to very low concentrations, its paralytic action can be used for good rather than evil. Several neuromuscular diseases are caused by overactive muscles and were previously untreatable or required surgery. Blepharospasm is a severe disease of the nerve endings near the eyelid. It causes people to blink so frequently that they can become functionally blind. Paralyzing these nerves by injections of Botox, the FDA-approved preparation of type A botulinal toxin, stops the blinking and lets the people see again. Strabismus, or cross-eye, is caused by unequal lengths of the muscles attached to the eye. In addition to the esthetic issue, the person stops using one eye and becomes functionally blind in that eye. The traditional cure is to surgically equalize the muscles. Injection of botulinal toxin has the same effect. The FDA approved both of these uses in 1989. In 2002, the FDA approved the use of Botox for treatment of cervical dystonia. In this disease, involuntary contractions of muscles in the neck and shoulder are so severe that the head is twisted to an abnormal position. Relaxing these muscles with Botox solves the problem. Botox can also be used cosmetically for removal of "frown lines" as an alternative to a face lift. Unfortunately, its effect lasts only ~4 months. Almost 1 million Americans per year receive Botox injections.

When Botox is injected under the skin near the wrinkles, flaccid paralysis of the underlying muscles causes the wrinkles to disappear. As toxicologists like to say, "The dose makes the poison."

Summary

- *C. botulinum* is an anaerobic sporeformer.
- Botulism is a rare but very serious disease. It can be caused by improper canning, temperature abuse, wounds, or, in the case of infants and at-risk adults, by colonization of the digestive tract.
- Symptoms of botulism poisoning are neurological and muscular, such as double vision, slurred speech, and paralysis.
- *C. botulinum* spores are the target of regulations for the canning of low-acid foods.
- Canned foods having a pH of >4.6 and a water activity of >0.86 must receive a "12-*D* bot cook."
- *C. botulinum* strains produce seven antigenically different neurotoxins.
- Botulinal toxin is a protein that is the most toxic substance known but is easily inactivated by heat.

Suggested reading

American Society for Microbiology. Home page (contains links to the latest legislation on select agents). http://www.asm.org.

Austin, J. W. 2001. *Clostridium botulinum*, p. 329–350. *In* M. P. Doyle, L. R. Beuchat, and T. J. Montville (ed.), *Food Microbiology: Fundamentals and Frontiers*, 2nd ed. ASM Press, Washington, D.C.

Hauschild, A. H. W., and K. L. Dodds. 1993. *Clostridium botulinum—Ecology and Control in Foods.* Marcel Dekker, Inc. New York, N.Y.

Richmond, J. Y., and R. B. McKinney (ed.). 1999. *Biosafety in Microbiological and Biomedical Laboratories*, 4th ed. Department of Health and Human Services, Washington, D.C.

Questions for critical thought

1. Why did the soup that made the elderly New York couple ill have to be vichyssoise? (If you know the ingredients of vichyssoise soup and how it is served, the answer will be obvious.) If creamy Cheddar cheese soup were similarly underprocessed, would you expect people to get sick from it?
2. Create another scenario for temperature-abuse botulism.
3. Why do most recipes for home-canned foods end with instructions to "boil for 10 min before serving?"
4. Does it matter that high-acid foods, such as tomato sauce, may contain viable botulinal spores after being processed?
5. Imagine that it is 2020 and you are a college professor researching botulism. Hundreds of wild geese have died on the college pond. The stench of their floating carcasses is hindering student recruitment efforts. The state Wildlife and Game Commission has bagged some of the dead geese and taken them to the lab. They have just called the dean with the results—type C botulism. The dean (your classmate, who got a C in food microbiology) remembers that botulinal toxin is the most

poisonous substance known. She calls you in panic: should she evacuate the campus? What do you tell her?

6. *Estimate* how many people might be killed if 1 ounce of botulinal toxin were placed in a warhead that dispersed it over a large area. State all your assumptions. Show all your work. Be sure to include an appropriate unit next to each numerical value.

7. You are the secretary of defense. How might you neutralize the threat described in question 6?

8. Comment on the fact that the most toxic toxin known to humans is a protein and "all natural." How does this make you feel about "all natural?" How does it make you feel about Mother Nature?

9. After reading the chapter, what questions do you have? Exchange your questions with a classmate, and research your classmate's question.

10. Read the following and respond:

    WHO: Two couples, Harry and Hilda Harrington; Ching-Ping and Pei-Ling Tseng

    WHAT: Shortness of breath, double vision, and slurred speech requiring hospitalization on a respirator and treatment with botulinal antitoxin

    WHEN: About 8 h after a party at the Harringtons'

    WHERE: At home

    FACTS OF THE INVESTIGATION: Both couples had consumed a homemade bean dip which contained garlic. The garlic label said, "Ollie's Olive Garden 'garlic in oil,' packaged in 100% virgin olive oil; keep refrigerated." Botulinal toxin was isolated from the remaining garlic in the jar.

    The Harringtons did not refrigerate the product but claimed that it was not refrigerated at Superlow Supermarket when they purchased it. A visit to Superlow revealed that the product was still being sold in the condiment isle without refrigeration. The manager says it was delivered on an unrefrigerated truck.

    Ollie's Olive Garden, Inc., claims that it is not responsible, since the product was actually packed by George's Great Gourmet Gifts. Ollie says that all he did was supply the olive oil. When asked if the distribution system was refrigerated, Ollie replies that the garlic is refrigerated at the place of manufacture, placed on refrigerated trucks, and delivered cold to food wholesalers (middlemen), but they have no control over what the wholesaler does with it.

    George's Great Gourmet Gifts is actually run by two men named Joe. They say that they are not responsible because the product was not refrigerated in the correct manner. They also point out that they were packing the garlic according to the specifications of Ollie's Olive Garden, Inc. A discussion with Joe and Joe reveals that they are investors who always wanted to run their own business. They did not know the reason for refrigerating the product but did so and kept it on the label because that is what George did when he sold them the company.

    At the beginning of the trial, the judge calls all parties into her chambers and tells them, "Someone is responsible for the suffering of these good people. We are going to find out who and make them pay."

    You are called on to be the expert witness. What do you tell the judge?

16

Clostridium perfringens*

LEARNING OBJECTIVES

The information in this chapter should enable the student to:
- use basic biochemical characteristics to identify *Clostridium perfringens*
- understand what conditions in foods favor *C. perfringens* growth.
- recognize, from symptoms and time of onset, a case of foodborne illness caused by *C. perfringens*
- choose appropriate interventions (heat, preservatives, and formulation) to prevent the growth of *C. perfringens*
- identify environmental sources of the organism
- understand the roles of spores and the sporulation cycle in *C. perfringens* foodborne illness

THE FOODBORNE ILLNESS

A Spore's-Eye View of *C. perfringens* Toxicoinfections

We spores are part of "the great circle of life." Most of the time, we just wait. Dormant. Patient. Then, there is a change. Heat kills our competitors, the vegetative cells. The temperature drops to that of a warm incubator, and we morph into cells and multiply. There are millions of us. A poor human eats us, and we're in perfringens heaven. We have warm temperatures, lots of nutrients, and anaerobic conditions. Our growth produces so much carbon dioxide that the poor human thinks she's having a gas attack. Oh, the cramps. Then, as nutrients run out, we sporulate back into our dormant form. During sporulation, as we cells break apart, we release a protein toxin that causes diarrhea in our victims. Expelled from our host, we return to the great circle of life. We wait. Dormant. Patient.

A Human View of *C. perfringens* Type A Foodborne Illness

Two outbreaks of *C. perfringens* type A foodborne illness occurred around St. Patrick's Day in 1994. The Cleveland, Ohio, outbreak involved 156 persons who ate corned beef prepared at a local delicatessen. The corned beef was prepared by boiling it for 3 h. It was then cooled slowly at room temperature before being put in the refrigerator. Four days later, the corned beef was warmed to 48.8°C and served. Sandwiches prepared with this corned beef were held at room temperature from late morning until later that afternoon, when they were consumed along with copious quantities of green beer.

---

*This chapter was originally written by Bruce A. McClane for *Food Microbiology: Fundamentals and Frontiers,* 2nd ed., and has been adapted for use in an introductory text.

The second St. Patrick's Day outbreak involved 86 people in Virginia who had attended a traditional corned beef dinner. The corned beef was a frozen, commercially prepared, brined product. It was thawed, cooked in large (4.54-kg) pieces, stored in a refrigerator, and held for 90 min under a heat lamp before being served. Samples of the food contained large populations of *C. perfringens.*

These two outbreaks illustrate the typical involvement of meat and temperature abuse in *C. perfringens* type A foodborne illness. In the Ohio outbreak, the corned beef was cooled too slowly after being cooked and was then reheated to a temperature insufficient to kill *C. perfringens* cells. In the Virginia outbreak, the beef was cooked in overly large portions and not reheated adequately before being served.

Incidence

C. perfringens type A toxicoinfection is the third most common cause of foodborne disease in the United States. From 1993 to 1997, 40 outbreaks (representing 4.1% of total bacterial foodborne disease outbreaks) of *C. perfringens* type A toxicoinfection occurred in the United States. These accounted for 2,772 cases (6.3% of total cases of bacterial foodborne diseases). However, like all foodborne illnesses, most cases of *C. perfringens* type A toxicoinfection are not recognized or reported. Conservative estimates suggest that 250,000 cases of *C. perfringens* type A foodborne illness occur in the United States each year, causing an average of seven deaths. The economic costs associated with *C. perfringens* type A toxicoinfections probably exceed $240 million.

Outbreaks of *C. perfringens* type A toxicoinfection are usually large (~50 to 100 cases) and often occur in institutional settings. The large size of outbreaks is due to two factors. First, large institutions often prepare food in advance and then hold it for later serving. This lets *C. perfringens* grow if the stored food is temperature abused. Second, given the relatively mild and routine symptoms of most *C. perfringens* toxicoinfection cases, public health officials usually become involved only when a lot of people get sick. *C. perfringens* foodborne illness occurs throughout the year, but like most foodborne illness, it is slightly more common during the summer months than at other times. The warmer weather may facilitate temperature abuse of foods during cooling and holding or provide more occasions (like picnics, fairs, and carnivals) on which food is prepared in casual settings where people are having too much fun to worry about hygienic practices.

Food Vehicles for *C. perfringens* Foodborne Illness

The most common vehicles for *C. perfringens* foodborne illness in the United States are meats and poultry. Meat-containing products (e.g., gravies and stews) and Mexican foods are also significant vehicles.

Factors Contributing to *C. perfringens* Type A Foodborne Illness

C. perfringens foodborne illness usually results from temperature abuse during the cooking, cooling, or holding of foods. The Centers for Disease Control and Prevention (CDC) reported that improper storage or holding temperature was a contributor to 100% of recent *C. perfringens* outbreaks for which factors were identified. Improper cooking was a factor in ~30% of these outbreaks. Contaminated equipment contributed to ~15% of the outbreaks.

The importance of temperature abuse in *C. perfringens* foodborne illness is not surprising. *C. perfringens* vegetative cells are relatively heat tolerant. More importantly, *C. perfringens* spores are much more heat resistant than the vegetative cells. Incomplete cooking can induce the germination of *C. perfringens* spores; if the food is improperly cooled or stored, the vegetative cells produced from the germinated spores can rapidly multiply to levels that cause illness.

Preventing *C. perfringens* Type A Foodborne Illness
Thoroughly cooking food is the best approach to preventing *C. perfringens* foodborne illness. This is particularly important for large roasts and turkeys. Because of their size, it is hard to reach the high internal temperatures needed to kill *C. perfringens* spores. The difficulty of cooking large pieces of meat to such high internal temperatures helps explain why the foods are such common vehicles for *C. perfringens* outbreaks. A second, and perhaps even more important, preventive step is to rapidly cool and store cooked foods to temperatures at which *C. perfringens* vegetative cells cannot grow (i.e., below 40 or above 140°F).

*Authors' note*_____
The U.S. Food Code suggests that food be held below 40 or above 140°F to prevent microbial growth.

Identification of *C. perfringens* Type A Foodborne Illness Outbreaks
Public health agencies traditionally use criteria such as incubation time and symptoms of illness or type and history of food vehicles (e.g., is temperature-abused meat or poultry involved?) for identifying *C. perfringens* outbreaks. However, the similarities between the onset times and symptomology of *C. perfringens* foodborne illness and other illnesses, such *Bacillus cereus* toxicoinfection, make it unwise to rely completely on clinical and epidemiologic features for identifying outbreaks. Laboratory analyses, such as the presence of large numbers of spores in the feces or detection of specific toxins, are more reliable in identifying outbreaks of *C. perfringens* illness. Bacteriologic criteria used by the CDC to identify an outbreak include the presence of either 10^5 *C. perfringens* organisms/g of stool from two or more ill persons or 10^5 *C. perfringens* organisms/g of the epidemiologically implicated food. Because *C. perfringens* is widely distributed in the environment, simply finding it in suspect food or feces is not enough to unequivocally identify an outbreak. Most strains of *C. perfringens* in food or feces do not carry the gene for toxin production and are unable to cause illness.

The CDC and the Food and Drug Administration now use the detection of *C. perfringens* enterotoxin in feces of ill individuals to identify *C. perfringens* poisoning outbreaks. Several commercially available serologic assays can detect *C. perfringens* enterotoxin.

Since *C. perfringens* enterotoxin can also be present in feces of people suffering from nonfoodborne gastrointestinal (GI) diseases (such as antibiotic-associated diarrhea), determining the presence of *C. perfringens* enterotoxin in the feces of a single individual is not sufficient to identify *C. perfringens* toxicoinfection. However, detection of *C. perfringens* enterotoxin in feces from several individuals provides strong evidence for an outbreak. This is particularly true when the individuals consumed a common food, developed illness within typical incubation times, and presented with the characteristic symptoms of *C. perfringens* toxicoinfection. The usefulness of fecal *C. perfringens* toxin detection for identifying *C. perfringens* outbreaks is limited, because fecal samples must be collected soon after the onset of symptoms. It is often difficult to recruit donors under these conditions.

CHARACTERISTICS OF *C. PERFRINGENS* TYPE A FOODBORNE ILLNESS

Symptoms of *C. perfringens* type A toxicoinfection develop 8 to 16 h after contaminated food is eaten. They resolve spontaneously within the next 12 to 24 h. Victims of *C. perfringens* toxicoinfection usually suffer only diarrhea and severe abdominal cramps. Vomiting and fever are not typical. While death rates from *C. perfringens* type A toxicoinfection are low, fatalities do occur in debilitated or elderly populations.

The typical pathogenesis of *C. perfringens* type A toxicoinfection is illustrated in Fig. 16.1. Initially, temperature abuse stimulates *C. perfringens* cells

Figure 16.1 Schematic representation of *C. perfringens* food poisoning. **(Top)** Vegetative cells or spores contaminate a meat product and multiply rapidly when food is incubated. **(Middle)** A person consumes the vegetative cells, which then multiply rapidly in the small intestine (producing a lot of gas) and sporulate, releasing the toxin at the same time as the spores. **(Bottom)** The victim is very sick with gas, cramps, and diarrhea but recovers in 24 to 48 h.

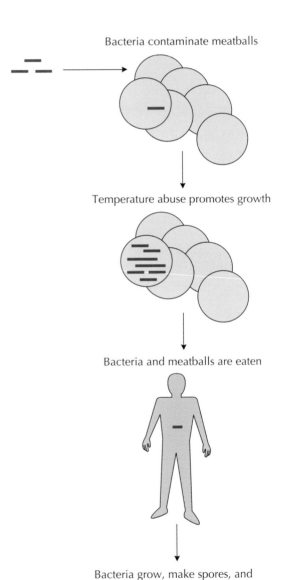

Bacteria contaminate meatballs

Temperature abuse promotes growth

Bacteria and meatballs are eaten

Bacteria grow, make spores, and release toxin, causing cramps and diarrhea

to rapidly multiply in food. Large numbers of the bacteria are consumed. Some vegetative cells survive passage through the stomach and remain viable when they enter the small intestine, where they multiply and sporulate. *C. perfringens* enterotoxin is made during the sporulation of *C. perfringens* cells in the small intestine. After being released in the intestine, *C. perfringens* enterotoxin quickly binds to intestinal epithelial cells and damages them. This *C. perfringens* toxin-induced intestinal damage initiates fluid loss, i.e., diarrhea.

Two factors help to explain why *C. perfringens* toxicoinfection is usually relatively mild and self-limited. First, the diarrhea probably flushes unbound *C. perfringens* toxin-bound cells from the small intestine. Second, *C. perfringens* toxin preferentially affects villus tip cells. These are the oldest intestinal cells and are rapidly replaced in young, healthy individuals by normal cell turnover.

INFECTIOUS DOSE FOR *C. PERFRINGENS* TYPE A FOODBORNE ILLNESS

C. perfringens cells are killed by stomach acidity. Hence, cases of *C. perfringens* toxicoinfection usually develop only when a heavily contaminated food (i.e., a food containing $>10^6$ to 10^7 *C. perfringens* vegetative cells/g of food) is eaten. The *C. perfringens* toxicoinfection toxin is produced in the victim when *C. perfringens* cells sporulate in the intestine. Therefore, this illness is considered a toxicoinfection.

Everyone is susceptible to *C. perfringens* toxicoinfection; however, the illness is more serious in elderly or debilitated individuals. Many people develop at least a transient serum antibody response to *C. perfringens* toxin following illness. There is, however, no evidence that previous exposure provides future protection.

THE ORGANISM

C. perfringens was first recognized as a foodborne pathogen in the 1940s and 1950s. It gradually became apparent that *C. perfringens* causes two quite different human enteric diseases that can be transmitted by food. These are *C. perfringens* toxicoinfection and necrotic enteritis. Since foodborne necrotic enteritis is rare in industrialized societies, it is not discussed here.

General

C. perfringens is a gram-positive, rod-shaped, encapsulated, nonmotile bacterium that causes a broad spectrum of human and veterinary diseases. *C. perfringens* virulence results from the organism's prolific toxin-producing ability, including at least two toxins, i.e., *C. perfringens* enterotoxin and β-toxin, that are active on the human GI tract.

In addition to producing GI tract-active toxins, *C. perfringens* has several other characteristics that contribute to its ability to cause foodborne disease.

- *C. perfringens* grows rapidly, doubling in number in <10 min. This allows the organism to quickly multiply in food.
- *C. perfringens* forms spores (Fig. 16.2) that are resistant to environmental stresses, such as radiation, desiccation, and heat.
- *C. perfringens* spores survive in incompletely cooked, or inadequately reheated, foods.

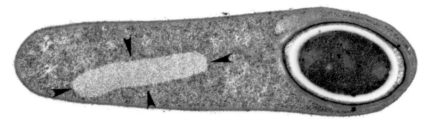

Figure 16.2 Transmission electron micrograph of sporulating *C. perfringens* cell. Note the terminal fully developed spore and the paracrystalline inclusion body (arrowheads). Courtesy of Ron Labbe, University of Massachusetts.

C. perfringens is considered an *anaerobe*, since it does not produce colonies on agar plates exposed to air. However, it tolerates moderate exposure to air. Compared to most other anaerobes, *C. perfringens* requires only relatively modest reductions in oxidation-reduction potential (E_h) for growth.

Classification: Toxin Typing of *C. perfringens*

There are at least 14 different *C. perfringens* toxins. However, no individual *C. perfringens* cell carries the genes encoding all of these toxins. This limitation provides the basis for a toxin-based typing system. This system classifies *C. perfringens* isolates into five types (A through E), depending upon an isolate's ability to make 4 (alpha, beta, epsilon, and iota) of the 14 toxins (Table 16.1). Toxin typing of *C. perfringens* has traditionally involved laborious toxin antiserum neutralization tests in mice, but the development of polymerase chain reaction (PCR)-based schemes permitting toxin genotyping of *C. perfringens* isolates made toxin typing easier.

The two foodborne diseases caused by *C. perfringens* are each associated with a distinct *C. perfringens* type. Necrotic enteritis is caused by type C isolates. The β-toxin produced by those isolates is the primary virulence factor for the disease. As implied by its name, *C. perfringens* type A toxicoinfection is associated with type A isolates of *C. perfringens*, even though other *C. perfringens* types sometimes express a toxin with similar properties. *C. perfringens* enterotoxin-producing type A isolates are more widely distributed than other types of *C. perfringens* and are thus responsible for most of the illness caused by the bacterium.

Table 16.1 Toxin typing of *C. perfringens*[a]

C. perfringens type	Toxin produced[b]			
	Alpha	Beta	Epsilon	Iota
A	+	−	−	−
B	+	+	+	−
C	+	+	−	−
D	+	−	+	−
E	+	−	−	+

[a]Adapted from B. A. McClane, p. 351–372, *in* M. P. Doyle, L. R. Beuchat, and T. J. Montville (ed.), *Food Microbiology: Fundamentals and Frontiers* (ASM Press, Washington, D.C., 2001).
[b]+, produced; −, not produced.

Susceptibility of *C. perfringens* to Preservation Methods

Pathogen growth in food is affected by factors such as temperature, E_h, pH, and water activity (a_w) levels. Therefore, the effects of those factors on *C. perfringens* growth are briefly discussed below.

Temperature

The heat resistance of *C. perfringens* spores allows their survival in incompletely cooked foods. Spore heat resistance is influenced by both environmental and genetic factors. The environment in which a *C. perfringens* spore is heated affects its heat resistance properties. Spores of some *C. perfringens* strains survive boiling for an hour or longer. There are also genetic variations in heat resistance properties. Perhaps due to selective pressure, spores of toxicoinfectious isolates generally have considerably greater heat resistance than spores of other *C. perfringens* isolates. Finally, incomplete cooking of foods not only may fail to kill *C. perfringens* spores in foods but can actually favor development of *C. perfringens* type A toxicoinfection by inducing the germination of *C. perfringens* spores.

Vegetative cells of *C. perfringens* are also somewhat heat tolerant. Although not truly thermophilic, they have a relatively high optimal growth temperature (43 to 45°C) and can often grow at 50°C. Vegetative cells of toxicoinfectious isolates are ~2-fold more heat resistant than vegetative cells of other *C. perfringens* isolates.

C. perfringens cells are not tolerant of refrigeration or freezing. *C. perfringens* growth rates rapidly decrease at temperatures below ~15°C. No growth occurs at 6°C. In contrast, spores of *C. perfringens* are cold resistant. Toxicoinfection can result if viable spores present in refrigerated or frozen foods germinate when that food is warmed for serving.

Other Factors

Growth of *C. perfringens* in food is also affected by the a_w, E_h, pH, and (probably) the presence of curing agents, such as nitrates. *C. perfringens* is less tolerant of low-a_w environments than is *Staphylococcus aureus*. The lowest a_w supporting vegetative growth of *C. perfringens* is 0.93 to 0.97, depending upon the solute used to control the a_w.

C. perfringens does not require an extremely reduced environment for growth. If the environmental E_h is suitably low for initiating growth, *C. perfringens* can then modify the E_h of its environment (by producing reducing molecules, such as ferredoxin) to produce more optimal growth conditions. The E_hs of many common foods (e.g., raw meats and gravies) are low enough to support *C. perfringens* growth.

C. perfringens growth is also pH sensitive. Optimal growth occurs at pH 6 to 7, whereas *C. perfringens* grows poorly, if at all, at pH values of ≤ 5 and ≥ 8.3.

Preservation factors, such as pH, a_w, and perhaps curing agents, control *C. perfringens* by inhibiting the outgrowth of *C. perfringens* spores. However, ungerminated spores may remain viable in foods in the presence of preservation factors that prevent cell growth. Those spores may germinate later if the growth-limiting factor(s) is removed during food preparation.

RESERVOIRS FOR *C. PERFRINGENS* TYPE A

C. perfringens is present throughout the natural environment, including soil (at levels of 10^3 to 10^4 colony-forming units/g), foods (e.g., ~50% of raw or frozen meats contain some *C. perfringens* organisms), dust, and the intestinal tracts of humans and domestic animals (e.g., human feces usually contain 10^4 to 10^6 *C. perfringens* organisms/g). The widespread distribution of *C. perfringens* was originally linked to its frequent occurrence in foodborne illness. However, recent studies have revealed that the ubiquitous distribution of *C. perfringens* in nature is not especially relevant for understanding the reservoir(s) for this foodborne pathogen. This is because <5% of all *C. perfringens* isolates harbor the *cpe* (for *C. perfringens* enterotoxin) gene required for toxin production. Only *C. perfringens* toxin-positive isolates cause foodborne illness. Because most *C. perfringens* isolates are *cpe* negative, it is important to find where the small minority of enterotoxigenic *C. perfringens* isolates exist in nature.

The *cpe* gene can be located on either the chromosome or a plasmid. However, the *cpe* gene in most toxicoinfectious isolates is located on the chromosome. This strong association between foodborne illness and isolates that possess a chromosomal *cpe* gene provides another criterion for identifying the locations of toxicoinfectious isolates in nature. By learning the location of the *cpe* gene in environmental isolates, microbiologists will be able to answer critical questions about the ecology of *C. perfringens* toxicoinfectious isolates. The questions to be answered include the following. Are the isolates present in low numbers in some healthy human carriers? Are they present in some food animals? Do they enter foods during processing? Do they enter foods during final handling, cooking, or holding?

VIRULENCE FACTORS CONTRIBUTING TO *C. PERFRINGENS* TYPE A FOODBORNE ILLNESS

Heat Resistance

Most cases of *C. perfringens* toxicoinfections are caused by isolates carrying a chromosomal *cpe* gene. This association between chromosomal-*cpe*-carrying isolates and toxicoinfections was initially puzzling, because *C. perfringens* isolates carrying chromosomal or plasmid *cpe* genes both make similar levels of identical toxins. The association between chromosomal *cpe* and toxicoinfections may be caused by the cells with chromosomal *cpe* genes having greater heat resistance than the cells and spores of isolates with plasmid-borne *cpe* genes. Since cooked meats cause most *C. perfringens* toxicoinfection outbreaks, the greater heat resistance of isolates with chromosomal *cpe* genes favors their survival during incomplete cooking.

C. perfringens Enterotoxin

Evidence that *C. perfringens* Enterotoxin Is Involved in Foodborne Illness

C. perfringens toxin is classified as an enterotoxin because it induces fluid and electrolyte losses from the intestinal tracts of mammals. There is strong epidemiological evidence that *C. perfringens* enterotoxin causes *C. perfringens* type A foodborne illness.

1. The presence of *C. perfringens* toxin in a victim's feces is strongly correlated with illness.

2. Toxin concentration in the feces causes serious intestinal effects in experimental animals.
3. Human volunteers fed purified *C. perfringens* toxin get sick and show the symptoms of *C. perfringens* foodborne illness.
4. The intestinal inflammation caused by *C. perfringens* toxin-positive isolates in experimental animals can be neutralized with *C. perfringens* toxin-specific antisera.

More recently, the importance of enterotoxin for the pathogenesis of *C. perfringens* toxicoinfectious isolates received compelling support from experiments that fulfilled Koch's postulates on a molecular basis. Those experiments revealed that sporulating (but not vegetative) culture lysates of a *C. perfringens* cpe+ strain induced fluid accumulation in experimental animals. This is consistent with *C. perfringens* toxin (whose expression is sporulation associated) being necessary for the activity of the toxicoinfectious isolate. Furthermore, neither vegetative nor sporulating culture lysates of a *cpe*-negative mutant made animals sick. The *cpe*-negative mutant's loss of virulence could be attributed to its lack of the *cpe* gene. Full virulence was restored when the isolate had the *cpe* gene reintroduced.

C. perfringens type A toxicoinfection is not the only disease involving *C. perfringens* toxin. *C. perfringens* toxin-producing isolates are also responsible for several nonfoodborne human GI illnesses. These include antibiotic-associated diarrhea, sporadic diarrhea, and some veterinary diarrheas. The *cpe*-positive *C. perfringens* isolates causing nonfoodborne GI diseases are genetically distinct from those causing toxicoinfections. That is, the *cpe* gene is located on the chromosomes of toxicoinfectious isolates but on a plasmid in nonfoodborne-disease-producing isolates.

Expression and Release of *C. perfringens* Enterotoxin
There are three interesting features of *C. perfringens* enterotoxin expression and release:

1. *C. perfringens* toxin is made by sporulating vegetative cells.
2. Many *C. perfringens* toxin-positive isolates make extremely large amounts of toxin.
3. *C. perfringens* toxin is not actually *secreted* by sporulating *C. perfringens* cells but is released when the mother cell lyses.

Synthesis of *C. perfringens* enterotoxin. Synthesis of *C. perfringens* toxin begins when cells sporulate and increases for the next 6 to 8 h. After 6 to 8 h of sporulation, *C. perfringens* toxin can represent up to 30% of the total cell protein. Why do some *C. perfringens* strains produce so much toxin during sporulation? The amount of *C. perfringens* toxin expressed by an isolate is not influenced by the location of the *cpe* gene, nor can the amount of toxin made be attributed to a gene dosage effect; all *cpe*-positive isolates carry only a single copy of the *cpe* gene. There is a general relationship between an isolate's sporulation ability and its toxin production, i.e., the better a *C. perfringens* isolate sporulates, the more *C. perfringens* toxin is produced. However, this correlation is not absolute.

Release of *C. perfringens* enterotoxin from *C. perfringens*. *C. perfringens* toxin accumulates in cells during its synthesis, often reaching high

Authors' note

Cells can have multiple copies of a given gene. One would expect that a higher gene copy number would result in a larger amount of the gene product (i.e., toxin). This is not always true.

enough concentrations to form paracrystalline inclusion bodies. Intracellular *C. perfringens* toxin is released into the intestines at the completion of sporulation, when the mother cell lyses to free its mature spore. The need for the mother cell to lyse and release the *C. perfringens* toxin explains, at least in part, why (despite *C. perfringens* toxin's quick intestinal action) symptoms develop only 8 to 24 h after ingestion. *C. perfringens* cells must complete sporulation before the toxin can be released into the intestine. This takes 8 to 12 h.

C. perfringens Enterotoxin Biochemistry
C. perfringens toxin is heat sensitive. Its biologic activity can be inactivated by heating it for 5 min at 60°C. The toxin is also quite sensitive to pH extremes but is resistant to some proteolytic treatments. In fact, limited treatment with the proteolytic enzyme trypsin or chymotrypsin increases *C. perfringens* toxin activity two- to threefold. This suggests that human intestinal proteases may activate *C. perfringens* toxin.

C. perfringens Enterotoxin Effects on the Gastrointestinal Tract
The principal target for *C. perfringens* enterotoxin is the small intestine. Several features differentiate the biological activity of *C. perfringens* toxin from those of cholera and *Escherichia coli* heat-labile toxins. *C. perfringens* toxin inhibits glucose absorption. *C. perfringens* toxin induces direct cellular damage to the small intestine. The villus tips are particularly sensitive. While other bacterial toxins (e.g., Shiga toxin and *Clostridium difficile* toxins) are also cytotoxic and cause intestinal-tissue damage, *C. perfringens* toxin is unique with respect to how quickly it damages intestinal tissue. *C. perfringens* toxin-induced intestinal damage can develop in 15 to 30 min.

Summary

- *C. perfringens* is an anaerobic spore-forming bacterium commonly found on meat and meat products.
- The toxicoinfection toxin is released as a paracrystalline protein when vegetative cells release their spores in the gut.
- Improper heating or cooling is the major cause of *C. perfringens* type A toxicoinfection.
- The *cpe* gene encodes the enteroxin; it is chromosomally located in toxicoinfectious isolates but is on the plasmids of isolates not associated with foods.
- Victims of *C. perfringens* type A toxicoinfections usually suffer only diarrhea and severe abdominal cramps.

Suggested reading

Labbe, R. G. 1989. *Clostridium perfringens*, p. 192–234. *In* M. P. Doyle (ed.), *Microbial Foodborne Pathogens*. Marcel Dekker, Inc., New York, N.Y.

McClane, B. A. 2001. *Clostridium perfringens*, p. 351–372. *In* M. P. Doyle, L. R. Beuchat, and T. J. Montville (ed.), *Food Microbiology: Fundamentals and Frontiers*, 2nd ed. ASM Press, Washington, D.C.

U.S. Food and Drug Administration. 2002. Foodborne pathogenic microorganisms and natural toxins. http://vm.cfsan.fda.gov/~mow/intro.html.

Questions for critical thought

1. Create an imaginary *C. perfringens* outbreak. Describe how the food was prepared and handled and steps that led to the outbreak. Also include the victims' symptoms.

2. Would you rather be a spore or a vegetative cell? Why?

3. You are an eminent professor of food microbiology, awaiting delivery of food for your son's graduation party. The caterer arrives with 10 trays of Swedish meatballs, lasagna, and chicken parmigiana. She assures you that "they are good and cold; just heat them up for a couple of hours on the steam tables." What do you do?

4. What is the difference between a case and an outbreak? Calculate the number of cases per outbreak for *C. perfringens* and some other foodborne pathogens. What can you infer from these numbers?

5. What is the biggest single causative factor in the occurrence of *C. perfringens* type A toxicoinfections?

6. What characteristics of *C. perfringens* contribute to its ability as a foodborne pathogen?

7. Both *C. perfringens* and *Clostridium botulinum* foodborne diseases are linked to spores. How are these diseases different? How are they similar? Why are their frequencies so different?

8. When foods are heated from 40 to 140°F or cooled from 140 to 40°F, it is important to get through this interval as rapidly as possible to prevent microbial growth. Rapid heating is easily achieved by obvious means. However, improper cooling causes as many outbreaks as improper heating and is much harder to control. Think of three ways to cool foods through the danger zone rapidly.

9. What is a paracrystalline inclusion body?

10. Here are several facts. Multiple copies of the *cpe* gene do not result in increased pathogenicity. Isolates from foodborne disease usually have a *cpe* gene on the chromosome rather than the plasmid. Spores from isolates with a chromosomal *cpe* gene are more heat resistant than those from isolates that have *cpe* on the chromosome. How do you relate these facts to the role of *cpe* in *C. perfringens* foodborne illness? What experiments could you do to refine this understanding?

17

Bacillus cereus*

LEARNING OBJECTIVES

The information in this chapter should enable the student to:
- use basic biochemical characteristics to identify *Bacillus cereus*
- understand what conditions in foods favor the growth of *B. cereus*
- recognize, from symptoms and time of onset, a case of foodborne illness caused by *B. cereus*
- choose appropriate interventions (heat, preservatives, and formulation) to prevent the growth of *B. cereus*
- identify environmental sources of *B. cereus*
- understand the roles of spores and toxins in causing foodborne illness

Outbreak

Even though fried rice is the classical example used to illustrate *B. cereus* food poisoning, it still causes problems. *Morbidity and Mortality Weekly Report* (the Centers for Disease Control and Prevention's weekly publication that lists how many people are getting sick or dying of various diseases) summarized an outbreak of *B. cereus* food poisoning that occurred at two day care centers in Virginia. Eighty-two children ate a special catered lunch. Eighty were interviewed, and 67 children who ate the lunch were identified. None of the 13 people who did not eat the lunch got sick. Fourteen of the people who ate the lunch became ill with nausea, abdominal cramps, and diarrhea. The median time of appearance of illness was 2 h after eating. The median time for recovery was 4 h after onset.

Chicken fried rice was associated with the outbreak. One-third of the people who ate the rice became ill. None of the people who did not eat the rice became ill. *B. cereus* at >10^5 organisms per g was isolated from the vomit of one child. It was isolated from the leftover chicken at >10^6 organisms per g.

The investigation revealed a classic pattern of food mishandling. The rice was cooked the night before and left to cool at room temperature before being refrigerated. The next morning, the rice and chicken were fried with oil, delivered to the day care center, held without refrigeration, and served without heating. When questioned, neither the restaurant staff nor day care center staff knew that this was a dangerous food-handling practice.

*This chapter was originally written by Per Einar Granum for *Food Microbiology: Fundamentals and Frontiers,* 2nd ed., and has been adapted for use in an introductory text.

INTRODUCTION

With "only" ~20,000 cases of illness per year in the United States, *B. cereus* is considered a minor foodborne pathogen. Britons, however, take it quite seriously. *B. cereus* produces two toxins, an emetic (vomit-inducing) toxin that can act rapidly (0.5 to 6 h) and a slower-acting (6 to 14 h) diarrheal type that causes diarrhea, cramps, and rectal tenesmus. The diarrheal illness is caused by an enterotoxin(s) produced during the vegetative growth of *B. cereus* in the small intestine. The emetic toxin is produced by cells growing in the food. Both types of illness are mild and brief and therefore underreported. In both cases, the food is usually cooked. This kills vegetative cells, but spores survive and become the source of the food poisoning. *B. cereus* is not a competitive microorganism but grows well after cooking and cooling of food. Outbreaks are associated with meats, gravies, fried rice, pasta, sauces, puddings, and dairy products.

CHARACTERISTICS OF THE ORGANISM

Genus and species designations were originally based on the basic physiology of the organism (as determined by biochemical tests) and what the organism looked like (under the microscope). Thus, *any* spore-forming, aerobic, rod-shaped bacterium is classified as a *Bacillus* species. The advent of molecular biology and genetic characterization has revealed that many of the phenotypic species-level identifications are inaccurate. The genus *Bacillus* as described in *Bergey's Manual of Systematic Bacteriology* (the dictionary of microbial classification) is too diverse for a single genus. An analysis of 16S ribosomal RNA (rRNA) sequences revealed at least five genera, or rRNA groups, within the genus *Bacillus.* With the subsequent identification of many new species, the number of genera has increased to approximately 16.

Bacillus anthracis, B. cereus, B. mycoides, B. thuringiensis, B. pseudomycoides, and *B. weihenstephanensis* are members of the *B. cereus* group. These bacteria have highly similar 16S and 23S rRNA sequences, indicating that they have diverged from a common evolutionary line. *B. anthracis* is the most distinctive member of the group, both in its highly virulent pathogenicity and in its notoriety as an agent for biological warfare. In addition, extensive studies of DNA from strains of *B. cereus* and *B. thuringiensis* have revealed that there is no scientific reason to give them separate species status. Indeed, *B. thuringiensis* strains used as natural bioinsecticides have caused cases of "*B. cereus*" food poisoning. Nevertheless, the name *B. thuringiensis* is retained for those strains that make a crystalline inclusion protein that is toxic to insects.

ENVIRONMENTAL SOURCES

B. cereus is widespread in nature and is frequently isolated from soil and growing plants. From this environment, it is easily spread to foods by cross contamination. *B. cereus* is spread from soil and grass to cows' udders and into raw milk. *B. cereus* spores survive milk pasteurization, and after germination, free from competition from other vegetative cells, they cause problems in many milk products. Most *B. cereus* strains are unable to grow below 10°C or in milk products stored between 4 and 8°C. However, psychotolerant strains grow at temperatures as low as 4 to 6°C.

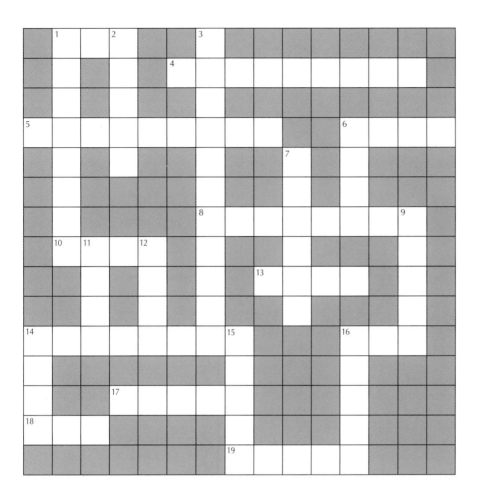

Across
1. Not good
4. *Bacillus* species used in terrorist attack
5. One of two *B. cereus* toxin types
6. *B. cereus* cells _____ in the small intestine
8. Rectal analog of "dry heaves"
10. Another source of spores; don't call it "dirt" at an agricultural college
13. Asian food source of spores
14. Order of magnitude for yearly number of *B. cereus* cases in the United States
16. Something you see with
17. Italian food source of spores
18. Baked dessert not usually associated with *B. cereus*
19. *B. cereus* rarely_____ people

Down
1. Most aerobic spore-forming rods belong to this genus
2. _____ products are frequently associated with *B. cereus*
3. Chemical that causes illness in the gut
6. Can be positive, negative, or a unit of mass
7. Vomit-inducing
9. Resistant life form made by *Bacillus* species
11. _____ man Empire
12. _____ of spores are required for illness to occur
14. Poor control of this extrinsic factor contributes to *B. cereus* outbreaks (abbreviation)
15. *B. cereus* was confirmed as a foodborne pathogen by scientists who_____ it
16. Answer to 9 Down spelled backwards

FOODBORNE OUTBREAKS

The number of outbreaks of *B. cereus* foodborne illness is highly underestimated. This can be attributed to the relatively short duration of the illness (usually <24 h), the similarity of symptoms to those of illnesses caused by other foodborne pathogens, and the likelihood that illness involving

B. cereus-contaminated milk is limited to one or two cases and is not identified as an outbreak. Toward the end of milk's shelf life, there frequently are enough *B. cereus* cells to cause illness. Fortunately, *B. cereus* proteases cause off flavors. One would hope that this spoilage would keep consumers from drinking highly contaminated milk.

It is not possible to compare the incidences of outbreaks in different countries because their methods of surveillance of foodborne illnesses differ greatly. The percentages of outbreaks and cases attributed to *B. cereus* in Japan, North America, and Europe vary from 1 to 47% for outbreaks and from ~0.7 to 33% for cases. The greatest numbers of reported *B. cereus* outbreaks and cases are from Iceland, The Netherlands, and Norway. In the Netherlands in 1991, *B. cereus* caused 27% of foodborne disease outbreaks where a causative agent was identified. However, the actual incidence of *B. cereus* illness was only 2.8% of the total number of cases because most cases of foodborne illness were of unknown etiology.

CHARACTERISTICS OF DISEASE

There are two types of *B. cereus* foodborne illness. The first type, caused by an emetic toxin, results in vomiting. The second type, caused by an enterotoxin(s), results in diarrhea. In a small number of cases, both toxins are produced and both types of symptoms occur. It appears that the incubation time (>6 h; average, 12 h) is too long for diarrheal illness to be caused by toxin premade in the food. However, enterotoxin(s) can be preformed in food when the *B. cereus* population is at least 100-fold higher than that necessary for causing food poisoning. Products with such large populations of *B. cereus* are no longer acceptable to the consumer, although food containing >10^7 *B. cereus* organisms/ml may not always appear spoiled. The two types of *B. cereus* foodborne illness are characterized in Table 17.1.

DOSE

Authors' note

Although the use of oneself as a human subject was accepted, and even considered courageous, in the early days of microbiology, this practice would be unethical by current standards.

After the first diarrheal outbreak of *B. cereus* foodborne illness in Oslo, Norway (from vanilla sauce), Professor S. Hauge isolated the causative agent, grew it to 4×10^6 cells/ml, and drank 200 ml of the culture. About 13 h later he developed abdominal pain and watery diarrhea that lasted for ~8 h. Counts of *B. cereus* ranging from 200 to 10^9/g (or ml) have been reported in foods implicated in outbreaks, indicating that the total dose (number of

Table 17.1 Characteristics of two types of illness caused by *B. cereus*[a]

Characteristic	Diarrheal syndrome	Emetic syndrome
Dose causing illness	10^5–10^7 (total cells)	10^5–10^8 (cells/g)
Toxin production	In small intestine of host	Preformed in foods
Type of toxin	Protein; enterotoxin(s)	Cyclic peptide; emetic toxin
Incubation period	8–16 h (occasionally >24 h)	0.5 to 5 h
Duration of illness	12–24 h (occasionally several days)	6–24 h
Symptoms	Abdominal pain, watery diarrhea, occasionally nausea	Nausea, vomiting, malaise (sometimes followed by diarrhea, due to production of enterotoxin)
Foods most frequently implicated	Meat products, soups, vegetables, puddings, sauces, milk and milk products	Fried and cooked rice, pasta, pastry, noodles

[a]Reprinted from P. E. Granum, p. 373–381, *in* M. P. Doyle, L. R. Beuchat, and T. J. Montville (ed.), *Food Microbiology: Fundamentals and Frontiers* (ASM Press, Washington, D.C., 2001).

cells per milliliter × number of milliliters consumed) ranges from ~5×10^4 to 1×10^{11}. The total number of *B. cereus* cells that must be present to produce enough toxin to produce illness is probably 10^5 to 10^8. Food containing >10^3 *B. cereus* cells/g cannot be considered safe for consumption.

VIRULENCE FACTORS AND MECHANISMS OF PATHOGENICITY

Very different types of toxins cause the two types of *B. cereus* foodborne illness. The emetic toxin, which causes vomiting, has been isolated and characterized, whereas the diarrheal disease is caused by one or more enterotoxins.

The Emetic Toxin

The emetic toxin (Table 17.2) causes emesis (vomiting) only. Its structure was once a mystery, because the only detection system involved living primates. However, the recent discovery that Hep-2 cells could detect the toxin has led to its isolation and structure determination. The emetic toxin has been named cereulide and consists of a ring structure of three repeats of four amino and/or oxy acids: (D-*O*-Leu-D-Ala-L-*O*-Val-L-Val)₃. The structure suggests that cereulide is an enzymatically synthesized peptide and not a gene product. The biosynthetic pathway and mechanism of action of the emetic toxin are unknown. The emetic toxin is resistant to heat, acid, and proteolysis. It is not antigenic.

Enterotoxins

Cloning and sequencing studies show that *B. cereus* produces at least five different proteins (or protein complexes) referred to as enterotoxins. The extents to which the different proteins are involved in foodborne illness can be determined through microslide diffusion tests (Fig. 17.1).

Three of the enterotoxins are probably involved in *B. cereus* foodborne illness. Two of the enterotoxins are multicomponent and related, whereas the third is a single protein of 34 kilodaltons. A three-component hemolysin (Hbl) with enterotoxin activity has dermonecrotic and vascular permeability activities and causes fluid accumulation in ligated rabbit ileal loops. It may be the virulence factor in *B. cereus* diarrhea. All three proteins of Hbl are transcribed from one operon (*hbl*).

Table 17.2 Properties of the emetic toxin cereulide[a]

Trait	Property, activity, or value
Molecular mass	1.2 kilodaltons
Structure	Ring-shaped peptide
Isoelectric point	Uncharged
Antigenic	No (?)
Biological activity in living primates	Vomiting
Cytotoxic	No
Heat stability	90 min at 121°C
pH stability	Stable at pH 2–11
Effect of proteolysis (trypsin, pepsin)	None
Conditions under which toxin is produced	In food: rice and milk at 25–32°C

[a]Adapted from P. E. Granum, p. 373–381, *in* M. P. Doyle, L. R. Beuchat, and T. J. Montville (ed.), *Food Microbiology: Fundamentals and Frontiers* (ASM Press, Washington, D.C., 2001).

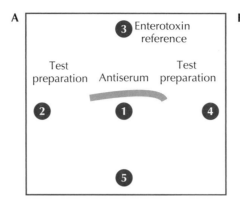
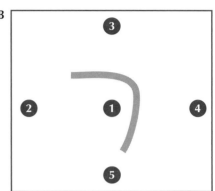
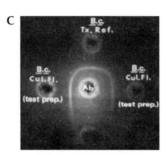

Figure 17.1 Microslide diffusion tests are used to detect *B. cereus* diarrheagenic and staphylococcal enterotoxins. Antiserum to the toxin is put in well 1, and reference toxin (a positive control) is put in well 3. Test samples are put in wells 2, 4, and 5. When the antiserum and toxin diffuse from their respective wells and meet, a precipitation line is formed. **(A)** Test where the food samples are negative but a line is formed with the known enterotoxin. **(B)** Test where the sample in well 4 contains enterotoxin. **(C)** Photograph of actual results, where samples in wells 2 and 4 (but not 5) are positive for toxin. Photo courtesy of R. Bennett, U.S. Food and Drug Administration; diagrams based on those by R. Bennett.

The Spore

The *B. cereus* spore is important in foodborne illness. It is more hydrophobic than any other *Bacillus* sp. spore. This enables it to adhere to several types of surfaces. It is difficult to remove during cleaning and is a difficult target for disinfectants. *B. cereus* spores also contain appendages and/or pili that are involved in adhesion. Not only can these properties enable the spores to withstand sanitation and remain on surfaces to contaminate foods, they also aid in adherence to epithelial cells.

Summary

- *B. cereus* is a sporeformer.
- It is a normal inhabitant of soil and is isolated from a variety of foods.
- Two different toxins cause an emetic or a diarrheal type of food-associated illness.
- Desserts, meat dishes, and dairy products are frequently associated with diarrheal illness.
- Rice and pasta are the most common vehicles of emetic illness.
- Some strains of the *B. cereus* group grow at refrigeration temperature.
- *B. cereus* foodborne illness is probably highly underreported.

Suggested reading

Granum, P. E. 2001. *Bacillus cereus*, p. 373–381. *In* M. P. Doyle, L. R. Beuchat, and T. J. Montville (ed.), *Food Microbiology: Fundamentals and Frontiers,* 2nd ed. ASM Press, Washington, D.C.

Questions for critical thought

1. Speculate as to why *B. cereus* is considered a larger problem in Britain than in the United States.
2. Some people have suggested that *B. cereus* be used as a surrogate to research the deadly bioagent *B. anthracis*. What are the pros and cons of this proposal?
3. Why are *B. cereus* foodborne illnesses more likely to be underreported than other foodborne illnesses?
4. Create a case study of a *B. cereus* foodborne outbreak.
5. What is rectal tenesmus?

SECTION *IV* Other Microbes Important in Food

18

Fermentative Organisms*

LEARNING OBJECTIVES

The information in this chapter will help the student to:
- understand the biochemical basis of food fermentation
- appreciate the similarities and differences among fermentations of dairy, vegetable, and meat products
- relate specific bacteria to specific fermentations
- outline the process for making fermented foods
- understand the benefits of using fermentation as a food-processing method

INTRODUCTION

Fermentation is a word with many meanings. Louis Pasteur used it to describe life in the absence of oxygen. A strict biochemical definition of fermentation is the process that bacteria use to make energy from carbohydrates in the absence of oxygen. However, fermentation also describes any biological process (for example, fermentations that make vinegar, antibiotics, monosodium glutamate, amino acids, and citric acid), whether oxygen is present or not. With a few exceptions, *food fermentations* are bioprocesses that change the properties of a food while the bacteria generate energy in the absence of oxygen. These changes go far beyond acid production. Fermentations produce flavor compounds and carbonation, alter texture and nutrient bioavailability, can detoxify, and add value to foods that are very different from their starting commodity.

This chapter simplifies fermentation biochemistry for students who have not had a biochemistry course. These biochemical principles are applied to the fermentation of dairy, vegetable, and meat products. The use of yeast to make wine, bread, and beer is explained. Finally, the use of fermentation to preserve food in other societies is examined.

THE BIOCHEMICAL FOUNDATION OF FOOD FERMENTATION

Energy is released when a compound is oxidized (i.e., an electron is lost; think back to Chemistry 101). While fermentation products such as alcohol are more oxidized than the starting product (usually a sugar), they are not

*This chapter is condensed and integrated from the seven chapters in the Food Fermentations section of *Food Microbiology: Fundamentals and Frontiers*, 2nd ed. They were written by Mark E. Johnson, James L. Steele, Herbert J. Buckenhüskes, Steven C. Ricke, Irene Zabala Díaz, Jimmy T. Keeton, Larry R. Beuchat, Sterling S. Thompson, Kenneth B. Miller, Alex S. Lopez, Iain Campbell, and Graham H. Fleet and edited by Larry R. Beuchat.

completely oxidized. For example, more energy can be released from alcohol by burning it. When one compound is oxidized and loses an electron, that electron must go somewhere. It goes to an electron acceptor, which becomes reduced. This can be expressed as follows:

Reduced electron donor (compound A) → Oxidized compound A
Oxidized electron acceptor (compound B) → Reduced compound B

When oxygen is the electron acceptor in the oxidation of sugars:

$$\frac{\begin{array}{l} \text{Sugar } (C_6H_{12}O_6) \rightarrow CO_2 \\ O_2 \rightarrow H_2O \end{array}}{\text{Sugar} + O_2 \rightarrow CO_2 + H_2O \ (+ \ \text{energy})}$$

This complete oxidation gives the cell a lot of energy. The energy is stored in the form of phosphorylated compounds, such as adenosine triphosphate (ATP). The energy can be transferred to other compounds by moving the phosphoryl group. In the example above, the complete oxidation of glucose produces 34 ATP molecules.

In fermented foods, there is no oxygen to serve as an electron acceptor, so part of the sugar serves as the electron acceptor. The end products are incompletely oxidized compounds like ethanol, acetic acid, and lactic acid. This incomplete oxidation does not yield very much energy (only one or two molecules of ATP from one molecule of glucose), but in the land of the blind, the one-eyed man is king.

Catabolic Pathways

Figure 2.5 shows the composite, step-by-step, biochemical pathways used by fermentative bacteria. The more generalized Fig. 18.1 allows the pathways to be understood with less biochemical detail. The Embden-Meyerhof-Parnas (EMP) pathway from glucose to pyruvic acid may be the most important catabolic pathway. One molecule of glucose is converted to two molecules of pyruvic acid. At this point, the oxidation-reduction reactions are not balanced. They can be balanced by oxidizing pyruvate to compounds like lactic acid or ethanol. Nonfermentative organisms (like us) completely oxidize the pyruvic acid to carbon dioxide and water, using oxygen as the terminal elec-

Figure 18.1 Simplified catabolic pathways used in fermented foods.

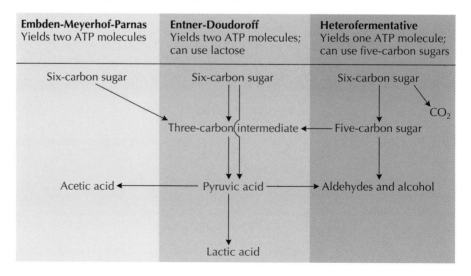

tron acceptor. In terms of molecules of ATP generated per mole of glucose used (i.e., two), the EMP pathway is the best of the fermentative pathways. Since its only product is lactic acid, organisms that use it are called *homolactic.*

The Entner-Doudoroff and heterofermentative pathways yield only 1 ATP molecule per mol of glucose. The Entner-Doudoroff pathway is important in dairy fermentations. Lactose (made up of two sugars, glucose and galactose), or *milk sugar,* is broken down to glucose and galactose during fermentation. However, the galactose portion of lactose cannot be used by the EMP pathway. Organisms that use the EMP pathway excrete galactose, wasting half of their energy source. Organisms using the Entner-Doudoroff pathway use this wasted galactose but produce only one three-carbon intermediate and therefore one ATP molecule.

The heterofermentative pathway releases a carbon dioxide molecule. It can make only one three-carbon intermediate and one ATP molecule. However, it allows organisms to use five-carbon sugars (*pentoses*) that other organisms cannot use. Thus, in environments where there are a lot of pentoses but no *hexoses* (six-carbon sugars that can yield two ATP molecules), heterofermentative organisms have a big advantage.

Which pathway an organism uses is determined by many things. These include the genes for specific enzymes, their regulation, and what sugars are available. Bacteria in the genera *Lactococcus* and *Pediococcus* are homofermentative, producing only lactic acid. Some *Lactobacillus* species are homofermentative, some are heterofermentative, and some can use both pathways.

DAIRY FERMENTATIONS

Many microbes are used to make fermented milk products (Table 18.1). The main bacteria used for acid production are the homofermentative lactic acid bacteria (LAB). Heterofermentative LAB contribute flavor compounds.

Table 18.1 Microorganisms involved in the manufacture of cheeses and fermented milks[a]

Product	Principal acid producer	Intentionally introduced secondary microbiota
Cheeses		
Colby, Cheddar, cottage, cream	*Lactococcus lactis* subsp. *cremoris* or *lactis*	None
Gouda, Edam, Havarti	*Lactococcus lactis* subsp. *cremoris* or *lactis*	*Leuconostoc* spp., *Lactococcus lactis* subsp. *lactis*
Brick, Limburger	*Lactococcus lactis* subsp. *cremoris* or *lactis*	*Geotrichum candidum, Brevibacterium linens, Micrococcus* spp.
Camembert	*Lactococcus lactis* subsp. *cremoris* or *lactis*	*Penicillium camemberti,* sometimes *B. linens*
Blue	*Lactococcus lactis* subsp. *cremoris* or *lactis*	*L. lactis* subsp. *lactis, Penicillium roqueforti*
Mozzarella, provolone, Romano, Parmesan	*Streptococcus thermophilus, Lactobacillus delbrueckii* subsp. *bulgaricus, Lactobacillus helveticus*	None; animal lipases added to Romano for piquant or rancid flavor
Swiss	*S. thermophilus, Lactobacillus helveticus, Lactobacillus delbrueckii* subsp. *bulgaricus*	*Propionibacterium freudenreichii* subsp. *shermanii*
Fermented milks		
Yogurt	*S. thermophilus, Lactobacillus delbrueckii* subsp. *bulgaricus*	None
Buttermilk	*Lactococcus lactis* subsp. *cremoris* or *lactis*	*Leuconostoc* spp., *Lactococcus lactis* subsp. *lactis*
Sour cream	*Lactococcus lactis* subsp. *cremoris* or *lactis*	None

[a]Adapted from M. E. Johnson and J. L. Steele, p. 651–664, *in* M. P. Doyle, L. R. Beuchat, and T. J. Montville (ed.), *Food Microbiology: Fundamentals and Frontiers* (ASM Press, Washington, D.C., 2001).

Because pasteurization kills most of the natural LAB in raw milk, manufacturers can control the rate and extent of acid development by adding specific bacteria to the milk. This gives greater process control and a more consistent product. The LAB added to start the manufacturing process are called *starter cultures.*

Other bacteria, referred to as *secondary microbiota,* are added to some fermented products to influence flavor and texture. *Leuconostoc* species and strains of *Lactococcus lactis* subsp. *lactis* that metabolize citric acid produce aroma compounds and carbon dioxide in cultured buttermilk and cheeses, such as Gouda, Edam, blue, and Havarti. Heterofermentative lactobacilli (*Lactobacillus brevis, Lactobacillus fermentum,* and *Lactobacillus kefiri*) are part of the varied microbiotas (including several yeast species) that produce ethanol, carbon dioxide, and lactic acid in more exotic cultured milks, such as kefir and koumiss. *Propionibacterium freudenreichii* subsp. *shermanii* is added to Swiss-type cheeses. Its scavenging metabolism converts L-lactic acid produced by LAB to propionic acid, acetic acid, and carbon dioxide. The carbon dioxide forms the holes (eyes) in Swiss cheese. Propionibacteria also ferment citric acid to glutamic acid, a natural flavor enhancer.

Fermented dairy products are not manufactured in a sterile environment. A generic scheme for making cheese is shown in Fig. 18.2. Environmental lactobacilli contribute to flavor development in aged cheeses. Rapid

Figure 18.2 Generic scheme for making fermented dairy products. U.S. sales of cheese are $1.5 billion per year.

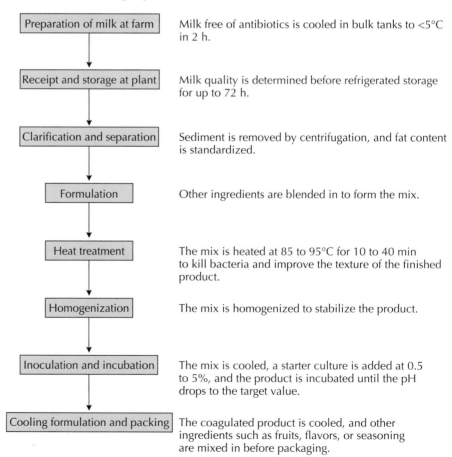

Preparation of milk at farm	Milk free of antibiotics is cooled in bulk tanks to <5°C in 2 h.
Receipt and storage at plant	Milk quality is determined before refrigerated storage for up to 72 h.
Clarification and separation	Sediment is removed by centrifugation, and fat content is standardized.
Formulation	Other ingredients are blended in to form the mix.
Heat treatment	The mix is heated at 85 to 95°C for 10 to 40 min to kill bacteria and improve the texture of the finished product.
Homogenization	The mix is homogenized to stabilize the product.
Inoculation and incubation	The mix is cooled, a starter culture is added at 0.5 to 5%, and the product is incubated until the pH drops to the target value.
Cooling formulation and packing	The coagulated product is cooled, and other ingredients such as fruits, flavors, or seasoning are mixed in before packaging.

(within 4 to 8 h) acid development by starter cultures lowers the pH to <5.3 in cheese and <4.6 in fermented milk products. At this pH, only acid-tolerant bacteria can grow. However, poor acidification allows contaminants to grow.

Starter Cultures

The key to the commercial success of fermented milk products is consistent and predictable rates of acid production. This is guaranteed by adding defined bacterial cultures, i.e., starter cultures, to initiate the fermentation. Acidity has profound effects on moisture control, protein retention, loss of minerals, hydration of proteins, and interactions between protein molecules. These, in turn, determine the product's sensory characteristics.

The flavor of cultured milk products is determined by the microorganisms used as starter cultures and the secondary microbiota. With cultured milks and some cheeses, such as mozzarella, cream, and cottage cheeses, the short time from processing to consumption (1 day to 4 weeks) precludes the development of "ripened" flavors by secondary organisms.

Production of Aroma Compounds

Lactic acid is the main end product in dairy fermentations. It contributes the acid taste but not the aroma. The main aromas and flavor of fermented milk products come from acetic acid, acetaldehyde, and diacetyl. In yogurt, these are formed by *Streptococcus thermophilus* and *Lactobacillus delbrueckii* subsp. *bulgaricus*. *Leuconostoc* species and citrate-utilizing *Lactococcus lactis* subsp. *lactis* are added to produce aroma compounds in buttermilk and some cheeses.

Milk contains 0.15 to 0.2% citric acid. The citric acid is metabolized by *Leuconostoc* species, citrate-metabolizing *Lactococcus lactis* subsp. *lactis*, and facultative heterofermentative lactobacilli to make diacetyl, acetic acid, and carbon dioxide. The carbon dioxide causes the holes in Gouda and Edam cheeses and the effervescent quality of buttermilk.

Several metabolic pathways form acetaldehyde. The enzyme threonine aldolase cleaves threonine to glycine and acetaldehyde. Acetaldehyde is also formed by citric acid-metabolizing *Lactococcus lactis* subsp. *lactis*. A yogurt or green-apple flavor defect is generated in fermented milks when the ratio of diacetyl to acetaldehyde is high. Excessive acetaldehyde in yogurt is caused by prolonged fermentation and is associated with high acid content. The use of *Leuconostoc* species that convert acetaldehyde to ethanol prevents excessive acetaldehyde levels. Off-flavor production is also prevented by rapid cooling and refrigerated storage.

Proteolysis, the enzymatic breakdown of milk proteins, generates flavors in ripened cheeses. The free amino acids and peptides produced in the cheese have positive or negative effects. A major negative effect of proteolysis is bitterness. This is caused by the proteolysis of *casein* (milk protein) into hydrophobic peptides. These can be hydrolyzed to nonbitter peptides and amino acids by bacterial peptidases. Therefore, the accumulation of bitter peptides depends on their relative rates of production and destruction.

Genetics of Lactic Acid Bacteria

Research on the genetics of LAB began in the early 1970s. It initially focused on plasmids and natural gene transfer in lactococci. The development of a detailed understanding of LAB genetics, natural gene transfer systems, and the development of tools required for genetic engineering have made this an important area of food microbiology.

In LAB, genes can be located on the chromosomes or on insertions into the chromosome (*introns*), move from one location to another (transposable elements), and/or be located on *extrachromosomal* (not on the chromosome) pieces of DNA called *plasmids*. The chromosomes of LAB are relatively small. Transposable genetic elements that move from one chromosomal site to another are important in some LAB. Insertion sequences, the simplest of transposable elements, also occur in many LAB. Their ability to cause molecular rearrangements and affect gene regulation can be good or bad for dairy fermentations. There are more complex transposable elements. These self-transmissible conjugal transposons code for nisin production, sucrose utilization, and a bacteriophage defense mechanism. They can be moved from one cell to another. Plasmids are independently replicating extrachromosomal circular DNA molecules. They are of particular importance in lactococci, where they encode characteristics essential for dairy fermentations, including lactose metabolism, proteinase activity, oligopeptide transport, bacteriophage resistance, exopolysaccharide production, and citric acid utilization. Lactococci, lactobacilli, and *S. thermophilus* can also gain new genes by using transduction, the transfer of bacterial genetic material by a bacteriophage.

Bacteriophages of Lactic Acid Bacteria

Viruses are host-specific invaders that take over the metabolic machinery of their host to do their work (see chapter 21). Humans are attacked by viruses specific for humans, horses are attacked by viruses specific for horses, and bacteria are attacked—you guessed it, by viruses specific for bacteria. The viruses that attack bacteria are called *bacteriophages*. Bacteriophages are bad news for fermentations. They infect the bacteria, split them open, kill them, and stop the fermentation. This is the most common cause of "stuck" fermentations.

Bacteriophages can be *lytic* or *temperate*. Lytic bacteriophages stick onto a sensitive cell through a mechanism that requires calcium. They then insert their DNA. The host's cellular machinery replicates the viral DNA, makes the proteins needed for virus assembly, and synthesizes other viral parts. Finally, the viral parts self-assemble, break open the host, and are released to infect more bacteria.

Temperate bacteriophages are subtler. They adsorb onto a sensitive cell and insert their DNA. The viral DNA becomes part of the bacterial chromosome. The viral DNA is reproduced with the bacterial DNA and is passed to all the host's descendents. It can remain unexpressed for generations. Eventually, the temperate bacteriophages go into a lytic cycle, taking over the host's cellular machinery to replicate viral DNA, make the proteins needed for virus assembly, and synthesize other viral parts. Once again, the viral parts self-assemble, break open the host, and are released to infect more bacteria. The fermentation is a failure.

Bacteriophage Resistance Mechanisms

LAB have several ways to defend against bacteriophages. They can interfere with bacteriophage adsorption, degrade phage DNA, or prevent the viral infection from progressing. These defense genes are carried by plasmids that can be transferred between bacteria. The details of biological and manufacturing methods to defend LAB against bacteriophages are shown in Table 18.2.

Table 18.2 Bacteriophage defenses of LAB

Defense	Mechanism
Biological	
Adsorption interference	LAB can prevent bacteriophage adsorption if they lack specific proteins or mask the receptors that the bacteriophages recognize.
Restriction/modification	Resistant LAB have enzymes that modify their own DNA to distinguish it from phage DNA, which they destroy using restriction enzymes.
Abortive infection	These systems inhibit infection after the bacteriophage have adsorbed, penetrated, and started the lytic process.
Commercial	
Phage-inhibitory media	Starter cultures are grown in media that lack Ca, which is required for phage adsorption.
Frozen concentrated starter cultures	Starters are grown in phage-inhibitory media, concentrated, and added to the fermentation at such high levels that no further growth is required.
Strain rotation	LAB strains are constantly rotated so that bacteriophage-sensitive strains are identified and replaced with bacteriophage-insensitive strains.
Good manufacturing practices	Minimize contamination with viruses.

FERMENTED VEGETABLES

Fermentation is an ancient process of preserving plants to retain their nutritive value. Many groups of microorganisms ferment vegetables. LAB and yeasts are preferentially used in the West (Europe and America). Many Eastern foods are fermented by molds. The most extensively used processes for biopreservation of vegetables involve lactic acid fermentation.

Lactic acid fermentation of vegetables was practiced by the Chinese in prehistoric times. However, the oldest written evidence dates from the first century A.D., when Plinius described the preservation of white cabbage in earthen vessels. This procedure results in the lactic acid fermentation of cabbage into sauerkraut. The development of heat sterilization and refrigeration systems have made fermentation less important as a preservation method in industrialized countries. Nonetheless, in developing countries, fermentation is still a critical means of food preservation. In addition to preservation, fermentation has many other functions (Table 18.3).

Vegetable fermentations involve complex microbiological, biochemical, chemical, and physical reactions. Fermentations are also influenced by many external factors. These are classified into four groups: technological factors, ingredients, raw material quality, and native microbiota.

Table 18.3 Uses of fermentation

1. To preserve vegetables and fruits
2. To develop characteristic sensory properties, i.e., flavor, aroma, and texture
3. To destroy naturally occurring toxins and undesirable components in raw materials
4. To improve digestibility, especially of some legumes
5. To enrich products with desired microbial metabolites, e.g., L-(+)-lactic acid or amino acid
4. To create new products for new markets
5. To enhance dietary value

Table 18.4 Steps in a typical vegetable fermentation

1. Select vegetables that are sound, undamaged, uniformly sized, and at the proper ripeness.
2. Pretreat, for example, by peeling, blanching, or cooking.
3. Place whole, pierced, shredded, or sliced vegetables in fermentation vessels. These can hold from 100 liters to 100 tons. Starter cultures can also be added.
4. Completely cover the vegetables with brine, and seal the fermentation vessels to exclude oxygen and ensure anaerobic conditions.
5. Let the fermentation take its natural course. The fermentation time depends on the temperature, they type of product, and the bacteria present.
6. Distribute the final fermented product fresh, unpackaged, packaged, or pasteurized.

Table 18.4 shows a generic process for manufacturing fermented vegetables. Because the botanical, physical, and chemical properties of various vegetables differ, the exact process must be tailored to specific products.

Ingredients and Additives Used during Fermentations

Salt is added to fermentations for many reasons. A major one is flavor. The amount of salt used depends on the vegetable and on consumer preference. In sauerkraut production, salt enhances fluid release from the shredded cabbage. It also helps create anaerobic conditions in fermentation vessels. Salt also has a selective effect on the vegetables' natural microbiota. Increasing amounts of salt favor the growth of LAB and inhibit undesirable bacteria and fungi. In sauerkraut fermentations, heterofermentative LAB are favored by a low salt concentration (~1%) and are greatly inhibited at 3%. Higher salt concentrations favor homofermentative species, accelerating the fermentation. Salt content below 0.8% often results in undesirable fermentation as well as in soft sauerkraut.

Other food additives are used in specific fermented vegetables and are regulated by law in most countries. The addition of ascorbic acid to sauerkraut prevents gray or brown discoloration. Citric acid or sulfur dioxide is also used for this purpose. Sorbic acid prevents the growth of yeasts, molds, and other microbes.

The overall microbial population, as well as the population of LAB, changes during the course of the fermentation. The fermentation of cabbage, for example, has four stages:

1. Fermentation starts as soon as the cabbage is placed into vessels. When the cabbage is tightly packed, the number of strictly aerobic bacteria decreases, while facultatively anaerobic enterobacteria grow for the first 2 or 3 days. During this period, dissolved oxygen is consumed by the microorganisms and by plant respiration. The formation of lactic, acetic, formic, and succinic acids lowers the pH. Carbon dioxide may generate foam.
2. Non-LAB are suppressed by anaerobic conditions and are overgrown by the LAB. Lactic acid fermentation is initiated by heterofermentative LAB, namely, *Leuconostoc mesenteroides* and then *Lactobacillus brevis*. This succession of microorganisms is complete after 3 to 6 days, and the lactic acid concentration increases to ~1%.

3. The third stage of fermentation is dominated by homofermentative LAB. Their growth is favored by the complete lack of oxygen, low pH, and high salt content. The homofermentative LAB increase the total acid content to 1.5 to 2.0%. Most sauerkraut is pasteurized when it reaches pH 3.8 to 4.1.

4. Only sauerkraut that is stored in the fermentation vessel and distributed fresh undergoes the final stage of fermentation. This stage is dominated by *Lactobacillus brevis* and some heterofermentative species that metabolize the pentoses released by the breakdown of cell walls. The acid content may increase to 2.5%.

In cucumber fermentation, excessive growth of heterofermentative *Leuconostoc mesenteroides* is undesirable, since carbon dioxide production causes gaseous spoilage and "floaters." This is also a problem with Spanish-style olives. To destroy naturally occurring oleuropein, an extremely bitter-tasting compound in olives, freshly harvested fruits are treated with dilute sodium hydroxide (lye) before fermentation. The lye is removed later by washing and neutralization. This causes extensive losses of water-soluble nutrients.

Floater formation is a serious problem in cucumber fermentation. Damage is worse with larger cucumbers, at higher fermentation temperatures, and at high dissolved carbon dioxide concentrations in the brine. The brine's carbon dioxide content can be reduced by controlling the initial microbial population that produces it, limiting the growth of heterofermentative LAB, and removing carbon dioxide from the brine.

MEAT FERMENTATIONS

Fermented meats are not as popular as fermented vegetable or dairy products, but they still constitute a major type of fermented food. Fermentation alters meat characteristics and prevents spoilage. The most common types of fermented meats are dry and semidry fermented sausages. The U.S. Department of Agriculture Food Safety and Inspection Service requires that shelf-stable sausages must contain nitrite and curing agents, be fermented to a pH of <5.0, and have a moisture/protein ratio of <3.1:1.0.

Sausage making is straightforward. Grinding the meat reduces its particle size. It is then blended with nitrites, curing agents, spices, a fermentable carbohydrate, and sometimes glucono-delta-lactone. The nitrite (40 to 50 parts per million) is added to inhibit *Clostridium botulinum*, contribute to the cured meat taste, and convert myoglobin to nitrosomyoglobin (which produces the pink color of cured meat). Fermentable sugars are added at 0.4 to 0.8% because there is not enough glucose in meat to produce much acid. Glucono-delta-lactone is converted to gluconic acid. This is converted by LAB to lactic and acetic acids, which accelerates the acidification. The starter cultures (10^7 colony-forming units [CFU]/g) are also added at this point. The mixture is stuffed into an oxygen-impermeable casing to ensure an anaerobic fermentation. The links are incubated to promote microbial growth. The starter cultures decrease the pH from 5.6 to 4.8 within 8 h. The lactic acid coagulates the protein, which facilitates drying. The finished sausage is sometimes given a heat treatment.

Defined starter cultures are relatively new to the meat industry. The benefits of using defined starters are shown in Table 18.5. The native bacteria of naturally fermented meat are lactobacilli. Unfortunately, they do not

Table 18.5 Benefits of using starter cultures in manufacture of fermented food

Rapid acid production
Decreased rate of contamination
Production of flavor compounds
More consistent product
Greater process predictability

tolerate the freeze-drying process used to make starter cultures, so in the 1950s, the industry started using *Pediococcus acidilactici.* It is also homofermentative, can tolerate 6% salt, and is not proteolytic. Freeze-dried cultures were hard to dissolve and had long lag times, so in the 1980s, manufacturers started using frozen cultures. These shorten the lag time and make acidification more rapid.

OTHER FERMENTED FOODS

Bread

As ancient societies evolved, their cereal food progressed from porridges and gruels to unleavened flat breads to breads leavened with yeast. Egyptians had baking ovens in 2700 B.C. In these early times, baking and brewing were both done with *Saccharomyces cerevisiae* (baker's yeast [Fig. 18.3]). Bakeries were often attached to breweries so that the yeast by-product of brewing could be used in bread making. Today, different specialized strains of yeast are used for each purpose.

The main role of the yeast is to produce carbon dioxide. This makes the bread rise. The first step in bread making is to mix the flour, sugar (as a fermentable carbohydrate), fat (for texture), salt, and other ingredients. The yeast is added at 1 to 6% (wt/wt). The yeast may be a dried powder, block, or cream. For sourdough breads, heterofermentative LAB are also added. Water is added, and the dough is kneaded so that gluten protein in the flour stretches and the dough forms a viscoelastic mass. The bread is fermented at 28 to 32°C. The yeast produces carbon dioxide and amylases that break down starch to the more fermentable glucose. During baking, carbon dioxide expansion causes a 40% increase in the bread volume.

Figure 18.3 Phase-contrast micrograph of yeast. Courtesy of George Carman, Rutgers University.

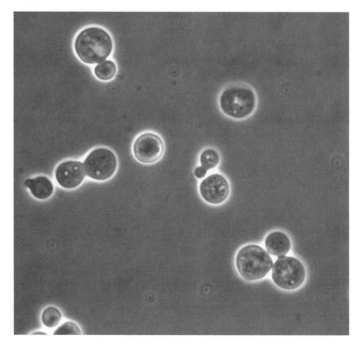

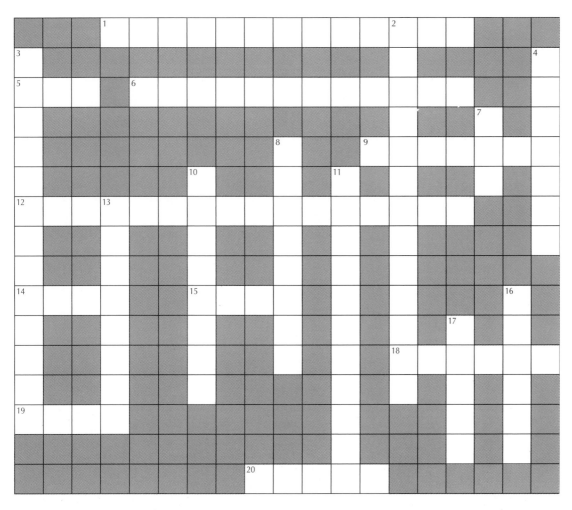

Across
1. Virus that attacks bacteria
5. Homofermentation yields 2, heterofermentation yields 1
6. A bioprocess is also called a _____
9. Used in cured meat for color, taste, and antimicrobial properties
12. Type of fermentation that makes only lactic acid
14. During Prohibition, amateur fermentation microbiologists were often the targets of a _____
15. Propionibacteria produce these in Swiss cheese
18. Fermentation can be described as life in the absence of _____
19. Analogous to the _____ planted to start the growth of a vegetable, starter cultures are used to _____ a new fermentation
20. Used to make beer, wine, and bread

Down
2. Fermented foods are microbially safe due to _____
3. Acids are made from fermentable _____
4. Oxidized alcohol
7. Abbreviation for citric
8. Heterofermentative, but not homofermentative, microbes use these
10. Their breakdown is responsible for bitter flavor in cheese
11. Genus consisting of only homofermentative species
13. Energy is made when compounds are _____
16. The liquid drained away in making chocolate sweets
17. Bacteriophage cycle that breaks open the host

Beer

Beer is proof that God loves us and wants us to be happy.

BENJAMIN FRANKLIN

Beer production can also be traced to ancient times. The essential ingredients of beer are hops (for flavor and antimicrobial activity), yeast (to produce alcohol and carbon dioxide), water (for obvious reasons), and malt (to provide the fermentable carbohydrate). German purity laws forbid the use of any other ingredients. Other countries allow less expensive cereals, grains, or corn to be used as *adjuncts*. The adjuncts provide up to 30% of the fermentable carbohydrates.

The brewing process consists of malting, mashing, wort boiling, fermentation, and postfermentation treatments (Fig. 18.4). *S. cerevisiae,* which grows on top of the fermentation mixture, is used for ales. *Saccharomyces carlsbergensis,* which settles to the bottom, is used for lagers. In both cases, the inoculum level is quite high ($\sim 10^7$ CFU) and increases eight-fold, i.e., growth consists of three doublings. The postfermentation processes include aging to remove "green" flavors and filtering to remove the yeast. Bottled beers are heated for 5 to 30 min at 60°C to kill the remaining yeast. Draft beer in kegs is not heated and must be refrigerated.

Wine

Wine making undoubtedly started in ancient times when the natural yeasts contaminated grape juice and fermented it to an alcoholic beverage. In the early 1900s, Pasteur discovered that microbes caused the fermentation. He also discovered the role of bacteria in wine spoilage. Pasteurization was invented to kill these bacteria in wine. It was only later applied to milk.

Figure 18.5 outlines the wine-making process. Grapes can be fermented in barrels, but large stainless steel tanks are more commonly used. To inoculate or not to inoculate grape juice with a defined inoculum of yeast is a subject of ongoing debate. As with other fermentations, using a defined

Figure 18.4 Schematic for making beer. Worldwide consumption tops 18 billion gallons per year. Adapted from G. H. Fleet, p. 747–772, *in* M. P. Doyle, L. R. Beuchat, and T. J. Montville (ed.), *Food Microbiology: Fundamentals and Frontiers,* 2nd ed. (ASM Press, Washington, D.C., 2001).

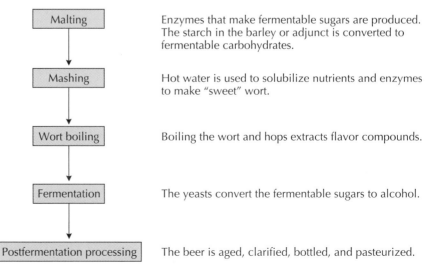

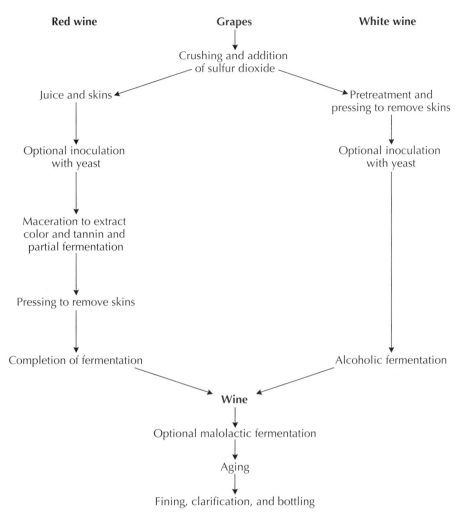

Figure 18.5 Schematic for making wines. In 2002, 595 million gallons of wine, worth $21 billion, was consumed in the United States. Adapted from G. H. Fleet, p. 747–772, *in* M. P. Doyle, L. R. Beuchat, and T. J. Montville (ed.), *Food Microbiology: Fundamentals and Frontiers*, 2nd ed. (ASM Press, Washington, D.C., 2001).

inoculum (invariably *S. cerevisiae*) gives a more consistent product and a more rapid process. However, *S. cerevisiae* does not naturally occur at high levels on grapes. The native grape yeasts that cause the fermentation are mostly *Kloeckera* and *Hansenia* species, with some *Candida*, *Pichia*, and *Hansenula* species. If the native yeasts are used, the process is less predictable but can be extraordinarily good, as well as very bad.

Malolactic fermentation is a secondary fermentation used to decrease the acidity of some wines. Malic acid occurs naturally in some grapes. *Oenococcus oeni* (formerly classified as a leuconostoc) can decarboxylate malic acid to lactic acid. This elevates the pH by 0.3 to 0.5 units and creates a mellower flavor.

Vinegar

Vinegar fermentation is almost as ancient as that of wine. It was undoubtedly discovered by accident when a *Gluconobacter* sp. contaminated wine and turned it to vinegar. This turned out to be not such a bad thing, since the vinegar could then be used to preserve other foods.

Vinegar is made by oxidizing ethanol to acetic acid:

$$CH_3CH_2OH + O_2 \rightarrow CH_3COOH + H_2O$$

This occurs in a two-step bioprocess. The first step in vinegar production is the alcoholic fermentation by yeast to produce alcohol. Ethanol from wine and hard cider are frequently used, but it can be made by the fermentation of almost any fruit or starchy material. In the second step, the ethanol is oxidized to acetic acid. Since *Gluconobacter oxidans,* used in making vinegar, is a strict aerobe, oxygen transfer is the rate-limiting step in the fermentation. The oldest methods involved simply leaving wine in wooden barrels exposed to the air, inoculating it with *Gluconobacter,* and waiting. In several weeks, or perhaps months, the wine became vinegar. The "trickling" fermentor radically reduced this time to a day or two by increasing the oxygen transfer rate. The *Gluconobacter* organisms grew as biofilms on wood shavings in a large wooden chamber. A rotating sprayer at the top of the chamber sprayed the alcoholic liquid on top of the chips, and it trickled down. The large surface area exposed to oxygen made this method very efficient. The trickling fermentor is probably the first bioreactor using immobilized-cell technology. The most modern method of vinegar production is the submerged culture reactor. By pumping oxygen through these very large fermentors, large quantities of vinegar can be made in a very short time.

Cocoa

Good cocoa flavor is the result of a complex fermentation involving microbial succession. Cocoa is made from the plant *Theobroma cacao.* It is native to South America and grows up to 15° north and south of the equator. The cacao beans, or seeds, are contained in pods about the size of a coconut. Without fermentation, the cocoa seeds would be bitter and astringent. The pods are filled with mucilage and seeds, much like a pumpkin. The seeds and mucilage are removed from the pods and put into heaps or containers and fermented for 2 to 8 days. They are turned periodically to introduce oxygen. The acetic acid and heat generated by the fermentation keep the seeds from germinating.

The mucilage contains sucrose, other sugars, and citric acid. Its pH is 3.4 to 4, and it contains 10 to 12% sugars. The microbes come from the environment, the plants, and the workers. During the first phase of the fermentation, which lasts 2 or 3 days, the yeasts are most important. They produce alcohols and aldehydes and create anaerobic conditions that favor the growth of LAB. This phase lasts up to 3 days. The LAB population can be as high as 10^7 CFU/ml. The LAB produce large amounts of hydrolytic enzymes, invertase, glycosidases, and proteases. There are some heterofermentative LAB in the population, but it is primarily homofermentative. The turning of the heaps or boxes introduces air that favors acetic acid bacteria in the final stage of the fermentation. The levels of *Gluconobacter* species and *Acetobacter* species can reach 10^6 CFU/g. The fermentation provides flavor precursors that are converted to flavors during drying and roasting.

There are two main methods of fermenting the cocoa. In Ghana and other small-scale producing regions, the seeds are put into heaps on the ground, covered with banana leaves, and turned by hand. All of this is very labor-intensive but produces high-quality seed. Large boxes from 1 by 1 by 1 to 7 by 5 by 1 m are used on plantations and estates. These are frequently grouped or arranged in tiers to make turning easier. In both methods, a liquid fermentate called "sweats" is produced and is drained away from the beans. This makes drying easier.

The development of good cocoa flavor is still very much an art. The complexity of the microbial succession prevents the use of starter cultures. The flavors are also dictated by the genetic potential of the plant. While a good fermentation cannot upgrade an inferior product, a bad fermentation can ruin a superior cocoa.

Coffee

Coffee beans are contained in a cherry-like fruit. The fleshy mucilage that surrounds the beans can be removed by dehydration or fermentation. When the beans are fermented, they are first mechanically depulped. The beans, covered with residual mucilage, are submerged in tanks of water. Native yeast, molds, LAB, and gram-negative bacteria ferment the mucilage to water-soluble products that can then be washed away. This takes 12 to 60 h. The beans are then dried, roasted, and ground.

Fermented Foods of Non-Western Societies

There are dozens of fermented foods that play important roles throughout Asia, Africa, and South America. The substrates are usually starchy local crops. The native microbiotas serve as the inocula. A few of these fermentations are shown in Table 18.6.

*Authors' note*_____

Many "traditional," "indigenous," and "non-Western" societies are more adept in the use of fermented foods than is mainstream American culture.

Table 18.6 Regionally important food fermentations[a]

Product	Organism(s)	Description	Use
Soy sauce	*Aspergillus oryzae, Aspergillus sojae*, yeast, pediococci	Soy and roasted wheat are fermented at 35–40°C for 2–4 mo (12–19% salt, pH 4.6–4.8) and pasteurized at 70–80°C.	Flavoring agent
Miso	*A. oryzae* and secondary fermentation of yeasts and pediococci	Rice and/or soybean paste is fermented for 1 wk to 1 yr at 25–30°C, 4–14% salt.	Soup base
Natto	*Bacillus natto (B. subtilis)*	Fermented at 25–40°C for 12 h to 2 wk	Whole fermented soybean, eaten as food
Sufu	*Mucor* spp., *Rhizopus chinensis, Actinomucor elegans*	Curd of soybean milk is pressed, cubed, boiled in brine, and inoculated and then aged for 1–12 mo.	Condiment
Lao-chao	*Rhizopus oryzae, R. chinensis, Endomycopsis* spp.	Incubation for 1 or 2 days at room temperature to yield juicy sweet slightly alcoholic product	Dessert
Tempeh	*Lactococcus* spp., *Lactobacillus* spp., *Rhizopus oligosporus*	Soybeans are pressed, inoculated with LAB, fermented for 2 days, and inoculated with *R. oligosporus*, whose mycelia bind it together into a cake.	Sliced, deep-fried, or used in fruit
Oncom	*Neurospora intermedia, R. oligosporus*	Deoiled peanut cakes are inoculated and incubated at 25–30°C for 1 or 2 days. Surface becomes covered with colored conidia or spores.	Fermented peanut press-cake
Poi	Initially lactobacilli and lactococci, followed by yeast	Corns of taro plant are ground, hydrated, and fermented by native microbiota for 1–3 days.	Eaten as food or side dish

[a]Adapted from L. R. Beuchat, p. 701–720, *in* M. P. Doyle, L. R. Beuchat, and T. J. Montville (ed.), *Food Microbiology: Fundamentals and Frontiers,* 2nd ed. (ASM Press, Washington, D.C., 2001).

Summary

- Fermentation is an incomplete oxidation of sugars in the absence of oxygen using an internal organic compound as an electron acceptor.
- "Fermentation" can also mean "bioprocess."
- Fermentative metabolism yields only 1 or 2 ATP molecules per mol of glucose, much less than oxidative metabolism.
- Homofermentative bacteria make only lactic acid.
- Heterofermentative bacteria make lactic acid, ethanol, carbon dioxide, and acetic acid.
- Homofermentative bacteria are used to make acid, whereas heterofermentative bacteria contribute flavor.
- Starter cultures increase the speed and consistency of fermentations.
- Lactic acid bacteria have several methods for transferring genes.
- Plasmids carry the genes for many traits important to fermentations.
- Bacteriophages are viruses that infect bacteria and can damage fermentations by killing the starter cultures.
- Vegetable fermentations rely on indigenous microbiota.
- Bread and beer both use yeast for fermentation.
- Vinegar production is an oxidative fermentation.
- Mucilage is fermented in coffee and cocoa fermentations.

Suggested reading

Beuchat, L. R. 2001. Traditional fermented foods, p. 701–720. *In* M. P. Doyle, L. R. Beuchat, and T. J. Montville (ed.), *Food Microbiology: Fundamentals and Frontiers,* 2nd ed. ASM Press, Washington, D.C.

Buckenhüskes, H. J. 2001. Fermented vegetables, p. 665–680. *In* M. P. Doyle, L. R. Beuchat, and T. J. Montville (ed.), *Food Microbiology: Fundamentals and Frontiers,* 2nd ed. ASM Press, Washington, D.C.

Campbell, I. Beer, p. 735–747. 2001. *In* M. P. Doyle, L. R. Beuchat, and T. J. Montville (ed.), *Food Microbiology: Fundamentals and Frontiers,* 2nd ed. ASM Press, Washington, D.C.

Fleet, G. H. 2001. Wine, p. 747–772. *In* M. P. Doyle, L. R. Beuchat, and T. J. Montville (ed.), *Food Microbiology: Fundamentals and Frontiers,* 2nd ed. ASM Press, Washington, D.C.

Fox, P. E. (ed.). 1993. *Cheese: Chemistry, Physics, and Microbiology,* vol. 1 and 2. Chapman and Hall, Ltd., London, England.

Johnson, M. E., and M. E. Steele. 2001. Fermented dairy products, p. 651–665. *In* M. P. Doyle, L. R. Beuchat, and T. J. Montville (ed.), *Food Microbiology: Fundamentals and Frontiers,* 2nd ed. ASM Press, Washington, D.C.

Ricke, S. C., I. Z. Díaz, and J. T. Keeton. 2001. Fermented meat, poultry, and fish products, p. 681–700. *In* M. P. Doyle, L. R. Beuchat, and T. J. Montville (ed.), *Food Microbiology: Fundamentals and Frontiers,* 2nd ed. ASM Press, Washington, D.C.

Salminen, S., and A. von Wright (ed.). 1998. *Lactic Acid Bacteria.* Marcel Dekker, Inc., New York, N.Y.

Thompson, S. S., K. B. Miller, and A. S. Lopez. 2001. Cocoa and coffee, p. 721–734. *In* M. P. Doyle, L. R. Beuchat, and T. J. Montville (ed.), *Food Microbiology: Fundamentals and Frontiers,* 2nd ed. ASM Press, Washington, D.C.

Questions for critical thought ————————————————

1. How is the saying, "In the land of the blind, the one-eyed man is king" relevant to energy production by fermentative metabolism? Can you think of another folk saying that captures the same concept?

2. If you were a *Lactobacillus* species able to use pentoses and hexoses and landed in an environment with both, which would you use? Why?

3. What are three general usages of the word "fermentation?"

4. Why does pH have profound effects on moisture control during cheese manufacture, retention of coagulants, and hydration of proteins?

5. Draw pictures reflecting the life cycles of lytic and temperate bacteriophages.

19

Spoilage Organisms*

LEARNING OBJECTIVES

The information in this chapter will enable the student to:

- gain knowledge about microorganisms responsible for spoilage of a wide range of food products
- discuss intrinsic mechanisms that inhibit spoilage of food
- understand the impact of food processing on microbial spoilage
- identify the sources of microorganisms responsible for spoilage of specific products
- discuss procedures that can be implemented to minimize contamination of raw materials

INTRODUCTION

The phone rings, and your friend from out of town asks if she can drop by for a visit. You go so far as to suggest that you will make dinner. You open the refrigerator to gather the necessary ingredients to make a delicious meal. Upon opening a jar of spaghetti sauce, you notice a greenish-white fuzzy mass on the surface of the sauce. The luncheon meat that you remove to make antipasto has a green sheen and a strange odor, and it feels slimy. For a salad, you take out lettuce, but it has turned brown, and the peppers have large dark spots. In order to gather your thoughts, you decide to sit down and have a glass of milk. When pouring the milk, you notice that it is lumpy and a bit smelly. You start to realize that you should have paid more attention in the food microbiology course that you attended as an undergraduate.

A product (meat, dairy, fruit, or seafood) is considered spoiled if sensory changes make it unacceptable to the consumer. Factors associated with food spoilage include color defects or changes in texture, the development of "off" flavors or odors, slime, or any other characteristic that makes the food undesirable for consumption. While enzymatic activity within a food contributes to changes during storage, *organoleptically* detectable (i.e., detectable through a change in odor or color) spoilage is generally a result of decomposition and the formation of metabolites resulting from microbial growth. This chapter discusses spoilage of many foods, from meat to dairy to produce. The types of microorganisms involved, conditions conducive to spoilage, and defects associated with spoilage are covered.

*This chapter contains information from chapters on meat, poultry, and seafood; milk and dairy products; and fruits, vegetables, and grains originally written by Timothy C. Jackson, Douglas L. Marshall, Gary R. Acuff, James S. Dickson, Joseph F. Frank, and Robert E. Brackett for *Food Microbiology: Fundamentals and Frontiers*, 2nd ed., and has been adapted for use in an introductory text.

MEAT, POULTRY, AND SEAFOOD PRODUCTS

Origin of Microflora in Meat

The levels of bacteria in muscle tissues of healthy live animals are extremely low. High numbers of bacteria are present on the hide, hair, and
hooves of red-meat animals, as well as in the gastrointestinal tract. Microorganisms on the hide include bacteria, such as *Staphylococcus, Micrococcus,*
and *Pseudomonas* species, and fungi, such as yeasts and molds, which are
normally associated with skin microflora, as well as species contributed by
fecal material and soil. The numbers and composition of this microflora are
influenced by environmental conditions.

The majority of bacteria on a dressed red-meat carcass originate from
the hide. During hide removal, bacteria are carried from the hide onto the
underlying tissue with the initial incision. Unlike those of cattle and sheep,
the skin of hogs is usually not removed but is scalded and left on the carcass. Recontamination can also occur during dehairing due to the presence
of debris in dehairing machines. Contamination can occur if the intestinal
tract is pierced or if fecal material is introduced from the rectum during
the removal of the abdominal contents. Handling can result in cross-
contamination of other carcasses. In addition to the hide and viscera, the
processing environment, such as floors, walls, contact surfaces, knives,
and workers' hands, can be a source of contamination in red meats.

Origin of Microflora in Poultry

The skin, feathers, and feet of poultry harbor microorganisms resident on
the skin, as well as from litter and feces. Although present on the skin, psychrotrophic bacteria, consisting primarily of *Acinetobacter* and *Moraxella*,
are primarily associated with the feathers. Contamination and cross-
contamination with fecal material may occur during transportation of birds
from growing houses to slaughter facilities and during processing, when
the birds are hung and bled. Following processing, carcasses are chilled
rapidly to limit the growth of microbes. Slush ice, continuous-immersion,
spray, air, and carbon dioxide chilling systems have been utilized or proposed for this purpose. Potable water should be used in these systems,
since bacteria present in untreated water can contribute to spoilage.

Origin of Microflora in Finfish

The numbers and composition of microflora on finfish are influenced by
the environment from which the fish are taken, the season, and the conditions of harvesting, handling, and processing. Water temperature has a significant influence on the initial number and types of bacteria on the surface
of the fish. Higher numbers of bacteria are generally present on fish from
warm subtropical or tropical waters than on fish from colder waters. Fish
taken from temperate waters harbor predominantly psychrotrophic bacteria, while mesophilic bacteria predominate on fish taken from tropical areas. Bacteria from the genera *Acinetobacter, Aeromonas, Cytophaga, Flavobacterium, Moraxella, Pseudomonas, Shewanella,* and *Vibrio* dominate on fish and
shellfish taken from temperate waters, while *Bacillus*, coryneforms, and *Micrococcus* species frequently predominate on fish taken from subtropical
and tropical waters. The initial microflora on fish is influenced by the
method of harvesting. Trawled fish generally have higher microbial levels
than those that are line caught. In trawling, the dragging of fish and debris
along the ocean bottom stirs up mud that contaminates the fish. In addi-

tion, the compaction of fish in trawling nets may cause expression of intestinal contents, with subsequent contamination of the fish surface. A delay in chilling fish also enhances the possibility of rapid microbial growth. In some Scrombridae and Scomberesocidae fish (tuna, mackerel, and skipjack), this can lead to the generation of toxic levels of histamine by *Morganella* (*Proteus*) *morganii* and related gram-negative bacteria that produce histidine decarboxylase. *Photobacterium phosphorum* may produce histamine during low-temperature storage of these fish.

Origin of Microflora in Shellfish

Unlike other crustacean shellfish (lobster, crabs, or crayfish) that are kept alive until they are heat processed, shrimp die soon after harvesting. Decomposition begins soon after death and involves bacteria on the shrimp surface that originate from the marine environment or from contamination during handling and washing. Molluscan shellfish (oysters, clams, scallops, and mussels) are stationary filter feeders, and thus, their microfloras depend greatly on the quality of the water in which they reside, the quality of the wash water, and other factors. Bacteria, including *Pseudomonas* spp., *Shewanella putrefaciens*, *Acinetobacter*, and *Moraxella*, that spoil finfish also cause spoilage of shellfish.

Bacterial Attachment to Food Surfaces

Spoilage of meat, poultry, and seafood generally occurs as a result of the growth of bacteria that have colonized muscle surfaces. The first stage in colonization and growth involves the attachment of microbial cells to the muscle surface. Bacterial attachment to muscle surfaces involves two stages. The first is a loose, reversible sorption that may be related to van der Waals forces or other physicochemical factors. One of the factors that influence attachment at this point is the population of bacteria in the water film. The second stage consists of an irreversible attachment to surfaces involving the production of an extracellular polysaccharide layer known as a *glycocalyx*. Many factors can influence bacterial attachment, including surface characteristics, growth phase, temperature, and the motility of the bacteria.

Microbial Progression during Storage

The initial microflora of muscle foods is highly variable. It comes from the microorganisms resident in and on the live animal; environmental sources, such as vegetation, water, and soil; ingredients used in meat products; workers' hands; and contact surfaces in processing facilities. Most perishable meat, poultry, and seafood products are stored at refrigeration temperatures to prolong their shelf life. As microbial growth occurs during storage, the composition of the microflora is altered so that it is dominated by a few, or often a single, microbial species, usually of the genera *Pseudomonas*, *Lactobacillus*, *Moraxella*, and *Acinetobacter* or the species *Brochothrix thermosphacta*. *Pseudomonas* species are able to compete successfully on aerobically stored refrigerated muscle foods, since they have a competitive growth rate, but *Moraxella* and *Acinetobacter* species are less capable of competing at refrigeration temperatures and lower pH. The growth of the aerobic spoilage microflora is suppressed during storage under vacuum and modified atmospheres. Under these conditions, lactic acid bacteria are favored because of their growth rate, their fermentative metabolism, and their ability to grow at the pH range of meat. At higher pH, *B. thermosphacta* and *S. putrefaciens* may grow and contribute to spoilage. The water activity

(a_w) of some types of processed meats is lowered by dehydration or the addition of solutes, such as salt or sugar. When low a_w restricts the growth of bacteria, growth of fungi may occur. Microbial growth does not occur on products with a_w values of <0.60.

Muscle Tissue as a Growth Medium

The composition (percent adipose, lean, and carbohydrate) of the meat influences the microbial growth and type of spoilage. Research suggests that spoilage defects in meat become evident when the number of spoilage bacteria on the surface reaches 10^7 colony-forming units (CFU)/cm^2. During aerobic spoilage, off odors are first detected when levels reach 10^7 CFU/cm^2. When numbers reach 10^8 CFU/cm^2, the muscle tissue surface will begin to feel tacky, the first stage in slime formation.

Composition and Spoilage of Red Meats

The a_w of red-meat lean muscle tissue is 0.99, with a corresponding water content of 74 to 80%. The protein content may vary from 15 to 22% on a wet-weight basis. The lipid contents of intact red meats vary from 2.5 to 37%, and the carbohydrate compositions range from 0 to 1.2%. Glycolysis leads to the accumulation of lactic acid, and as a result, the pH of the muscle tissue decreases. The spoilage of meats stored at ambient temperature results from the growth of mesophiles, predominantly *Clostridium perfringens* and members of the family *Enterobacteriaceae*. Spoilage deep within muscle tissue, known as "sours" or "bone taint," has been attributed to a slow cooling of carcasses, resulting in the growth of anaerobic mesophiles thought to be already present in the muscle tissues. If the surfaces of whole carcasses or fresh meat cuts become dry, bacterial growth may be restricted and fungal spoilage may occur. *Thamnidium, Mucor,* and *Rhizopus* may produce a whiskery, airy, or cottony gray-to-black growth on beef due to the presence of mycelia. Black spot has been attributed to the growth of *Cladosporidium,* white spot has been attributed to the growth of *Sporotrichum* and *Chrysosporium,* and green patches have been attributed to *Penicillium.* Molds will not grow on beef held at temperatures below −5°C.

Composition and Spoilage of Poultry Muscle

The mechanism of microbial spoilage of poultry muscle is similar to that of red meat. Spoilage is generally restricted to the outer surfaces of the skin and cuts and has been characterized by off odors and sliminess, as well as various types of discoloration. Skin may provide a barrier to the introduction of spoilage microorganisms to the underlying muscle tissue. Although the pH of breast muscle (pH 5.7 to 5.9) differs from that of leg muscle (pH 6.4 to 6.7), the organisms that cause spoilage are similar and include *Pseudomonas, Aeromonas,* and *S. putrefaciens.* For poultry carcasses packaged in oxygen-impermeable films, spoilage may be caused by *Shewanella, B. thermosphacta,* and atypical lactobacilli. The ability to make sulfide compounds, such as hydrogen sulfide, dimethyl sulfide, and methyl mercaptan, make *S. putrefaciens* an important component of the spoilage microflora.

Composition and Spoilage of Finfish

The internal muscle tissue of a healthy live fish is generally sterile. Bacteria are present on the outer slime layer of the skin and gill surfaces and in the intestines in the case of feeding fish. The composition of fish muscle is highly variable between species and may fluctuate widely, depending upon size, season, fishing grounds, and diet. The average composition of nonfatty fish, such as cod, has been characterized as 18% protein and <1% lipid, whereas in fatty fish, such as herring, the lipid content may range from 1 to 30%, with the water content varying so that fat and water comprise ~80% of the muscle tissue. As with other muscle foods, the spoilage microflora of fresh ice-stored fish consists largely of *Pseudomonas* spp. *S. putrefaciens* may also contribute to the spoilage of seafood, and *Acinetobacter* and *Moraxella* may comprise a smaller portion of the spoilage microbes. The spoilage characteristics of fresh fish can be divided into four stages. Stage I occurs from 0 to 6 days after death and involves shifting of bacterial populations without odor. Stage II occurs from 7 to 10 days after death, with bacterial growth becoming apparent and the development of a slightly fishy odor. Stage III occurs from 11 to 14 days after death and is characterized by rapid bacterial growth, a sour and fishy odor, and the start of slime formation on the skin. Stage IV occurs >14 days after death, at which point bacterial numbers are stationary, proteolysis begins, the skin is extremely slimy, and the odor is offensive.

Composition and Spoilage of Shellfish

Crustacean and molluscan shellfish generally contain larger amounts of free amino acids than finfish. Trimethylamine oxide is present in crustacean shellfish, with the exception of cephalopods, scallops, and cockles, and is absent in molluscan tissue. Crustaceans possess potent cathepsin-like enzymes, which rapidly break down proteins, leading to tissue softening and the development of volatile off odors. Removal of the head after harvest can extend shelf life by eliminating an organ that releases degradative enzymes. Some crustacean meats (primarily shrimp) suffer from a visual defect known as black spot melanosis, which is due to polyphenol oxidase activity and not to microbial action.

Molluscan shellfish contain a lower total nitrogen concentration in their flesh than do finfish or crustacean shellfish and much more carbohydrate, mostly in the form of glycogen. As a result, the spoilage pattern of molluscan shellfish differs from that of other seafoods and is generally fermentative, with the pH of tissues declining as spoilage progresses, yielding a predominance of lactobacilli and streptococci.

Factors Influencing Spoilage

Proteolytic and Lipolytic Activities

Although *Pseudomonas* and other aerobic spoilage bacteria are able to produce proteolytic enzymes, their production is delayed until the late logarithmic phase of growth. Proteolysis occurs only in populations of $>10^8$ CFU/cm^2, when spoilage is well advanced and the bacteria are approaching their maximum cell density.

Oxidative rancidity of fat occurs when unsaturated fatty acids react with oxygen from the storage environment. Stable compounds, such as aldehydes, ketones, and short-chain fatty acids, are produced, resulting in the eventual development of rancid flavors and odors. Autoxidation, independent of microbial activity, occurs in muscle foods stored in aerobic environments. The rate is influenced by the proportion of unsaturated fatty acids in the fat. Generally, lipase production is restricted while carbohydrate substrates in muscle tissue are being utilized; it is unlikely that microbial lipolytic activity would occur until glucose on the muscle surface is depleted. At this point, amino acids would also be degraded, and the resulting spoilage characteristics would possibly mask the effects of rancidity.

Spoilage of Adipose Tissue

Adipose tissue consists predominantly of insoluble fat, which cannot be used for microbial growth until it is broken down and emulsified. While the spoilage processes and growth rates of spoilage bacteria are similar for adipose and muscle tissues, the low level of carbohydrates in adipose tissue means that spoilage odors are detected when lower numbers of bacteria are present and most, if not all, available glucose is depleted. This occurs when populations exceed 10^6 CFU/cm^2. The growth rates of some psychrotrophic microorganisms, such as *Hafnia alvei*, *Serratia liquefaciens*, and *Lactobacillus plantarum*, are higher on fat than on lean beef and pork tissues. Spoilage bacteria, such as *S. putrefaciens*, may also grow. In practice, spoilage of fat before lean tissue is unlikely, given the presence of muscle tissue fluids in vacuum-packaged cuts and the restriction of bacterial growth due to the drying of carcass surfaces.

Spoilage under Anaerobic Conditions

Lactic acid bacteria dominate the spoilage microflora of muscle foods when oxygen is excluded from the storage environment. If the pH of the muscle tissue is high or residual amounts of oxygen are present, other microorganisms, such as *B. thermosphacta* and *S. putrefaciens*, may cause spoilage. The growth rate of bacteria under anaerobic conditions is considerably reduced compared to growth under aerobic conditions. In addition, the maximum cell density achieved under anaerobic conditions ($\sim$$10^8$ CFU/cm^2) is considerably less than that achieved under aerobic conditions ($>$$10^9$ CFU/cm^2). The sour, acid, cheesy odors or cheesy and dairy flavors that develop in muscle tissue under anaerobic conditions can be attributed, at least in part, to the accumulation of short-chain fatty acids and amines. The presence of *B. thermosphacta* in significant numbers causes more rapid spoilage than that of lactic acid bacteria.

Other Meat Characteristics Associated with Spoilage

Animals subjected to excessive stress or exercise before slaughter can have depleted levels of muscle glycogen. This results in a condition known as dark, firm, and dry. Spoilage of dark, firm, and dry meat occurs more quickly than spoilage of meat with normal pH. Rapid spoilage is caused by the absence of glucose, and hence lactic acid, in the tissues. Pale, soft, exudative muscle tissue is a condition that occurs in pork and turkey, and to a lesser extent in beef, in which sugar utilization decreases the muscle pH to its ultimate level while the muscle temperature is still high. There is debate over whether pale, soft, exudative meats spoil more slowly than meat with normal pH.

Comminuted Products

The limited shelf life of comminuted (ground and blended) muscle foods has been attributed to (i) a higher initial microbial load due to use of a poorer-quality product for grinding, (ii) contamination during processing, and (iii) the effects of the comminution of the muscle tissue. On the surfaces of aerobically stored comminuted products, *Pseudomonas*, *Acinetobacter*, and *Moraxella* species are the main microflora, while in the interior, due to the limited availability of oxygen, lactic acid bacteria are dominant. Occasional contaminants, such as *Aeromonas* spp. or members of the family *Enterobacteriaceae*, occur more often on comminuted products than on intact tissue.

Cooked Products
Cooking muscle foods destroys vegetative bacterial cells, although endospores may survive. For perishable, cooked, uncured meats, spoilage is caused by bacteria that survive heat processing or by postprocessing contaminants. Microorganisms responsible for spoilage of these products include psychrotrophic micrococci, streptococci, lactobacilli, and *B. thermosphacta*.

Processed Products
Microbial spoilage of processed products depends upon the nature of the product and the ingredients used. Spoilage of processed meats may be characterized as slimy spoilage, souring, or greening. *Slimy spoilage,* which is usually confined to the outside surfaces of product casings, results from the growth of some yeasts; two genera of lactic acid bacteria, *Lactobacillus* and *Enterococcus;* and *B. thermosphacta. Souring* occurs, typically beneath casings, when bacteria, such as lactobacilli, enterococci, and *B. thermosphacta,* utilize lactose and other sugars to produce acids. While *greening* of fresh meats may be the result of hydrogen sulfide production by certain bacteria, greening of cured meat products may also develop in the presence of hydrogen peroxide, which may form on the surfaces of vacuum- or modified-atmosphere-packaged meats when they are exposed to air. *Lactobacillus viridescens* is the most common cause of this type of greening, but species of *Streptococcus* and *Leuconostoc* may also produce the defect. The presence of these bacteria is often a result of poor sanitation.

Cured meats are spoiled by microorganisms that tolerate low a_w, such as lactobacilli or micrococci. If sucrose is added to cured meats, a slimy dextran layer may form due to the activity of *Leuconostoc* spp. or other bacteria, such as *L. viridescens. B. thermosphacta* may also be involved in the spoilage of these products. If dried meats are properly prepared and stored, their low a_w renders them microbiologically stable. To restrict the growth of some species of microbes that can grow at low a_w, the water content of dried meat products must be sufficiently reduced. Microbial spoilage of these products should not occur unless exposure to high relative humidity or other high-moisture conditions results in an uptake of moisture. The addition of antifungal agents may retard the growth of microbes in these products.

Control of Spoilage of Muscle Foods
Modifying intrinsic characteristics of products or extrinsic characteristics of the storage environment can control the growth of spoilage microorganisms on muscle foods. The shelf life of processed meats can be extended by processing procedures and ingredients that either prevent the growth of spoilage microorganisms or select for a less offensive spoilage microflora. For fresh meats, the extension of shelf life has become especially important as the food industry moves toward the centralization of processing activities, with products distributed to more distant domestic and international markets. Specific methods to prevent spoilage and prolong the shelf life of muscle foods include effective good manufacturing practices at lower levels for products entering storage and distribution; rinsing by water, immersion, and spray systems to remove physical and microbial contaminants from carcasses; and incorporation of antimicrobial compounds, such as chlorine and organic acids, in wash water. The use of lactic, acetic, propionic, and citric acids has been investigated, although lactic and acetic acids are most often used to reduce microbial levels on carcass surfaces. The

effectiveness of organic acids is influenced by factors such as the type of acid, concentration, temperature, and point of application in processing.

Temperature is the most important environmental parameter influencing the growth of microorganisms in muscle foods. As the temperature is decreased below the optimum for the growth of microorganisms, generation times and lag times are extended, and growth is therefore slowed (Fig. 19.1). Under aerobic storage conditions, the comparatively high growth rate of *Pseudomonas* spp. allows successful competition with mesophiles and other psychrotrophs at temperatures below 20°C. Likewise, under anaerobic conditions at temperatures below 20°C, the rapid growth rate of psychrotrophic lactobacilli allows successful competition with other psychrotrophic spoilage microorganisms. Storage of foods at temperatures at which mesophiles will grow allows these microorganisms to contribute to spoilage. Higher storage temperatures may also allow the growth of psychrotrophic spoilage microorganisms that would otherwise be restricted by the pH or lactic acid concentration of muscle tissue at reduced temperatures. Likewise, the minimum temperature for the growth of psychrotrophic spoilage microorganisms may be increased by other factors that influence growth, e.g., pH, a_w, or oxidation-reduction potential, or by inhibitory agents, such as food additives or increased carbon dioxide.

The shelf life of muscle foods can be extended by storage under vacuum or modified atmospheres. Modified-atmosphere packaging (MAP) involves the storage of products under a high-oxygen barrier film with a headspace with a different gas composition than air. Typically, this headspace contains elevated amounts of carbon dioxide, and it may also contain nitrogen and oxygen in various proportions. Much of the extended shelf life of such products results from a modification of the spoilage microflora from an aerobic psychrotrophic population, consisting of bacteria such as

Figure 19.1 Predicted populations of total aerobic bacteria in ground beef as affected by storage temperature. Reprinted from T. C. Jackson, D. L. Marshall, G. R. Acuff, and J. S. Dickson, p. 91–109, *in* M. P. Doyle, L. R. Beuchat, and T. J. Montville (ed.), *Food Microbiology: Fundamentals and Frontiers* (ASM Press, Washington, D.C., 2001).

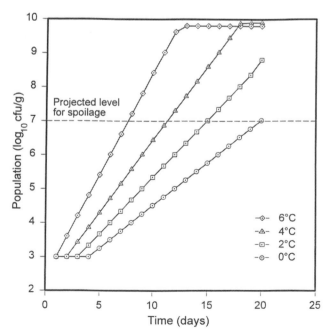

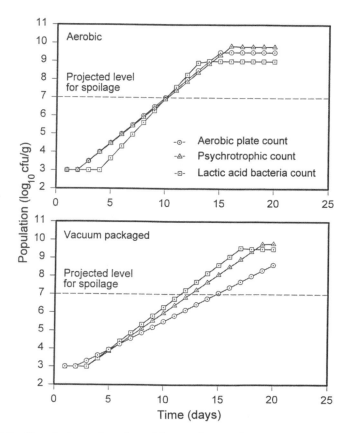

Figure 19.2   Populations of total aerobic, psychrotrophic, and lactic acid bacteria in aerobically packaged and vacuum-packaged ground pork stored at 0°C. The growth curves are based on the derived values for lag phase duration and generation time. Reprinted from T. C. Jackson, D. L. Marshall, G. R. Acuff, and J. S. Dickson, p. 91–109, *in* M. P. Doyle, L. R. Beuchat, and T. J. Montville (ed.), *Food Microbiology: Fundamentals and Frontiers* (ASM Press, Washington, D.C., 2001).

Pseudomonas, Moraxella, and *Acinetobacter,* to one consisting predominantly of lactic acid bacteria and *B. thermosphacta* (Fig. 19.2). Carbon dioxide is commonly used in modified-atmosphere environments due to its bacteriostatic activity. Carbon dioxide increases the lag phases and generation times of many microorganisms. Oxygen is occasionally included in modified atmospheres used to package fresh red meats to maintain the oxymyoglobin responsible for the fresh red, or "bloomed," color desired by the consumer. However, if significant amounts of oxygen are present, aerobic microorganisms and *B. thermosphacta* may grow, resulting in spoilage. Nitrogen may also be included to displace oxygen and delay oxidative rancidity and to inhibit the growth of aerobic microorganisms.

Irradiation of meats enhances microbiological safety and quality by significantly reducing populations of pathogenic and spoilage bacteria. At present, relatively low doses of irradiation, i.e., <5 kilograys (kGy), are receiving the most interest. The doses of radiation required to inactivate 90% of the vegetative cells of pathogenic bacteria vary with processing conditions (temperature and presence or absence of air) but typically range from a low of <0.2 kGy for *Aeromonas* and *Campylobacter* to as high as 0.77 kGy for *Listeria* and *Salmonella.* Although there is some variation among species, the doses required to inactivate 90% of the vegetative cells of spoilage bacteria typically fall in the range of 0.2 to 0.8 kGy. In contrast, the dose

Table 19.1   Some defects of fluid milk that result from microbial growth[a]

Defect	Associated microorganisms	Type of enzyme	Metabolic product(s)
Bitter	Psychrotrophic bacteria, *B. cereus*	Protease Peptidase	Bitter peptides
Rancid	Psychrotrophic bacteria	Lipase	Free fatty acids
Fruity	Psychrotrophic bacteria	Esterase	Ethyl esters
Coagulation	*Bacillus* spp.	Protease	Casein destabilization
Sour	Lactic acid bacteria	Glycolytic	Lactic and acetic acids
Malty	Lactic acid bacteria	Oxidase	3-Methyl butanal
Ropy	Lactic acid bacteria	Polymerase	Exopolysaccharides

[a]Adapted from J. F. Frank, p. 111–126, *in* M. P. Doyle, L. R. Beuchat, and T. J. Montville (ed.), *Food Microbiology: Fundamentals and Frontiers* (ASM Press, Washington, D.C., 2001).

required for 90% inactivation of *Clostridium* spores is as high as 3.5 kGy. Low-dose irradiation (<5 kGy) is effective in extending the shelf life of fresh meats. Because of significant reductions in both pathogenic and spoilage bacteria, low-dose irradiation has been suggested as a pasteurization process for fresh meats. One of the major drawbacks to irradiation is consumer hesitation to purchase irradiated foods.

MILK AND DAIRY PRODUCTS

This section focuses on spoilage defects in dairy products that result from microbial growth. Modern dairy processing utilizes pasteurization, heat sterilization, fermentation, dehydration, refrigeration, and freezing as preservation treatments. Combining a preservation method with component separation processes (i.e., churning, filtration, centrifugation, and coagulation) results in an assortment of dairy foods having vastly different tastes and textures and a complex variety of spoilage microflora. Spoilage of dairy foods is manifested as off flavors and odors and changes in texture and appearance. Some defects of milk and cheese caused by microorganisms are listed in Tables 19.1 and 19.2.

Milk and Dairy Products as Growth Media

Milk

Milk is a good growth medium for many microorganisms because of its high water content, near-neutral pH, and variety of available nutrients. Although milk is a good growth medium, the addition of yeast extract or protein hydrolysates often increases growth rates, suggesting that milk is not

Table 19.2   Some defects of cheese which result from microbial growth[a]

Defect	Associated microorganisms	Metabolic product
Open texture, fissures	Heterofermentative lactobacilli	Carbon dioxide
Early gas	Coliforms, yeasts	Carbon dioxide, hydrogen
Late gas	*Clostridium* spp.	Carbon dioxide, hydrogen
Rancidity	Psychrotrophic bacteria	Free fatty acids
Fruity	Lactic acid bacteria	Ethyl esters
White crystalline surface deposits	*Lactobacillus* spp.	Excessive D-lactate
Pink discoloration	*L. delbrueckii* subsp. *bulgaricus*	High redox potential

[a]Adapted from J. F. Frank, p. 111–126, *in* M. P. Doyle, L. R. Beuchat, and T. J. Montville (ed.), *Food Microbiology: Fundamentals and Frontiers* (ASM Press, Washington, D.C., 2001).

an ideal growth medium. Even though milk has a high fat content, few spoilage microorganisms utilize it as a carbon or energy source. This is because the fat is in the form of globules surrounded by a protective membrane composed of glycoproteins, lipoproteins, and phospholipids. Moreover, many microorganisms cannot utilize lactose and therefore must rely on proteolysis or lipolysis to obtain carbon and energy. Finally, freshly collected raw milk contains various microbial growth inhibitors.

The major nutritional components of milk are lactose, fat, protein, minerals, and various nonprotein nitrogenous compounds. Carbon sources in milk include lactose, protein, and fat. The citrate in milk can be utilized by many microorganisms but is not present in sufficient amounts to support significant growth. Sufficient glucose is present in milk to allow the initiation of growth by some microorganisms, but for fermentative microorganisms to continue growth, they must have the transport system and hydrolytic enzymes for lactose utilization. Other spoilage microorganisms may oxidize lactose to lactobionic acid. The lactose in milk is present in large quantities and is more than sufficient to support extensive microbial growth.

There are primarily two types of proteins in milk, caseins and whey proteins. Caseins are present in the form of highly hydrated micelles and are readily susceptible to proteolysis. Whey proteins (β-lactoglobulin, α-lactalbumin, serum albumin, and immunoglobulins) remain soluble in milk after the precipitation of casein. They are less susceptible to microbial proteolysis than caseins. Milk also contains nonprotein nitrogenous compounds, such as urea, peptides, and amino acids, that are readily available for microbial utilization, but these compounds are present in insufficient quantity to support the extensive growth required for spoilage.

The major microbial inhibitors in raw milk are lactoferrin and the lactoperoxidase system. Natural inhibitors of less importance include lysozyme, specific immunoglobulins, folate, and vitamin B_{12} binding systems. Lactoferrin, a glycoprotein, acts as an antimicrobial agent by binding iron. Psychrotrophic aerobes that commonly spoil refrigerated milk are inhibited by lactoferrin, but the presence of citrate in cow's milk limits its effectiveness, as the citrate competes with lactoferrin for binding the iron. The most effective natural microbial inhibitor in cow's milk is the lactoperoxidase system. Lactoperoxidase catalyzes the oxidation of thiocyanate and the simultaneous reduction of hydrogen peroxide, resulting in the accumulation of hypothiocyanite. Lactic acid bacteria, coliforms, and various pathogens are inhibited by this system.

The minimum heat treatment required for fluid milk to be sold for consumption in the United States is 72°C for 15 s, though most processors use slightly higher temperatures and longer holding times. Pasteurization affects the growth rate of spoilage microflora by destroying inhibitor systems. More severe heat treatments affect microbial growth by increasing available nitrogen through protein hydrolysis and by the liberation of inhibitory sulfhydryl compounds. Lactoperoxidase is only partially inactivated by normal pasteurization treatments.

Dairy Products

Dairy products are very different growth environments than fluid milk because they have nutrients removed or concentrated and may have a lower pH or a_w. Yogurt is essentially acidified milk and therefore provides a nutrient-rich low-pH environment. Cheeses are less acidified than yogurt, but they have added salt and less water, resulting in lower a_w. In addition, the solid nature of cheeses limits the mobility of spoilage microorganisms. Liquid milk concentrates, such as evaporated skim milk, do not have sufficiently low a_w to inhibit spoilage and must be canned or refrigerated for preservation. Milk-derived powders have sufficiently low a_w to completely inhibit microbial growth. Butter is a water-in-oil emulsion, so microorganisms are trapped within serum droplets. If butter is salted, the mean salt content of the water droplets will be 6 to 8%, sufficient to inhibit gram-negative spoilage organisms that could grow during refrigeration. Unsalted butter is usually made from acidified cream and relies on low pH and refrigeration for preservation.

Psychrotrophic Spoilage

The preservation of fluid milk relies on effective sanitation, pasteurization, timely marketing, and refrigeration. Immediately following collection, raw milk is rapidly cooled and is refrigerated until it is consumed (obviously, during pasteurization the milk is heated). There is often sufficient time between milk collection and consumption for psychrotrophic bacteria to grow. Many of the flavor defects detected by milk consumers result from this growth. Pasteurized milk is expected to have a shelf life of 14 to 20 days, so contamination of the contents of a container with even one rapidly growing psychrotrophic microorganism can lead to spoilage.

Psychrotrophic Bacteria in Milk

Psychrotrophic bacteria that spoil raw and pasteurized milk are primarily aerobic gram-negative rods in the family *Pseudomonadaceae*, with occasional representatives from the family *Neisseriaceae* and the genera *Flavobacterium* and *Alcaligenes*. Representatives of other genera, including *Bacillus, Micrococcus, Aerococcus,* and *Staphylococcus,* and the family *Enterobacteriaceae* may be present in raw milk and may be psychrotrophic, but they are usually outgrown by the gram-negative obligate aerobes (especially *Pseudomonas* spp.) when milk is held at its typical 3 to 7°C storage temperature. The psychrotrophic spoilage microflora of milk is generally proteolytic, with many isolates able to produce extracellular lipases, phospholipase, and other hydrolytic enzymes but unable to utilize lactose. The bacterium most often associated with flavor defects in refrigerated milk is *Pseudomonas fluorescens,* with *Pseudomonas fragi, Pseudomonas putida,* and *Pseudomonas lundensis* also commonly encountered. Psychrotrophic bacteria commonly found in raw milk are inactivated by pasteurization.

Sources of psychrotrophic bacteria in milk.   Soil, water, animals, and plant material constitute the natural habitat of psychrotrophic bacteria found in milk. Plant materials may contain $>10^8$ psychrotrophs per g, and farm water usually contains low levels of psychrotrophic microorganisms. Its use to clean and rinse milking equipment provides a direct means for their entry into milk. Psychrotrophic bacteria isolated from water are often very active producers of extracellular enzymes and grow rapidly at refrigeration temperatures. Milking equipment, utensils, and storage tanks are the major source of psychrotrophic contamination of raw milk. Pasteurized milk products become contaminated with psychrotrophic bacteria by exposure to contaminated equipment or air. Although the levels of psychrotrophic bacteria in air are generally quite low, only one viable cell per container can spoil the product.

Growth characteristics and product defects. Generation times in milk of the most rapidly growing psychrotrophic *Pseudomonas* spp. isolated from raw milk are 8 to 12 h at 3°C and 5.5 to 10.5 h at 3 to 5°C. These growth rates are sufficient to cause spoilage within 5 days if the milk initially contains only one cell per milliliter. However, most psychrotrophic pseudomonads present in raw milk grow much more slowly, causing refrigerated milk to spoil in 10 to 20 days. Defects of fluid milk associated with the growth of psychrotrophic bacteria are caused by the extracellular enzymes they make. Sufficient enzyme to cause defects is usually present when the population of psychrotrophs reaches 10^6 to 10^7 CFU/ml. Bitter, putrid flavors and coagulation result from proteolysis. Lipolysis leads to the formation of rancid and fruity flavors.

Proteases and their contribution to spoilage. *P. fluorescens* and other psychrotrophs that may be present in raw milk generally produce protease during the late-exponential and stationary phases of growth. The temperature for optimum protease production by psychrotrophic *Pseudomonas* spp. is lower than the temperature for optimum growth. Relatively large amounts of protease are produced at temperatures as low as 5°C, and production is inhibited in milk held at 2°C. The effects of calcium and iron ions on protease production by *Pseudomonas* spp. are relevant to dairy spoilage. Ionic calcium is required for protease synthesis. Iron, which may be at a growth-limiting concentration in milk, will repress protease production by *Pseudomonas* spp. when added to milk.

The properties of the proteinases produced by *P. fluorescens* that are most relevant to dairy product spoilage include temperature optima from 30 to 45°C with significant activity at 4°C and a pH optimum that is near neutral or alkaline, and all are metalloenzymes containing either Zn^{2+} or Ca^{2+}. Perhaps the most important characteristic of these enzymes in regard to dairy product spoilage is their extreme heat stability: they retain significant activity after ultrahigh-temperature (UHT) milk processing.

Proteases of psychrotrophic bacteria cause product defects either at the time they are produced in the product or as a result of the enzyme surviving a heating process. Degradation of casein in milk by enzymes produced by psychrotrophs results in the liberation of bitter peptides. Bitterness is a common off flavor in pasteurized milk that has been subject to postpasteurization contamination with psychrotrophic bacteria. Continued proteolysis results in putrid off flavors associated with low-molecular-weight degradation products, such as ammonia, amines, and sulfides. Bitterness in UHT (commercially sterile) milk develops when sufficient psychrotrophic bacterial growth occurs in raw milk (estimated at 10^5 to 10^7 CFU/ml), leaving behind residual enzymes after heat treatment. Low-level protease activity in UHT milk can also result in coagulation or sediment formation.

Lipases and their contribution to spoilage. Lipases of psychrotrophic pseudomonads, like proteases, are produced in the late-logarithmic or stationary phase of growth. As with protease, optimal synthesis of lipase generally occurs below the optimum temperature for growth. Milk is an excellent medium for lipase production by pseudomonads. Ionic calcium is required for lipase activity, as the activity inhibited by ethyldiaminetetraacetic acid (EDTA) is reversed by the addition of Ca^{2+}. Supplementation of milk with iron delays the onset of lipase production by *P. fluorescens* in raw milk.

Temperatures for optimal activity of lipases produced by *Pseudomonas* spp. range from 22 to 70°C, with most between 30 and 40°C. The optimal pH for activity is from 7.0 to 9.0, with most lipases having optima between 7.5 and 8.5. *P. fluorescens* lipase in milk is active at refrigeration temperatures and has significant activity at subfreezing temperatures and low a_w. The heat stability of lipases from psychrotrophic pseudomonads is similar to that of their proteases. Low-temperature inactivation is markedly affected by milk components, with milk salts increasing susceptibility and casein increasing stability.

The triglycerides in raw milk are present in globules that are protected from enzymatic degradation by a membrane. Milk becomes susceptible to lipolysis if this membrane is disrupted by excessive shear force (from pumping, agitation, freezing, etc.). Raw milk contains components that degrade the fat globule membrane and cause rancidity in raw and pasteurized milk. This process is independent of microorganisms. The rancid flavor and odor resulting from lipase action are usually due to the liberation of C-4 to C-8 fatty acids. Fatty acids of higher molecular weight produce a "soapy" flavor. Low levels of unsaturated fatty acids liberated by enzymatic activity may be oxidized to ketones and aldehydes to produce an oxidized, or "cardboardy," off flavor. *P. fragi* produces a fruity off flavor in milk by esterifying free fatty acids with ethanol. Lipolytic off flavors in UHT products generally take several weeks to months to develop due to the small amounts of lipase present or appear in products made from raw milk with excessively high populations of psychrotrophs ($>10^6$ CFU/ml).

A rancid defect in butter may result from the growth of lipolytic microorganisms, residual heat-stable microbial lipase, or milk lipase activity in the raw milk. The typical odor of rancid butter is associated with low-molecular-weight fatty acids (C-4 to C-8). Microbial lipases in butter are active even if the product is stored at −10°C. Growth of psychrotrophic bacteria in butter occurs only if the product is made from sweet rather than ripened (sour) cream. Sweet cream butter is preserved by salt and refrigeration. Butter is a water-in-fat emulsion, so moisture and salt will not equilibrate during storage. If salt and moisture are not evenly distributed in the product during manufacture, then lipolytic psychrotrophs will grow in pockets of high a_w present in the product.

Cheese is more susceptible to defects caused by bacterial lipases than to those caused by proteases. This is because most proteases are concentrated along with the fat in the curd. The acidic environment of most cheeses limits, but may not eliminate, lipase activity. Some cheeses, such as Camembert and Brie, increase in pH to near neutrality during ripening. More acidic cheeses, e.g., Cheddar, are susceptible if cured for several months or if large amounts of lipase are present. Psychrotrophic bacteria do not grow in cured cheeses because of the low pH and salt content, but they may grow in high-moisture fresh products, such as cottage cheese.

Control of Product Defects Associated with Psychrotrophic Bacteria
Preventing defects caused by psychrotrophic bacteria in raw milk involves limiting contamination levels, rapid cooling immediately after milking, and maintenance of cold storage temperatures. Rapid cooling of milk after collection is important. Fresh milk from the cow enters the farm storage tank (e.g., a bulk tank) at 30 to 37°C. Sanitary standards in the United States require raw milk to be cooled to 7°C within 2 h after milking, but most farm systems achieve rapid cooling to <4°C. Since milk is often picked up from

the farm every 48 h, three additional milkings will be added to the previously collected milk. The second milking should not warm the previously collected milk to >10°C.

Preventing contamination of pasteurized dairy products with psychrotrophic bacteria is primarily a matter of equipment cleaning and sanitation, although airborne psychrotrophs may also limit product shelf life. Even when filling equipment is effectively cleaned and sanitized, it can still become a source of psychrotrophic microorganisms, which accumulate during the normal hours of continuous use.

Spoilage by Fermentative Nonsporeformers

Fluid milk, cheese, and cultured milks are the major dairy products susceptible to spoilage by non-spore-forming fermentative bacteria. Non-spore-forming bacteria responsible for fermentative spoilage of dairy products are mostly in either the lactic-acid-producing or coliform group. The genera of lactic acid bacteria involved in spoilage of milk and fermented products include *Lactococcus, Lactobacillus, Leuconostoc, Enterococcus, Pediococcus,* and *Streptococcus*. Coliforms can spoil milk, but this is seldom a problem, since either the lactic acid or psychrotrophic bacteria usually outgrow them. Members of the genera *Enterobacter* and *Klebsiella* are most often associated with coliform spoilage, while *Escherichia* spp. only occasionally exhibit sufficient growth to produce a defect. Lactic-acid-producing bacteria are normal inhabitants of the skin of the cow's teat and in the environment. Coliform bacteria are present on udder skin and are also associated with milk residue buildup on inadequately cleaned milking equipment.

Defects of Fluid Milk Products

The most common fermentative defect in fluid milk products is souring. This is caused by the growth of lactic acid bacteria. Lactic acid by itself has a clean, pleasant acid flavor and no odor. The unpleasant sour odor and taste of spoiled milk result from small amounts of acetic and propionic acids. Sour odor can be detected before a noticeable acid flavor develops. Other defects may occur in combination with acid production. A malty flavor results from growth of *Lactococcus lactis* subsp. *lactis* bv. maltigenes. Malty flavor is primarily due to 3-methylbutanal.

Another defect associated with the growth of lactic acid bacteria in milk is "ropy" texture. Most dairy-associated species of lactic acid bacteria produce exocellular polymers that increase the viscosity of milk, causing the ropy defect. The polymer is a polysaccharide containing glucose and galactose with small amounts of mannose, rhamnose, and pentose. Some of polymer-producing strains are used to produce high-viscosity fermented products, such as yogurt and Scandinavian ropy milk.

Defects in Cheese

Some strains of lactic acid bacteria produce flavor and appearance defects in cheese. Lactobacilli are a normal part of the dominant microflora of aged Cheddar cheese. If heterofermentative lactobacilli predominate, the cheese is prone to develop an open texture or fissures, a result of gas production during aging. Gassy defects in aged Cheddar cheese are more often associated with the growth of lactobacilli than with the growth of coliforms, yeasts, or sporeformers. *Lactobacillus brevis* and *Lactobacillus casei* subsp. *pseudoplantarum* have been associated with gas production in retail mozzarella cheese. *Lactobacillus casei* subsp. *casei* produces a soft-body defect in mozzarella cheese. Some cheese varieties occasionally exhibit a pink discoloration, which is due to the growth of pigmented strains of propionibacteria and certain strains of *Lactobacillus delbrueckii* subsp. *bulgaricus*. Fruity off flavor in Cheddar cheese is usually a result of the growth of lactic acid bacteria (usually *Lactococcus* spp.) that produce esterase. The major esters contributing to fruity flavor in cheese are ethyl hexanoate and ethyl butyrate.

Growth of coliform bacteria usually occurs during the cheese-manufacturing process or during the first few days of storage and is therefore referred to as early gas (or early blowing) defect. In hard cheeses, such as Cheddar, this defect occurs when slow lactic acid fermentation fails to rapidly lower the pH or when highly contaminated raw milk is used. Soft, mold-ripened cheeses, such as Camembert, increase in pH during ripening, with a resulting susceptibility to coliform growth. Coliform growth in retail cheese often manifests as swelling of the plastic package. Approximately 10^7 CFU of coliform bacteria/g are needed to produce a gassy defect in the product.

Control of Defects Caused by Lactic Acid and Coliform Bacteria
Defects in fluid milk caused by coliforms and lactic acid bacteria are controlled by good sanitation practices during milking, maintaining raw milk at temperatures below 7°C, pasteurization, and refrigeration of pasteurized products. These microorganisms seldom grow to significant levels in refrigerated pasteurized milk because of their low growth rates compared to psychrotrophic bacteria. Control of coliform growth in cheese is achieved by using pasteurized milk, encouraging rapid fermentation of lactose, and good sanitation during manufacture. Controlling defects produced by undesirable lactic acid bacteria in cheese and fermented milks is more difficult, since the growth of lactic acid bacteria must be encouraged during manufacture and the final products often provide suitable growth environments. Undesirable strains of lactic acid bacteria are readily isolated from the manufacturing environment, so their control requires attention to plant cleanliness and protecting the product during manufacture.

Spore-Forming Bacteria
Spoilage by spore-forming bacteria can occur in low-acid fluid milk products that are preserved by substerilization heat treatments and packaged with little chance for recontamination with vegetative cells. Products in this category include aseptically packaged milk and cream and sweetened and unsweetened concentrated canned milks. Nonaseptic packaged refrigerated fluid milk may spoil due to growth of psychrotrophic *Bacillus cereus* and *Bacillus polymyxa* in the absence of more rapidly growing gram-negative psychrotrophs. Hard cheeses, especially those with low interior salt concentrations, are also susceptible to spoilage by spore-forming bacteria. Spore-forming bacteria that spoil dairy products usually come from the raw milk. The spore-forming bacteria in raw milk are predominantly *Bacillus* spp., with *Bacillus licheniformis*, *B. cereus*, *Bacillus subtilis*, and *Bacillus megaterium* most commonly isolated. *Clostridium* spp. are present in raw milk at such low levels that enrichment and most-probable-number techniques must be used for their quantification. Populations of spore-forming bacteria in raw milk vary seasonally; in temperate climates, *Bacillus* and *Clostridium* spp. are at higher levels in raw milk collected in the winter than in the summer.

Defects in Fluid Milk Products

Pasteurized milk can spoil due to the growth of psychrotrophic *B. cereus*. Germination of spores in raw milk occurs soon after pasteurization, indicating that they were heat activated. The defect produced by subsequent growth is described as *sweet curdling*, since it first appears as coagulation without significant acid or off flavor being formed. Coagulation is caused by a chymosin-like protease. Eventually, the enzyme degrades casein sufficiently to produce a bitter-flavored product. Growth may become visible as "buttons" at the bottom of the carton; these are actually bacterial colonies. Psychrotrophic *B. cereus* also produces phospholipase C (lecithinase), which degrades the fat globule membrane, resulting in the aggregation of the fat in cream. The result is described as *bitty cream defect*. Psychrotrophic *Bacillus* spp. other than *B. cereus*, including *Bacillus circulans* and *Bacillus mycoides*, are also capable of spoiling heat-treated milk. The major heat-resistant species in milk is *Bacillus stearothermophilus*. Other, less heat-resistant *Bacillus* spp. have been isolated from UHT milk, especially *B. subtilis* and *B. megaterium*.

Defects in canned condensed milk. Canned condensed milk may be either sweetened with sucrose and glucose to lower the a_w or left unsweetened. The unsweetened product must be sterilized by heat treatment. *Sweet coagulation* is caused by growth of *Bacillus coagulans*, *B. stearothermophilus*, or *B. cereus*. This defect is similar to the sweet-curdling defect caused by psychrotrophic *B. cereus* in pasteurized milk. Protein destruction, in addition to curdling, can also occur and is usually caused by the growth of *B. subtilis* or *B. licheniformis*. Swelling or bursting of cans can be caused by the growth of *Clostridium sporogenes*. *Flat sour defect* (acidification without gas production) can result from the growth of *B. stearothermophilus*, *B. licheniformis*, *B. coagulans*, *Bacillus macerans*, and *B. subtilis*. Sweetened condensed milk should have sufficiently low a_w to inhibit bacterial spore germination. However, if the a_w is not well controlled, *Bacillus* spp. may produce acid or acid-proteolytic spoilage. Methods for controlling the growth of spore-formers in fluid products mainly involve the use of appropriate heat treatments, including UHT treatments.

Defects in Cheese

The major defect in cheese caused by spore-forming bacteria is gas formation, usually resulting from the growth of *Clostridium tyrobutyricum* and occasionally *C. sporogenes* and *Clostridium butyricum*. This defect is often called late blowing or late gas, because it occurs after the cheese has aged for several weeks. Emmental, Swiss, Gouda, and Edam cheeses are most often affected because of their relatively high pH and moisture content, in addition to their low interior salt levels. The defect can also occur in Cheddar and Italian cheeses. Processed cheeses are susceptible to late blowing because spores are not inactivated during heat processing. Late gas defect results from the fermentation of lactate to butyric acid, acetic acid, carbon dioxide, and hydrogen gas.

Yeasts and Molds

Growth of yeasts and molds is a common cause of spoilage of fermented dairy products, because these microorganisms grow well at low pH. Yeast spoilage is manifested as fruity or yeasty odor and/or gas formation. Hard (or cured) cheeses, when properly made, have very small amounts of lactose, thus limiting the potential for yeast growth. Cultured milks, such as yogurt and buttermilk, and fresh cheeses, such as cottage cheese, normally contain fermentable levels of lactose. They are prone to yeast spoilage. A fermented or yeasty flavor observed in Cheddar cheese spoiled by growth of a *Candida* sp. is associated with elevated ethanol, ethyl acetate, and ethyl butyrate. The affected cheese has a high moisture content (associated with low starter activity and therefore high residual lactose) and a low salt content, which contribute to yeast growth. Yeast spoilage can also occur in dairy foods with low a_w, such as sweetened condensed milk and butter. The most common yeasts present in dairy products are *Kluyveromyces marxianus* and *Debaromyces hansenii* and their counterparts *Candida famata*, *Candida kefyr*, and other *Candida* spp. Also prevalent are *Rhodotorula mucilaginosa*, *Yarrowia lipolytica*, and *Torulospora* and *Pichia* spp. *Candida* yeasts have been isolated from vacuum-packaged cheese. Mold spores do not survive pasteurization. Fermented dairy products provide a highly specialized ecological niche for yeasts, selecting for those that can utilize lactose or lactic acid and that tolerate high salt concentrations. Yeasts able to produce proteolytic or lipolytic enzymes may also have a selective advantage for growth in dairy products.

The most common molds found on cheese are *Penicillium* spp. Others, such as *Aspergillus, Alternaria, Mucor, Fusarium, Cladosporium, Geotrichum,* and *Hormodendrum*, are occasionally found. Mold species commonly isolated from processed cheese include *Penicillium roqueforti, Penicillium cyclopium, Penicillium viridicatum,* and *Penicillium crustosum*. Vacuum-packaged Cheddar cheese supports the growth of *Cladosporium cladosporioides, Penicillium commune, Cladosporium herbarum, Penicillium glabrum,* and *Phoma* spp. Antimycotic chemicals, such as sorbate, propionate, and natamycin (pimaracin), are used to control mold growth. Some *Penicillium* spp. not only are resistant to sorbate but can degrade it by decarboxylation, producing 1,3-pentadiene. This imparts a kerosene-like odor to the cheese.

SPOILAGE OF PRODUCE AND GRAINS

This section focuses on changes in color, flavor, texture, or aroma brought about by the growth of microorganisms on fruits, vegetables, and grains. For the purpose of this chapter, plant pathology as it relates to the degradation of plant materials is restricted to spoilage problems that arise before harvest, whereas food microbiology deals with spoilage after harvest. Although, according to scientific definitions, there are differences between fruits and vegetables, some confusion does occasionally occur, especially in the minds of consumers. Fruits are defined as the seed-bearing organs of plants and include not only well-known commodities, such as apples, citrus fruits, and berries, but also items sometimes thought of as vegetables, such as tomatoes, bell peppers, and cucumbers. In contrast, vegetables are defined as all other edible portions of plants, including leaves, roots, and seeds.

Types of Spoilage

In general, three broad types of spoilage exist in plant products. The first type is active spoilage, caused by plant-pathogenic microorganisms actually initiating infection of otherwise healthy and uncompromised products. This reduces sensory quality. A second type of spoilage is passive, or wound-induced, spoilage, in which opportunistic microorganisms gain access to internal tissues via damaged epidermal tissue, i.e., peels or skins. This type of spoilage often occurs soon after the product has been damaged by harvesting, processing equipment, or insects. Similarly, passive spoilage can occur when opportunistic spoilage microorganisms gain entry into internal tissues via lesions caused by plant pathogens or via natural openings, such as stomata.

Spoilage of plant products can be manifested in a variety of ways, depending on the specific product, the environment, and the microorganisms involved. Traditionally, spoilage has been described by the symptoms most often associated with a particular product. Listed in Table 19.3 are examples of some common types of fungal spoilage of fruits and vegetables and the molds normally associated with them. Caution should be exercised in using this system, since more than one type of microorganism may produce identical or similar symptoms. For example, the best-known type of spoilage in vegetables is soft rot, which is usually evidenced by obvious softening of the plant tissue. The disease can be caused by various plant-pathogenic bacteria, most notably *Erwinia carotovora*; however, bacteria, such as *Bacillus* and *Clostridium* spp., and yeasts and molds are also

Table 19.3 Types of fungal spoilage of fruits and vegetables[a]

Product involved	Type of spoilage	Mold responsible
Fruits		
Citrus fruits	*Alternaria* rot	*Alternaria*
Bananas	Anthracnose (bitter rot)	*Colletotrichum musae*
Onions, sweet potatoes	Black rot	*Aspergillus niger, Ceratocystis fimriata*
Peaches	Brown rot	*Monilinia fructicola*
Bananas	Crown rot	*Colletotrichum musae, Fusarium roseum, Verticillium theobromae, Ceratocystis paradoxa*
Grapes	Gray mold rot	*Botrytis cinerea*
Pineapples	Pineapple black rot	*Ceratocystis paradoxa*
Tomatoes, citrus fruits	Sour rot	*Geotrichum candidum*
Apples, pears	Lenticel rot	*Cryptosporiopsis malicorticus, Phlyctaena vagabunda*
Citrus fruits	Green mold rot	*Penicillium digitatum*
Oranges	Blue rot	*Penicillium*
Peaches, cherries	*Cladosporium* rot	*Cladosporium herbarum*
Vegetables		
Onions	Black mold rot	*Aspergillus*
Carrots, cauliflower	Black rot	*Alternaria*
Lettuce, spinach	Downy mildew	*Bremia, Phytophthora*
Asparagus	*Fusarium* rot	*Fusarium*
Cabbage	Gray mold rot	*Botrytis*
Green beans	*Rhizopus* soft rot	*Rhizopus*
Onions	Smudge (anthracnose)	*Colletotrichum*
Potatoes	Tuber rot	*Fusarium*
Celery	Watery soft rot	*Sclerotinia*
Green beans	Wilt	*Pythium*
Oranges	Blue rot	*Penicillium*
Eggplant	Blight	*Phomopsis*
Bananas	Finger rot	*Pestalozzia, Fusarium, Gloeosporium*
Peaches	Pink rot	*Trichothecium*

[a]Adapted from J. M. Jay, *Modern Food Microbiology,* 4th ed. (Van Nostrand Reinhold, New York, N.Y., 1992).

occasionally implicated in soft-rot spoilage. Types of spoilage are also known by the names of the microorganisms that cause them. For example, *Fusarium* rot and *Rhizopus* soft rot describe not only the symptoms of the spoilage but also the mold that causes them.

Mechanisms of Spoilage

Intact healthy plant cells possess a variety of defense mechanisms to resist microbial invasion. Thus, before microbial spoilage can occur, these defense mechanisms must be overcome. Fruits, vegetables, grains, and legumes have an epidermal layer of cells, i.e., skin, peel, or testa, that provides a protective barrier against the infection of internal tissues. The compositions of the epidermal tissues vary, but the walls of cells in the tissue usually consist of cellulose and pectic materials, and the outermost cells are covered by a layer of waxes (*cutin*). Damage due to relatively uncontrollable factors, such as insect infestation, windblown sand, or rubbing against neighboring surfaces, can occur before harvest. Postharvest damage is often the result of poorly designed or maintained processing equipment. Some microorganisms, especially plant-pathogenic molds, possess mechanisms to penetrate external tissues of plants. Once external barriers are penetrated, not only do these microorganisms quickly invade internal tissues, but other opportunistic microorganisms often take advantage of the availability of nutrients present in damaged tissue and cause additional spoilage. Once microorganisms penetrate the outer tissues of fruits, vegetables, or grains, they still must gain access to internal areas and individual cells to extract nutrients. Plant tissues are held together by the middle lamella, which is composed primarily of pectic substances. The cell wall consists of the primary layer, composed of cellulose and pectates, and the secondary layer, which consists almost entirely of cellulose.

Degradative enzymes play an important role in the postharvest spoilage of plant products. Five classes of microbial enzymes are primarily responsible for the degradation of plant materials. These are pectinases, cellulases, proteases, phosphatidases, and dehydrogenases. Because pectin and cellulose constitute the main structural components of plant cells, pectinases and cellulases are the most important degradative enzymes involved in spoilage. Pectinases are enzymes that cause depolymerization of the pectin chain. Although pectinases are often discussed as a single enzyme, three main types are recognized for their roles in plant spoilage. Pectinases are produced by several plant pathogens, including *Botrytis cinerea*, *Monilinia fructicola*, *Penicillium citrinum*, and *E. carotovora*. Some pectinases are chain splitting and reduce the overall length of the pectin chain. The ultimate degradation of the pectin chain results in liquefaction of the pectin and complete breakdown of plant tissues. Cellulases are the second major class of degradative enzymes that can lead to spoilage. Cellulases function by degrading cellulose (glucose polymer) to glucose. Like pectinases, several types of cellulases exist, some of which attack native cellulose by cleaving cross-linkages between chains, while others act by breaking the cellulose into shorter chains. Cellulases are of less importance in postharvest spoilage of plant products than pectinases.

Influence of Physiological State

The physiological state of plant products, especially those consisting of fruits or vegetables, can have a dramatic effect on susceptibility to microbiological spoilage. Fruits, vegetables, and grains usually possess some sort of defense mechanism to resist infection by microorganisms. Usually these mechanisms are most effective when the plant is at peak physiologic health. Once plant tissues begin to age or are in a suboptimal physiologic state, resistance to infection diminishes. Fruits and vegetables differ in the ways in which they change physiologically after detachment from the plant. Nonclimacteric (i.e., no longer exhibiting a burst in respiration rate) fruits and vegetables, such as strawberries, beans, and lettuce, cease to ripen once they have been harvested. In contrast, climacteric fruits and vegetables, such as bananas and tomatoes, continue to mature and ripen after harvest. Ripening can continue to the point where normal cell integrity begins to diminish and tissues deteriorate. This process, known as *senescence,* arises from the accumulation of degradative enzymes produced by the fruit or vegetable and is unrelated to microbial decay. However, the loss of cellular integrity brought about by senescence makes fruits and vegetables even more susceptible to microbial infection and spoilage. Consequently, climacteric fruits and vegetables are usually among the most perishable of plant products.

Microbiological Spoilage of Vegetables

Although vegetables, fruits, grains, and legumes are all plant-derived products, they possess inherent differences that influence both the natural microflora and the type of spoilage encountered. Important intrinsic factors which influence the microfloras that develop on plant products include the pH and a_w. In general, the natural microfloras of vegetables include bacteria, yeasts, and molds representing many genera. Examples of the wide variety of microorganisms associated with fruit and vegetable products are shown in Table 19.4. However, microfloras can vary considerably, depending on the type of vegetable, environmental considerations, seasonality, and whether the vegetables were grown in close proximity to the soil. Both gram-positive and gram-negative bacteria are normally present on vegetables at the time of harvest.

Spoilage Microflora

A variety of microorganisms can cause spoilage of vegetables. Table 19.3 lists some molds that cause spoilage and the types of spoilage associated with them. Yeasts, molds, and bacteria cause spoilage of vegetables; however, bacteria are more frequently isolated from initial spoilage defects. The reason for this is that bacteria are able to grow faster than yeasts or molds in most vegetables and therefore have a competitive advantage, particularly at refrigeration temperatures.

Spoilage can be influenced by the history of the land on which vegetables are grown. For example, repeated planting of one type of vegetable on the same land over several seasons can lead to the accumulation of plant pathogens in the soil and increased potential for spoilage. Soil can also be contaminated by floodwater or poor-quality irrigation water.

Virtually all vegetables receive at least some processing or handling before they are consumed. It is important to realize that many common processing steps increase the likelihood of spoilage. In most cases, processors of vegetables destined to be sold as fresh produce add chlorine to wash water to achieve a concentration of 5 to 250 mg/liter. Chlorine is an effective antimicrobial agent. However, washing with chlorinated water and other disinfectants has only a limited antimicrobial effect on the microflora of the produce. Many processors consider washing fresh produce with chlorinated water a disinfection step, but this is not the real purpose

Table 19.4 Psychrotrophic and lactic acid bacteria and yeasts identified in stored vegetable salads[a]

Organism[b]	% of isolates
Psychrotrophic bacteria	
Enterobacter intermedium	1
Pasteurella haemolytica biovar T	1
Pseudomonas spp.	1
Pseudomonas aeruginosa	1
Pseudomonas picketti biovar 1	1
Pseudomonas putrefaciens	1
Pseudomonas stutzeri/P. mendocina group	1
Salmonella enterica serovar Choleraesuis	1
Acinetobacter	3
Enterobacter agglomerans	3
Pasteurella ureae	3
Plesiomonas shigelloides	3
Pseudomonas maltophilia	3
Staphylococcus cohnii	4
Acinetobacter antitratus	5
Chromobacterium violaceum	7
Pseudomonas putida	8
Pseudomonas fluorescens	19
ND	33
Lactic acid bacteria	
Lactobacillus curvatus	2
Lactobacillus lactis	2
Lactobacillus fermentum	4
Lactobacillus plantarum	7
Lactobacillus paracasei	11
Leuconostoc mesenteroides	16
Lactobacillus brevis	22
ND	36
Yeasts	
Candida valida	4
Geotrichum sp.	4
Hansenula anomala	4
Rhodotorula splutinis	4
Torulopsis spp.	4
Candida lambica	11
Trichosporon sp.	18
ND	50

[a]Adapted from R. M. Garcia-Gimeno and G. Zurera-Cosano, *Int. J. Food Microbiol.* **36**:31–38, 1997.
[b]ND, not determined.

of the inclusion of chlorine. Rather, chlorine is more effective in killing microorganisms in the water and minimizing contamination of the vegetables by the rinse water.

Cutting, slicing, chopping, and mixing are other important processing steps for fresh vegetables. These operations are becoming even more important as the demand for ready-to-eat products increases. The processes can result in increases in populations of microorganisms on fresh vegetables through the transfer of microorganisms from the equipment to the product.

Storage, packaging, MAP, and transportation are other important processing steps that can influence the development of microbiological spoilage of vegetables. Most modified-atmosphere techniques involve reducing the concentration of oxygen while increasing the concentration of carbon dioxide. MAP functions by reducing the respiration and senescence processes of fresh fruits and vegetables, thereby delaying undesirable changes in sensory quality. The reduction of senescence in produce usually requires carbon dioxide concentrations of at least 5%; however, concentrations in excess of 20% can be detrimental to food quality. In general, aerobic gram-negative bacteria are most sensitive to carbon dioxide. Obligate and facultative anaerobic microorganisms are more resistant. Likewise, molds are more sensitive to carbon dioxide than are fermentative yeasts.

More extensive processing techniques, such as thermal processing (canning) and freezing, have been used for years to preserve vegetables. Improper storage temperatures primarily cause microbiological spoilage of frozen vegetables. The microfloras in frozen vegetables are essentially the same as in raw products. Therefore, the spoilage patterns can be expected to be similar to those of raw products if the products are not properly maintained in a frozen state.

Preservation of vegetables by canning is accomplished by placing the vegetables in hermetically sealed containers and then heating them sufficiently to destroy microorganisms. For most vegetables, processing temperatures exceed 120°C. This eliminates all but the most heat-resistant bacterial spores. Consequently, spoilage of canned vegetables is usually caused by thermophilic spore-forming bacteria, unless the container integrity has been compromised in some way. Acidification without gas production is one of the most common types of spoilage observed in canned vegetables. This defect, called flat sour, is caused by *B. stearothermophilus* or *B. coagulans*. The defect ordinarily occurs when cans receive a marginally adequate thermal treatment or if they are stored at temperatures above 40°C. Another type of spoilage seen with canned vegetables is gas production and consequent swelling of cans. Swelling is caused by thermophilic spore-forming anaerobes, such as *Clostridium thermosaccharolyticum,* which grow and produce large amounts of hydrogen and carbon dioxide. The production of hydrogen sulfide in some canned vegetables can lead to a spoilage problem known as "sulfide stinker." This type of spoilage usually occurs in the absence of can swelling.

Microbiological Spoilage of Fruits

Fresh fruits are similar to vegetables in that they usually have a high enough a_w to support the growth of all but the most xerophilic (grows best at low a_w) or osmophilic (grows best at high a_w) fungi. However, most fruits differ from vegetables in that they have a more acidic pH (<4.4), the exception being melons. Fruits also have a higher sugar content. In addition, fruits usually possess more effective defense mechanisms, such as thicker epidermal tissues and higher concentrations of antimicrobial organic acids.

Normal and Spoilage Microfloras

As with vegetables, the normal microfloras of fruits are varied and include both bacteria and fungi. The sources of microorganisms include all those mentioned for vegetables, such as air, soil, and insects. Unlike that of vegetables, however, spoilage of fruits is most often due to yeasts or molds, although bacteria are also important. Examples of bacterial spoilage are *Erwinia* rots of pears or the production of buttermilk-like flavors in orange juice by lactic acid bacteria. Many of the same handling and processing techniques for fresh vegetables also apply to fresh fruits. Fruits are usually washed or rinsed immediately after harvest and then are usually packed into shipping cartons. The washing step can play an important role in reducing microbial contamination by as much as 3 log CFU/g.

Fresh-cut or minimally processed fruit products have become increasingly popular in recent years and will likely become more popular due to their convenience. Both molds and yeasts have the ability to cause spoilage; the latter are more often associated with spoilage of cut fruits due to their ability to grow faster than molds. However, bacteria also play a role in the spoilage of fruits.

Several types of fruit are dried to yield intermediate-moisture products, which normally rely on low a_w as the main mechanism for preservation. Raisins and prunes, and usually dates and figs, are consumed as dried fruit products. Dried apples, apricots, and peaches are also popular with consumers. Depending on the means by which fruits are dehydrated, moisture levels can range from <5 to 35%. In general, most dried fruits have sufficiently low a_w to inhibit bacterial growth. Spoilage is usually limited to osmophilic yeasts or xerotolerant molds. Populations of yeasts and molds on dried fruit usually average <10^3 CFU/g. Yeasts normally associated with spoilage of dried fruits include *Zygosaccharomyces rouxii* and *Hanseniaspora, Candida, Debaryomyces*, and *Pichia* species. Molds capable of growth below an a_w of 0.85 include several *Penicillium* and *Aspergillus* species (especially the *Aspergillus restrictus* series), *Eurotium* spp. (the *Aspergillus glaucus* series), and *Wallemia sebi*.

Fruit concentrates, jellies, jams, preserves, and syrups are resistant to spoilage due to their low a_w. Unlike that of dried fruits, however, the reduced a_w of these products is achieved by adding sufficient sugar to achieve a_w values of 0.82 to 0.94. In addition to reduced a_w, these products are also usually heated to temperatures of 60 to 82°C, which kills most xerotolerant fungi. Consequently, spoilage of these products usually occurs when containers have been improperly sealed or after consumers open them.

Various heat treatments are used to preserve fruit products. Fruits generally require less severe treatment than vegetables due to the enhanced lethality brought about by their acidic pH. Many canned fruits, whether halved, sliced, or diced, are processed by heating the products to a can center temperature of 85 to 90°C. Some processed fruit juices and nectars are rapidly heated to 93 to 110°C and then aseptically placed into containers. In either case, the processes are sufficient to kill most vegetative bacteria, yeasts, and molds. However, several genera of molds produce heat-resistant ascospores or sclerotia. Molds typically associated with spoilage of thermally processed fruits products include *Byssochlamys fulva, Byssochlamys nivea, Neosartorya fischeri*, and *Talaromyces flavus*. Signs of spoilage for heat-processed fruit products include visible mold growth, off odors, breakdown of fruit texture, or solubilization of starch or pectin in the suspending medium.

In recent years, food microbiologists have become aware that a thermotolerant spore-forming bacterium, *Alicyclobacillus acidoterrestris*, can be a problem in pasteurized juices. This bacterium is sufficiently acid and heat tolerant to survive the pasteurization conditions that kill most other bacteria. *A. acidoterrestris* produces 2-methoxyphenol (guaiacol), which imparts a medicine-like or phenolic off flavor to several types of juices, most notably apple and orange juices.

Microbiological Spoilage of Grains and Grain Products

Although grains are similar to fruits and vegetables in that they are of plant origin, they differ in many ways. Unlike fruits and vegetables, grains are typically thought of as primarily agronomic or field crops rather than horticultural crops. Grains destined for human food are ground into flour or meal for bakery or pasta products or further processed into snacks or breakfast cereals. The final products, except doughs, often have a_w values of <0.65, below which most microorganisms will not grow.

Natural Microfloras

Grains and grain products normally contain several genera of bacteria, molds, and yeasts. The specific species depends on the conditions encountered during production, harvesting, storage, and processing. Although the low a_w of grains and grain products might lead one to believe that fungi, especially molds, are the predominant natural microflora, this is not always the case. Indeed, depending on the storage conditions, type of grain, and moisture content, bacterial levels may far exceed mold populations. Of the many different types of microorganisms present on grains, only a few invade the kernel itself. Molds such as *Alternaria, Fusarium, Helminthosporium,* and *Cladosporium* are primarily responsible for invading wheat in the field. However, infection by these molds is minimal and does little damage unless the grains are allowed to become too moist.

Effects of Processing

Grains intended for human consumption are rarely used in their native state but undergo various processing treatments. Each step in the pretreatment and milling operations reduces populations of microorganisms so that the flours usually contain smaller populations than the grains from which they are made. Aside from fewer microorganisms being present in flour, the profile of the microflora in milled grains is similar to that of whole grains. The final processing step to which most grain products, specifically flours, are subjected is the addition of liquids (e.g., water or milk) and other ingredients to produce doughs.

Types of Spoilage

Properly dried and stored grains and grain products are inherently resistant to spoilage due to their low a_w. However, despite attempts to protect grains from the uptake of water during storage, the a_w can increase to a level that enables xerotolerant molds to grow. Moreover, it should be kept in mind that reduced a_w often will only slow the growth of fungi and that spoilage will ultimately occur given enough time. A subtle change in temperature in large bulk quantities of grains can cause condensation to develop in some areas, increasing moisture levels sufficiently to allow mold growth. In such cases, molds such as *Aspergillus, Eurotium* spp., and *W. sebi* may grow and cause spoilage. Molds are the primary spoilage microorganisms in grains due to their ability to grow at reduced a_w. The primary concern with mold spoilage is the potential production of mycotoxins. Molds can also cause adverse changes in the color, flavor, and aroma of grain products due to the production of aromatic volatile compounds. The production of volatile metabolites depends more on the fungal species than on the grain type. Another important adverse affect of mold growth on grains is an increase in the free fatty acid value (FAV). Increases in the FAV can result from fungal lipase activity. Because some of these lipases are heat stable, the FAV is sometimes used as an indication of fungal spoilage of grains.

Most grain-based foods receive some type of heat treatment before they are eaten. These treatments are usually sufficient to inactivate all but the most heat-resistant microorganisms. However, development of a spoilage condition known as *ropiness,* caused by the germination of spores of *B. subtilis* or *B. licheniformis,* and a defect referred to as *bloody bread* resulting from the growth of the red-pigmented bacterium *Serratia marcescens* may occur, albeit rarely, in commercially produced breads.

Summary

- *Pseudomonas* spp. are involved in spoilage of dairy, meat, poultry, and produce.
- Attachment of bacteria to muscle tissue is a two-step process.
- Processing equipment can be a major source of spoilage microorganisms.
- Good manufacturing practices are essential in limiting populations of spoilage microorganisms associated with raw materials and the finished product, thereby extending shelf life.
- Control of temperature is critical to prevent spoilage and extend shelf life.
- Proteolytic spoilage occurs prior to lipolytic spoilage.
- Food composition (percent adipose, carbohydrate, and protein) influences microbial growth and ultimately spoilage.
- Growth of one type of microorganism may influence the growth of other microorganisms and result in spoilage.

Suggested reading

Jay, J. M. 1992. *Modern Food Microbiology,* 4th ed., p. 187–198. Van Nostrand Reinhold, New York, N.Y.

Whitfield, F. B. 1998. Microbiology of food taints. *Int. J. Food Sci. Technol.* **33:**31–51.

Questions for critical thought

1. The microbial flora of fish depends on the temperature of the waters that the fish are harvested from. What types of microorganisms might be expected to be associated with fish harvested from temperate waters? From tropical waters? What others factors influence the spoilage of fish?
2. Raw milk is protected from spoilage by at least three intrinsic systems. Name the three systems and the bases on which they inhibit spoilage.
3. Chlorine is often added to water used to rinse vegetables. What impact does this treatment have on the microbial load of the treated product?

4. The a_ws of many products are such that the product is resistant to spoilage. Explain how a_w affects spoilage.

5. Why is it likely that a product will undergo proteolytic spoilage before lipolytic spoilage?

6. What properties of fruits and vegetables support spoilage by yeast and molds rather than by bacteria?

7. What is the cause of rancid defect in butter? How does the spoilage of ripened-cream butter differ from that of sweet-cream butter?

8. Comment on the use of irradiation to enhance the microbiological safety of meats. Can this technique be used for fresh produce (fruits and vegetables)? What are its disadvantages?

9. What types of microorganisms are involved in the contamination and spoilage of canned dairy products? Investigate methods that can be used to control sporeformers in canned foods.

10. How does spoilage differ in climacteric and nonclimacteric fruits? What is senescence, and how can it be controlled?

20

Molds*

LEARNING OBJECTIVES

The information in this chapter will enable the student to:

- gain knowledge about common food spoilage molds and those associated with disease
- understand methods used for differentiating and identifying molds
- associate specific mycotoxins with molds that produce toxins of concern
- identify foods that are commonly associated with mold problems
- know conditions that contribute to the growth of molds and toxin production in foods
- discuss methods used for screening of raw and processed materials for the presence of mycotoxins

INTRODUCTION

We have known for centuries that eating certain mushrooms can cause illness or even be fatal. That some common food spoilage molds cause disease and death is a newer realization. Indeed, in 1940, a *Penicillium* species was identified as the source of toxicity in "yellow" rice in Japan. Mycotoxins came to the attention of scientists in the Western world in the early 1960s with the outbreak of turkey "X" disease in England. This disease killed ~100,000 turkeys and other farm animals. The cause of the disease was traced to peanut meal in the feed, which was heavily contaminated with *Aspergillus flavus*. Analysis of the feed revealed that a group of fluorescent compounds, later named aflatoxins, were responsible for the outbreak. Aflatoxins also killed large numbers of ducklings in Kenya and caused hepatoma in hatchery-reared trout in California. This chapter focuses on molds that are commonly associated with the spoilage of foods.

ISOLATION, ENUMERATION, AND IDENTIFICATION

Bacteria grow much faster than molds and are often found in contaminated samples. Therefore, media for the enumeration of fungi contain antibacterial compounds and compounds to inhibit mold colony spreading, such as dichloran rose bengal chloramphenicol agar or dichloran 18% glycerol agar. Czapek agar, a defined medium based on mineral salts, or a derivative such as Czapek yeast extract agar and malt extract agar are used for the

---

*This chapter is based on chapters written by Ailsa D. Hocking, John I. Pitt, and Lloyd B. Bullerman for *Food Microbiology: Fundamentals and Frontiers*, 2nd ed., and has been adapted for use in an introductory text.

identification of *Aspergillus* species from foods. Growth on Czapek yeast extract 20% sucrose agar can also be a useful aid in identifying species of *Aspergillus*. The most effective medium for rapid detection of aflatoxigenic molds is *A. flavus* and *Aspergillus parasiticus* agar. Under incubation at 30°C for 42 to 48 h, *A. flavus* and *A. parasiticus* produce a bright orange-yellow colony reverse, which is diagnostic and readily recognized. The structures of *A. flavus* and *A. parasiticus* are shown in Fig. 20.1.

Unlike *Penicillium* species, *Aspergillus* species are conveniently "color coded," and the color of the *conidia* (nonmotile spores) serves as a very useful starting point in identification. The microscopic morphology is also important in identification. *Phialides* (cells producing conidia) may be produced directly from the swollen apex (*vesicle*) of long stalks (*stipes*), or there may be an intermediate row of supporting cells (*metulae*). Correct identification of *Aspergillus* species is an essential prerequisite to assessing the potential for mycotoxin contamination in a commodity, food, or feedstuff.

General enumeration procedures suitable for foodborne molds are effective for enumerating all common *Penicillium* species. Many antibacterial media give satisfactory results. However, some *Penicillium* species grow rather weakly on very dilute media, such as potato dextrose agar. Therefore, dichloran rose bengal chloramphenicol and dichloran 18% glycerol agars are recommended.

Identification of *Penicillium* isolates to the species level is not easy and is preferably carried out under carefully standardized conditions of medium, incubation time, and temperature. In addition to the microscopic morphology, gross physiological features, including colony diameters and the colors of conidia and colony pigments, are used to distinguish species. A selective medium, dichloran rose bengal yeast extract sucrose agar, is used for the enumeration of *Penicillium verrucosum* and *Penicillium viridicatum*. *P. verrucosum* produces a violet brown reverse coloration on this agar. Figure 20.2 shows penicillia of four *Penicillium* subgenera.

Figure 20.1 Aflatoxigenic fungi. **(A)** *A. flavus* head (magnification, ×300); **(B)** *A. flavus* conidia (magnification, ×1,875); **(C)** young *A. parasiticus* heads (magnification, ×300); **(D)** *A. parasiticus* conidia (magnification, ×1,875). Reprinted from A. D. Hocking, p. 451–465, *in* M. P. Doyle, L. R. Beuchat, and T. J. Montville (ed.), *Food Microbiology: Fundamentals and Frontiers*, 2nd ed. (ASM Press, Washington, D.C., 2001).

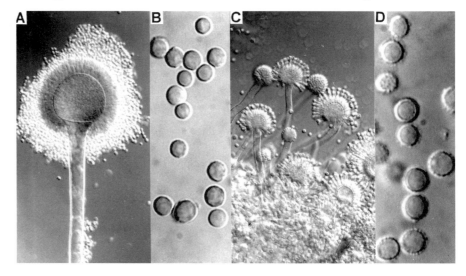

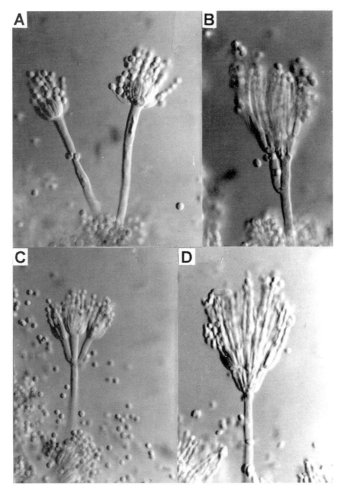

Figure 20.2   Penicillia (magnification, ×750) representative of the four *Penicillium* subgenera. **(A)** *Penicillium* subgenus *Aspergilloides* (*P. glabrum*); **(B)** *Penicillium* subgenus *Penicillium* (*P. expansum*); **(C)** *Penicillium* subgenus *Furcatum* (*P. citrinum*); **(D)** *Penicillium* subgenus *Biverticillium* (*P. variabile*). Reprinted from J. I. Pitt, p. 467–480, *in* M. P. Doyle, L. R. Beuchat, and T. J. Montville (ed.), *Food Microbiology: Fundamentals and Frontiers,* 2nd ed. (ASM Press, Washington, D.C., 2001).

Fusarium species are most often associated with cereal grains; seeds; milled cereal products, such as flour and corn meal; barley malt; animal feeds; and dead plant tissue. To isolate *Fusarium* species from these products, it is necessary to use selective media. The basic techniques for the detection and isolation of *Fusarium* employ plating techniques, either plate counts of serial dilutions of products or the placement of seeds or kernels of grain directly on the surface of agar medium in petri dishes, i.e., *direct plating.* Since many *Fusarium* species are plant pathogens and are found in fields where crops are grown, these molds respond to light. Growth, pigmentation, and spore production are most typical when cultures are grown in alternating light and dark cycles of 12 h each. Fluorescent light or diffuse sunlight from a north window is best. Fluctuating temperatures, such as 25°C (day) and 20°C (night), also enhance growth and sporulation.

Several culture media have been used to detect and isolate *Fusarium* species. These include Nash-Snyder medium, modified Czapek Dox agar,

Czapek iprodione dichloran (CZID) agar, potato dextrose iprodione dichloran agar, and dichloran chloramphenicol peptone agar. Nash-Snyder medium and modified Czapek Dox agar contain pentachloronitrobenzene, a known carcinogen, and are not favored for routine use in food microbiology laboratories. However, these media can be useful for evaluating samples that are heavily contaminated with bacteria and other fungi. The most commonly used medium for isolating *Fusarium* from foods is CZID agar, but rapid identification of *Fusarium* isolates to the species level is difficult, if not impossible, on this medium. Isolates must be subcultured on other media, such as carnation leaf agar (CLA), for identification. However, CZID agar is a good selective medium for *Fusarium*. Most molds are inhibited on CZID agar, and *Fusarium* species can be readily distinguished.

Identification of *Fusarium* species is based largely on the production and morphology of macroconidia and microconidia (Fig. 20.3). *Fusarium* species do not readily form conidia on all culture media, and conidia formed on high-carbohydrate media, such as potato dextrose agar, are often more variable and less typical. CLA is a medium that supports abundant and consistent spore production. Carnation leaves from actively growing, disbudded young carnation plants free of pesticide residues are cut into small pieces (5 mm²), dried in an oven at 45 to 55°C for 2 h, and sterilized by irradiation. CLA is prepared by placing a few pieces of carnation leaf on the surface of 2.0% water agar. *Fusarium* isolates are then inoculated onto the agar and leaf interface, where they form abundant and typical conidia and conidiophores in sporodochia (Fig. 20.4). CLA is low in carbohydrates and rich in other complex, naturally occurring substances that apparently stimulate spore production.

ASPERGILLUS SPECIES

Aspergillus was first described nearly 300 years ago and is an important genus in foods. Although a few species have been used to make food (e.g., *Aspergillus oryzae* in soy sauce manufacture), most *Aspergillus* species occur in foods as spoilage or biodeterioration fungi. They are extremely common in stored commodities, such as grains, nuts, and spices, and occur more frequently in tropical and subtropical than in temperate climates. The genus *Aspergillus* contains several species capable of producing mycotoxins, although the mycotoxin literature of the past 35 years has been dominated by papers on aflatoxins. Since foods contain many toxin-producing *Aspergillus* species, it is important to identify them and determine if they can produce toxin in a particular food.

The genus *Aspergillus* is large, containing >100 recognized species, most of which grow well in laboratory culture. Within the genus *Aspergillus*, species are grouped into six subgenera that are subdivided into sections. There are a number of teleomorphic (sexual-spore-producing) genera that have *Aspergillus* conidial states (*anamorphs*, i.e., asexual states or stages), but the only two of real importance in foods are the xerophilic (growing at low water activity [a_w]) *Eurotium* and *Neosartorya* species. They produce heat-resistant ascospores and cause spoilage in heat-processed foods, mainly fruit products. The most widely used taxonomy for *Aspergillus* is based on traditional morphological taxonomic techniques. However, chemical techniques, such as isoenzyme patterns, secondary metabolites, and ubiquinone systems, and molecular techniques have been used to clarify relationships within the genus.

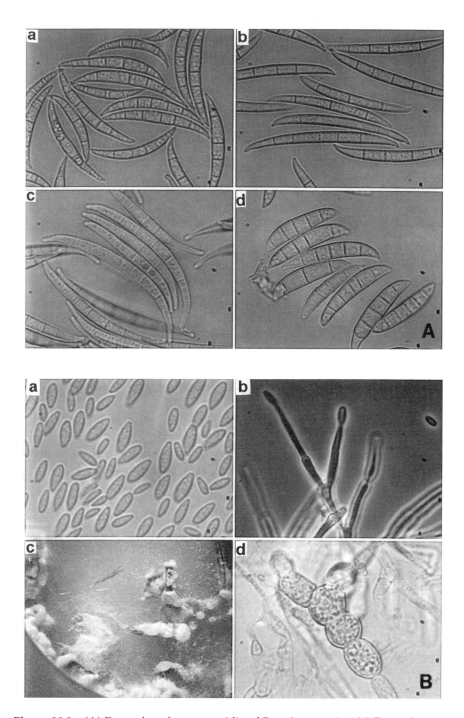

Figure 20.3 **(A)** Examples of macroconidia of *Fusarium* species. **(a)** *F. graminearum;* **(b)** *F. moniliforme;* **(c)** *F. equiseti;* **(d)** *F. culmorum* (magnification, ×1,000). **(B)** Examples of micro- and macroscopic structures of *Fusarium* species. **(a)** Microconidia; **(b)** monophialides; **(c)** sporodochia; **(d)** chlamydospores (magnification, ×1,000 [a, b, and d] and ×10 [c]). Reprinted from L. B. Bullerman, p. 481–497, *in* M. P. Doyle, L. R. Beuchat, and T. J. Montville (ed.), *Food Microbiology: Fundamentals and Frontiers,* 2nd ed., (ASM Press, Washington, D.C., 2001).

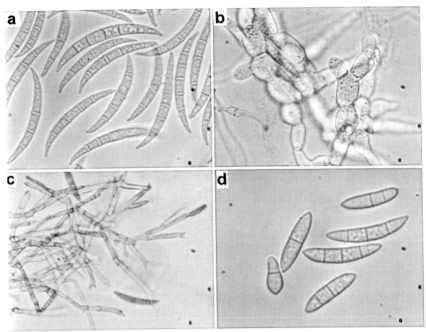

A. Fusarium sporotrichioides.

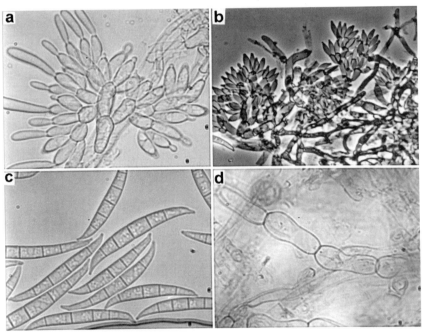

B. Fusarium graminearum.

Figure 20.4   **(A)** Microscopic structures of *F. sporotrichioides*. **(a)** Macroconidia; **(b)** chlamydospores; **(c)** phialides; **(d)** microconidia (magnification, ×1,000 [a, b, and d] and ×550 [c]). **(B)** Microscopic structures of *F. graminearum*. **(a)** Conidiophores (monophialides); **(b)** monophialides in sporodochia; **(c)** macroconidia; **(d)** chlamydospores (magnification, ×1,000 [a, c, and d] and ×550 [b]).

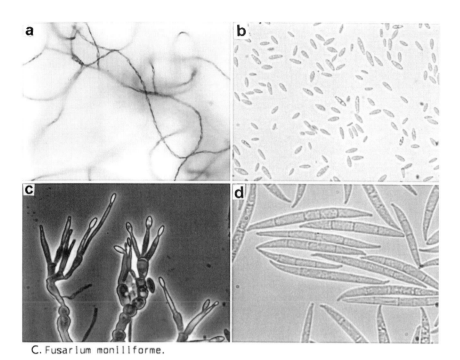

C. Fusarium moniliforme.

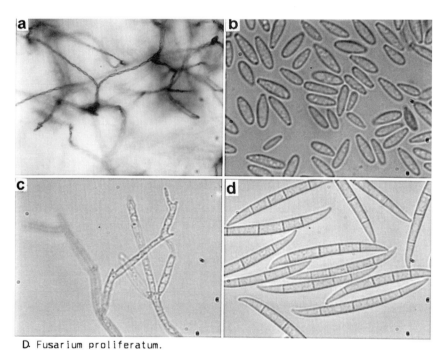

D. Fusarium proliferatum.

Figure 20.4 *(continued)*   **(C)** Microscopic structures of *F. moniliforme.* **(a)** Microconidia in chains; **(b)** microconidia; **(c)** monophialides producing microconidia; **(d)** macroconidia (magnification, ×165 [a], ×550 [b], and ×1,000 [c and d]). **(D)** Microscopic structures of *F. proliferatum.* **(a)** Microconidia in chains; **(b)** microconidia; **(c)** polyphialides; **(d)** macroconidia (magnification, ×165 [a] and ×1,000 [b to d]). Reprinted from L. B. Bullerman, p. 481–497, *in* M. P. Doyle, L. R. Beuchat, and T. J. Montville (ed.), *Food Microbiology: Fundamentals and Frontiers,* 2nd ed., (ASM Press, Washington, D.C., 2001).

Table 20.1 Significant mycotoxins produced by *Aspergillus* species and their toxic effects[a]

Mycotoxin(s)	Toxicity	Species
Aflatoxins B_1 and B_2	Acute liver damage, cirrhosis, carcinogenic (liver), teratogenic, immunosuppressive	*A. flavus, A. parasiticus, A. nomius*
Aflatoxins G_1 and G_2	Effects similar to B aflatoxins; G_1 toxicity is less than B_1 but greater than B_2	*A parasiticus, A. nomius*
Cyclopiazonic acid	Degeneration and necrosis of various organs, tremorgenic, low oral toxicity	*A. flavus, A. tamarii*
Ochratoxin A	Kidney necrosis (especially pigs), teratogenic, immunosuppressive, probably carcinogenic	*A. ochraceus* and related species *A. carbonarius* *A. niger* (occasional)
Sterigmatocystin	Acute liver and kidney damage, carcinogenic (liver)	*A. versicolor, Emericella* spp.
Fumitremorgens	Tremorgenic (rats and mice)	*A. fumigatus*
Territrems	Tremorgenic (rats and mice)	*A. terreus*
Tryptoquivalines	Tremorgenic	*A. clavatus*
Cytochalasins	Cytotoxic	*A. clavatus*
Echinulins	Feed refusal (pigs)	*Eurotium chevalieri, Eurotium amstelodami*

[a]Reprinted from A. D. Hocking, p. 451–465, *in* M. P. Doyle, L. R. Beuchat, and T. J. Montville (ed.), *Food Microbiology: Fundamentals and Frontiers*, 2nd ed. (ASM Press, Washington, D.C., 2001).

Almost 50 *Aspergillus* species can produce toxic metabolites, but the *Aspergillus* mycotoxins (Table 20.1) of greatest significance in foods and feeds are aflatoxins (produced by *A. flavus, A. parasiticus,* and *Aspergillus nomius*); ochratoxin A from *Aspergillus ochraceus* and related species and from *Aspergillus carbonarius* and occasionally *Aspergillus niger;* sterigmatocystin, produced primarily by *Aspergillus versicolor* but also by *Emericella* species; and cyclopiazonic acid (*A. flavus* is the primary source, but it is also reported to be produced by *Aspergillus tamarii* [Fig. 20.5]). *Tremorgenic* (i.e., producing involuntary trembling of the body) toxins are produced by *Aspergillus terreus* (territrems), *Aspergillus fumigatus* (fumitremorgens), and *Aspergillus clavatus* (tryptoquivaline). Citrinin, patulin, and penicillic acid may also be produced by certain *Aspergillus* species.

Aspergillus species produce toxins that exhibit a wide range of toxicities and cause significant long-term effects. Aflatoxin B_1 is perhaps the most potent liver carcinogen known for a wide range of animal species, including humans. Both ochratoxin A and citrinin affect kidney function. Cyclopiazonic acid has a wide range of effects. Tremorgenic toxins, such as territrems, affect the central nervous system. Table 20.1 lists the most significant toxins produced by *Aspergillus* species and their toxic effects.

A. flavus and A. parasiticus
Undoubtedly, the most important group of toxigenic aspergilli are the aflatoxigenic molds, *A. flavus, A. parasiticus,* and the recently described but much less common species *A. nomius,* all of which are classified in *Aspergillus* section *Flavi.* Although these three species are closely related and show many similarities, many characteristics may be used in their differentiation (Table 20.2). The toxins produced by these three species are species specific. *A. flavus* can produce aflatoxins B_1 and B_2 and cyclopiazonic acid, but only some isolates are toxigenic. *A. parasiticus* produces aflatoxins B_1, B_2, G_1, and G_2 but not cyclopiazonic acid, and almost all isolates are toxigenic. *A. nomius* is morphologically similar to *A. flavus,* but like *A. parasiticus,* it produces B and G aflatoxins without cyclopiazonic acid. Because this species appears to be uncommon, it has been little studied. The poten-

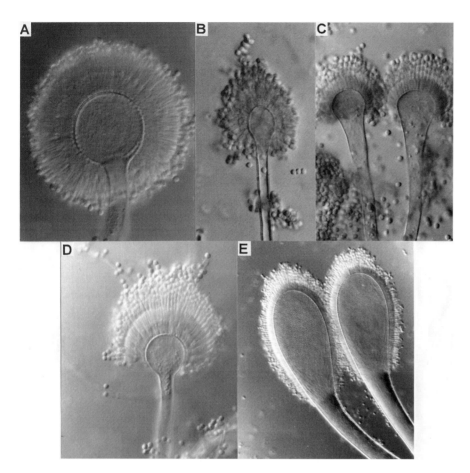

Figure 20.5 Some common mycotoxigenic *Aspergillus* species. **(A)** *A. ochraceus* (magnification, ×750); **(B)** *A. versicolor* (magnification, ×750); **(C)** *A. fumigatus* (magnification, ×750); **(D)** *A. terreus* (magnification, ×750); **(E)** *A. clavatus* (magnification, ×300). Reprinted from A. D. Hocking, p. 451–465, *in* M. P. Doyle, L. R. Beuchat, and T. J. Montville (ed.), *Food Microbiology: Fundamentals and Frontiers*, 2nd ed. (ASM Press, Washington, D.C., 2001).

tial toxigenicities of isolates are not known, and the practical importance of the species is hard to assess. *A. oryzae* and *Aspergillus sojae* are related to *A. flavus* and *A. parasiticus* and are used to make fermented foods, but they do not produce toxins. *A. tamarii* produces aflatoxin B and cyclopiazonic acid. Obviously, accurate differentiation of related species within section *Flavi* is important in order to determine the potential for toxin production and the types of toxins likely to be present.

Table 20.2 Distinguishing features of *A. flavus*, *A. parasiticus*, and *A. nomius*[a]

Species	Conidia	Sclerotia	Toxins
A. flavus	Smooth to moderately roughened, variable in size	Large, globose	Aflatoxins B_1 and B_2, cyclopiazonic acid
A. parasiticus	Conspicuously roughened, little variation in size	Large, globose	Aflatoxins B and G
A. nomius	Similar to *A. flavus* (bullet shaped)	Small, elongated	Aflatoxins B and G

[a]Reprinted from A. D. Hocking, p. 451–465, *in* M. P. Doyle, L. R. Beuchat, and T. J. Montville (ed.), *Food Microbiology: Fundamentals and Frontiers*, 2nd ed. (ASM Press, Washington, D.C., 2001).

Detection and Identification

A. flavus and *A. parasiticus* (Fig. 20.1) are easily distinguished from other *Aspergillus* species by using appropriate media and methods. The texture of the conidial walls is a reliable differentiating feature: conidia of *A. flavus* (Fig. 20.1B) are usually smooth to finely roughened, while those of *A. parasiticus* (Fig. 20.1D) are clearly rough when observed under an oil immersion lens. The combination of characteristics most useful in differentiating among the three aflatoxigenic species are summarized in Table 20.2.

Aflatoxins

Aflatoxins are difuranocoumarin derivatives. Aflatoxins B_1, B_2, G_1, and G_2 are produced in nature by the molds discussed above. The letters B and G refer to the fluorescent colors (blue and green, respectively) observed under long-wave ultraviolet light, and the subscripts 1 and 2 refer to their separation patterns on thin-layer chromatography plates. Aflatoxins M_1 and M_2 are produced from their respective B aflatoxins by hydroxylation in lactating animals and are excreted in milk at a rate of ~1.5% of ingested B aflatoxins.

Aflatoxins are synthesized through the polyketide pathway, beginning with condensation of an acetyl unit with two malonyl units and the loss of carbon dioxide. Intermediate steps result in the formation of norsolorinic acid. Norsolorinic acid undergoes several metabolic conversions to form aflatoxin B_1. G group aflatoxins are formed from the same substrate as B group aflatoxins, namely, *O*-methylsterigmatocystin, but by an independent pathway. Most of the genes involved in the biosynthesis of aflatoxins are contained within a single gene cluster in the genomes of *A. flavus* and *A. parasiticus*, and their regulation and expression are now relatively well understood.

Toxicity. Aflatoxins are both acutely and chronically toxic in animals and humans, producing acute liver damage, liver cirrhosis, and tumor induction. Even more important to human health are the immunosuppressive effects of aflatoxins, either alone or in combination with other mycotoxins. Immunosuppression can increase susceptibility to infectious diseases, particularly in populations where aflatoxin ingestion is *chronic* (continuous), and can interfere with the production of antibodies in response to immunization in animals, and perhaps also in children.

Acute aflatoxicosis in humans is rare; however, several outbreaks have been reported. In 1967, 26 people in two farming communities in Taiwan became ill with apparent food poisoning. Nineteen were children, three of whom died. Although postmortems were not performed, rice from the affected households contained ~200 µg of aflatoxin B_1/kg, which was probably responsible for the outbreak. An outbreak of hepatitis in India in 1974 that affected 400 people, 100 of whom died, was almost certainly caused by aflatoxins. The outbreak was traced to corn heavily contaminated with *A. flavus* and containing up to 15 mg of aflatoxins per kg. It was calculated that the affected adults may have consumed 2 to 6 mg on a single day, implying that the acute lethal dose for adult humans is on the order of 10 mg. The deaths of 13 Chinese children in the northwestern Malaysian state of Perak were reportedly due to ingestion of contaminated noodles. Aflatoxins were confirmed in postmortem tissue samples. A case of systemic aspergillosis caused by an aflatoxin-producing strain of *A. flavus* has been reported in which aflatoxins B_1, B_2, and M_1 were detected in lung lesions and were considered to have played a role in damaging the immune system of the patient.

The greatest direct impact of aflatoxins on human health is their potential to induce liver cancer. Human liver cancer has a high incidence in central Africa and parts of Southeast Asia. Studies in several African countries and Thailand have shown a correlation between aflatoxin intake and the occurrence of primary liver cancer. No such correlation could be demonstrated for populations in rural areas of the United States, despite the occurrence of aflatoxins in corn. However, evidence supports the hypothesis that high aflatoxin intakes are causally related to high incidences of cancer, even in the absence of hepatitis B. In animals, aflatoxins cause various syndromes, including cancer of the liver, colon, and kidneys. Regular low-level intake of aflatoxins can lead to poor feed conversion, low weight gain, and poor milk yields in cattle.

Mechanism of action.   Aflatoxin B_1 is metabolized in the liver, leading to the formation of highly reactive intermediates, one of which is 2,3-epoxy-aflatoxin B_1. Binding of these reactive intermediates to DNA results in disruption of transcription and abnormal cell proliferation, leading to mutagenesis or carcinogenesis. Aflatoxins also inhibit oxygen uptake in the tissues by acting on the electron transport chain and inhibiting various enzymes, resulting in decreased production of adenosine triphosphate (ATP).

Toxin detection.   Mycotoxins can be detected by chemical or biological methods. Chemical determination of aflatoxins has become fairly standardized. Samples are extracted with organic solvents, such as chloroform or methanol, in combination with small amounts of water. The representative sample size(s) depends on the particle size; therefore, a 500-g sample is acceptable for oils or milk powder, but 3 kg of flour or peanut butter is necessary. Extracts can be further cleaned up by passage through a silica gel column, concentrated, and then separated by thin-layer chromatography or high-performance liquid chromatography. Following thin-layer chromatography, aflatoxins are visualized under ultraviolet light and quantified by visual comparison with known concentrations of standards. High-performance liquid chromatography with detection by ultraviolet light or absorption spectrometry provides a more readily quantifiable (although not necessarily more accurate or sensitive) technique.

Immunoassay techniques, including enzyme-linked immunosorbent assays and dipstick tests, for aflatoxin detection have been developed. These kits are commercially available. Immunoaffinity and fluorimetric detection have been combined in developing biosensors able to detect aflatoxin down to 50 μg/kg in as little as 2 min. The polymerase chain reaction (PCR) has been applied to detect the presence of aflatoxigenic fungi in foods, targeting aflatoxin biosynthetic genes.

Occurrence of aflatoxigenic molds and aflatoxins.   While *A. flavus* is widely distributed in nature, *A. parasiticus* is less widespread. The actual extent of *A. parasiticus* occurrence in nature is likely greater than reported, since both species are reported indiscriminately as *A. flavus*. *A. flavus* and *A. parasiticus* have a strong affinity for nuts and oilseeds. Corn, peanuts, and cottonseed are the most important crops invaded by these molds, and in many instances, invasion takes place before harvest, not during storage as was once believed. Peanuts are invaded while still in the ground if the crop suffers drought stress or related factors. In corn, insect damage to developing kernels allows the entry of aflatoxigenic molds, but invasion can also occur through the silks of developing ears. Cottonseeds are invaded through the nectaries.

Significant amounts of aflatoxins can occur in peanuts, corn, and other nuts and oilseeds, particularly in some tropical countries, where crops may be grown under marginal conditions and where drying and storage facilities are limited. *A. flavus* grows well on cereals and spices, but aflatoxin production in these commodities can be prevented with proper drying, handling, and storage.

Control and inactivation.   Control of aflatoxins in commodities generally relies on screening techniques that separate the affected nuts, grains, or seeds. In corn, cottonseed, and figs, screening for aflatoxins can be done by examination under ultraviolet light: those particles which fluoresce may be contaminated. All peanuts fluoresce when exposed to ultraviolet light, so this method is not useful for detecting kernels that are contaminated with aflatoxins. Peanuts containing aflatoxins are removed by electronic color-sorting machines, which detect discolored kernels.

Aflatoxins can be partially destroyed by chemical treatments. Oxidizing agents, such as ozone and hydrogen peroxide, remove aflatoxins from contaminated peanut meals. Although ozone is effective in destroying aflatoxins B_1 and G_1 after exposure at 100°C for 2 h, there is no effect on aflatoxin B_2. The treatment also decreases the lysine content of the meal. Treatment with hydrogen peroxide destroys 97% of aflatoxin in defatted peanut meal. The most practical chemical method of aflatoxin destruction is the use of anhydrous ammonia gas at elevated temperatures and pressures, with a 95 to 98% reduction in total aflatoxin in peanut meal. This technique is used commercially for detoxification of animal feeds in Senegal, France, and the United States.

The ultimate method for the control of aflatoxins in commodities, particularly peanuts, is to prevent the plants from becoming infected with the molds. Progress toward this goal is being made through a biological control strategy, which is the early infection of plants with nontoxigenic strains of *A. flavus* to prevent the subsequent entry of toxigenic strains.

Aflatoxins are among the few mycotoxins covered by legislation. Statutory limits are imposed by some countries on the amounts of aflatoxin that can be present in particular foods. The limit imposed by most Western countries is 5 to 20 μg of aflatoxin B_1/kg (parts per billion) in several human foods, including peanuts and peanut products. The amount allowed in animal feeds varies, but up to 300 μg/kg is allowed in feedstuffs for beef cattle and sheep in the United States. There are no statutory limits for aflatoxins in foods or feeds in Southeast Asia.

Other Toxigenic Aspergilli

Although aflatoxins are of greatest concern, there are other toxins produced by other *Aspergillus* species. *A. ochraceus* (Fig. 20.2A) is a widely distributed mold that is particularly common in dried foods, such as nuts (e.g., peanuts and pecans), beans, dried fruit, biltong (a South African dried meat), and dried fish. *A. ochraceus* produces ochratoxin. Ochratoxin A is the major toxin, with minor components of lower toxicity designated ochratoxin B and C. Other reported toxic metabolites are xanthomegnin and viomellein. Other *Aspergillus* species closely related to *A. ochraceus* can also produce ochratoxin A. *Aspergillus sclerotiorum*, *Aspergillus alliaceus*, *Aspergillus melleus*, and *Aspergillus sulphureus* have all been reported to produce ochratoxins, but in *A. ochraceus* and related species, only a portion of isolates are toxigenic. Production of ochratoxin A by *Aspergillus niger* var. *niger* was reported in 1994. *A. carbonarius* also produces ochratoxin A.

A. versicolor (Fig. 20.5B) is the most important food spoilage and toxigenic species in *Aspergillus* section *Versicolores*. *A. versicolor* is the major producer of sterigmatocystin, a carcinogenic compound that is a precursor of the aflatoxins. Surprisingly, *A. versicolor* does not produce aflatoxins, but it produces ochratoxin A. Sterigmatocystin has been reported to occur naturally in rice in Japan, wheat and barley in Canada, cereal-based products in the United Kingdom, and Ras cheese. Sterigmatocystin has low acute oral toxicity because it is relatively insoluble in water and gastric juices. As a liver carcinogen, sterigmatocystin is ~1/150 as potent as aflatoxin B_1 but is still much more potent than most other known liver carcinogens.

A. fumigatus, section *Fumigati* (Fig. 20.5C), is best recognized as a human pathogen that causes aspergillosis of the lung. Its prime habitat is decaying vegetation. *A. fumigatus* is isolated frequently from foods, particularly stored commodities, but is not regarded as a serious spoilage mold.

A. fumigatus is capable of producing several toxins that affect the central nervous system, causing tremors. Fumitremorgens A, B, and C are toxic cyclic dipeptides that are produced by *A. fumigatus* and *Aspergillus caespitosus*. Verruculogen, also produced by these species, has a structure similar to that of the fumitremorgens but with a different side chain. Verruculogen is tremorgenic and appears to cause inhibition of the alpha motor cells of the anterior horn.

A. terreus occurs commonly in soil and in foods, particularly stored cereals and cereal products, beans, and nuts, but is not regarded as an important spoilage mold. *A. terreus* can produce a group of tremorgenic toxins known as territrems. *A. clavatus* (Fig. 20.5E) is found in soil and decomposing plant materials. Although *A. clavatus* has been reported to be present in various stored grains, it is especially common in malting barley, an environment particularly suited to its growth and sporulation. *A. clavatus* produces patulin, cytochalasins, and the tremorgenic mycotoxins tryptoquivaline, tryptoquivalone, and related compounds.

PENICILLIUM SPECIES

The discovery of penicillin in 1929 provided impetus to search for other *Penicillium* metabolites with antibiotic properties and ultimately led to the recognition of citrinin, patulin, and griseofulvin as "toxic antibiotics," later called mycotoxins. The literature on toxigenic penicillia is now quite vast. In a comprehensive review of the literature on fungal metabolites, ~120 common mold species were demonstrably toxic to higher animals. Forty-two toxic metabolites are produced by one or more *Penicillium* species. At least 85 *Penicillium* species are listed as toxigenic. Classification within the genus *Penicillium* is based primarily on microscopic morphology (Fig. 20.2). The genus is divided into subgenera based on the number and arrangement of phialides and metulae and *rami* (elements supporting phialides) on the stipes.

Significant *Penicillium* Mycotoxins

As already noted, >80 *Penicillium* species are reported to be toxin producers. The range of mycotoxin classes produced by *Penicillium* species is broader than for any other fungal genus. Toxicity due to *Penicillium* species is very diverse. However, most toxins can be placed in two broad groups, those that affect liver and kidney function and those that are neurotoxins. Generally, the *Penicillium* toxins that affect liver or kidney function are asymptomatic or can cause generalized debility in humans or animals. In contrast, the toxicity of the neurotoxins in animals is often characterized by sustained trembling. However, individual toxins show wide variations from these generalizations.

In fact, the most important toxin produced by a *Penicillium* species, ochratoxin A, is derived from isocoumarin linked to the amino acid phenylalanine. The target organ of toxicity in all mammalian species is the kidney. The lesions can be produced by both acute and chronic exposure. Ochratoxin A is produced by *P. verrucosum* (Fig. 20.6A), which grows in barley and wheat crops in cold climates, especially Scandinavia, central Europe, and western Canada. *P. verrucosum* grows most strongly at relatively low temperatures, down to 0°C, with a maximum near 31°C. The major source of ochratoxin A in foods is bread made from barley or wheat in which *P. verrucosum* has grown. Because ochratoxin A is fat soluble and not readily ex-

Figure 20.6 Some toxigenic *Penicillium* species (magnification, ×750). **(A)** *P. verrucosum;* **(B)** *P. crustosum;* **(C)** *P. roqueforti;* **(D)** *P. oxalicum.* Reprinted from J. I. Pitt, p. 467–480, *in* M. P. Doyle, L. R. Beuchat, and T. J. Montville (ed.), *Food Microbiology: Fundamentals and Frontiers,* 2nd ed. (ASM Press, Washington, D.C., 2001).

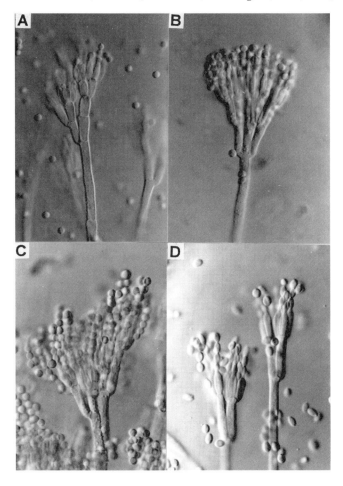

creted, it also accumulates in the depot fat of animals that eat feeds containing the toxin, and from there it is ingested by humans eating, for example, pig meats. It has been suggested that ochratoxin A is a causal agent of Balkan endemic nephropathy, a kidney disease with a high mortality rate in certain areas of Bulgaria, Yugoslavia, and Romania.

Citrinin is a significant renal toxin affecting domestic animals, such as pigs and dogs. It is also an important toxin in domestic birds, where it produces watery diarrhea and an increase in water consumption and reduced weight gain due to kidney degeneration. The importance of citrinin in human health is difficult to assess. Primarily recognized as a metabolite of *Penicillium citrinum,* citrinin is produced by more than 20 other fungal species. *P. citrinum* is a *ubiquitous* (i.e., found in many environments) mold and has been isolated from nearly every kind of food surveyed for fungi. The most common sources are cereals, especially rice, wheat, and corn; milled grains; and flour. Citrinin is produced over most of the growth temperature range, but the effect of a_w is unknown.

Patulin is a lactone, and it produces teratogenic effects (that is, it causes fetal malformations [birth defects]) in rodents, as well as neurological and gastrointestinal effects. The most important *Penicillium* species producing patulin is *Penicillium expansum,* best known as a fruit pathogen but also widespread in other fresh and processed foods. *P. expansum* produces patulin as it rots apples and pears. Patulin appears to lack chronic toxic effects in humans, and therefore, low levels in juices are perhaps of little concern. However, because apple juice is widely consumed by children as well as adults, some countries have set an upper limit of 50 μg/liter for patulin in apple juice and other apple products. It is also important as an indicator of the use of poor-quality raw materials in juice manufacture. Patulin is produced by several other *Penicillium* species besides *P. expansum* (e.g., *Penicillium griseofulvum* and *Penicillium roqueforti*), but the potential for production of unacceptable levels in foods appears to be much lower. Patulin is quite stable in apple juice during storage; pasteurization at 90°C for 10 s caused <20% reduction.

Cyclopiazonic acid, a highly toxic compound, causes fatty degeneration and hepatic cell necrosis in the liver and kidneys of domestic animals. Cyclopiazonic acid is produced by *A. flavus* and at least six *Penicillium* species (e.g., *Penicillium camemberti* Thom, *P. commune,* and *P. griseofulvum*). Along with *A. flavus, P. commune,* a frequent cause of cheese spoilage, is probably the most common source of cyclopiazonic acid in foods. Other species producing this toxin have a wide range of habitats.

Acute cardiac beriberi was a common disease in Japan in the second half of the 19th century and has been linked to citreoviridin. The symptoms are heart distress, labored breathing, nausea, and vomiting, followed by anguish, pain, restlessness, and sometimes maniacal behavior. In extreme cases, progressive paralysis leading to respiratory failure occurred. The major source of citreoviridin is *Penicillium citreonigrum* (synonyms, *Penicillium citreoviride* and *Penicillium toxicarium*), a species which usually occurs in rice, less commonly in other cereals, and rarely in other foods. Citreoviridin is produced at temperatures from 10 to 37°C, with a maximum near 20°C.

Several tremorgenic mycotoxins are produced by *Penicillium* species. The most important is the highly toxic penitrem A. Verruculogen, which is equally toxic, is not produced by species common in foods. Less toxic compounds include fumitremorgen B, paxilline, verrucosidin, and janthitrems.

At doses of ~1 mg/kg of body weight, penitrem A causes brain damage and death in rats, but the mechanism remains unclear. The only common foodborne species producing penitrem A is *P. crustosum* (Fig. 20.6B). Nearly all isolates of *P. crustosum* produce penitrem A at high levels, so the presence of this species in food or feed is a warning signal. Penitrem A appears to be produced only at high a$_w$. *P. crustosum* is a ubiquitous spoilage mold, occurring almost universally in cereal and animal feed samples. *P. crustosum* causes spoilage of corn, processed meats, nuts, cheese, and fruit juices, as well as being a weak pathogen on pomaceous fruits and cucurbits.

P. roqueforti (Fig. 20.6C), a species which is used in cheese manufacture but which can also be a spoilage mold, produces two toxins of interest. Roquefortine has a relatively high 50% lethal dose (340 mg/kg intraperitoneally in mice), but it has been reported to be the cause of the deaths of dogs in Canada, with symptoms similar to strychnine poisoning. PR toxin (produced by *P. roqueforti*) is apparently much more toxic (50% lethal dose, 6 mg/kg intraperitoneally in mice) than roquefortine, but it has not been implicated in animal or human disease. Due to their potential occurrence in staple foods, PR toxin and roquefortine are of considerable public health significance.

Secalonic acids are dimeric xanthones produced by a range of taxonomically distantly related molds. Secalonic acid D, the only secalonic acid produced by *Penicillium* species, has significant animal toxicity. Secalonic acid D is produced as a major metabolite of *Penicillium oxalicum* (Fig. 20.6D). A major habitat for *P. oxalicum* is corn at harvest. However, the role of secalonic acid D in human or animal disease remains unclear.

FUSARIA AND TOXIGENIC MOLDS OTHER THAN ASPERGILLI AND PENICILLIA

The most important group of mycotoxigenic molds other than *Aspergillus* and *Penicillium* species are species of the genus *Fusarium*. *Fusarium* species are most often found as contaminants of plant-derived foods, especially cereal grains. Many *Fusarium* species are plant pathogens, while others are saprophytic; most can be found in the soil. As such, these molds and their toxic metabolites (mycotoxins) find their way into animal feeds and human

Authors' note_____

Saprophytic organisms live on dead plant tissue. They do not affect the health of a plant.

Table 20.3 Major mycotoxins that may be produced by *Fusarium* species of importance in cereal grains and grain-based foods[a]

Section	Species	Potential mycotoxin(s)
Sporotrichiella	F. poae	Type A trichothecenes, T-2 toxin, diacetoxyscirpenol
	F. sporotrichioides	T-2 toxin
Gibbosum	F. equiseti	Unknown
Discolor	F. graminearum	Deoxynivalenol (vomitoxin), 3-acetyldeoxynivalenol, 15-acetyldeoxynivalenol, zearalenone, possibly others
	F. culmorum	Nivalenol, zearalenone
Liseola	F. verticillioides	Fumonisins and others
	F. proliferatum	Fumonisins, moniliformin, and others
	F. subglutinans	Moniliformin and others

[a]Reprinted from L. B. Bullerman, p. 481–497, in M. P. Doyle, L. R. Beuchat, and T. J. Montville (ed.), *Food Microbiology: Fundamentals and Frontiers*, 2nd ed. (ASM Press, Washington, D.C., 2001).

foods (Table 20.3). In terms of human foods, *Fusarium* species are most often encountered as contaminants of cereal grains, oil seeds, and beans. Another mold genus, other than *Aspergillus* and *Penicillium,* whose members can produce mycotoxins is *Alternaria. Alternaria* species are widely distributed in the environment and can be found in soil, decaying plant materials, and dust.

Toxigenic *Fusarium* Species

The most common toxic species, discussed below, belong to the sections *Sporotrichiella (Fusarium sporotrichioides* and *Fusarium poae), Gibbosum (Fusarium equiseti), Discolor (Fusarium graminearum* and *Fusarium culmorum),* and *Liseola (Fusarium verticillioides, Fusarium moniliforme, Fusarium proliferatum,* and *Fusarium subglutinans).* The major mycotoxins produced by these species are summarized in Table 20.3. The genus *Fusarium* is characterized by production of septate *hyphae* that generally range in color from white to pink, red, purple, or brown due to pigment production. The most common characteristic of the genus is the production of large septate, crescent-shaped, fusiform, or sickle-shaped spores known as *macroconidia.* The macroconidia exhibit a foot-shaped basal cell and a beak-shaped or snout-like apical cell (Fig. 20.3A). Some species also produce smaller one- or two-celled conidia known as *microconidia* (Fig. 20.3B), and some produce swollen, thick-walled chlamydospores in the hyphae or in the macroconidia (Fig. 20.3B).

Authors' note

Hyphae are branched or unbranched filaments. A mass of hyphae is referred to as a mycelium.

F. poae and *F. sporotrichioides* are in section *Sporotrichiella. F. poae* is widespread in soils of temperate regions and is found on grains, such as wheat, corn, and barley. It exists primarily as a saprophyte but may be weakly parasitic. It is most commonly found in temperate regions of Russia, Europe, Canada, and the northern United States. *F. poae* has also been found in warmer regions, such as Australia, India, Iraq, and South Africa. *F. sporotrichioides* is found in soil and a wide variety of plant materials. The mold is found in the temperate to colder regions of the world, including Russia, northern Europe, Canada, the northern United States, and Japan. It can grow at low to very low temperatures, e.g., at $-2°C$ on grain overwintering in the field.

Of the species in section *Gibbosum,* only *F. equiseti* will be discussed here. *F. equiseti* has been found in soils from Alaska to tropical regions and has been isolated from cereal grains and overwintered cereals in Europe, Russia, and North America. However, *F. equiseti* is particularly common in tropical and subtropical areas. For the most part, *F. equiseti* is saprophytic, but it may be pathogenic to plants such as bananas, avocados, and cucurbits. *F. equiseti* may contribute to leukemia in humans by affecting the immune system.

F. graminearum, in section *Discolor,* is a plant pathogen found worldwide in the soil and is the most widely distributed toxigenic *Fusarium* species. It causes various diseases of cereal grains, including gibberella ear rots in corn and fusarium head blight or scab in wheat and other small grains. These two diseases are important to food microbiology and food safety because the mold and its major toxic metabolites, deoxynivalenol and zearalenone, may contaminate grain and food products made from the grain.

F. culmorum is also widely distributed in the soil and causes diseases of cereal grains, among which ear rot in corn is important in food microbiology and food safety, since the mold and its toxins may contaminate corn-based foods. *F. culmorum* produces the mycotoxins deoxynivalenol, zearalenone, and acetyldeoxnivalenol.

The *Liseola* section contains four species of the genus *Fusarium* that are of interest: *F. verticillioides*, *F. moniliforme*, *F. proliferatum*, and *F. subglutinans*. *F. verticillioides* is a soilborne plant pathogen that is found in corn growing in all regions of the world. It is the most prevalent mold associated with corn. It often produces symptomless infections of corn plants but may infect the grain as well and has been found worldwide on food-grade and feed-grade corn. The presence of *F. verticillioides* in corn is a major concern because of the possible widespread contamination of corn and corn-based foods with its toxic metabolites, especially the fumonisins. The main human disease associated with *F. verticillioides* is esophageal cancer. Mycotoxins produced by *F. verticillioides* include fumonisins, fusaric acid, fusarins, and fusariocins.

F. proliferatum is closely related to *F. verticillioides*, yet less is known about this species, possibly because of its frequent misidentification as *F. moniliforme*. It is also frequently isolated from corn, where it probably occurs in much the same way as *F. verticillioides*. *F. proliferatum* is capable of producing fumonisins but has not been associated with animal or human diseases.

F. subglutinans is very similar to *F. verticillioides* and *F. proliferatum*. It is widely distributed on corn and other grains. Little information is available about this mold, probably because of its misidentification as *F. moniliforme*. *F. subglutinans* has not been specifically associated with any reported animal or human diseases, but it has been found in corn from regions with high incidences of human esophageal cancer.

Detection and Quantitation of *Fusarium* Toxins

Fusarium species produce several toxic or biologically active metabolites. The toxins are trichothecenes, a group of closely related compounds. The trichothecenes are divided into three groups: type A (diacetoxyscirpenol, T-2 toxin, HT-2 toxin, and neosolaniol), type B (deoxynivalenol, 3-acetydeoxynivalenol, 15-acetyldeoxnyivalenol, nivalenol, and fusarenon-X), and type C (satratoxins). Of these, the toxin most commonly found in cereal grains and most often associated with human illness is deoxynivalenol. Other *Fusarium* toxins associated with diseases are zearalenone and the fumonisins. The most common chromatographic separation techniques used to identify these toxins are thin-layer chromatography and high-performance liquid chromatography. Gas chromatography also has some applications, particularly when coupled with mass spectrometry. Immunoassays have been developed for *Fusarium* toxins. Qualitative kits for screening, as well as kits for quantitative analysis, are available for deoxynivalenol, zearalenone, T-2 toxin, and fumonisins.

Other Toxic Molds

Other potentially toxic molds that may contaminate foods include species of the genera *Acremonium*, *Alternaria*, *Chaetomium*, *Cladosporium*, *Claviceps*, *Myrothecium*, *Phomopis*, *Rhizoctonia*, and *Rhizopus*, as well as the species *Diplodia maydis*, *Phoma herbarum*, *Pithomyces chartarum*, *Stachybotrys chartarum*, and *Trichothecium roseum*. However, most of these molds are more likely to be present in animal feeds, and their significance for food safety may be minimal. Some have been shown to produce toxic secondary metabolites in vitro which have yet to be found to occur naturally.

The ergot mold, *Claviceps purpurea,* is the cause of the earliest recognized human mycotoxicosis, known as ergotism. Convulsive ergotism may have been the cause of the behavior that led to the witchcraft trials of 1692 in Salem, Mass. *Alternaria* species infect plants in the field and may contaminate wheat, sorghum, and barley. *Alternaria* species also infect various fruits and vegetables, including apples, pears, citrus fruits, peppers, tomatoes, and potatoes.

Summary

- *Aspergillus, Penicillium,* and *Fusarium* are the mold genera most commonly associated with foods.
- Toxigenic molds are most often found as contaminants of plant-derived foods.
- The characteristics used to differentiate and identify molds are microscopic morphology and gross physiological features, including colony diameters and the colors of conidia and colony pigments.
- Specific growth media that inhibit the growth of bacteria are required for molds.
- Hydrogen peroxide is used commercially for detoxification of grains and other commodities.
- Molds can grow at low to very low temperatures, e.g., at $-2°C$ on grain overwintering in the field.

Suggested reading

Bullerman, L. B. 1996. Occurrence of *Fusarium* and fumonisins on food grains and in foods, p. 27–38. *In* L. Jackson, J. DeVries, and L. Bullerman (ed.), *Fumonisins in Foods.* Plenum Publishing Corp., New York, N.Y.

El-Banna, A. A., J. I. Pitt, and L. Leistner. 1987. Production of mycotoxins by *Penicillium* species. *Syst. Appl. Microbiol.* **10:**42–46.

Ellis, W. O., J. P. Smith, B. K. Simpson, and J. H. Oldham. 1991. Aflatoxins in food: occurrence, biosynthesis, effects on organisms, detection, and methods of control. *Crit. Rev. Food Sci. Nutr.* **30:**403–439.

Flannigan, B., and A. R. Pearce. 1994. *Aspergillus* spoilage: spoilage of cereals and cereal products by the hazardous species *A. clavatus,* p. 115–127. *In* K. A. Powell, A. Renwick, and J. F. Peberdy (ed.), *The Genus Aspergillus. From Taxonomy and Genetics to Industrial Application.* Plenum Press, New York, N.Y.

Pitt, J. I. 2000. *A Laboratory Guide to Common Penicillium Species,* 3rd ed. CSIRO Food Science Australia, North Ryde, New South Wales, Australia.

Samson, R. A., E. S. Hoekstra, J. C. Frisvad, and O. Filtenborg (ed.). 1995. *Introduction to Food-Borne Fungi.* Centraalbureau voor Schimmelcultures, Baarn, The Netherlands.

Questions for critical thought

1. The presence of patulin in apple juice is a human health risk. Why? Detection of patulin may indicate quality control problems. Explain.
2. Which aflatoxins are covered by legislation and why?
3. How would you identify and differentiate *Aspergillus* and *Penicillium?*
4. Food microbiologists are concerned with the presence of *F. verticillioides* in corn because of the possible widespread contamination of corn and

corn-based foods with its toxic metabolites. What toxic metabolites does this microorganism produce and what is the main disease of concern for humans?

5. In identifying members of the genus *Fusarium,* what common (distinct) characteristics of the microorganism are used?

6. Compare the characteristics and properties of sterigmatocystin and aflatoxin.

7. List several ways to prevent the growth of mold in grains. List methods that can be used to detoxify grains.

8. Investigate a recent outbreak of foodborne illness related to the growth of toxigenic mold in the product. Describe the characteristics of the product, including processing, handling, and storage practices. Formulate a strategy that could have been implemented to prevent the outbreak.

21

Viruses and Prions*

LEARNING OBJECTIVES

The information in this chapter will help the student to:
- tell the difference between viruses, bacteria, and prions
- compare and contrast ways to control viruses, bacteria, and prions in foods
- recognize the signs that a virus has caused foodborne illness
- explain what a prion is and how it is transmitted
- appreciate that science is constantly changing

INTRODUCTION

Most of food microbiology, and the bulk of this book, are devoted to bacteria. However, if one could count the number of people who get sick with diarrhea per year, viruses would be just as important. Viruses cause 50 to 66% of the cases of foodborne illness in the United States. Prions, which cause mad cow disease and new-variant Creutzfeldt-Jakob disease (CJD) (the "old" CJD is hereditary; new-variant CJD, or vCJD, is caused by eating certain types of infected meat), have not been proven to have made anyone sick in the United States (to the best of our knowledge at the time this chapter was written). However, prions are biological, have recently entered the U.S. food supply, and are causing people to panic, so they are discussed here even though they are not microbes.

Bacteria, viruses, and prions are fundamentally different (Table 21.1). Bacteria are complex. There are many types of bacteria in food, and they are easily cultured in the laboratory. The ability of bacteria to grow in food creates special problems. For example, a food having a low "safe" level of 10 *Staphylococcus aureus* cells when it is made may have a disease-producing 10 million when it is consumed. Viruses and prions do not grow in foods, cannot easily be grown in the laboratory, and are hard to detect. Foodborne viruses consist of RNA and a few proteins. There are only two major foodborne viruses, noroviruses and hepatitis A virus. Food is a carrier for viruses, which move from feces to food due to poor hygiene. When the food is eaten, they are carried to the intestine, where they attack intestinal cells and cause illness. A *prion* is even simpler; it is a single "nonconformist" protein molecule. This type of protein is part of certain meat tissues. Heat does not inactivate it. Bacteria and viruses cause millions of cases of foodborne illness per year. No cases of prion-related CJD have been confirmed in the United States through the late months of 2004.

*This chapter is based on one written by Dean O. Cliver for *Food Microbiology: Fundamentals and Frontiers,* 2nd ed. It has been adapted and augmented with material on prions for use in an introductory text.

Table 21.1   Comparison of bacteria, viruses, and prions

Trait	Bacteria	Viruses	Prions
Composition	Complex structures of lipid, carbohydrates, proteins, DNA, and RNA	Single strand of RNA in a simple protein structure	Specific type of protein
Types important in food	More than a dozen gram-positive and gram-negative organisms	Two most important are hepatitis A virus and Norwalk virus, or noroviruses	Only one that we know of
Control in food	Control growth by changing intrinsic or extrinsic factors; kill by heat, irradiation, etc.	Do not grow in food; inactivated by heat; prevent entry by good hygiene	Do not grow in food; not inactivated by heat; prevent consumption of meat from mad cows
Disease	Infection or intoxication	Infection	Protein shape morphing
Transmission	Eating food where bacteria have grown	Fecal-oral	Eating meat that contains prions
Detection method(s)	Relatively easy to culture in microbiology laboratory	Electron microscopy or human cell culture	Examination of tissue at autopsy
No. of U.S. cases/yr	$\geq$30 million	Up to 45 million	None yet

VIRUSES

The Challenge

Viruses infect humans and grow in intestinal cells, releasing billions of their progeny into the feces. There can easily be 10^8 viruses per g of a food handler's feces. Imagine that the food handler does not wash his or her hands after using the toilet (Box 21.1). A 0.01-g piece of feces hiding under a fingernail would contain 10^6 (1 million) virus particles. If the infectious dose were 1,000 particles, the feces could get mixed into 1,000 servings of salad dressing, donut glaze, or whipped cream, causing illness in 1,000 people (1,000 particles/serving × 1,000 servings = the million virus particles originally under the fingernail). (This example gives new meaning to the phrase "finger-licking good.")

Such incidents occurred in:

- St. Paul, Minn., when 3,000 people got sick after eating butter cream frosting prepared by a food handler infected with a Norwalk-like virus
- cruise ships, where hundreds of people became sick from eating fruits and ice contaminated with enteroviruses
- a central school lunch kitchen in Japan, where enteroviruses infected 3,200 children and 120 teachers
- Michigan schools, where frozen strawberries from the school lunch program caused 230 cases of hepatitis

Eating raw shellfish from water polluted with feces is a classic cause of viral food poisoning. Shellfish concentrate the virus from the water because they are filter feeders. An outhouse draining into an oyster bed caused the first reported outbreak of hepatitis A originating from oysters. Approximately *300,000* people who ate raw clams harvested from polluted waters in the People's Republic of China came down with hepatitis A. Clearly, "dilution is not the solution" when it comes to feces entering shellfish beds.

BOX 21.1

Don't get caught dirty-handed

One of the first things we learned (or should have learned) was to wash our hands after going to the bathroom. Nonetheless, the fecal-oral route is a frequent source of foodborne illness. The failure of health care workers to wash their hands is an important cause of infections acquired in the hospital. To combat this public health threat, the American Society for Microbiology (ASM) has started a Clean Hands Campaign (see figure on left and http://www.washup.org). Food safety educators are also encouraging consumers to take responsibility for food safety through educational programs, such as the FightBac campaign (see figure on right).

When asked, "Do you always wash your hands after going to the bathroom?" >95% of us respond "yes." However, when the ASM commissioned a survey to observe public behavior in six airports, the results were quite different (see table). The observers discreetly watched >7,000 people in six cities to see who washed their hands and who didn't. There were significant gender and geographical differences (what's going on in San Francisco?). The high rate of hand washing in Toronto was due to the increased awareness caused by the SARS virus.

City	% of people who did not wash their hands	
	Male	Female
New York City	28	13
Chicago	27	15
San Francisco	21	41
Dallas	31	8
Miami	30	21
Toronto	6	4
Avg	26	17

Washing one's hands for 15 to 20 s with warm soapy water is the first line of defense against infectious disease. However, an ASM member who did a ministudy of ~300 colleagues' bathroom habits during the annual meeting found that no one washed his hands this well (although >95% of these microbiologists did wash their hands).

Gloves are not a substitute for hand washing. An undergraduate research project (R. Montville, Y. Chen, and D. W. Schaffner, *J. Food Prot.* **64**:845–849, 2001) that examined the transfer of bacteria from hands to food without gloves found a 10% transfer rate. The use of gloves reduced the transfer rate to 0.01%. This 1,000-fold difference in the transfer of bacteria from hands to food is good, but it can be overcome by dirty hands. If a food handler has 10^6 bacteria on her hand and transfers 0.01% of them, 10^4 bacteria will end up on the food. You can't check out the kitchen in a restaurant before you eat, but you can check out the bathroom. If there is no soap, eat elsewhere.

Courtesy of the American Society for Microbiology.

Courtesy of the Partnership for Food Safety Education.

Virus Biology

Viruses are *host specific,* i.e., they grow only on specific host organisms. Humans are the *only* host for human *enteric* (intestinal) viruses. Human enteric viruses infect only humans who eat them. They attach to and infect intestinal cells. After they multiply inside, they break open the cell and are shed into the feces.

Enteric viruses are simple. They are just a strand of RNA surrounded by a few proteins. The RNA makes more RNA without getting DNA involved. In fact, these *RNA viruses* have no DNA! The RNA contains the information required to make a few proteins and the enzymes needed for viral self-replication. There is some debate as to whether viruses are alive, since they cannot multiply outside of their specific hosts. Thus, they are frequently referred to as virus particles.

Intervention

In theory, it is very simple to prevent viral diseases transmitted by foods.

- Keep the viruses out of the food by using good food handler hygiene and preventing the harvest of shellfish from polluted water.
- Heat the food to kill the viruses.
- Use potable water in all areas of food processing.
- Sanitize food contact surfaces.

Nonetheless, there are 50 million cases of foodborne viral disease per year. This suggests that these simple measures are not so simple to follow.

Although heat kills viruses, other antibacterial food-processing methods may or may not kill viruses. The abilities of new, nonthermal methods, such as high pressure, pulsed electric light, and electromagnetic waves, to kill viruses are largely unknown. Viruses are relatively acid resistant and survive stomach acid. More radiation is needed to kill viruses than bacteria.

A process called *depurination* is used to eliminate viruses from shellfish harvested from contaminated waters. In depurination, the shellfish are moved from the polluted water to clean water and held for some period. The shellfish should wash the viruses out with the clean water. However, the viruses are frequently retained after the bacteria used to monitor the process have been washed out. Thus, it is hard to tell when depurination is complete or if it is successful.

Noroviruses

Noroviruses are a major cause of foodborne disease. Outbreaks can have thousands of cases. Noroviruses occupy a perennial position among the top 10 agents of foodborne illness, and they occasionally win the coveted number 1 spot. They, like all viruses, are difficult to study because they cannot be grown in the laboratory. Electron microscopes, antibodies, and genetic methods are used to detect noroviruses.

These small, round, structured caliciviruses contain a single strand of RNA. They are transmitted by a fecal-oral route. Raw shellfish, especially oysters, concentrate the virus from polluted waters. The viral particles are hearty, surviving storage, refrigeration, freezing, and passage through stomach acid. Upon reaching the small intestine, the virus infects mucosal cells, multiplies inside, kills the cells, and is released to contaminate more feces. Fecal shedding of the virus can continue for more than a week after the infected person has gotten better.

Hepatitis A Virus

Food-related cases of hepatitis A comprise 5% of all hepatitis cases and are not very important in the overall control of hepatitis. From the food safety point of view, however, this 5% equals ~2,000 cases per year. Hepatitis A is transmitted in foods by the feces of infected food handlers.

Hepatitis A virus has an incubation period of up to 50 days. This is troublesome, since it is shed in the feces of an infected person for 10 to 14 days *before* the person gets sick. This provides a large window of opportunity for unwitting virus transmission. Hepatitis is a nasty disease. It causes *jaundice* (yellowing of the skin), anorexia, vomiting, and a profound melancholy. These symptoms can last for weeks. The shedding continues for a few more weeks after the infected person gets better. Since there is no way to culture the virus, the diagnosis can be made only by finding hepatitis-specific antibodies in the person's blood. Hepatitis can cause kidney problems. When the virus infects liver cells, the body's immune system responds with T cells. These kill the virus *and* the liver cells. In countries where fecal contamination is common, most children get a mild case (and acquire immunity) by the time they are 5 years old. In more developed countries, hepatitis A is more common among adults, who get more serious cases.

Norovirus and hepatitis A virus are compared in Table 21.2.

Food Relatedness of Other Viruses

Astroviruses cause as many as 40,000 illnesses/year in the United States. Named for the star patterns on their coats, they carry a single RNA. Diarrhea is the main symptom after an incubation period of 3 to 4 days. Recovery usually occurs in 2 to 3 days but may require up to 15 days.

Half of the diarrhea cases in hospitalized children are caused by rotaviruses. They, too, are transmitted in food and water by the fecal-oral route. Fecal-oral transmission is especially troublesome in day care centers and nursing homes. This double-stranded RNA virus kills ~1 million children worldwide each year. The death toll in the United States is ~100/year. The normal symptoms are diarrhea, and sometimes vomiting, for 4 to 8 days.

Table 21.2 Characteristics of two major foodborne viruses

Characteristic	Norovirus	Hepatitis A virus
Disease symptoms	Nausea, vomiting, diarrhea due to action on intestinal cells; lasts for 24–48 h	Jaundice, anorexia, vomiting, prolonged sickness; lasts for weeks to months
Incubation period	Usually 24–48 h but can be as quick as 10 h	15–50 days; median = 29 days
Shedding of viral particles in feces	Simultaneous with symptoms; continues for a few weeks after illness	Shed for 10–14 days *before* symptoms appear
Immunity	Exposure does not confer immunity	Exposure confers durable immunity[a]
Type of virus	Calicivirus; spherical; 28-nm diam	Picornavirus; spherical; 32-nm diam; depressions on surface
Estimated no. of foodborne cases/yr	50 million	2,000

[a]However, there can be a relapse years later. It is noteworthy that children get less severe symptoms and can have no symptoms at all. These children have lifelong immunity. Thus, in developing countries, where most people are exposed as children, adult cases are rarer. In industrialized countries, adults are at greater risk.

Polio was one of the first foodborne viral diseases. Fortunately, vaccination and pasteurization have almost conquered the virus. The World Health Organization has certified the Americas, Europe, China, and most of Asia as polio free. With <2,000 cases per year worldwide, poliovirus may soon be totally eradicated. Tick-borne encephalitis virus, another single-stranded RNA virus, can be transmitted in raw milk. It causes diarrhea. Foodborne virus infections are entirely preventable. They could be eliminated by improving the sanitation of water and food and by breaking the fecal-oral linkage.

Many viruses are *not* transmitted by foods. These include:

- rabies virus
- hantaviruses (a group of viruses that includes an extremely dangerous pulmonary virus carried by rodents in the southwestern United States)
- rhinoviruses (which cause the common cold)
- human immunodeficiency virus (HIV)

HIV-positive food handlers pose no risk of transmitting HIV through the food they prepare. Even people symptomatic with AIDS are not a threat to food safety if they do not have diarrhea. (However, no one with active diarrhea should be preparing or handling food.)

PRIONS

The prion story tells us a lot about the nature of science.

- Science-based public policy can be based only on what is known.
- The fact that something has not yet been discovered does not mean it does not exist.
- Established scientific "fact" is often theory, subject to modification.
- New findings can overthrow the most established dogma.

A Short History of the Prion

Most people have heard of mad cow disease. It has brought the British meat industry to its knees, and regulators dread its appearance in the United States. There are many prion-related diseases. However, mad cow disease did not appear (or perhaps even exist) until 1986. The prion story starts earlier, in 1972. A young doctor, Stanley Prusiner, became fascinated with a patient whose brain was being eaten away even though the rest of her body was untouched. Prusiner's studies of this disease earned him ridicule and almost cost him his job but ultimately won him the Nobel Prize.

Transmissible spongiform encephalopathies (TSEs) (medical terminology for brains that look decayed and spongy) are a symptom of scrapies in sheep, kuru in cannibals, and CJD in elderly humans. In all three diseases, autopsies reveal decayed, spongy brains. However, until Prusiner, no one had connected these illnesses to a common mechanism. They were thought to be caused by mutations, genetic predispositions, or "slow" (since the incubation period was 3 to 5 years) undetectable viruses. Prusiner proposed the prion concept (coined by combining proteinaceous and infectious) to explain scrapies. Prusiner defined a prion as an infectious protein particle without any nucleic acid. It is a specific protein normally found in healthy brain and backbone membranes. When it is touched by the infectious mod-

Toles © 2000 *The Buffalo News.* Reprinted with permission of Universal Press Syndicate. All rights reserved.

ified form of the protein (the prion), the specific protein refolds into the prion shape. The prion concept contradicts the dogma that only DNA and RNA transmit biological information. It was met with ridicule and derision. Furthermore, the idea that prion proteins can exist in two forms (the normal shape and the infectious shape) broke the central dogma of molecular biology. Everyone who has taken even one course in biochemistry *knows* that:

1. DNA sequence is used to make RNA sequence.
2. RNA sequence is used to make a protein with a specific amino acid sequence.
3. The protein sequence dictates its (only possible) structure.

The two holy mantras of molecular biology are "DNA makes RNA makes protein" and "protein sequence determines structure determines function." Prusiner showed that these facts are not always true.

This would be an obscure piece of scientific history except for one thing—in 1986, bovine spongiform encephalopathy (BSE) broke out in British cattle. Although it was attributed to slow viruses at the time, prions had jumped the species barrier from sheep to cows. The prion is transmitted from animal to animal when they are fed offal, or the rendered remains of infected animals. (The United States banned the use of offal in animal feed in 1989.) Over the next 10 years, a new form of CJD (vCJD), found in young people rather than the elderly, became statistically correlated with beef consumption. The British beef industry reacted by changing feeding practices and killing all animals in herds afflicted with BSE. There was a diplomatic crisis as countries banned importation of British beef. Tens of

Authors' note

Students taking biochemistry or genetics courses should be warned that these mantras are almost always correct. The exception proves the rule.

thousands of cattle were killed as a preventive measure. All hell broke lose when, in 1996, the British government finally admitted that BSE could cause vCJD in humans. The dust still has not settled. BSE has recently appeared in the United States.

In 1997, Stanley Prusiner was awarded the Nobel Prize in Medicine for discovering the prion.

Prion Biology

The prion concept, especially the "without any nucleic acid" part, remains controversial. It is hard to prove that something (in this case, the nucleic acid) does not exist. Maybe it exists but cannot be found. The argument against prions is that there *must* be a small, undetectable amount of nucleic acid. However, treatments that destroy nucleic acids do not destroy prions, and nucleic acids have not been found in purified prions.

The concept of protein shape morphing explains all of the TSEs. However, protein morphing is only one of three mechanisms that cause spongiform encephalopathies. A genetic defect in the normal protein causes hereditary CJD in certain ethnic populations. A mutation in the normal protein causes sporadic CJD in the elderly. Finally, there is the morphing of normal protein into prions by prions consumed in the diet. It is very difficult to determine the cause of CJD in a specific person.

You need to understand a bit of protein chemistry to understand how prions work. The order in which amino acids are linked in a protein (*primary sequence*) determines its local shape (*secondary structure*). Common secondary structures are helices and pleated sheets. A protein molecule may contain several helices, pleated sheets, and unordered sections. These arrange themselves to form the protein's overall shape. The normal protein has a shape with several helices. When touched by the prion, the normal protein refolds to match the prion shape. This has large runs of pleated sheets (Fig. 21.1). In this shape, the prion protein destroys brain tissue.

Prions are not inactivated by heat, irradiation, or food preservatives. Once they are in a food, they are there for good. The only way to prevent TSEs is to prevent the converted form from touching the normal form. This is done by banning the use of animal by-products in animal feed. However, wild animals, such as mink and deer, also carry TSEs. Theoretically, these prions might jump the species barrier into domestic cattle.

*Authors' note*_____

The term "morphing" is conceptually correct, but it is not the correct technical term. If you are interested in the nomenclature, read Prusiner's paper.

Figure 21.1 Structures of normal protein **(left)** and infectious prion form **(right).** Note the changes in the secondary structures (coils and pleated sheets [arrows]), as well as the change in the overall shape. Both forms have the same amino acid sequence. Used with permission of Sarit Helman.

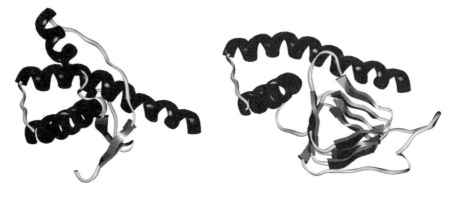

Summary

- Foodborne viruses are as important as bacteria in causing foodborne disease.
- Viruses grow only in specific hosts and cannot grow in foods.
- Fecal-oral transmission is the main route of infection for enteric viruses.
- Heat and good hygiene should keep foods safe from viruses.
- Noroviruses and hepatitis A virus are the major foodborne viruses.
- Prions are infectious proteins devoid of RNA and DNA that were discovered by Stanley Prusiner in the 1980s.
- Prions cause "transmissible spongiform encephalopathies," such as scrapies, kuru, mad cow disease, and Creutzfeldt-Jakob disease.
- TSEs are caused when normal protein is converted into an infectious prion by inherited or spontaneous mutation or by contact with an ingested prion molecule.

Suggested reading

Cliver, D. O. 2001. Foodborne viruses, p. 501–511. *In* M. P. Doyle, L. R. Beuchat, and T. J. Montville (ed.), *Food Microbiology: Fundamentals and Frontiers,* 2nd ed. ASM Press, Washington, D.C.

Cliver, D. O. 1997. Virus transmission via food (an IFT Scientific Status Summary). *Food Technol.* **51**:71–78.

Prusiner, S. B. 1998. Prions. *Proc. Natl. Acad. Sci. USA* **95**:13363–13383.

Questions for critical thought

1. This chapter's chronology of mad cow disease is accurate up to late 2004. What has happened with mad cow disease in North America since that time?
2. How are bacteria, viruses, and prions different?
3. What intervention can be used to make food safe from all of them?
4. There are no routine laboratory tests to detect viruses or prions in foods. How does this make their control different from controlling foodborne pathogens?
5. What is a prion? What three ways can it be formed?
6. Much of this book deals with how to prevent microbial growth in food. How can viruses be prevented from growing in foods?
7. Think of three terms that could be used to convey to your less-sophisticated peers the concept of fecal-oral transmission.
8. Why is dilution *not* the solution for preventing viral contamination by shellfish? Consider the case of the outhouse over the oyster bed. Assume 50 people, all infected at 10^8 particles per g of feces, depositing 2 kg of fecal material per day into a body of water that is 1 by 1 km and 10 m deep. How many viral particles per milliliter would there be? Why might you be able to swim in the water without getting sick but get sick from the oysters that live in the water?

SECTION V

Control of Microorganisms in Food

22

Antimicrobial Chemicals*

LEARNING OBJECTIVES

The information is this chapter will allow the student to:
- link chemical preservatives to their ability to prevent food spoilage
- distinguish positive and negative aspects of food preservatives
- characterize traits that distinguish "natural" from "chemical" preservatives
- describe factors that affect antimicrobial preservation systems and how they interact
- understand how organic acids work and the influence of pK_a and pH on their action
- identify enzymes and inorganic food preservatives, how they work, and foods they are used in

INTRODUCTION

Preservatives are good. Without them, food spoils. Microbiological, enzymological, chemical, and physical changes cause spoilage, and preservatives prevent these changes. Most antimicrobials inhibit growth without killing the microbe. Chilling, freezing, water activity (a_w) reduction, acidification, fermentation, energy, and antimicrobial compounds all inhibit microbial growth.

Antimicrobials are sometimes called preservatives. However, preservative is a broader term than antimicrobial. It includes antibrowning agents, antioxidants, stabilizers, etc. The major antimicrobial targets are foodborne pathogens and spoilage microorganisms. They cause off odors, off flavors, liquefaction, and discoloration and produce slime in food. Spoilage can be a good thing. It warns people not to eat foods that may contain pathogens. However, pathogens can grow without obvious signs of spoilage. (If pathogens always spoiled food, no one would eat the food and there would be no foodborne illness.) Preservation systems that allow pathogen growth without spoilage are inherently dangerous.

Most antimicrobials only *inhibit* growth. They are *bacteriostatic* rather than *bactericidal*. Bacteriostatic compounds inhibit microbes; bactericidal compounds kill them. Antimicrobials act against specific metabolic targets. Antimicrobials having different mechanisms can be combined to give a synergistic effect. This is called *hurdle technology*. The cell wall, cell membrane, enzymes, and genes are all targets for antimicrobial action. Because

---

*This chapter was originally written by P. Michael Davidson for *Food Microbiology: Fundamentals and Frontiers*, 2nd ed., and has been adapted for use in an introductory text.

metabolism is an interlocking web, it is hard to determine an antimicrobial's exact mechanism of action.

Antimicrobial classification is arbitrary. This chapter divides antimicrobials into two classes, traditional "chemical" antimicrobials and naturally occurring ones.

Chemical antimicrobials:

- have been used for years
- are approved by many countries
- are made by synthetic means
- are considered "bad" by consumers

Chemical antimicrobials are also found in nature. They include acetic acid from vinegar, benzoic acid from cranberries, and sorbic acid from rowan berries.

Naturally occurring food preservatives are compounds that:

- are extracted from natural sources
- are "organic"
- are not usually subject to regulatory approval as food additives
- look "friendly" on the label (label friendly)

FACTORS THAT AFFECT ANTIMICROBIAL ACTIVITY

Foods are complex systems. Many variables influence microbial growth and inhibition. For instance, the food's pH, a_w, storage temperature, and gas composition affect antimicrobial effectiveness. These factors must be considered when developing preservation systems. Antimicrobial activity is affected by microbiological, intrinsic, extrinsic, and process factors (Table 22.1).

There has been an explosion of knowledge about how cells respond to stress. Tolerance gained by exposure to one stressor, such as cold, may make cells more tolerant of another, such as acid. This is called *adaptation, stress response,* or *acquired tolerance.* Special proteins are produced when cells are stressed. They are responsible for the acquired tolerance. Some stress responses are specific to a given stressor; others confer resistance to a broad range of stressors.

Intrinsic factors are part of the food itself. These include composition, pH, buffering capacity, oxidation-reduction potential, and a_w. They can be specified during product development to formulate "food safety" into the food. When a product is reformulated to reduce salt, acidity, sugar, or other

Table 22.1 Factors that influence antimicrobial activity

Microbiological	Intrinsic	Extrinsic	Process
Genetic resistance	pH	Time	Alteration of food microstructure
Vegetative cell or spore	Oxidation-reduction potential	Temperature	Alteration of microbial ecology
Cell number	a_w	Atmosphere	Alteration of food composition
Growth phase	Buffering capacity	Relative humidity	Induced responses, such as injury
Growth rate	Food composition		
Inactivation mechanisms			
Interaction with other microbes			
Cell physiology			

potentially inhibitory substances, the impact on microbiological safety should be considered.

Extrinsic factors also influence antimicrobial activity. These variables are external to the food. They include temperature, storage time, atmosphere, and relative humidity. The time-temperature combination is perhaps the most important. It is arguably the most difficult to control. Extrinsic factors also need consideration during product development. A manufacturer of frozen foods expanding into refrigerated entrees faces new microbiological challenges.

The microbiology of a processed food is often different from that of an unprocessed food. Processing can change food composition. For example, drying changes the a_w. Heating foods can eliminate vegetative cells and allow spores to thrive. Food processing can also change food microstructure.

*Authors' note*_____

The shopper who forgets that he left his refrigerated entrée in the trunk of his car and then eats it the next day creates a special challenge for food safety.

ORGANIC ACIDS

The short-chain organic acids (Table 22.2) have similar mechanisms and applications. They are all found in nature but are often made chemically for use in foods. They have "generally recognized as safe" status and are label friendly. Due to their mechanism of action (see below), organic acids are most useful below their pK_a (explained below). They are rarely used in foods with pH values of >5.5. This is a major limitation.

To be effective, organic acids must cross the bacterial cell membrane and get into the cytoplasm (Fig. 22.1). Organic acids exist in charged and uncharged forms but cannot cross the membrane when they are charged, so only the uncharged (*protonated*) acid can get into the bacterium and inhibit its growth. The pH at which the concentrations of charged and uncharged acids are equal is the pK_a. Since pH is a log scale, at 1 pH unit above the pK_a, there is 10-fold less protonated acid. This low concentration is not inhibitory. This is why organic acids are most useful in acidic foods near or below their pK_as. At low pH, the protonated acid crosses the cell membrane, and the neutral cytoplasmic pH then causes the acid to dissociate. The release of the proton acidifies the cytoplasm. The cell uses adenosine triphosphate (ATP) to pump protons out of the cell. This maintains the neutral pH needed for viability but uses up the ATP energy needed for growth.

Table 22.2 Short-chain organic acids

Organic acid	Formula	pK_a	Where found	Target of activity	Applications	Concentration (%)
Acetic	CH_3COOH	4.5	Vinegar	Bacteria and yeast	Baked goods, cheese, olives, gravies, sauces, meat and poultry	<0.1–0.8
Benzoic	⬡—COOH	4.2	Cranberries	Fungi	Beverages, jams, jellies	<0.1
Lactic	$CH_3OHCOOH$	4.8	Fermented food	Bacteria	Used mainly for pH control and flavor; also used for sanitation, sprayed on meat	<0.4–2.0
Propionic	CH_3CH_2COOH	4.9	Swiss cheese	Yeast, mold, rope-forming bacteria	Baked goods and cheeses	0.3–no limit
Sorbic	$CH_3CH=CHCH=CHCOOH$	4.7	Rowanberries	Fungi and some bacteria	Baked goods, cheese, wine	0.1–0.3

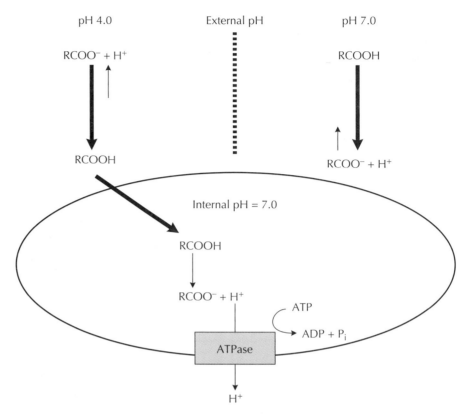

Figure 22.1 Organic acids are effective inhibitors only at low pH, where their protonated forms can cross the cell membrane (oval). At the higher pH inside the cell, the acid dissociates and the cell has to use energy to pump out the proton that would acidify the cytoplasm. At high external pH, the dissociated acid cannot cross the membrane and get into the cell.

NITRITES

Nitrites are used in cured meats, such as bacon, hot dogs, and bologna. Curing solutions contain nitrite, salt, sugar, spices, and ascorbate or erythorbate. Nitrite has many uses in cured meats. Nitrosomyoglobin makes cured meat pink. In addition to being antimicrobials, nitrites contribute to cured meat flavor and texture and serve as antioxidants. Nitrite's effectiveness depends on many factors. Nitrites work better at lower pH and are more inhibitory in the absence of air. The reducing agents ascorbate and isoascorbate enhance nitrite's antibotulinal action. Temperature, salt, and microbial load also influence nitrite effectiveness.

Sodium nitrite inhibits *Clostridium botulinum* in cured meats. Nitrite also inhibits bacteria, such as *Achromobacter, Enterobacter, Escherichia coli, Flavobacterium, Micrococcus, Pseudomonas,* and *Listeria monocytogenes.* Bacon, bologna, corned beef, frankfurters, luncheon meats, ham, fermented sausages, shelf-stable canned cured meats, and ham usually contain nitrite. Governmental regulations limit its use to 156 parts per million (ppm) for most products and 100 to 120 ppm in bacon. Sodium erythorbate or isoascorbate is required as a cure accelerator and to inhibit nitrosamine formation.

Nitrosamines are carcinogens formed by reactions of nitrite with amines. The discovery that nitrosamines are produced from nitrite and meat protein at high temperature triggered extensive research on nitrite substitutes. However, no other compound could replace nitrite's antimicrobial, taste, and color functions. Concern about nitrosamines decreased when it was learned that vegetables were a larger source of dietary nitrites.

PARABENZOIC ACIDS

Benzoic acid has a pK_a similar to those of the organic acids. However, the carbon chains present between the benzene and the acid parts of the molecule increase the pK_a. These parabenzoic acids (Fig. 22.2), commonly called *parabens*, remain undissociated up to their pK_a of 8.5. This gives them an effective pH range of 3.0 to 8.0. The methyl, propyl, and heptyl parabens are legal for use as antimicrobials in most countries.

The antimicrobial activity of parabens increases with increasing chain length. Unfortunately, solubility in water decreases with increasing chain length. The oppositional influence of chain length and solubility determines which paraben can be used in a given application. Methyl and propyl parabens are often used in a 2:1 to 3:1 ratio to capitalize on their solubility and increased activity, respectively. The *n*-heptyl ester is used in beers, noncarbonated soft drinks, and fruit-based beverages. Parabens are generally more active against molds and yeast than against bacteria (Table 22.2). Parabens are used in baked goods, beverages, fruit products, jams and jellies, fermented foods, syrups, salad dressings, wine, and fillings.

Phenolics and parabens act at microbial membranes, but the specific mechanisms are unclear. The membrane's barrier properties may be degraded, letting intracellular compounds leak out. Respiration may also be inhibited, perhaps due to changes in membrane fluidity.

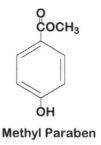

Methyl Paraben

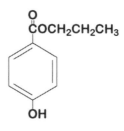

Propyl Paraben

Figure 22.2 Structures of alkyl esters of *p*-hydroxybenzoic acid (parabens). Reprinted from P. M. Davidson, p. 593–627, *in* M. P. Doyle, L. R. Beuchat, and T. J. Montville (ed.), *Food Microbiology: Fundamentals and Frontiers*, 2nd ed., (ASM Press, Washington, D.C., 2001).

SODIUM CHLORIDE

Sodium chloride (NaCl), or common salt, is the oldest preservative known. A few foods, such as raw meats and fish, are preserved solely by high salt concentrations, i.e., *salting*. More often, salt is combined with processes such as canning, pasteurization, or drying. Foodborne pathogens are usually inhibited by an a_w of ≤ 0.92 (equivalent to 13% [wt/vol] NaCl) under otherwise-optimum conditions. The exception is *Staphylococcus aureus*, which can grow at an a_w of 0.83 to 0.86 under aerobic conditions. Under anaerobic conditions, the minimum a_w is >0.90. *L. monocytogenes* is also salt tolerant. It survives in saturated salt solutions at low temperatures.

Sodium chloride's primary lethal mechanism is *plasmolysis* (shrinkage of the cell wall due to water loss). Salt reduces the a_w, creating unfavorable growth conditions. As the a_w decreases, cells undergo osmotic shock and rapidly lose water through plasmolysis. During plasmolysis, a cell dies or remains dormant. To resume growth, the cell must concentrate specific solutes to equilibrate with the external a_w.

PHOSPHATES

Sodium acid pyrophosphate, tetrasodium pyrophosphate, sodium tripolyphosphate, sodium tetrapolyphosphate, sodium hexametaphosphate (SHMP), and trisodium phosphate are active against microbes in foods.

Gram-positive bacteria are more susceptible to phosphates than gram-negative bacteria. Tetrasodium pyrophosphate, sodium tripolyphosphate, and SHMP at 0.5% inhibit *Bacillus subtilis*, *Clostridium sporogenes*, and

Clostridium bifermentans. Sodium acid pyrophosphate inhibits *C. botulinum* toxin production but not growth. Outgrowth of *Bacillus* spores is prevented by 0.2 to 1.0% SHMP.

The ability of polyphosphates to chelate metal ions is responsible for their antimicrobial action. Polyphosphates probably inhibit gram-positive bacteria and fungi by removing essential cations from the cell wall binding sites. Divalent cations can reverse polyphosphate inhibition. Inhibition is also reduced at lower pH due to protonation of the chelating sites. Orthophosphates have no chelating ability and thus no inhibitory activity.

SULFITES

Sulfur dioxide (SO_2) has been used as a disinfectant since ancient times. Antimicrobial salts of sulfur dioxide include potassium sulfite (K_2SO_3), sodium sulfite (Na_2SO_3), potassium bisulfite ($KHSO_3$), sodium bisulfite ($NaHSO_3$), potassium metabisulfite ($K_2S_2O_5$), and sodium metabisulfite (NaS_2O_5). Sulfites are used primarily in fruit and vegetable products to control spoilage yeasts and molds, acetic acid bacteria, and malolactic bacteria. Sulfites are also antioxidants that inhibit enzymatic browning.

The most important factor for the antimicrobial activity of sulfites is pH. In water, sulfur dioxide and its salts exist as a pH-dependent equilibrium mixture:

$$SO_2 \cdot H_2O \leftrightarrow HSO_3^- + H^+ \leftrightarrow SO_3^{2-} + H^+$$

Once again, dissociation of the acid plays a major role. As the pH decreases, the proportion of $SO_2 \cdot H_2O$ increases and the bisulfite (HSO_3^-) ion concentration decreases. The pK_a values for sulfur dioxide are ~1.8 and 7.2. Sulfite is most inhibitory when the acid or $SO_2 \cdot H_2O$ is undissociated. Therefore, their most effective pH range is <4.0. $SO_2 \cdot H_2O$ is 100 to 1,000 times more active than HSO_3^- or SO_3^{2-} against *E. coli*, yeast, and *Aspergillus niger*. Sulfites, especially the bisulfite ion, are very reactive. They form addition compounds (α-hydroxysulfonates) with aldehydes and ketones.

Low concentrations of sulfur dioxide are fungicidal. The inhibitory concentration of sulfur dioxide against *Saccharomyces*, *Zygosaccharomyces*, *Pichia*, *Hansenula*, and *Candida* species is 0.1 to 20.2 ppm. Sulfur dioxide at 25 to 100 ppm inhibits *Byssochlamys nivea* growth and patulin production in juices. Sulfites may inhibit spoilage bacteria in wines, fruits, and meats.

Sulfur dioxide controls microbial spoilage in fruits, fruit juices, wines, sausages, fresh shrimp, and pickles. It is added at 50 to 100 mg/liter to grape juices used for making wines to inhibit molds, bacteria, and yeasts. If the right concentration is used, sulfur dioxide does not interfere with wine yeasts or wine flavor. During fermentation, sulfur dioxide also serves as an antioxidant, clarifier, and dissolving agent. Sulfur dioxide (50 to 75 mg/liter) also prevents postfermentation spoilage.

The targets for sulfite inhibition are the cytoplasmic membrane, DNA replication, protein synthesis, and enzymes. In *Saccharomyces cerevisiae*, *Saccharomyces ludwigii*, and *Zygosaccharomyces bailii*, sulfites diffuse into the cell. Other fungi have active transport systems for sulfites.

NATURALLY OCCURRING ANTIMICROBIALS

Many foods contain natural antimicrobials. They may extend the food's shelf life, but their use as direct food additives presents many challenges. Ideally, they should work when added as food ingredients without purification. The cost of using the natural antimicrobial should be as low as that for chemical preservatives. They should not affect the food's taste, nutrition, or safety. However, most natural antimicrobials are normally present in amounts that are too small to work when mixed into foods. If larger amounts were used, taste and smell would be affected.

Lysozyme

Lysozyme is an antimicrobial enzyme produced in eggs, milk, and tears. Commercial lysozyme is made from dried egg white. The lysozyme in hen's eggs has 129 amino acids. It is stable to heat (100°C) at low pH (<5.3). It is inactivated at lower temperatures when the pH is increased. The enzyme breaks the β-1,4 glycosidic bonds between N-acetylmuramic acid and N-acetylglucosamine of peptidoglycan. This breaks down the bacterial cell wall.

Lysozyme is most active against gram-positive bacteria, probably because their peptidoglycan is more exposed. It inhibits *C. botulinum, Clostridium thermosaccharolyticum, Clostridium tyrobutyricum, Bacillus stearothermophilus, Bacillus cereus,* and *L. monocytogenes*. Lysozyme is the main antimicrobial in egg albumen. Ovotransferrin, ovomucoid, and alkaline pH enhance its activity in eggs. Lysozyme is less effective against gram-negative bacteria because they have less peptidoglycan and a protective outer membrane.

The chelator ethylenediaminetetraacetic acid (EDTA) enhances lysozyme activity. EDTA helps lysozyme penetrate to the peptidoglycan. The susceptibility of gram-negative cells can be increased by chelators (e.g., EDTA) that bind Ca^{2+} or Mg^{2+}. These cations are needed to maintain a functional lipopolysaccharide layer. Gram-negative cells are also sensitized to lysozyme by pH shock, heat shock, osmotic shock, drying, and freeze-thaw cycling.

Lysozyme is one of the few natural antimicrobials subject to regulatory approval for food use. In Europe, it is used to prevent gas formation (blowing) in cheeses. Lysozyme is used in Japan to preserve seafood, vegetables, pasta, and salads.

Lactoferrin and Other Iron-Binding Proteins

Milk and eggs contain antimicrobial iron-binding proteins. Lactoferrin is the main iron-binding protein in milk. Milk also contains low levels of the iron-binding protein transferrin. There are two iron-binding sites per lactoferrin molecule. One bicarbonate (HCO_3^-) is required for each Fe^{3+} bound by lactoferrin. Citrate, another chelator in milk, inhibits lactoferrin's activity, while bicarbonate reverses the inhibition. To be effective, lactoferrin must be in a low-iron environment where bicarbonate is present. Milk provides these conditions.

Microorganisms with a low iron requirement are not inhibited by lactoferrin. Iron stimulates growth in many genera, including *Clostridium, Escherichia, Listeria, Pseudomonas, Salmonella, Staphylococcus, Vibrio,* and *Yersinia*. Lactoferrin's depletion of iron may cause inhibition.

Egg albumen has another iron-binding molecule, ovotransferrin. This is sometimes called conalbumin. It makes up 10 to 13% of the total egg white protein. Ovotransferrin has an amino acid sequence that is very similar to that of lactoferrin. Each ovotransferrin molecule has two iron-binding sites. Like lactoferrin, it binds anions, such as bicarbonate, with each bound ferric iron molecule. To be inhibitory, there must be more ovotransferrin molecules than iron molecules and the pH must be alkaline. Ovotransferrin inhibits gram-positive and gram-negative bacteria. Gram-positive bacteria are generally more sensitive. *Bacillus* and *Micrococcus* species are very sensitive, as are some yeasts. Ovotransferrin is bacteriostatic. Like lactoferrin, the primary inhibitory mechanism is probably iron depletion.

Avidin

Avidin is another egg albumen protein. It makes up ~0.05% of the total egg albumen protein. It is stable to heat and a wide pH range. Avidin binds biotin at a ratio of four biotin molecules per avidin molecule. Biotin activates enzymes in some metabolic cycles. Avidin inhibits bacteria and yeasts that require biotin for growth. However, avidin also binds transport proteins in the *E. coli* outer membrane. This suggests that avidin may inhibit bacteria by interfering with transport.

Spices and Their Essential Oils

Spices are usually added to foods as flavoring agents. However, they also have some antimicrobial activity. The ancient Egyptians used spices for food preservation and embalming around 1550 B.C. Cloves, cinnamon, oregano, and thyme, and to a lesser extent sage and rosemary, are the strongest antimicrobial spices. Table 22.3 lists their active chemical components and required concentrations, and the organisms sensitive to them. Not all spices have antimicrobial activity.

There are not many studies of the inhibitory mechanism(s) of spices. The terpenes in spice oils are the primary antimicrobials. Many of the most active terpenes, e.g., eugenol, thymol, and carvacrol, are phenolic. Therefore, their modes of action are probably related to those of other phenolic compounds. They interfere with membrane function.

Onions and Garlic

Onions and garlic contain the best-characterized plant antimicrobials. These compounds inhibit *B. subtilis*, *Serratia marcescens*, *Mycobacterium* spp., *B. cereus*, *C. botulinum* type A, *E. coli*, *Lactobacillus plantarum*, *Leuconostoc*

Table 22.3 Antimicrobial activities of spices

Spice	Antimicrobial compound	Concentration	Target of activity
Cloves Cinnamon	Eugenol Cinnamon aldehyde	Spices are 10 to 20% volatile oil. Clove is active at 1/100 dilution.	Clove inhibits *B. subtilis*, *E. coli*, *Salmonella*, *L. monocytogenes*, and *S. aureus*
Oregano Thyme	Carvacrol Thymol	MIC[a], 0.02–0.05%; MBC[b], 0.03–0.1% of essential oil	*E. coli*, *Salmonella*, *L. monocytogenes*, *S. aureus*, *B. cereus*, some yeast and molds
Sage Rosemary	Terpenes, borneol, piene, camphor, thujone	MIC, 0.3%; MBC, 0.5%	More effective against gram-positive bacteria: *B. cereus*, *S. aureus*, *L. monocytogenes*
Vanillin	4-Hydroxy-3-methoxy-benzaldehyde	500–2,000 µg/ml	Molds and nonlactic gram-positive bacteria

[a]MIC, minimum inhibitory concentration.
[b]MBC, minimum bactericidal concentration.

mesenteroides, Salmonella spp., *Shigella* spp., and *S. aureus*. The fungi *Aspergillus flavus, Aspergillus parasiticus, Candida albicans,* and *Cryptococcus, Penicillium, Rhodotorula, Saccharomyces, Torulopsis,* and *Trichosporon* spp. are also inhibited.

The major antimicrobial in garlic is allicin. It is formed by allinase from alliin when garlic cells are disrupted. Allicin probably inhibits sulfhydryl-containing enzymes in bacteria. A similar reaction occurs in onions, forming thiopropanal-*S*-oxide. Onions also contain the antimicrobial phenolic compounds protocatechuic acid and catechol.

Isothiocyanates

Isothiocyanates (R—N=C=S) are potent antimicrobials. They are made from glucosinolates in plant cells from the Cruciferae, or mustard, family. This family contains cabbage, kohlrabi, Brussels sprouts, cauliflower, broccoli, kale, horseradish, mustard, turnips, and rutabagas. Fungi, yeasts, and bacteria are inhibited by 0.016 to 0.062 µg/ml in the vapor phase or 10 to 600 µg/ml in liquid. Isothiocyanates may inhibit cells by reacting with disulfide bonds or inactivating sulfhydryl enzymes.

Phenolic Compounds

Phenolic compounds have an aromatic ring with one or more hydroxyl groups. Phenolic compounds are classified as simple phenols and phenolic acids, hydroxycinnamic acid derivatives, and the flavonoids. Many of the inhibitors discussed above are phenolics.

Simple phenolic compounds include monophenols (e.g., *p*-cresol), diphenols (e.g., hydroquinone), and triphenols (e.g., gallic acid). Gallic acid occurs in plants as quinic acid esters or hydrolyzable tannins (tannic acid). The only simple phenols used as preservatives are those in wood smoke. Smoking meats, cheeses, fish, and poultry imparts both a desirable flavor and a preservative effect. Phenol and cresol contribute to the smoke's flavor and antioxidant and antimicrobial actions. Several commercial smoke preparations at 0.25 and 0.5% reduce *L. monocytogenes* viability in buffer. The phenolic glycoside oleuropein, or its aglycone, inhibits *L. plantarum, L. mesenteroides, Pseudomonas fluorescens, B. subtilis, Rhizopus* spp., and *Geotrichum candidum.*

Summary

- Microbiological spoilage is just one type of spoilage.
- Foods that permit pathogens, but not spoilage organisms, to grow are especially dangerous.
- Antimicrobials can be "cidal" or "static" but are usually the latter.
- Antimicrobial activity is influenced by microbiological, intrinsic, and extrinsic factors.
- Organic acids are effective below their pK_as because they cross the bacterial membrane, dissociate, and acidify the cytoplasm.
- Parabenzoic acids work similarly to organic acids but have higher pK_a values.
- Lysozyme is an enzyme that kills gram-positive cells by breaking down their walls.
- Nitrites prevent *C. botulinum* growth in cured meats and confer color and flavor.

- Sodium chloride lowers a_w and causes cellular plasmolysis.
- Phosphates and lactoferrin chelate metal ions that microbes need to grow.
- Sulfites are used primarily on fruits and vegetables to inhibit yeasts and molds.
- The sulfur-containing compounds in onions and garlic inhibit microbes.

Suggested reading

Ash, M., and I. Ash. 1995. *Handbook of Chemical Preservatives.* Grower, Brookfield, Vt.

Branen, A. L., P. M. Davidson, and S. Salminen (ed.). 1990. *Food Additives.* Marcel Dekker, New York, N.Y.

Davidson, P. M. 2001. Chemical preservatives and natural antimicrobial compounds, p. 593–627. *In* M. P. Doyle, L. R. Beuchat, and T. J. Montville (ed.), *Food Microbiology: Fundamentals and Frontiers,* 2nd ed. ASM Press, Washington, D.C.

Questions for critical thought

1. If an antimicrobial were bacteriostatic and inhibited microbes forever, would foods last forever? Explain.

2. You are the director of microbiology at Fine and Fat Foods. The marketing department has directed that all those nasty preservatives be removed from all products. How can you maintain the safety and quality of your products?

3. Diagram the process by which organic acids work in high-acid foods but not in low-acid foods.

4. If you were assigned the task of developing the ideal antimicrobial, what criteria would it have to meet?

5. Calculate the amount of protonated acetic acid at pH 3.5 and 5.5. How much acetic acid would you have to add at pH 5.5 to get as much protonated acetic acid as there is at pH 3.5?

6. You are an extension specialist at Big State University. A local inventor approaches you for advice on marketing a new antimicrobial. Called Superlator, it chelates all cations, preventing the growth of bacteria that require any cations. What advice do you give her?

7. What concepts from Chemistry 101 are important for understanding how preservatives work?

23

Biologically Based Preservation and Probiotic Bacteria*

LEARNING OBJECTIVES

The information in this chapter will help the student to:

- understand that biological methods can be used to "naturally" enhance food safety without changing the nature of the food (as fermentations do)
- appreciate the potential antimicrobial uses of the small proteins called bacteriocins
- relate the organization of bacteriocin genes to the way bacteriocins are made
- describe the characteristics of probiotic bacteria, suggest possible health benefits, and critically evaluate commercial claims for probiotic bacteria
- link the scarcity of validated data about the health benefits of probiotic bacteria to the human gastrointestinal tract's complex ecosystem

INTRODUCTION

It is easy to tell if foods are processed or contain preservatives. Changes in the way the food looks or information on the label make it obvious. However, this information is often misused. Consumers are wary of preservatives and "processed" foods, even though they give us a safe and diverse diet. As a result, consumers increasingly rely on refrigeration to ensure the safety of "fresh" foods. (Note well that the opposite of *fresh* is not "processed" or even "frozen," it is *spoiled*.) There are two reasons that refrigeration should not be used as the sole preservation method. The first is that temperatures in 20% of refrigerators are >10°C (50°F). The second reason is that *Listeria monocytogenes* can grow at refrigeration temperatures of <10°C. Because of this, additional means of keeping refrigerated foods safe are required. Lactic acid bacteria (LAB) can provide added protection to refrigerated foods. LAB are accepted by consumers as natural and health promoting. They may, through fermentation, be the oldest form of food preservation (after drying). Biologically based preservation methods that do not ferment foods are among the newest forms of food preservation.

This chapter provides an overview of *biopreservation*. Biopreservation is the use of LAB, their metabolic products, or both to improve or ensure the safety and quality of foods that are not seen as typical fermented foods.

Some LAB produce antimicrobial proteins called *bacteriocins*. Bacteriocins may inhibit spoilage and pathogenic bacteria without changing the

---

*This chapter is based on the one written by T. J. Montville and Karen Winkowski for the first edition of *Food Microbiology: Fundamentals and Frontiers*. It has been adapted and augmented with material from Todd Klaenhammer's chapter on probiotics in the second edition.

food (i.e., acidifying it or producing gas, solidification, etc.). The use of bacteriocins is a new and emerging area of food microbiology.

The food industry is market driven. Advertisers try to turn technical needs that might be seen as marketing negatives (i.e., the need to preserve a food) into something that will help sell the food (i.e., the preservation method improves consumer health). Thus, because consumers want foods that are more than nutritious, foods should be distinctly health promoting. Since LAB are natural and are perceived as health promoting, foods preserved by biological techniques may have an edge if marketed as "probiotic."

BIOPRESERVATION BY CONTROLLED ACIDIFICATION

Organic acids (such as acetic, lactic, or citric acid) can inhibit microbial growth. The acids are usually added to the food. LAB, however, can produce lactic acid *in* the food. Many factors determine the effectiveness of acidification in place. These include the food's initial pH, its buffering capacity, the specific pathogen, the nature of the fermentable carbohydrate, ingredients that affect LAB growth, and the growth rates of the LAB and pathogen at refrigerated and abuse temperatures. Biopreservation can require customization and research on each product use. Bacteriocins, diacetyl, and hydrogen peroxide may enhance acid inhibition. For example, MicroGard, a "generally recognized as safe" (GRAS) cultured milk product, is frequently added to cottage cheese in the United States as a biopreservative. MicroGard is made by fermenting milk using *Propionibacterium shermanii* to produce acetic acid, propionic acid, low-molecular-weight proteins, and a bacteriocin.

The idea of using LAB acid production to prevent botulinal poisoning dates back to the 1950s. This technology exploits the inability of *Clostridium botulinum* to grow at a pH of <4.8 to prevent its growth if bacon is left unrefrigerated. LAB and a fermentable carbohydrate (such as glucose or lactose) are added to the food. The LAB grow and produce acid only when the food is temperature abused. Under proper refrigeration, the LAB cannot grow, no acid is formed, and the preservation system is invisible to the consumer.

Nitrites are powerful inhibitors of botulinal spores. When it was found that the nitrites used to cure meats form cancer-causing chemicals called *nitrosamines*, research was initiated to find nitrite substitutes with antibotulinal activity. Nobi Tanaka at the University of Wisconsin reduced nitrite in bacon by using controlled acidification. When bacon was inoculated with 10^3 botulinal spores/g and incubated at 28°C, toxin was produced in 58% of the bacon samples prepared with the standard 120 parts per million nitrite. When the nitrite was reduced to 80 or 40 parts per million and supplemented with sucrose and starter cultures, the pH dropped and ≤2% of the bacon became toxic. The U.S. Department of Agriculture approved the "Wisconsin process" for bacon manufacture in 1986.

BACTERIOCINS

General Characteristics

LAB bacteriocins are a relatively heterogeneous group of small proteins. They act against closely related bacteria but not against the bacteria that make the bacteriocins. The term "closely related" can cover a wide range of gram-positive bacteria. Chelating agents, hydrostatic pressure, or injury can make gram-negative bacteria sensitive to bacteriocin. Bacteriocins are not enzymes but may have some unknown cellular function.

Authors' note

The toxicity level (<2%) of Wisconsin-process bacon may not sound good, but in the control experiment using conventional bacon, more than half of the samples became toxic. If you like bacon, be sure to refrigerate it!

The term "bacteriocins" was originally used for only those plasmid-mediated proteins produced by bacteria late in the stationary phase of growth that are bactericidal to a narrow range of closely related bacteria having specific binding sites for that bacteriocin. It is now clear that few LAB bacteriocins meet all of these criteria. Bacteriocins differ in which bacteria they kill, their biochemical traits, and whether their genes are located on the chromosome or a plasmid. Most bacteriocins are small (3 to 10 kDa), have a high isoelectric point, and contain both hydrophobic (repelled by water) and hydrophilic (water-loving) sections.

The lantibiotics and the pediocin-like bacteriocins are the two most important bacteriocin groups. If you know the basic biochemistry of amide linkages in proteins, you can see that the lantibiotic nisin is very strange (Fig. 23.1). Nisin contains unusual amino acids, other than the "normal" 20. After the protein is made, some amino acids react with cysteine to form thioether (single-sulfhydryl) lathionine rings. Bacteriocins containing lanthionine rings are called *lantibiotics*. There are many structurally similar lantibiotics. Nisin, the first and best-characterized LAB bacteriocin, is produced in two related forms. Nisin A contains a histidine at position 27, whereas nisin Z has an asparagine. Subtilin, produced by *Bacillus subtilis*, also contains five lanthionine rings and has a conformation similar to that of nisin.

Pediocin-like bacteriocins are small heat-stable proteins made of the usual 20 amino acids. They all have the same leader amino acid sequence containing glycine-glycine ↓ (any amino acid) (the arrow indicates the cleavage site). The upstream amino acids are cut off during the process of excreting the bacteriocin. Pediocin-like bacteriocins act against *L. monocytogenes*. They have a -tyrosine-glycine-asparagine-glycine-valine-(any amino acid)-cysteine amino-terminal sequence. Pediocin PA-1, pediocin AcH, sakacins A and P, leucocin A, bavaricin MN, and curvacin A are members of this group.

Bacteriocin Applications in Foods

The bacteriocin or the bacteria that make it can be added to foods to inhibit pathogens. Only nisin is commercially available as an ingredient, but pediocin addition is also effective. Many applications add the LAB rather than the pure bacteriocin. Bacteriocins can also be used to improve the

*Authors' note*_____
A plasmid is an extra piece of DNA separate from the chromosome.

*Authors' note*_____
Proteins are made by linking amino acid amide and carboxyl groups. There is an amine left over at the starting end and an unlinked carboxyl group at the terminus.

Figure 23.1 Structure of nisin showing positions of unusual amino acids [dehydroalanine (Dha), dehydrobutyrine (Dhb), lanthionine (Ala-S-Ala), and methyl lanthionine (ABA-S-Ala)] in addition to regular amino acids.

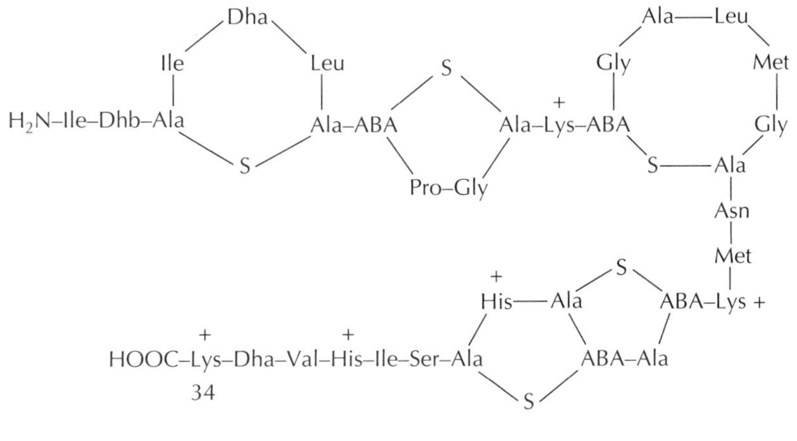

quality of fermented foods by inhibiting unwanted bacteria and promoting the growth of the starter culture. There are many benefits to using defined starter cultures to make fermented foods. However, the native bacteria of the food usually outcompete the starter cultures. Using starter cultures that produce bacteriocins to kill the native bacteria solves this problem.

Nisin is added to milk, cheese, and dairy products; many canned foods; mayonnaise; and baby foods throughout the world. It is GRAS as an antibotulinal agent in many food products, and it can prolong shelf life. Nisin also sensitizes spores to heat so the severity of thermal treatments can be reduced. However, this application is not approved in the United States because it might mask poor process control.

Nisin is often used with other inhibitors and may be used with food packaged in a modified atmosphere. Nisin increases the shelf life and delays toxin production by type E botulinal strains in fresh fish packaged in a carbon dioxide atmosphere. The combination of nisin and modified atmosphere to prevent *L. monocytogenes* growth in pork is more effective than either treatment used alone. When nisin is added to liquid whole eggs at 5 mg/liter before pasteurization, their refrigerated shelf life doubles.

Pediocins inhibit *L. monocytogenes* vegetative cells but not clostridial spores. European patents cover the use of pediocin PA-1 as a dried powder or culture liquid to extend the shelf life of salads and salad dressing and as an antilisterial agent in cream, cottage cheese, meats, and salads.

Pediocins are more effective than nisin in meat. Dipping meat in pediocin PA-1 decreases the viability of attached *L. monocytogenes* organisms 100- to 1,000-fold. Pretreating meat with pediocin reduces subsequent *L. monocytogenes* attachment. Pediocin AcH can cause a 10- to 1,000,000-fold reduction in listeria populations in ground beef, sausage, and other products. Emulsifiers, such as Tween 80, or entrapment in other lipids makes pediocin work better in fatty foods. In most cases, pediocin kills listeriae rapidly and delays the growth of the survivors.

Bacteriocin-Producing Starter Cultures Can Improve the Safety of Fermented Foods

Foods that are usually made by LAB are easily improved by using bacteriocin-producing starter cultures. For example, including a nisin-producing starter culture to make Cheddar cheese increases the shelf life from 14 to 87 days at 22°C.

Pediococci that make pediocins are especially effective in fermented meats. Pediocin production by *Pediococcus acidilactici* PAC 1.0 during the manufacture of fermented dry sausage reduces *L. monocytogenes* viability >10-fold. When *P. acidilactici* H is used to ferment summer sausage, *L. monocytogenes* viability drops >1,000-fold.

Organization of Bacteriocin Genes: a Generic Operon

The DNA or amino acid sequences of many bacteriocins are known. There is a general model for the genetic organization of bacteriocin genes. The organization of a generic *operon* is shown in Fig. 23.2. An operon clusters the genes containing the structural information near the genes involved in immunity, maturation, processing, and export of the bacteriocin molecules. The genes that control bacteriocin biosynthesis are also on the operon. Not all bacteriocin operons contain all of these genes, nor are the organizations of the genes in each operon identical for every bacteriocin. The generic operon shows many of the similarities.

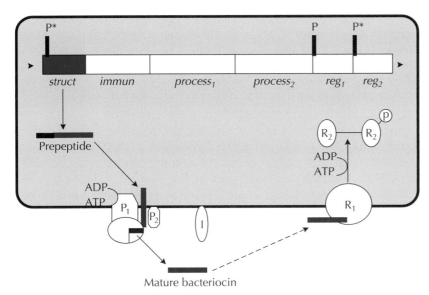

Figure 23.2   A generic bacteriocin operon. The structural gene (*struct*) codes for a prepropeptide, which is modified and excreted by the products (P_1 and P_2) of the processing gene (*process$_1$* and *process$_2$*). The operon also contains the immunity gene (*immun*). The operon may be regulated by a signal transduction pathway coded for by *reg$_1$* and *reg$_2$*, producing a histidine kinase (R_1) and response regulator (R_2) subject to phosphorylation (R_2–p). The regulatory molecules P_1 and P_2 affect transcription of the operon. ADP, adenosine diphosphate. Adapted from T. J. Montville, K. Winkowski, and M. L. Chikindas, p. 629–647, *in* M. P. Doyle, L. R. Beuchat, and T. J. Montville (ed.), *Food Microbiology: Fundamentals and Frontiers,* 2nd ed. (ASM Press, Washington, D.C., 2001).

The *structural gene* usually encodes a prepeptide. The prepeptide is made up of the excreted bacteriocin plus an N-terminal leader sequence that is cut off during transport. Pediocin-like bacteriocin genes code for a prepeptide containing a glycine-glycine cleavage site. The leader sequence may be needed for recognition by export machinery. The *immunity gene* protects cells that make bacteriocins from being killed by it. Immunity is specific and is coordinated with bacteriocin production. The *processing and export genes* code for at least two proteins that mature the bacteriocin and export it from the cell. The *regulatory genes* code for proteins that turn bacteriocin synthesis on and off.

Bacteriocin action.   LAB bacteriocins target the cytoplasmic membrane of sensitive bacteria. They disrupt the membrane by making pores. This increases its permeability to small compounds, causing a rapid efflux of preaccumulated ions, amino acids, and, in some cases, adenosine triphosphate (ATP) molecules. The equilibration of compounds across the membrane destroys the gradients required for many vital cellular functions.

Legal status.   The legal status of bacteriocins and the bacteria that make them is unclear. LAB are GRAS for the production of fermented foods. GRAS status, which is given by the U.S. Food and Drug Administration (FDA), is especially desirable because it lets a compound be used in a specific application without additional approval. Some people overlook the linkage of GRAS status to a specific application. The GRAS status of LAB

Authors' note

For bacteriocins, "immunity" means "protection from" and has nothing to do with antigens, antibodies, or the immune response.

Authors' note

GRAS status can be conferred in three ways: if the product was in use before 1958, if the FDA affirms that a compound is GRAS in response to an industry request, or if a company "self-affirms" that its compound is GRAS. (Welcome to the wild world of food law!)

for the production of fermented foods does not automatically make LAB or their metabolic products GRAS for other uses, such as the preservation of foods that are not fermented. Nisin is the only bacteriocin that has GRAS status. The 1988 GRAS affirmation for the use of nisin in pasteurized processed cheese was supported by toxicological data. This affirmation is the foundation for many additional GRAS affirmations. In the nonregulatory sense, bacteriocins are widely considered safe.

Purified bacteriocins produced by GRAS organisms are not automatically GRAS themselves. Bacteriocins that are not GRAS are regulated as food additives and require premarket approval by the FDA. A food fermented by bacteriocin-producing starters can be used as an ingredient in a second food product. Its use *as an ingredient* might coincidentally extend the shelf life of the product without the need for listing it as a preservative on the label. However, if the ingredient was added *for the purpose of extending shelf life*, the FDA would probably consider it an additive and require both premarket clearance and label declaration. Purified bacteriocins used as preservatives definitely require premarket approval by the FDA.

PROBIOTIC BACTERIA

The idea that eating LAB improves health first appeared in print in 1907. In his book *Prolongation of Life*, the Nobel laureate Elie Metchnikoff attributed the longevity of people in the Balkans to the bacteria in the yogurt they ate. He hypothesized that the bacteria in the yogurt suppress "bad" bacteria in the gut. The term "probiotic" was rediscovered in the 1970s and was used to describe the healthful effects of feeding microbial feed supplements to animals. In 1989, R. Fuller proposed a narrower definition in which viable "good" bacteria are consumed and act in the gastrointestinal (GI) tract to benefit the host organism (i.e., us). While we know that fermentation improves digestibility, generates amino acids, and produces vitamins in food, there are not many peer-reviewed data showing that probiotic bacteria promote "intestinal well-being" or are otherwise health promoting. There are many reports of the probiotic effect of LAB. Some of the species involved are listed in Table 23.1. Applications are given in Table 23.2. However, because there are so few well-controlled studies in humans, the therapeutic use of LAB has not gained acceptance in mainstream medicine (Box 23.1).

Cultures used to make fermented products are usually chosen for their technological traits rather than their health-promoting attributes. For example, *Lactobacillus delbrueckii* subsp. *bulgaricus* and *Streptococcus thermophilus*, traditionally used to make yogurt, are hearty and acidify rapidly

Authors' note_____

There is no agreement as to what constitutes "intestinal well-being."

Table 23.1 Examples of human probiotic species and strains with research documentation[a]

Lactobacillus acidophilus	*Lactobacillus rhamnosus*	*Bifidobacterium lactis*
Lactobacillus casei	*Lactobacillus plantarum*	*Bifidobacterium longum*
Lactobacillus johnsonii	*Lactobacillus reuteri*	*Bifidobacterium breve*
Lactobacillus fermentum	*Lactobacillus salivarius*	*Lactobacillus paracasei*
Saccharomyces boulardii	*Streptococcus thermophilus*	*Lactobacillus delbrueckii* subsp. *bulgaricus*

[a]Adapted from Table 39.2 of T. R. Klaenhammer, p. 797–811, *in* M. P. Doyle, L. R. Beuchat, and T. J. Montville (ed.), *Food Microbiology: Fundamentals and Frontiers*, 2nd ed. (ASM Press, Washington, D.C., 2001). Information in source table was compiled from M. E. Sanders and J. Huis in't Veld, *Antonie Leeuwenhoek* **76:**293–315, 1999.

Table 23.2 Possible health benefits and mechanisms of probiotic bacteria

Health effect	Possible mechanism(s)
Promotes ability to digest lactose	Probiotic cultures make lactose-using enzymes
Fights foodborne disease	Occupies colonization sites, promotes immunity, decreases severity of diarrhea
Anticancer	Binds cancer-causing compounds, decreases levels of natural enzymes that promote cancer, stimulates immune function
Enhances immune system	Strengthens defenses against infections and tumors, increases antigen-specific immune responses, lowers inflammatory responses
Fights heart disease	Increases bile activity, may reduce cholesterol
Reduces blood pressure	Bacterial action on milk proteins produces a compound that lowers blood pressure in animals
Reduces ulcers	Production of inhibitor against *Helicobacter pylori*[a]
Limits urogenital infections	Adhesion to urinary tract and vaginal cells, competitive exclusion, production of inhibitors (surfactants, hydrogen peroxide)

[a]It is now widely accepted that ulcers are caused by *H. pylori,* not stress.

BOX 23.1

Probiotics in the marketplace

"You can eliminate painful and embarrassing Candida *yeast infections with probiotic bacteria!"*

"Probiotic bacteria inhibit harmful organisms, boost immune function, and increase resistance to infection."

"Promoting a proper balance of bacteria is critical to good health."

These are just a few claims from dozens of Internet sites that promote the consumption of probiotic bacteria. Despite the scientific uncertainty that surrounds dietary supplements, probiotics are big business. Neither foods nor drugs, dietary supplements are legally defined as dietary ingredients that are taken by mouth to supplement the diet, are not labeled as a food, and are not the sole item of a meal (for then it would be a *food*) or taken to prevent, cure, or treat a disease (for then it would be a *drug*). (Therefore, the first claim above is illegal.) Supplements are not regulated as foods, nor are they regulated as drugs. They are hardly regulated at all! There are few government regulations governing the purity, potency, or effectiveness of supplements.

Only the manufacturer's reputation ensures the identity and viability of probiotic cultures. Nonetheless, in 2002 U.S. probiotic sales reached $160 million, and they are expected to grow at 10% per year for the next 10 years. There are even probiotic products for dogs!

The area of probiotic bacteria is a difficult one in which to demonstrate cause and effect. Its secrets may lie beyond classical pure-culture microbiology, for probiotics are supposed to affect the microbial members of the human intestinal ecosystem. The largest producer of probiotics in the world is Yakult. This company was founded by Minoru Shirota, who developed a strain of *Lactobacillus casei* that was exceptionally hearty and resistant to stomach acid. With its modest product claims of containing "friendly bacteria" that "promote intestinal well being," the company grossed $1.5 billion in 23 countries. There is increasing clinical support for the use of probiotics in the management of many gastric disorders. Nestlé, Dannon, Rhodia, and other "mainstream" food companies have recently introduced probiotic products. It's hard to be anti*probiotics.

but have poor resistance to acid and bile salts and do not survive stomach passage. To capitalize on the health-promoting potential, many yogurt manufacturers now use *Lactobacillus acidophilus* and *Bifidobacterium* spp. that are resistant to acid and bile salts.

The Human GI Tract Is a Microbial Ecosystem

The complexity of the human GI tract makes research on probiotic bacteria difficult. The average human contains over 400 species of bacteria. Thirty or 40 of these comprise 99% of the bacteria in healthy humans. It is existentially noteworthy that we contain as many bacterial cells as human cells ($\sim 10^{13}$ to 10^{14}). Because it has so many organisms and environments, the GI tract must be studied as an ecosystem. Humans, like other ecosystems, experience a progression of inhabitants. We are born with insides that are more or less sterile. We are colonized by bifidobacteria when breast-fed or by lactobacilli when fed with cow milk. Strict anaerobes come with the introduction of solid food. Once established, the composition and distribution of our microbial component are very hard to change. In ecological terms, the microbiota of the human GI tract is a *climax community*. The metabolic activity of the climax community is quite variable and is strongly influenced by the host physiology and by everything that affects the host physiology. (Think about what happens when you eat too many beans.)

The human GI tract is ~ 350 cm long from the oral to the anal orifices. The GI tract is divided into three major sections, and each has its own distinct microbiota. The stomach is highly acidic (pH 1 to 3) and is populated by $<10^3$ colony-forming units (CFU) of aerobic gram-positive organisms per g. The intestinal pH (6.8 to 8.6) favors microbial growth. The small intestine is a transitional zone inhabited by 10^3 to 10^4 CFU of *Lactobacillus*, *Bifidobacterium*, *Bacteroides*, and *Streptococcus* organisms per g. Microbial growth in the large intestine is luxuriant. There are 10^{11} to 10^{12} CFU/g, with anaerobes outnumbering aerobes 100- to 10,000-fold. A diverse microbial population populates the large intestine. Bacteria from the genera *Bacteroides*, *Fusobacterium*, *Lactobacillus*, *Bifidobacterium*, and *Eubacterium* are present in large numbers. *L. acidophilus* is especially important. The *Enterobacteriaceae* are present at relatively lower levels.

The normal human biota contains two different bacterial populations. There are indigenous bacteria that have colonized the host by adhering to the intestine, and there are also foodborne bacteria that are just passing through. Dietary factors, such as a carnivorous or vegetarian diet or even starvation, make surprisingly little difference in the distribution of bacteria in the GI tract.

The complexity of the GI ecosystem makes colonization hard to study. Colonization is influenced by gastric acidity, bile salt concentration, peristalsis, digestive enzymes, and the immune response. Feces are 33 to 50% bacteria on a dry-weight basis and are their own ecosystem. The fecal ecosystem is subject to dehydration and has high levels of enzyme activities.

The future use of probiotic bacteria is difficult to predict. *Something* happens in the GI tract when LAB are consumed in large numbers. Exactly *what* happens and *why* are hard to determine. It is very important that these benefits be studied in the context of diet, well-characterized probiotic bacteria, individual human intestinal ecosystems, and the established principles of microbial ecology. Much of this lies outside the scope of pure-culture microbiology and may be the work of future food microbiologists.

*Authors' note*_____

A baby's feces change as diet and microbiota change. I was fascinated to watch my first child's poop change from black meconium to mustardy yellow cream to foul brown solid as his diet progressed.

*Authors' note*_____

A climax community is an ecosystem in which the progression of species has stopped, all niches are filled, and a characteristic group of organisms is maintained.

Summary

- Refrigeration alone is insufficient to ensure the microbial safety of food.
- LAB may be used to produce acid in place when a food is held at temperatures that are too warm. The acidification prevents pathogens from growing and making toxins.
- Bacteriocins are small proteins that kill sensitive bacteria by making pores in them.
- Lantibiotic bacteriocins have unusual amino acids and single sulfur rings.
- Pediocin-like bacteriocins have a common leader sequence and cleavage site and are active against *L. monocytogenes.*
- Bacteriocins can be produced by GRAS LAB in the food or added to the food as natural preservatives if they are approved as food additives or have GRAS affirmation.
- The genes for bacteriocins, their regulation, and the enzymes that process them are grouped together in an operon.
- Probiotic bacteria are thought to confer health benefits by improving the balance of bacteria in the intestines.
- Possible health benefits of probiotics include decreased cholesterol, increased ability to utilize lactose, decreased symptoms of diarrheal illness, and anticancer action.
- The human GI tract is a complex ecosystem. This makes the study of probiotics difficult.

Suggested reading

Cleveland, J., T. J. Montville, and M. L. Chikindas. 2001. Bacteriocins: safe food preservatives of the future. *Int. J. Food Microbiol.* **71:**1–20.

Klaenhammer, T. R. 2001. Probiotics and prebiotics, p. 797–811. *In* M. P. Doyle, L. R. Beuchat, and T. J. Montville (ed.), *Food Microbiology: Fundamentals and Frontiers,* 2nd ed. ASM Press, Washington, D.C.

Montville, T. J., K. Winkowski, and M. L. Chikindas. 2001. Biologically based preservation systems, p. 629–647. *In* M. P. Doyle, L. R. Beuchat, and T. J. Montville (ed.), *Food Microbiology: Fundamentals and Frontiers,* 2nd ed. ASM Press, Washington, D.C.

Sanders, M. E. 1999. Probiotics—scientific status summary. *Food Technol.* **52:**67–77.

Questions for critical thought

1. What is the temperature of your refrigerator? (If you are living in a dorm room without a refrigerator, check the temperature of the refrigerator at the store where you buy food.)
2. Why does the pH drop in Wisconsin process bacon?
3. What were the original six characteristics used to determine if an inhibitor was a bacteriocin?
4. Why is it a good thing that a bacteriocin does not act against the organism that makes it?
5. Why does it benefit a bacterium to have all the genes involved in bacteriocin synthesis and regulation clustered in an operon?

6. Make a narrow and a wide column on a piece of paper. List the genes involved in bacteriocin production in the narrow column. In the wide column, write what would happen if each gene were deleted. Choose one of the genes and describe how it might be altered for commercial benefit.

7. Choose one health benefit of probiotic bacteria. Find, read, and evaluate three research articles that relate to it.

8. If humans contain 10^{13} to 10^{14} human cells and 10^{13} to 10^{14} bacterial cells, why don't we look like slimy bacterial colonies?

9. List and discuss factors that influence the conclusions of studies of the health benefits of probiotics.

24

Physical Methods of Food Preservation*

LEARNING OBJECTIVES

The information in this chapter will help the student to:

- use the water activity (a_w) concept to distinguish between bulk water and water that is unavailable for microbial growth
- understand the effect of a_w on microbial growth and microbial ecology
- appreciate the roles of temperature, controlled atmosphere, and freezing in microbial growth
- differentiate among different types of heat processing, their objectives, and their industrial uses
- quantitatively determine, using D and z values, the lethality of a given heat process for an organism
- make a technical and social judgment about the use of irradiation preservation in foods
- gain familiarity with new nonthermal methods of food preservation

INTRODUCTION

Conditions that are too stressful for microbial growth can damage bacterial cells. Minor stresses inhibit cell growth; major stresses kill cells. Thus, food environments can be manipulated to inhibit or kill microbes. Physical manipulation of foods can inhibit microbial growth, kill the cells, or mechanically remove microbes from the food. Dehydration, refrigeration, and freezing inhibit microbial growth. Microbes can be killed by heating or by ultraviolet (UV) or ionizing radiation. Nonthermal energy is also lethal, and membrane filtration removes microbes from liquid foods. This chapter discusses physical food preservation methods, except filtration.

PHYSICAL DEHYDRATION PROCESS

Water is the major factor in controlling food spoilage. However, it is the *availability* of the water, not its amount, that determines inhibition. Water is chemically bound to food molecules. (The water is bound by ionic interactions, hydrogen bonding, etc., with chemical constituents of the food.) The bound water is not available for chemical reactions or microbial growth. The amount of unbound, or available, water determines if microbes can grow.

---

*This chapter was originally written by József Farkas for *Food Microbiology: Fundamentals and Frontiers*, 2nd ed., and has been revised for use in an introductory text.

Table 24.1   Moisture contents of various dry or dehydrated food products when the a_w is 0.7 at 20°C[a]

Food	% Moisture content
Grains	4–9
Milk powder	7–10
Cocoa powder	7–10
Whole-egg powder	10–11
Skim milk powder	10–15
Dried fat-free meat	10–15
Rice and legume seeds	12–15
Dehydrated vegetables	12–22
Dried soups	13–21
Dried fruits	18–25

[a]Based on J. Farkas, p. 567–592, *in* M. P. Doyle, L. R. Beuchat, and T. J. Montville (ed.), *Food Microbiology: Fundamentals and Frontiers,* 2nd ed. (ASM Press, Washington, D.C., 2001).

The measure of available water in foods is *water activity* (a_w), defined as the ratio of the vapor pressure of water in a food, P, to the vapor pressure of pure water, P_0, at the same temperature:

$$a_w = \frac{P}{P_0}$$

The movement of water vapor from a food to the air depends on the moisture content, the food composition, the temperature, and the humidity. At constant temperature, the water in the food equilibrates with water vapor in the air. This is the food's *equilibrium moisture content.* At the equilibrium moisture content, the food neither gains nor loses water to the air. The relative humidity of the surrounding air is thus the equilibrium relative humidity (ERH [%]). ERH is defined as follows:

$$\text{ERH (\%)} = a_w \times 100$$

Foods at the same a_w may have different moisture contents due to their chemical compositions and different water binding capacities (Table 24.1.).

Dehydration preserves food by removing available water, i.e., reducing the a_w. Hot air removes water by evaporation. Freeze-drying removes water by sublimation after freezing. These processes reduce the a_w of the food to levels that inhibit growth.

The a_ws of various foods are compared in Table 24.2. High-moisture foods, such as fruits, vegetables, meats, and fish, have a_ws of ≥0.98. Intermediate-moisture foods (jams, sausages, etc.) have a_w levels of 0.7 to 0.85. Additional preservative factors (e.g., reduced pH, preservatives, and pasteurization) are required for their microbiological stability. The relationship among a_w, growth, and storage stability is shown in Fig. 24.1.

Various microbes have different a_w requirements. Decreasing the a_w increases the lag phase of growth, decreases the growth rate, and decreases the number of cells at stationary phase. Foodborne microbes are grouped by their minimal a_w requirements in Table 24.3. Gram-negative species usually require the highest a_w. Gram-negative bacteria, such as *Pseudomonas* spp. and most *Enterobacteriaceae*, usually grow only above an a_w of 0.96 and 0.93, respectively. Gram-positive non-spore-forming bacteria are less sensitive to reduced a_w. Many *Lactobacillaceae* have minimum a_ws near 0.94.

Authors' note

Sublimation is the conversion of ice to vapor without passing through the liquid stage.

Table 24.2   Typical a_w of various foods[a]

Food(s)	a_w
Fresh raw fruits, vegetables, meat, fish	≥0.98
Cooked meat, bread	0.95–0.98
Cured meat products, cheeses	0.91–0.95
Sausages, syrups	0.87–0.91
Rice, beans, peas	0.80–0.87
Jams, marmalades	0.75–0.80
Candies	0.65–0.75
Dried fruits	0.60–0.65
Dehydrated vermicelli, spices, milk powder	0.20–0.60

[a]Adapted from J. Farkas, p. 567–592, *in* M. P. Doyle, L. R. Beuchat, and T. J. Montville (ed.), *Food Microbiology: Fundamentals and Frontiers,* 2nd ed. (ASM Press, Washington, D.C., 2001).

Some *Micrococcaceae* grow below an a_w of 0.90. Staphylococci are unique among foodborne pathogens because they can grow at a minimum a_w of ~0.86. However, they do not make toxins below an a_w of 0.93. Most spore-forming bacteria do not grow below an a_w of 0.93. Spore germination and outgrowth of *Bacillus cereus* are prevented at an a_w of 0.97 to 0.93. The minimum a_w for *Clostridium perfringens* spore germination and growth is between 0.97 and 0.95.

Several yeast species grow at a_w levels lower than those of bacteria. Salt-tolerant species, such as *Debaryomyces hansenii, Hansenula anomala,* and *Candida pseudotropicalis,* grow well on cured meats and pickles at NaCl concentrations of up to 11% (a_w = 0.93). Some xerotolerant species (such as *Zygosaccharomyces rouxii, Zygosaccharomyces bailii,* and *Zygosaccharomyces bisporus*) grow and spoil foods, such as jams, honey, and syrups, that have high sugar content (and correspondingly low a_w).

Authors' note

Xerotolerant organisms can grow at low a_w. Halotolerant organisms are relatively insensitive to salt.

Figure 24.1   Relationship among a_w, growth, and storage stability.

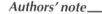

Organisms inhibited	a_w	Foods	Shelf life
	1.00		Days
Clostridium botulinum		Most fresh foods, meat, fish, poultry, fruit, and vegetables	
Salmonella, most bacteria			
Staphylococcus aureus (anaerobic)	0.90		
Most yeasts		Cured meat products	
Staphylococcus aureus (aerobic)	0.85		
			Weeks
Most molds	0.80		
		Syrups, salted foods	
Halophilic bacteria			Months
	0.70		
		Dried foods	
			Years
Osmophilic yeasts and molds	0.60		

Table 24.3 Minimal a_w required for growth of foodborne microbes at 25°C[a]

Group of microorganisms	Minimal a_w required
Most bacteria	0.91–0.88
Most yeasts	0.88
Regular molds	0.80
Halophilic bacteria	0.75
Xerotolerant molds	0.71
Xerophilic molds and osmophilic yeasts	0.62–0.60

[a]Adapted from J. Farkas, p. 567–592, *in* M. P. Doyle, L. R. Beuchat, and T. J. Montville (ed.), *Food Microbiology: Fundamentals and Frontiers,* 2nd ed. (ASM Press, Washington, D.C., 2001).

Molds generally grow at lower a_ws than foodborne bacteria. The most common xerotolerant molds belong to the genus *Eurotium.* Their minimal a_w for growth is 0.71 to 0.77, while the optimal a_w is 0.96. True xerophilic molds, such as *Monascus* (*Xeromyces*) *bisporus,* do not grow at an a_w of >0.97 to 0.99. The relationship of a_w to mold growth and toxin formation is complex. Molds can grow, but not make toxins, under many conditions.

The varied a_w growth limits of bacteria and fungi reflect the mechanisms that help them grow at low a_w. Bacteria protect themselves from osmotic stress by accumulating *compatible solutes* intracellularly. Compatible solutes equilibrate the cells' intracellular a_w to that of the environment but do not interfere with cellular metabolism. Some bacteria accumulate K^+ ions and amino acids, such as proline. Halotolerant and xerotolerant fungi concentrate polyols, such as glycerol, erythritol, and arabitol.

Drying

When foods are dried, hot air evaporates the water and carries it away. Both the drying temperature and the decreased a_w affect microbes. Factors such as the size and composition of food pieces and the time-temperature combinations influence the microbiological effects of drying.

During the initial period of drying, the temperature is still low and the relative humidity is high. The size of the food particles determines the length of this phase. If the phase is long, microbes may grow. During later phases of drying, the temperature is high but the relative humidity is low. There is no opportunity for growth, but there is not much lethality either. The higher temperatures result in lower moisture content, and dry heat is less lethal than wet heat. If the drying lasts long enough (at least 30 min), there are time-temperature combinations that kill microbial cells. The temperature and moisture distribution within the food determines the lethality. With certain drying methods, the surface temperature may reach 100°C, while the internal temperature remains lower. When microbes are mainly on the surface, their inactivation is enhanced. Lethal combinations of high temperature and high humidity are rare during drying. In fact, the major microbial die-off is due to the high temperature-low humidity conditions, which last long after the actual drying. Cell populations decrease during storage because injured cells that cannot recover at low a_w gradually die.

The a_w of dried foods is usually much lower than the microbial growth minimum. Therefore, dried products are microbiologically stable. If the ERH increases or the temperature drops, condensation on the product surface permits mold growth.

Freeze-Drying

Freeze-drying (*lyophilization*) combines two preservation methods. It freezes the food and then dehydrates the frozen food through vacuum sublimation of the ice. That is, water moves from the ice state to the vapor state without becoming liquid. This is a gentle way to remove water. In other drying methods, the solution moves to the food surface, where the water leaves and the solutes concentrate. During freeze-drying, the sublimation front moves into the food and the ice sublimates where it is formed. Thus, solutes remain inside the food at their original locations, and the food retains its original structure. As a result, freeze-dried foods rehydrate rapidly to 90 to 95% of their original moisture content. Unfortunately, the large surface area of freeze-dried foods makes them vulnerable to oxidation during storage. Packaging freeze-dried foods in inert atmospheres delays oxidation. Vapor-impermeable packaging prevents their rehydration.

Freeze-drying foods influences microbes by decreasing the a_w during freezing and during sublimation of the ice. The extents of cell damage in these two phases may be different. Microbial survival also depends on the composition of freeze-dried food. Carbohydrates, proteins, and collodial substances are usually protective. Cell viability slowly decreases during storage of freeze-dried foods.

Freeze-drying is optimized for cell survival during the preservation of stock cultures. Low dehydration temperature, protective additives (such as glycerol or nonfat dry milk solids) in the microbial suspension, and storage of lyophilized cultures under vacuum increase viability. Gram-positive bacteria survive freeze-drying better than gram-negative bacteria.

COOL STORAGE

Cool (or chilled) storage refers to storage at temperatures from ~16 to −2°C. While pure water freezes at 0°C, most foods remain unfrozen until ≤−2°C. However, many fruits and vegetables suffer from chill injury when kept at <4 to 12°C. The length of cool storage may vary from a few days to several weeks (Table 24.4).

Chemical reaction rates decrease as temperatures decrease. This decreases microbial growth rates. Refrigeration temperature is below the minimal growth temperatures of most foodborne microbes (see chapter 2). Even psychrotrophic microbes, such as *Yersinia enterocolitica*, *Vibrio parahaemolyticus*, *Listeria monocytogenes*, and *Aeromonas hydrophila*, grow very

Authors' note_____

While beyond the scope of this book, the preservation of bacteria in culture collections is a neglected aspect of food microbiology. Many microbiologists are amateur naturalists, maintaining their own private culture collections and trading cultures with their colleagues like baseball cards. However, bacterial traits can drift over years of repeated transfer. It is better to get cultures in the freeze-dried form from a recognized culture collection, such as the American Type Culture Collection. Secure forms of culture preservation are essential for process security and research reproducibility.

Authors' note_____

While the importance of refrigeration is well recognized, 20% of refrigerators have temperatures that are >50°F! Don't count on them to ensure food safety.

Table 24.4 Shelf life extension of raw foods by cool storage[a]

Food	Avg useful storage life (days) at:	
	0°C (32°F)	22°C (72°F)
Meat	6–10	1
Fish	2–7	1
Poultry	5–18	1
Fruits	2–180	1–20
Leafy vegetables	3–20	1–7
Root crops	90–300	7–50

[a]Reprinted from J. Farkas, p. 567–592, *in* M. P. Doyle, L. R. Beuchat, and T. J. Montville (ed.), *Food Microbiology: Fundamentals and Frontiers,* 2nd ed. (ASM Press, Washington, D.C., 2001); adapted from N. N. Potter, *Food Science,* 4th ed. (AVI-Van Nostrand Reinhold, New York, N.Y., 1986).

Across

1. Trade name often used as a synonym for photocopier
3. If there is a _____, it's not free
5. Amount of time at a given temperature required to kill 90% of a population
7. Phase change of water to solid state
8. Water (prefix)
9. Canning helps prevent growth of this
10. Bad for sun-dried food
12. Oxidation-reduction potential (abbreviation)
14. Mild heat treatment that kills pathogens but not spoilage bacteria
17. Method of preservation that involves an electron beam
18. Hemolytic-uremic syndrome (abbreviation)
19. Opening that serves as an outlet
21. Period from midnight to noon (abbreviation)
22. Unit of pressure
23. This kills by heat but has no special nonthermal effect

Down

1. Prefix meaning "dry"
2. Able to grow at low water activity
3. A good method for preserving microbes
4. Adding or removing this affects microbial growth
5. Removal of water
6. Augment
9. Emergency signal (abbreviation)
11. Type of preservation that does not add heat
13. Hydrostatic _____ is a promising nonthermal method of food preservation
15. Water _____, not water content, governs microbial growth
16. Type of extruded pasta
20. Prefix meaning 10^{-12}
21. Water activity (abbreviation)
23. Molecular weight (abbreviation)

slowly at low temperature. A temperature increase of only a few degrees can dramatically increase microbial growth rates. Refrigerated foods can spoil rapidly if their initial microbial populations are high. Since the temperature requirements for various microbes differ, refrigeration may change the composition of the microbial population.

An organism's growth temperature range depends on how well it regulates its membrane fluidity. Psychrotrophs have more unsaturated fatty acid residues in their lipids at low temperatures. This leads to a decrease in the lipid melting point. The process by which cells alter their fatty acid compositions to maintain membrane fluidity is called *homeoviscous adaptation.* The fluid state allows membrane proteins to function.

Controlled-Atmosphere Storage

The minimal growth temperature is lowest when other factors are optimal. When other factors are unfavorable, the minimal growth temperature increases. The most important factors are pH, a_w, and oxygen concentration. The suboptimal combination of these factors with refrigeration extends food stability. Controlled-atmosphere storage is widely used for certain fruits and vegetables. The oxygen content is reduced (to 2 to 5%), and the carbon dioxide content is increased (to 8 to 10%). These conditions are maintained in airtight chilled storage rooms. Modified-atmosphere storage slows the respiration of the fruit or vegetable. This limits quality deterioration and microbial growth.

Carbon dioxide inhibits growth for several reasons. When dissolved in water, carbon dioxide reduces the pH. However, the primary mechanism is direct inhibition of microbial respiration (see chapter 2). The oxygen reduction contributes to carbon dioxide's inhibitory effect. Psychrotrophic spoilage bacteria, such as *Pseudomonas* and *Acinetobacter,* are particularly sensitive to carbon dioxide, while lactic acid bacteria and yeasts are not.

Modified-Atmosphere Packaging

Modified-atmosphere packaging (MAP) is similar to controlled-atmosphere storage. In the case of vacuum-packaged products, the air pressure rather than the composition is changed. The residual air pressure of only 0.3 to 0.4 bar (1 bar = 10^5 pascals) reduces the available oxygen. If packaging films with very low gas permeability are used for fresh foods, the carbon dioxide concentration may change. This is because oxygen is used and carbon dioxide is produced during respiration. This modified atmosphere is very inhibitory to certain microbes and increases the keeping quality of foods. However, *Clostridium botulinum* (and other pathogens) grows well in the absence of O_2 and may be a hazard. Because of this, some vacuum-packed foods, especially pasteurized foods, should be kept below 3.3°C and may require secondary barriers to microbial growth.

MAP for live (respiring) products, like fresh fruit and vegetables, is complicated. The packaging film must have a gas permeability that equilibrates the atmosphere in the package. The package atmosphere is affected by the rates at which respiration and gas permeability change with temperature. MAP could extend the shelf life of many perishable products, but its adoption is hindered by concern about the growth of pathogens at refrigeration temperatures. Fish and fish products pose a risk for *C. botulinum* growth in MAP products. The U.S. National Academy of Sciences recommends that fish not be packed under modified atmospheres until the safety of the process has been proven.

FREEZING AND FROZEN STORAGE

Freezing lowers a food's temperature to −18°C or below. Many convenience foods are frozen to maintain their high quality. Freezing is, however, a highly energy-intensive process. Commercially, foods are frozen in cold air, by contact with a cooled surface, by submersion in cold refrigerant liquid, or by spraying the refrigerant onto the food.

Foods do not freeze at a well-defined temperature; they freeze over a broad temperature range. Depending on the food's composition, water in the food starts freezing at −1 to −3°C. The freezing of the water increases the solute concentration in the water that is not yet frozen. This further decreases the freezing point. When the temperature reaches the *eutectic point,* the rest of the material freezes as a solution without further separation of water and solute. The totally frozen state is a complex system of ice crystals and crystallized soluble substances. The totally frozen state occurs at −15 to −20°C for fruits and vegetables and at ≤40°C for meats.

Freezing produces osmotic shock in microbes. Intercellular ice crystal formation causes mechanical injury. The concentration of cellular liquids changes the pH and ionic strength. This then inactivates enzymes, denatures proteins, and inhibits metabolic processes. Cell membranes suffer major damage. In addition, thawing reexposes surviving cells to these effects. The injury of microbial cells may be reversible or irreversible. The extents of injury, repair, death, and survival vary according to the freezing, frozen-storage, and thawing conditions.

The freezing rate has a major impact on microbial viability. During slow freezing, crystallization occurs extracellularly, and the cell's cytoplasm becomes supercooled at −5 to −10°C. Cells lose water that then freezes extracellularly. Water crystallization increases the concentration of the external solution. This also removes water from the cells. Overall, microbes are exposed to osmotic effects for a relatively long time. This increases injury. Increased freezing rates decrease the duration of the osmotic effects. This increases microbial survival. When freezing rates are high, crystal formation also occurs intracellularly. This injures the cells drastically and decreases survival rates. Freezing rates also impact food quality. Faster freezing results in higher quality. However, while food manufacturers can optimize freezing, they have little control over thawing, which has an equally important effect.

The food environment is the most important influence on the microbial effects of freezing. Certain compounds enhance, while others diminish, freezing's lethal effects. Sodium chloride reduces the freezing point of solutions, thereby extending the time during which cells are exposed to high solute concentrations before freezing occurs. Compounds such as glycerol, saccharose, gelatin, and proteins generally protect cells from freeze damage.

Microbes do not grow at temperatures lower than about −8°C. The fate of microbes that survive freezing may change during storage. Often the death of survivors is fast initially and slows gradually, and finally, the survival level stabilizes. Death during frozen storage is probably due to the unfrozen, very concentrated residual solution. The residual solution may change during storage. At fluctuating temperatures, the size of the ice crystals may increase. Fluctuating temperatures are more lethal than stable ones, but they adversely affect food quality. Gram-positive microbes survive frozen storage better than gram-negative bacteria.

Although freezing and frozen storage reduce microbial viability, freezing is not considered a lethal process. Under some conditions, survivors can grow during thawing. Their levels may then equal or exceed the level before freezing. Thawing releases a nutrient-rich solution from the food cells. During thawing, microbes can penetrate damaged food tissue more easily, and liquid condenses on the food surface. These conditions favor microbial growth. Thawed products are especially vulnerable to rapid microbial spoilage. Refreezing thawed products can be dangerous.

PRESERVATION BY HEAT TREATMENTS

Heat is the most widely used method for killing microbes. For a food microbiologist, a microbe's heat resistance is one of its most important traits.

Pasteurization (named after Louis Pasteur [Box 24.1]) is a relatively mild heat treatment. The purpose of pasteurization is to kill non-spore-forming *pathogenic* bacteria. It also inactivates enzymes and kills many (99 to 99.9%) spoilage organisms. Because of this, pasteurized foods take

---

BOX 24.1

Louis Pasteur, the first food microbiologist

Louis Pasteur in the laboratory.

Louis Pasteur (see figure) is one of the mythic giants of science, but he was also a man of great humility and integrity. He was not a particularly good student as a child. He'd rather go fishing or paint. But his life bent toward science when, at 26, he "discovered" stereochemistry, i.e., that some acids could exhibit "handedness" by bending light to the left or to the right. He concluded that life was asymmetric and abandoned chemistry to study biology.

Pasteur the biologist fathered several more disciplines. Until this time, many people thought that life "popped up" spontaneously and that fermentation was a chemical process. Pasteur disproved this using a simple but elegant experiment with a curved-neck flask (see figure). It was full of culture medium that would support the growth of bacteria but prevented them from entering. These flasks, some of which are still on exhibit at the Institut Pasteur in Paris, remained sterile. By showing that "life comes from life," he killed the idea of spontaneous generation. From this, he postulated that fermentation was a biological rather than a chemical process. Milk, wine, and beer spoilage caused by "ferments" could be stopped by heating the products for a few minutes at relatively low temperatures; a process we now know as pasteurization. Pasteur was the first food microbiologist!

Pasteur is also the father of immunology, having developed animal vaccines for cholera and anthrax. He

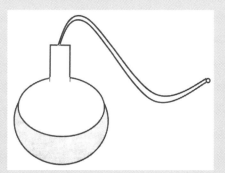

Curved-neck flask of the type used by Pasteur.

did not want to test the vaccination concept on humans, fearing the guilt and ignominy that causing a human death would confer if the vaccine failed. His hand was forced, however, when Madame Meister appeared at his laboratory with her 9-year-old son, Joseph. Joseph had been mauled by a rabid dog 2 days earlier and would certainly die without the vaccine. With fear in his heart, Pasteur treated the boy, who recovered and worked at the Institut Pasteur for the rest of his life. Joseph Meister's loyalty to Pasteur is legendary. There is a story that, when the Germans overran France in 1940, they sought to disinter Pasteur in retribution for his refusal to accept an honorary degree from the University of Bonn. Fifty years after his cure, Meister took his own life rather than open the crypt of his savior.

We live in an age where there is debate about the worth of "basic" versus "applied" science. Some feel that basic science is "better" science. Others see no need to pay for research unless it addresses some defined problem. We would do well to remember the words of Louis Pasteur: "There does not exist . . . a science one can give the name applied science. There are science and the application of science, bound together as the fruit and the tree which bears it."

longer to spoil, especially when they are refrigerated to delay the growth of surviving organisms.

Sterilization kills *all* microbes. The industrial goal of sterilization is freedom from pathogens and shelf stability rather than absolute sterility. A product may contain a viable spore that cannot grow (due to low pH, low a_w, etc.) and still be "commercially sterile." *C. botulinum* spores, for example, may be present in high-acid foods. Because they cannot germinate and grow, they do not present a hazard.

Technological Fundamentals

Foods can be heat processed after or before being packaged. The most common method is to heat process after packaging by *canning*. Canning is the commercial sterilization of food in *hermetically sealed* (airtight) containers. Nicholas Appert invented the basic canning process in the early 1800s. His research was done in response to Napoleon's offer of a prize for the development of a food preservation method that would extend the range of his troops. Canned foods, in the form of C rations or today's MRE (meal, ready to eat), remain the staple of soldiers around the world. When foods are packaged after being heated in the process of *aseptic packaging*, the food and the packaging material are sterilized separately. The food is then placed *aseptically* (in a sterile environment) into the package. Aseptic technology is widely used for fruit juices, dairy products, creams, sauces, and soups that would suffer quality loss if they were heated for long periods. Postprocess contamination can be a problem if bacteria are not rigorously excluded from the packaging area.

High-temperature, short-time heating (HTST) (rapid heating to temperatures of ~140°C, holding for several seconds, and then rapid cooling) produces shelf-stable food. HTST processes improve product quality and process efficiency. HTST preserves nutrients and sensory attributes better than conventional processing, which subjects foods to lower temperatures for long periods. Juice boxes, drink pouches, and a variety of condiments are made using HTST technology.

The mild heating of packaged foods combined with well-controlled refrigeration produces a very high quality product. These products, also known as REPFEDs (i.e., refrigerated processed foods of extended durability), cook-chill products, or *sous vide* meals (foods mildly heated within vacuum packs), are becoming very popular. However, some *C. botulinum* strains can grow at temperatures as low as 3.0°C and are a hazard in these foods. The Food and Drug Administration (FDA) requires a second hurdle (i.e., a preservative) in addition to refrigeration to ensure the safety of these foods.

Thermobacteriology

Wet heat (e.g., steam) kills microbes by denaturing nucleic acids, proteins, and enzymes. DNA damage may be the key event in killing vegetative cells. In spores, the germination system is the most heat-sensitive component. During mild heating, cytoplasmic membranes are a major site of injury. Dry heat is less lethal than wet heat and kills microbes by dehydration and oxidation. Dry heat needs higher temperatures and longer times to cause the same lethality as wet heat.

The death of a bacterial population by constant-temperature heating has logarithmic kinetics. That is, an equal time at a given temperature kills an equal percentage of the bacteria, regardless of the number of bacteria present. For example, if 5 min of boiling kills 90% of the bacteria, 10 min

---

Authors' note

Note that foods can be "canned" in glass jars, foil pouches, and other containers besides cans.

Authors' note

Should food be thought of as "durable"?

will kill 99% and 15 min will kill 99.9%. In theory, the number of organisms killed can never reach 100%. Another way to look at this is that if the initial number of bacteria were 10,000, after 5 min the number would be reduced to 1,000. Similarly, if the initial number were 10,000,000, after 5 minutes 1,000,000 would still be alive. It should be evident that, when many bacteria are present, a 90 or 99% reduction may not ensure the safety of the product. Figure 24.2 shows arithmetic and log transforms of the same data.

This negative exponential curve is expressed by the following equation:

$$N = N_0 e^{-k \cdot t}$$

In this equation, N is the number of viable cells after any time (t) of heating. N_0 is the initial number of viable cells, and k is the rate constant (time^{-1}) for destruction at temperature T. If N_0, k, and t are known, N can be calculated. Similarly, if N_0 and k are known, the time t required to reduce a population to N bacteria can be determined.

The discussion above may seem complicated. Fortunately, if the number of viable bacteria after a given time of heating is expressed as the $\log_{10}$, a linear curve results (Fig. 24.3). In this curve, the time it takes to reduce the number of viable bacteria 10-fold (i.e., 90%) is called the *D value*, or *decimal reduction value*. In food microbiology, the *D* value replaces the concept of the rate constant, k. The *D* value is the time required to destroy 90% (1 log unit) of the population, as shown in Fig. 24.3 (survival curve). Survival curves show the rate of destruction of a specific organism in a specific medium or food at a specific temperature. Under given conditions, the death rate at any given temperature is constant and is independent of the initial cell number. The logarithms of *D* values plotted against the heating temperatures give the thermal-death time curve (Fig. 24.4). The thermal-death time is the time necessary to kill a given number of organisms at a specified temperature.

*Authors' note*_____

The D value is inversely related to k by the equation $D_T = 2.3/k_T$.

Figure 24.2 Graph illustrating linear and logarithmic plots of the same data for viability of a bacterial population that is heated at a constant temperature. The black plot should be read using the linear (left) y axis. The blue plot corresponds to $\log_{10}$ values and should be read using the $\log_{10}$ (right) y axis.

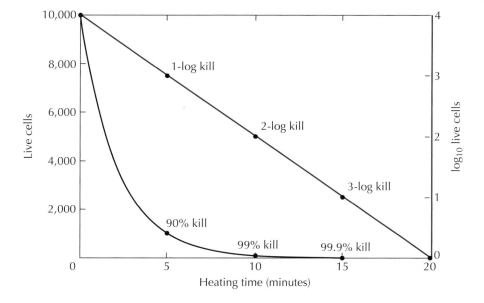

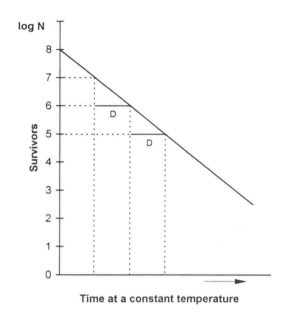

Figure 24.3 The bacterial survival curve shows the logarithmic order of bacterial death from which the *D* value is derived. Reprinted from J. Farkas, p. 567–592, *in* M. P. Doyle, L. R. Beuchat, and T. J. Montville (ed.), *Food Microbiology: Fundamentals and Frontiers,* 2nd ed. (ASM Press, Washington, D.C., 2001).

The *D* value gives the microbe's heat resistance at a single temperature. However, in heat processing, microbes are exposed to many temperatures as the product heats up to the processing temperature, holds it, and then cools off. Higher temperatures have greater lethality (smaller *D* values) than lower temperatures. The thermal-death time curve shows how heat resistance changes at different temperatures. This enables one to calculate the cumulative lethality of each time-temperature combination encountered during the process.

The *z* value describes how lethality changes with temperature. The *z* value is the temperature increase required to reduce the *D* value by a factor of 10. Graphically, it is equal to the slope of the thermal-death time curve. The *z* value can also be calculated from the equation

$$t_2 = t_1 \times 10^{(T_1 - T_2)/z}$$

where T_1 is the higher of the two temperatures, at which the *D* value is t_1; T_2 is the lower temperature, at which the *D* value is t_2; and *z* is the number of degrees required to change the *D* value by a factor of 10. For example, if the *D* value at 250°F were 2 min and the *z* value were 18°F (typical for many spores), then the *D* value at 232°F would be 20 min. It takes an 18°F change in temperature to cause a 10-fold change in the *D* value. What would be the *D* value at 268°F?

Know the difference:

The *D* value is the time (in minutes) it takes at a given temperature to decrease viability by a factor of 10.

The *z* value is the number of degrees it takes to change the *D* value by a factor of 10.

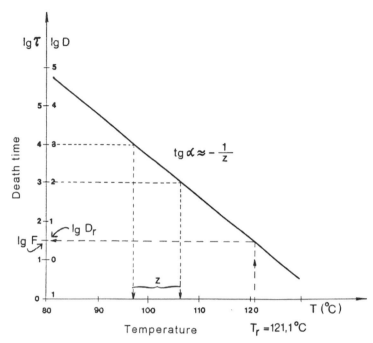

Figure 24.4 The thermal-death time curve illustrates the z and F values. T_r, reference temperature; τ, death time. D_r, D value at reference temperature; tg, slope of the thermal-death time curve. Reprinted from J. Farkas, p. 567–592, *in* M. P. Doyle, L. R. Beuchat, and T. J. Montville (ed.), *Food Microbiology: Fundamentals and Frontiers,* 2nd ed. (ASM Press, Washington, D.C., 2001).

The heat resistances (D values) and how they change with temperature (z value) are different for different microbes and are influenced by many factors. There are inherent differences among species, among strains within the same species, and between spores and vegetative cells. Bacterial spores are more heat resistant than vegetative cells of the same species. Some spores survive for minutes at 120°C or for hours at 100°C. Less heat-resistant spores have ($D_{100°C}$) of <1 min. Vegetative cells of bacteria, yeasts, and molds can have D values of <0.1 min at 70 to 80°C.

The age of the cells, their growth phase, the growth temperature, the growth medium composition, and exposure to prior stressors also affect heat resistance, as does the environment in which microbes are heated. Thus, unlike chemical constants such, as pK_a or molecular mass, D and z values depend on the situation of use. Since they depend on so many factors, great caution should be used in applying values obtained in one environment to another food environment.

Heat resistance data for vegetative cells and spores are given in Tables 24.5 and 24.6. Vegetative cells of spore-forming bacteria are not especially heat resistant. The high heat resistance of spores is caused by their specific structures, which result in a relatively dehydrated spore core. (Remember that dry heat is much less lethal than wet heat.) The molecular basis of heat resistance in spores is covered in chapter 3.

Vegetative forms of yeasts are killed by 10- to 20-min heat treatment at 55 to 60°C. Increasing the temperature 5 to 10°C gives similar lethality for yeast ascospores. Vegetative propagules of most molds and conidiospores

Table 24.5   Relative heat resistances of selected vegetative bacteria[a]

Organism	Heating medium	D value (min) at:				z value (°C)
		70°C	65°C	60°C	55°C	
Lactobacillus plantarum	TSB[b] + sucrose (a_w = 0.95)		4.7–8.1			
	Tomato juice (pH 4.5)	11.0				12.5
Pseudomonas fluorescens	Nutrient agar				1–2	
Salmonella enterica serovar Senftenberg	Skim milk			10.8		6.0
	Phosphate buffer (0.1 M; pH 6.5)		0.29			
	Same as above plus sucrose (% [wt/vol]), 30:70		1.4–43			
	Milk chocolate	440				18.0
Staphylococcus aureus	Custard or pea soup			7.8		4.5

[a]Adapted from J. Farkas, p. 567–592, *in* M. P. Doyle, L. R. Beuchat, and T. J. Montville (ed.), *Food Microbiology: Fundamentals and Frontiers*, 2nd ed. (ASM Press, Washington, D.C., 2001).
[b]TSB, tryptic soy broth.

are inactivated by 5 to 10 min of wet-heat treatment at 60°C. Several mold species that form sclerotia or ascospores are much more heat resistant. They cause serious spoilage problems. Mold spores survive dry-heat treatment for 30 min at 120°C and cause spoilage of baked foods.

The death rates of bacteria are also influenced by heating rates and the prior history of temperature exposure. When cells are exposed to sublethal temperatures above their growth maxima (i.e., to a sublethal heat shock), their heat resistance at higher temperatures increases. This is because the

Table 24.6   Approximate heat resistances of selected bacterial spores[a]

Type of food and typical organisms	D value (min) at:		z value (°C)
	121°C	100°C	
Low-acid foods (pH > 4.6)			
Thermophilic aerobe			
Bacillus stearothermophilus	4.0–4.5	3,000	7
Thermophilic anaerobes			
Clostridium thermosaccharolyticum	3.0–4.0		12–18
Desulfotomaculum nigrificans	2.0–3.0		
Mesophilic anaerobes			
Clostridium sporogenes	0.1–1.5		9–13
C. botulinum types A and B	0.1–0.2	50	10
Clostridium perfringens		0.3–20	10–30
Mesophilic aerobes			
Bacillus licheniformis		13	6
Bacillus subtilis		11	7
B. cereus		5	10
Acid foods (pH ≤ 4.6)			
Thermotolerant aerobe			
Bacillus coagulans	0.01–0.1		
Mesophilic aerobes			
Bacillus polymyxa		0.1–0.5	
Clostridium butyricum		0.1–0.5	

[a]Adapted from J. Farkas, p. 567–592, *in* M. P. Doyle, L. R. Beuchat, and T. J. Montville (ed.), *Food Microbiology: Fundamentals and Frontiers*, 2nd ed. (ASM Press, Washington, D.C., 2001).

production of heat shock proteins is induced by the sublethal temperature. In addition to heat, other environmental stresses, chemicals, or mechanical treatments can trigger the production of "heat shock" proteins. This confers cross-resistance among stressors. For example, cells that form heat shock proteins in response to sublethal temperature may become more resistant to acid. The heat shock response caused by industrial processes can protect the microbes.

Calculating Heat Processes for Foods

Foods in containers do not reach processing temperatures instantly. Furthermore, all temperatures (above a minimum) encountered on the way to the process temperature contribute to microbial inactivation. Because of this, many time-temperature combinations can give the same process lethality. Six-tenths of a minute at 268°F can be as lethal as 6 min at 250°F. There must be some common denominator or reference value used to compare these lethalities. The F value is used for this purpose. The F value is the time at a specific temperature required to kill a specific number of cells having a specific z value.

Since the F value is the time required to decrease a population with a specific z value at a specific temperature, some temperature must be (arbitrarily) set for the universal reference F value. This universal reference is the F_0 value, the temperature is 121°C (250°F), and the z value is 10°C (18°F). The F_0 value of a heat treatment is called its *sterilization value*. F_0 measures the lethality of a given heat treatment and relates it to the time required for the same lethality at some other temperature T. Thus, if one knows the time it takes to kill a population of cells with a z value of 10°C at 121°C (the reference temperature), one can get the same result at other times or temperatures by using the following equation:

*Authors' note*_____
The reference temperature is actually not so arbitrary, since it is the temperature produced by low-pressure (15 pounds per square inch) steam in canneries.

$$\text{Time}_{\text{at new temperature}} = F_0/10^{(\text{new temperature} - 121)/z}$$

Various foods have different F_0 requirements. Heat sensitivity and thermal-death curves are affected by many factors, so thermal-death curves should be made for the actual food being processed. Because the spores of C. botulinum types A and B are the most heat-resistant spores of a foodborne pathogen, commercial sterilization of low-acid (pH >4.6) foods must kill these spores. A *botulinum cook* is a heat process that reduces the population of C. botulinum spores by an arbitrarily established factor of 12 decimal ($\log_{10}$) values. This 12-D concept provides a large safety margin in low-acid canned foods. (If there were one spore in each of 10^{12} cans that received a 12-D botulinum cook, there would be only 1 can with a surviving spore.) Since botulinal spores have a D value of 0.20 min at 121°C, a botulinum cook is equal to 2.4 min (12 D × 0.20 min/D). Any combination of time and temperature that yields an F_0 of 2.4 min is acceptable for canned foods. Spoilage organisms may have spores that are more heat resistant than botulinal spores. The higher heat resistance of the spoilage organisms often determines the commercial process.

Microwave Heat Treatment

When foods are exposed to radio frequency (1- to 500-MHz) or microwave (500-MHz to 10-GHz) energy, they heat up. The heat is generated by the water molecules. Water molecules are electric dipoles, having negatively charged oxygen atoms and positively charged hydrogen atoms. When water molecules encounter a rapidly oscillating radio frequency or microwave field, they reorient with each change in the field direction. This creates intermolecular friction and produces heat. Claims of nonthermal lethal effects from microwaves have been investigated experimentally and largely dismissed. The ability of microwaves to kill bacteria can be explained solely by the heat produced.

Microwave heating can reduce process times and energy usage in food processing. However, the geometry of the food; its thermal, physical, and dielectric properties; its mass; the power input; and the radiation frequency all influence the electromagnetic waves in a food. This can cause differential microwave heating of food constituents (i.e., fat versus lean in meat) or of different parts of a meal (the salty noodles may heat faster than the fried fish). This nonuniform heating limits the use of microwave heating for microbial inactivation. Nonuniform temperatures within a food might allow localized survival of pathogens.

PRESERVATION BY IRRADIATION

UV Radiation

UV radiation at wavelengths of 240 to 280 nm is destructive. It damages nucleic acids by cross-linking thymine dimers in DNA. This prevents repair and reproduction. Gram-negative bacteria are most easily killed by UV radiation, while bacterial endospores and molds are more resistant. Viruses are more UV resistant than bacterial cells. Pigment formation also influences UV resistance. Cocci that form colored colonies are less susceptible to UV radiation than those with colorless colonies. Dark conidia of certain molds are highly UV resistant.

The very low penetration of UV radiation and the difficulty in attaining an even exposure level over the food surface limit its use in food preservation. UV sources are used mainly for disinfection of air, e.g., in aseptic filling of liquid food, in packaging sliced bread, in the ripening rooms of cheese and dry sausages, or for water disinfection.

Ionizing Radiation

Ionizing radiation has high energy and penetrating power. Its lethality is caused by damaging DNA and generating free radicals. Free radicals damage the membrane and other cellular structures. Proteins and enzyme activity are not affected, so irradiated food can undergo enzymatic spoilage. Ionizing radiation can pasteurize or sterilize foods without quality loss. By preventing microorganisms, insects, and plants from reproducing, irradiation stabilizes foods (Table 24.7).

Table 24.7 Preservative efforts of ionizing radiation[a]

Effect	Results
Inhibits sprouting	Increased shelf life of root crops
Decreases "after ripening"	Increased shelf life of fruits and vegetables
Kills insects	Insects; disinfestation of food
Reduces microbial populations	Decreased contamination of food; increased shelf life of foods; prevention of food poisoning

[a]Adapted from J. Farkas, p. 567–592, *in* M. P. Doyle, L. R. Beuchat, and T. J. Montville (ed.), *Food Microbiology: Fundamentals and Frontiers,* 2nd ed. (ASM Press, Washington, D.C., 2001).

In parts of the world where transportation and refrigeration infrastructure are lacking, irradiation can increase the food supply. In countries where refrigeration is widespread but organisms like *L. monocytogenes, Salmonella enteritidis,* and *Escherichia coli* O157:H7 are problematic, raw meat can be pasteurized by radiation. There was initial consumer resistance to irradiated meats, probably because consumers saw a potential hazard with no offsetting benefit to them. As consumers come to understand the very real benefit of irradiation to microbial safety, the hypothetical risks become less important and irradiated foods become more widespread in the marketplace.

High-energy electromagnetic radiation (gamma rays of ^{60}Co or X rays) with energies up to 5 mega-electron volts (MeV) or electrons from electron accelerators with energies up to 10 MeV are used to irradiate food. This irradiation increases the stability of the food, does not make food radioactive, and is commercially available.

Electron beams and X or gamma rays penetrate differently. The practical depth limit for 10-MeV electron beams in high-moisture foods is 3.9 cm. In contrast, gamma rays at 5 MeV penetrate 23.0 cm. Except for the penetration difference, electromagnetic radiation and electrons are equivalent and are used interchangeably in foods. Although the use of electron beams may calm the fears of consumers who are afraid of "radioactivity," food treated with either method must be labeled "treated by irradiation" in the United States.

Microbiological Fundamentals

Ionizing radiation is used primarily to kill microbes. Ionizing radiation acts on cells directly by breaking DNA, the most critical target. There is also indirect action caused by radiolytic products of other molecules, particularly free radicals formed from water. An organism's radiation sensitivity is inversely proportional to its size and complexity (Table 24.8). Larger organisms are more radiation sensitive because they are more complex and have bigger DNA targets than smaller organisms. The radiation sensitivities of similar organisms may differ due to differences in their chemical and physical structures and in their abilities to repair radiation injury.

Ionizing radiation kills microbes by damaging DNA. Ionizing radiation can affect DNA directly via the ionizing ray or indirectly via the primary water radicals H, OH, and e_{aq}^-. The most important of these radicals is the OH radical. The OH radicals formed in the hydration layer around the DNA molecule are responsible for 90% of DNA damage. Thus, the indirect effects of radiation cause the most damage in living cells. Ionizing

Table 24.8   Approximate lethal radiation doses for various organisms[a]

Organism	Dose (kGy)
Higher animals	0.1
Insects	1.0
Non-spore-forming bacteria	10
Bacterial spores	50
Viruses	100

[a]Adapted from J. Farkas, p. 567–592, *in* M. P. Doyle, L. R. Beuchat, and T. J. Montville (ed.), *Food Microbiology: Fundamentals and Frontiers,* 2nd ed. (ASM Press, Washington, D.C., 2001).

radiation causes chemical damage to the organism's DNA. In addition, physicochemical damage breaks the phosphodiester backbone in one or, less frequently, both DNA strands.

Most microorganisms can repair single-strand DNA breaks. Radiation-sensitive organisms, such as *E. coli,* cannot repair double-strand breaks. Highly resistant species (e.g., *Deinococcus* spp.) can repair them. The low water content of bacterial spores is a major factor in their radiation resistance. During germination, the spore's water content increases and the radiation resistance disappears.

Gram-negative bacteria, including spoilage organisms and pathogenic species, are generally more sensitive than vegetative gram-positive bacteria. Bacterial spores are even more resistant. More highly resistant vegetative forms, typified by *Deinococcus* (formerly *Micrococcus*) *radiodurans* and *Deinobacter* spp., are rare. The radiation sensitivity of many molds is similar to that of vegetative bacteria. However, fungi with pigmented hyphae have radiation resistance similar to that of bacterial spores. Yeasts are as resistant as the more resistant bacteria. Since they are very small and simple, viruses are highly radiation resistant.

Radiation lethality is represented by plotting the logarithm of the number of surviving organisms against the radiation dose. Typical radiation lethality curves are shown in Fig. 24.5. Type 1 is an exponential survival curve, quite common for radiation-sensitive organisms. Type 2 is characterized by an initial shoulder, indicating that equal increments of radiation are more effective at high doses than at low doses. The shoulder may be explained by the microbes' ability to repair low-dose damage. Type 3 is characterized as concave, due to a small population of resistant organisms that have survived.

The radiation sensitivity of microbial populations is expressed by the decimal reduction dose (D_{10} value). This is conceptually similar to the mea-

Figure 24.5 Radiation survivor curve as described in the text. Reprinted from J. Farkas, p. 567–592, *in* M. P. Doyle, L. R. Beuchat, and T. J. Montville (ed.), *Food Microbiology: Fundamentals and Frontiers,* 2nd ed. (ASM Press, Washington, D.C., 2001).

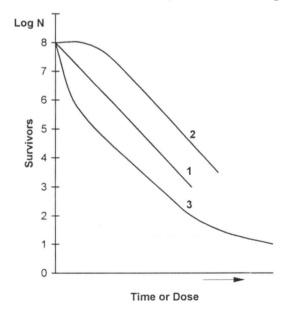

surement of heat resistance. When the dose-survival curve is linear, D_{10} is the reciprocal of its slope:

$$D_{10} = \frac{\text{radiation dose}}{\log N_0 - \log N}$$

where N_0 is the initial number of organisms, and N is the number of organisms surviving the radiation dose. Survival curves of type 2 and type 3 are often best represented by quoting an inactivation dose, for example, the dosage to inactivate 90 or 99% of the initial viable-cell count.

Technological Fundamentals

Radiation requires a minimum dose to be effective. In some cases, only the food's surface needs to be irradiated. In others, the entire food must be treated. Guidelines for dose requirements are given in Table 24.9. Like D values in heating, these are specific for a given food.

Food irradiation is sometimes called *radurization* in analogy to pasteurization. It eliminates pathogens but also reduces the number of spoilage microorganisms in food, thereby extending the shelf life. Relatively high doses, approaching 5 kilograys (kGy), can eliminate spoilage organisms. More resistant and metabolically less active species, e.g., resistant *Moraxella* spp., lactic acid bacteria, and yeasts, remain viable. During refrigerated storage, the survivors grow more slowly than the normal spoilage organisms, increasing shelf-life three- or fourfold.

Radappertization applies ionizing radiation to prepackaged, enzyme-inactivated foods to reduce microbial activity and make the food microbiologically stable. It is analogous to canning, although any packaging material can be used. The irradiation dose is determined by the most radiation-resistant microorganism associated with the food. For nonacid, low-salt foods, this is the *C. botulinum* type A spore. By analogy with canning, the safety of the radappertization process is based on a 12-*D* reduction in the viability of botulinal spores. The dose required is ~45 to 50 kGy for low-salt, nonacid foods. Since this is above the 10 kGy generally permitted for all foods and can produce off flavors, radappertization is not a commercial process.

Radiation treatment does not kill by heating the product. The largest dose used is 50 kGy. This amount of energy equals ~12 calories (50 joules [J]).

Authors' note

The term "radappertization" is a tribute to Nicolas Appert, the inventor of canning.

Table 24.9 Dose requirements for applications of food irradiation[a]

Application	Dose requirement (kGy)
Inhibition of sprouting	0.03–0.12
Insect disinfestation of seed products, flours, fresh and dried fruits, etc.	0.2–0.8
Parasite disinfestation of meat	0.1–3.0
Radurization of perishable food items	0.5–10
Radicidation of frozen meat, poultry, eggs, and other foods and feeds	3.0–10
Reduction or elimination of microbial population in dry food ingredients	3.0–20
Radappertization of meat, poultry, and fish	25–60

[a]Adapted from J. Farkas, p. 567–592, *in* M. P. Doyle, L. R. Beuchat, and T. J. Montville (ed.), *Food Microbiology: Fundamentals and Frontiers,* 2nd ed. (ASM Press, Washington, D.C., 2001).

Figure 24.6 The radura symbol for food treated by irradiation.

Hence, if all the ionizing radiation degrades to heat, the food temperature rises ~12°C. For this reason, irradiation has been called a cold process. In an attempt to gain public acceptance of food irradiation, it is sometimes referred to as cold pasteurization. The FDA allows cold pasteurization (accompanied by the radura [Fig. 24.6]) to be used in place of irradiation, but use of this term is controversial. Consumer activists claim that it is deceptive.

Radiation can damage fruit and vegetable texture by degrading polysaccharides, such as pectin. This often is accompanied by calcium release, which causes tissue softening. Softening can reduce the natural microbial resistance of the plant. This accelerates spoilage if postirradiation contamination occurs.

Consumer Acceptance of Food Irradiation

Food irradiation can hardly be considered new technology. The foundational science was done in the middle of the last century. Progressively more applications have gained FDA approval over the last 30 years (Table 24.10). Even though consumer activists now concede that eating irradiated foods will not make you radioactive, consumer acceptance has been slow. Consumers will not take even hypothetical risks if they see no personal benefit. The arguments being used against irradiated foods are amusingly similar to those used against the mandatory pasteurization of milk in the early 1900s:

- allows use of inferior raw materials
- reduces quality
- destroys nutrients
- creates new toxins
- high cost of equipment squeezes out small processors
- no multigenerational studies have been done in humans
- unknown effects may hurt future generations
- increases consumer costs

With the current emphasis on microbiological food safety, consumers are beginning to realize that "raw" food can contain listeriae, salmonellae, and other enteric pathogens. This makes the benefit of food irradiation more tangible. Studies suggest that it costs 2 to 7 cents per pound to irradiate meat and that consumers are willing to pay a 35- to 70-cent-per-pound premium for pathogen-free meat. Perhaps some day unpasteurized meat will be as rare as unpasteurized milk.

Table 24.10 Approval of irradiated foods

Yr	Application	Dose (kGy)
1963	Insect control, wheat	0.5
1964	Sprouting, potatoes	0.1
1983	Spices	30
1986	Insect control, fruit maturation, and killing of trichinae[a]	1.0
1990	Pathogen control in poultry	1.5–3.0
1997 (FDA)	Pathogen control in red meat	4.5 (fresh)
1999 (USDA[b])		7.0 (frozen)
2000	Shell eggs	3

[a]Trichinae in pork, if not killed by heat or irradiation, develop into tapeworms in humans.
[b]USDA, U.S. Department of Agriculture.

Table 24.11 Nonthermal methods of microbial inactivation in food

Method	Principle	Energy	Applications	Advantages	Disadvantages
UV light	Cross-links DNA through thymine dimers	100–400 J/cm^2	Surface and air disinfection, juices	Dry, no "heated" characteristics in food quality	Dose changes with age of lamp, limited penetration
High-intensity light	Contains UV energy and also creates photochemical reactions	0.01–50 J/cm^2	Surfaces	Dry, no residue, no heat generation	Limited penetration
Pulsed electric field	Creates pores and degrades membrane integrity	20–80 kW^a/cm	Liquids, eggs, milk	No protein denaturation, heat-induced products	Limited by electrical properties of product
Hydrostatic pressure	Instantaneous transmission of energy throughout food causes structural damage	100–800 MPa^b	Liquids, jams, juices, guacamole	No heat-induced defects; high-quality fresh product	Damages food structure; not a continuous process

^{a}kW, kilowatts.
^{b}MPa, megapascals.

Other Nonthermal Processes

Nonthermal methods of food preservation have the potential to produce products with superior quality. They all work by applying different forms of energy to the food. The basics of some nonthermal processes are shown in Table 24.11.

Summary

- Physical treatments of food can inhibit, kill, or remove microbes.
- The amount of available water (a_w) determines which organisms can grow in a food and influences the lag time, growth rate, and final cell density.
- An a_w of 0.85 is the lowest a_w that allows the growth of pathogens. This makes it a cardinal value in food microbiology.
- Drying does not necessarily kill microbes but prevents their growth.
- Refrigeration prevents the growth of many pathogens and slows the growth of all microbes.
- Freezing prevents microbial growth but cannot be considered lethal.
- Controlled and modified atmospheres can extend the shelf life of fruits and vegetables but can create conditions favorable for specific pathogens.
- Pasteurization is designed to kill all pathogens but may leave viable spoilage organisms.
- Commercial sterilization aims to kill all microbes that can grow in a product.
- Heat kills microbes with negative exponential kinetics.
- The *D* value is the *time* required for a 90% (1-log-unit) reduction in the viability of a population.
- The *z* value is the *number of degrees* required to change the *D* value 10-fold.
- The lethality of microwave energy is due to the heat produced.

- Ionizing radiation from electron beams or radioactive sources kills microbes by DNA breaks and the generation of free radicals.
- Ionizing radiation kills microbes with negative exponential kinetics.

Suggested reading _____

Farkas, J. 2001. Physical methods of food preservation, p. 567–592. *In* M. P. Doyle, L. R. Beuchat, and T. J. Montville (ed.), *Food Microbiology: Fundamentals and Frontiers*, 2nd ed. ASM Press, Washington, D.C.

Institute of Food Technologists. 2000. *Kinetics of Microbial Inactivation for Alternative Food Processing Technologies. Journal of Food Science Supplement.* Institute of Food Technologists, Chicago, Ill. (Also available at http://vm.cfsan.fda.gov/~comm/ift-toc.html.)

Novak, J. S., G. M. Sapers, and V. K. Juneja (ed.). 2003. *Microbial Safety of Minimally Processed Foods.* CRC Press, Boca Raton, Fla.

Olsen, D. G. 1998. Irradiation of food. *Food Technol.* **51:**56–62.

Questions for critical thought _____

1. Consider the list of consumer objections to food irradiation. Use the World Wide Web (put "food irradiation" as a search engine keyword) to obtain facts that refute five of these objections.

2. Microbiological kinetic constants have their counterparts in chemical kinetic constants. If the D value is analogous to the chemical rate constant k, what chemical constant is analogous to the z value?

3. You may remember that low-acid canned foods must receive a thermal process equivalent to F_0 value of 2.4 min. This is based on a 12-D reduction of *C. botulinum* spores, which have a D_{250} of 0.2 min. Imagine two process deviations. In deviation A, the time is 10% less than it should be (i.e., the product receives 2.16 min at 250°C). In deviation B, the temperature is 10% too low (i.e., the product receives 2.40 min at 225°F). Which process deviation represents a greater threat to public safety? Be quantitative in your answer (i.e., calculate the log reduction in *C. botulinum* spores as the result of each deviation, assuming a z value of 18°F). Show all of your work.

4. Identify (circle) the factual errors in the paragraph below, which was written by an anti-irradiation food activist. Rewrite the paragraph in a factual manner.

   > **Unite Against Radioactive Food**
   > All consumers should unite against the irradiation of food. Unlike other forms of food preservation, it can never make the food 100% sterile. It robs foods of essential nutrients and generates free radicals that accelerate aging and cause cancer. This technology is untried, and its safety is untested. Don't let them use you as a lab rat: ban the irradiation of food!

5. Define z value. If an organism has a $D_{80°C}$ of 20 min, a $D_{95°C}$ of 2 min, and a $D_{110°C}$ of 0.2 min, what is its z value?

6. The year is 2007, and you are the director of research for Peter's Premium Preservative-Phree Phoods. One of your product lines is a tomato sauce. Your procurement department has identified a very inexpensive source of tomatoes from Israel, but your microbiology department re-

ports that they typically contain *Bacillus citricum* spores at ~100 colony-forming units (CFU)/g. This organism can metabolize citric acid and thereby increase the pH of the product. Anticipating your every need, the microbiologists scan the World Wide Web for data on *B. citricum* heat resistance and present you with the information in the table below:

Time (min)	CFU/g at:		
	80°C	**85°C**	**90°C**
0	2.2×10^6	2.2×10^6	3.3×10^6
8			2.1×10^5
10	7.0×10^5	1.6×10^5	
16			2.0×10^4
20	4.4×10^5	5.9×10^4	
24			3.5×10^3
30		1.8×10^4	
32			1.7×10^3
40	2.0×10^5	3.2×10^3	
60	4.1×10^4		

a. You decide to show those microbiologists that you are still technically competent, so you determine the $D_{90°C}$ and the z value (in degrees Celsius) for *B. citricum* from the data. (Each survivor curve has five data points; ignore the blank cells.) What values do you get?

b. Your current process requires that the jars of sauce be held for 7 min at an internal temperature of 100°C. Will this process be adequate for tomatoes contaminated with *B. citricum?*

c. Do you have any other concerns about this product or the data upon which you are basing your judgment? Are there other factors you should consider?

NOTE: You should document your decision in case the FDA wants to inspect your records. Show all your work. If you solve the problem graphically, attach your graph to your answer. Justify your "yes or no" answers to parts b and c.

7. You have taken a position as the director of microbiology with Angelica's Premium Potted Products in equatorial Africa. One of your product lines is canned meat. The main spoilage problem is a gram-positive spore-forming rod. The person who had the job before you decided that product quality could be vastly improved by using a higher temperature for a shorter period. (The product is currently given a 7-D process of 791 min at 220°F.) The previous microbiologist had prepared spores and determined that the $D_{250°C}$ was 2.3 min. She had also conducted thermal-resistance studies at 230 and 240°F but then ate some bad fugu fish and tragically died. You find her data in the table below. How long would it take to achieve a 7-D process at 260°F? Would this process be sufficient to meet the 12-D botulinum cook requirement if you wanted to export the product to the United States?

NOTE: You should document your decision in case the FDA wants to inspect your records. Show all your work, including the calculation of the D value at 230 and 240°F, and the z value determination. If you solve the problem graphically, attach your graph to your answer. Justify your "yes or no" answers to the botulinum cook.

*Authors' note*_____

"Potted" is an old-fashioned word for canned.

Data on heat resistance of spores that cause product spoilage

Time (min)	CFU/g at:	
	230°F	**240°F**
0	1.0×10^4	12,500
5		3,400
10	4.7×10^3	650
15		190
20	2.4×10^3	45
35	7.0×10^2	

8. Your company makes a pasteurized juice product with an initial (post-pasteurization) microbial load of 10^5 CFU/g. Storage studies show that after 5 days at 35°C, the microbial load is 10^8 CFU/g. Your process engineer finds that by increasing the pasteurization temperature by a few degrees, the initial (postpasteurization) microbial load can be lowered to 10^3 CFU/g. If this process were used, what would the final number of CFU be after 5 days at 35°C? (Assume that all conditions except the initial number are identical.) Justify your answer. (Hint: you may wish to consider the equation for the exponential growth of microbes.)

9. Why do sugar syrup and salad dressing that are both 50% water present different challenges for microbial stability?

25

Industrial Strategies for Ensuring Safe Food

LEARNING OBJECTIVES

The information in this chapter will enable the student to:

- discuss the benefits of sanitation, good manufacturing practices (GMPs), and the Hazard Analysis Critical Control Point (HACCP) system with respect to human safety and food quality
- outline the process for development of a HACCP program
- recognize the limitations of HACCP
- outline the basic concepts of GMPs
- discuss the proper application of sanitizing agents
- integrate sanitation, GMPs, and HACCP
- identify practices that can result in the introduction of chemical, biological, or physical hazards into food

INTRODUCTION

Most people in the food industry are aware of Hazard Analysis Critical Control Point (HACCP) systems. Indeed, few students majoring in food science have not heard of HACCP by the end of the freshman year. This chapter focuses primarily on HACCP; however, before a HACCP program can be introduced or developed, basic building blocks must be in place. Specifically, in an industrial setting, implementation of good manufacturing practices (GMPs) and a sound sanitation program are key elements. The implementation of GMPs and of sanitation and HACCP systems is perhaps even more important now to address public concern that food supplies are safe and secure from intentional contamination by terrorists.

GMPs

In 1969, the Food and Drug Administration (FDA) published regulations on GMPs that provide basic rules for food plant sanitation. In 1986, the FDA published a revised version that expanded the original regulations. The GMPs can be found in the *Code of Federal Regulations* under Title 21, part 110, as "Current Good Manufacturing Practices in Manufacturing, Packing, or Holding Human Food" (referred to here as CGMPs). To clarify what the *Code of Federal Regulations* contains, the following explanation may help. Proposed regulations are first published in a daily publication, the *Federal Register,* setting forth rulings by the executive branch and agencies of the federal government. These are then codified and published as the *Code of Federal Regulations.* The Code of Federal Regulations is divided into 50 titles that represent areas that are subject to federal regulation.

The CGMPs are divided into seven subparts, each detailing requirements related to various operations or areas in food-processing facilities. The requirements are established to prevent contamination of a product from direct and indirect sources. Subparts D and F are reserved; therefore, only subparts A to C, E, and G are discussed here.

General Provisions (Subpart A)

Subpart A contains a list of definitions, the underlying reason for CGMPs linked to the Food, Drug and Cosmetic Act (referred to here as the Act); responsibilities imposed on plant management regarding personnel; and exclusions. Definitions are provided to maintain consistency; however, they are often vague and open to interpretation. The criteria and definitions used to determine whether a food is *adulterated* (made unfit for human consumption by the addition of impurities) are covered under sections 402(a)(3) and 402(a)(4) of the Act. A food is considered adulterated if it contains any filth or putrid and/or decomposed material or is otherwise unfit for use as food. The Act goes on to state that food that has been prepared, packed, or held under unsanitary conditions; that may become contaminated with filth; or that has been rendered injurious to human health is adulterated.

Criteria for disease control, cleanliness, and the education and training of personnel are included. These requirements are designed to prevent the spread of disease-causing microorganisms from workers to food contact surfaces and the food itself. Workers in direct contact with food or areas that indirectly contact food must conform to hygienic practices. These include maintaining personal cleanliness, wearing clean clothing, removal of jewelry, washing hands, and wearing appropriate protective garments (gloves, hair nets, and beard covers). Personnel responsible for plant sanitation should receive education and exhibit a level of competency so that they can pass the information on to workers in a facility.

The exclusions section holds exempt food establishments that are engaged solely in the harvesting, storage, or distribution of raw agricultural commodities [section 201(r) of the Act defines this as any food in its raw or natural state, including all fruits that are washed, colored, or otherwise treated in their unpeeled natural form prior to being marketed] that are ordinarily cleaned, prepared, treated, or otherwise processed before being marketed.

Buildings and Facilities (Subpart B)

Subpart B covers regulations for plant and grounds, which state that the grounds of a food plant should be maintained in a condition that will protect against the contamination of food. The plant design and construction must be such that operations within a plant can be separated and that materials used for walls, floors, etc., facilitate cleaning and sanitizing. The sanitary operations section covers general requirements related to general maintenance, cleaning and sanitizing compounds, toxic-material storage, pest control, sanitation of food contact surfaces, and storage and handling of cleaned equipment and utensils. The last section covers sanitary facilities and controls and provides minimum requirements for sanitary facilities and accommodations. Facilities and accommodations include, but are not limited to, the water supply, plumbing, sewage disposal, toilet facilities, hand-washing facilities, and disposal of rubbish and *offal* (animal parts that are considered waste, generally applied to meat- and poultry-processing plants).

Equipment (Subpart C)

The equipment and utensils section specifies that the materials used and the design of plant equipment and utensils must allow adequate cleaning and proper maintenance. Emphasis is placed on preventing the adulteration of food by microbial contamination or by other contaminants, such as lubricants or metal fragments. Moreover, equipment used in the manufacturing, holding, or conveying of food must be designed and constructed for easy cleaning and maintenance. Cooling equipment (freezers and cold storage) and control instruments (e.g., pH meters and thermometers) must be accurate and adequately maintained. Finally, compressed gases that are introduced into food or used to clean food contact surfaces must be free of contaminants that are considered unlawful indirect food additives.

Production and Process Controls (Subpart E)

Subpart E is the most detailed subpart and is broken down into several sections and subsections that contain rules to ensure the fitness of raw materials and ingredients, to maintain manufacturing operations, and to protect finished products from deterioration. The processes and controls section stresses that all operations must be conducted in accordance with adequate sanitation principles. Quality control programs must be in place to ensure that food is suitable for human consumption and that packaging materials are safe and suitable. Quality control sanitation programs are required to ensure compliance with the Act.

The subsection on raw materials and ingredients contains descriptions for how those materials should be inspected, handled, washed, and segregated to minimize contamination and deterioration. Materials must not contain microorganisms or must be treated during manufacture to ensure that the levels of microorganisms that produce food poisoning or other human illness no longer present a problem and do not render the food adulterated. Compliance with regulations for natural toxins and other *deleterious* (harmful) substances is of the utmost importance and should be verifiable. Holding of raw materials must be adequate to protect against contamination and must be consistent with requirements for that material. Therefore, frozen materials must be kept frozen, dry materials must be protected against moisture and temperature extremes, and material scheduled for rework must be identified as such.

The handling of food in an appropriate way to prevent contamination is covered in the subsection on manufacturing operations. Required conditions are listed for the handling of foods that support the rapid growth of undesirable microbes, particularly those that cause foodborne illness. Requirements to protect several types of foods, for example, batters, sauces, and dehydrated foods, from contamination or adulteration are stated. Various operations, such as mechanical manufacturing (washing, cutting, and mashing), heat blanching, and filling, assembling, and packaging must be performed to protect against contamination. To facilitate the tracking process, companies should code individual consumer packages with appropriate information (most companies routinely code each package). This is also important in the voluntary recall of products that are in violation of food laws and regulations.

The final section of subpart E, warehousing and distribution, is brief but nonetheless extremely important. The section states that the storage and transport of finished food must be conducted under conditions that prevent physical, chemical, and microbial contamination. The conditions must also prevent deterioration of the food and the container.

DALs (Subpart G)

The section of subpart G covering natural or unavoidable defects in food for human use that present no health hazards was established because it is not possible, and never has been possible, to grow (in open fields), harvest, and process crops that are totally free of natural defects. Certain defects that present no human health hazard, particularly in raw agriculture commodities, may be carried through to the finished product. The defect action level (DAL) represents the limits at or above which the FDA may take legal action to remove a product from the consumer market. The fact that the FDA has established DALs does not mean that a manufacturer need only stay below that level. Manufacturers cannot mix or blend a product containing any amount of defective food at or above the current DAL with another lot of the same or another product. This is not permitted by the FDA and renders the food unlawful regardless of the DAL of the finished food. Companies cannot use DALs to mask poor manufacturing practices. The FDA clearly states that the failure to operate under GMPs will result in legal action even though products might contain natural or unavoidable defects at levels lower than the currently established DALs.

An extensive list of DALs is available for products from allspice to wheat flour. The defect may be mold, rodent filth, insects, rot or mammalian excreta. Below are a few examples of DALs (listed as product, defect, and action level):

- apple butter: rodent filth; average of four or more rodent hairs per 100 g of apple butter
- cocoa beans: mammalian excreta; mammalian excreta at 10 mg or more per pound
- fig paste: insects; >13 insect heads per 100 g of fig paste in each of two or more subsamples
- peaches, canned and frozen: moldy or wormy; average of 3% or more of fruit by count is wormy or moldy
- potato chips: rot; average of 6% or more of pieces by weight contain rot

Although not specified or required in the CGMPs, maintaining accurate records to document compliance with laws and regulations is implied. Records from all phases of the food operation should be maintained to verify that raw materials, food-packaging materials, and the finished product comply with the provisions of the Act. Records related to processes used to destroy or prevent the growth of microorganisms of public health concern are also important. To facilitate tracking of finished products and recalls, records related to distribution must be maintained. Recordkeeping will also aid in the development of an HACCP program.

SANITATION

Many students, and for that matter food plant workers, incorrectly consider sanitation to be only the removal of trash and the cleaning of equipment. However, sanitation includes a host of activities that are ultimately designed to prevent the adulteration of a product during processing. The true goal of any sanitation program is to protect the health of the consumer. In addition, a sound sanitation program increases the shelf life of a product, thereby minimizing economic loss, and prevents contamination of a product with materials that may offend the consumer (e.g., pieces of wood, rocks, or machinery). During the past 50 years, significant advances have been made in productivity and the ways that foods are processed. These advances have enhanced the safety and shelf life of products, but they require strict sanitation programs to ensure that contamination does not occur.

Before sanitation can be discussed, it must be defined with respect to the food industry. *Sanitation* is the creation of a comprehensive program designed to control microorganisms in a food-processing plant. A sanitation program must be based on sound science to ensure that healthy workers handle food in a clean environment. The goal of the program is to prevent contamination of food with pathogens that cause human illness and to limit the introduction and growth of microorganisms that result in spoilage. To achieve this goal, personal and environmental hygiene must be practiced. Sanitation is considered an applied science because of the human health issues that could arise should a breakdown in sanitation practices occur.

Within the sanitary-operations section of the CGMPs, general requirements for the maintenance of buildings, fixtures, and other physical facilities; pest control; sanitation of food contact surfaces; and storage and handling of cleaned equipment and utensils are covered. Depending on the product, the U.S. Department of Agriculture (USDA) may have jurisdiction over a food-processing facility, specifically those associated with meat, poultry, and eggs (according to the Federal Meat Inspection Act, Poultry Products Inspection Act, and Egg Products Inspection Act). Other agencies may also become involved. For instance, the Environmental Protection Agency may be involved if issues arise over water pollution (Federal Water Pollution Control Act) or air pollution (Clean Air Act) or through the use of pesticides (Federal Insecticide, Fungicide, and Rodenticide Act).

The benefits of sanitation can be numerous, but first and foremost they aid in protecting the consumer from illness. An effective sanitation program facilitates the development of an effective HACCP program. Inspections of food-processing plants, not only by federal inspectors but also by other companies (company A purchases a product from company B for use in a company A product) are becoming more stringent. An effective sanitation program can limit the number of negative consumer contacts related to food spoilage (a product has off odor or flavor) and foodborne illness. These types of problems weaken consumer confidence, causing decreased sales and increased claims. A stringent sanitation program can increase the storage life and shelf life of food, minimizing economic loss. Regular cleaning and maintenance of processing equipment and heating and cooling systems permit more efficient operation and reduce the microbial load in the facility and the product. In fact, the spread of microbes from cooling systems has been linked to the contamination of food and outbreaks of foodborne illness. Overall, a sound sanitation program can increase the quality of the product, improve customer relations, and increase the trust of inspectors associated with regulatory agencies.

Before sanitation standard operating procedures (SSOPs) can be developed, a sanitation audit must be conducted. A review of the facilities will aid in identifying areas of noncompliance and establishing corrective actions. A systematic approach should be used when conducting an audit. Starting at the receiving area, check food materials (ingredients) for temperature, odor, color, and appearance. Next, move to the general plant area and inspect walls, floors, drains, and footbaths. Processing equipment, tanks, pumps, pasteurization charts, and retort charts can also be inspected. While moving through the plant, observe employee practices and determine compliance. Check for employees who appear ill, have open lesions, are wearing jewelry, or are walking in restricted areas; check for the use of protective devices (hair nets, beard nets, and gloves) and footbaths. The building and the area around the outside of the building must also be checked for signs of pests, broken windows, a damaged roof, open doors and windows, standing water, and trash. Recall that sanitation is more than trash removal.

Microbes can be brought into processing facilities on raw materials or picked up by food during processing as it passes through or over handling equipment. Employees can spread microbes to equipment and to food. Microbes from process waters during washing, conveying, or preparation can contribute to the microbial load of the product. Microorganisms can also be present in ingredients that may be added to the product. This is particularly important if ingredients are added after a process designed to reduce or eliminate microbes has been completed.

Sanitization (using germicidal or sanitizing agents) of food contact surfaces is the key to maintaining a sanitary food-processing environment, thereby reducing the number of microorganisms associated with a food. Sanitization involves the treatment of a food contact surface so that the process effectively destroys vegetative cells of microorganisms capable of causing human illness and adequately reduces populations of other microbes. The process does not need to reduce populations of bacterial spores. The treatment should not have an adverse impact on the food (impart an off odor or color) or the safety of the consumer.

The main criterion for selecting a chemical sanitizing agent is the ability of the agent to kill microorganisms of concern within a specified time. While chlorine or other chemical sanitizing agents can be extremely effective in reducing the microbial load, they must be used appropriately. In food-processing facilities, chlorine compounds are the most widely used sanitizing agents. A number of factors can reduce or effectively inactivate sanitizers. Sanitizers can react with organic materials (soil, food, bacteria, oils, etc.) that have not been removed from equipment surfaces, reducing the effectiveness of the sanitizer. Impurities in water, including iron, manganese, nitrites, and sulfides, can react with a sanitizer and decrease its effectiveness. Although temperature is generally not a factor, temperatures below 10°C and above 90°C can dramatically influence the availability and action of some sanitizers. Generally, increasing the concentration of a sanitizer increases the rate of destruction of the microorganisms. However, in the case of hypochlorites (e.g., sodium hypochlorite), this is not the case. For example, the addition of hypochlorite to water to give 1,000 parts per million (ppm) of chlorine produce a solution that requires three times as long to kill the same number of bacteria as a solution containing 25 ppm. This is because hypochlorites are produced by combining chlorine with hydroxides. The addition of hypochlorite results in the addition of a proportional amount of alkali that raises the pH of the solution. A 1,000-ppm solution of chlorine has a pH of ~11.0, whereas a 25-ppm solution has a pH of 8.0. In general, the effectiveness of chlorine and iodine compounds decreases as the pH increases.

Chlorine sanitizers may be the most popular sanitizers, but a number of other chemical sanitizers are available that are more useful than chlorine compounds for certain applications. Iodophors are good as skin disinfectants, since they do not irritate the skin at recommended levels of use. They can be used to clean equipment but may stain epoxy and polyvinylchloride and may cause discoloration of starchy food. Hydrogen peroxide is effective against biofilms and may be used on all types of surfaces (equipment, floors, and walls). Ozone, glutaraldehyde, and quaternary ammonium are also effective. The efficacies of these compounds against certain microbes differ compared to chlorine compounds. For example, quaternary ammonium sanitizers are effective against molds and gram-positive bacteria but are ineffective against most gram-negative bacteria.

Sanitizing can be accomplished without the use of chemical agents. Ultraviolet irradiation is used to destroy microorganisms in hospitals. In Europe, ultraviolet light units are commonly used to disinfect drinking water and food-processing waters. Steam can be used, but it is costly and not particularly effective. In fact, condensation from the application of steam may complicate the cleaning process and provide moisture for surviving microorganisms. Hot water may be useful for sanitizing small components (eating utensils and small containers), but like steam, the process is expensive because of high energy costs.

A sound sanitation program results in the manufacture of foods that are safe and wholesome and that meet legal obligations. Sanitation is the responsibility of every person in a food-processing plant. Sanitation must be practiced every day, a policy the food company must reinforce through proper training and encouragement of employees. If every person understands his or her responsibility in a sanitation program, individual efforts will ensure that a consistent product is produced and consumer confidence is maintained. If executed properly, sanitation should reduce (perhaps eliminate) concern about the spreading of human diseases and food poisoning. Moreover, if sanitation is maintained, a product with a reduced microbial population will be produced and waste and spoilage will be minimized.

SSOPs

SSOPs are guided by CGMPs detailing a specific sequence of events necessary to perform a task to ensure sanitary conditions. To prevent direct product contamination, SSOPs should contain a description of procedures that a food-processing facility will follow to address the elements of preoperational and postoperational sanitation. A set of written SSOPs must be developed and maintained by federally and state-inspected meat and poultry plants.

The USDA Food Safety and Inspection Service (FSIS) felt that SSOPs would permit each company to consistently follow effective sanitation procedures, thereby reducing the likelihood of direct product contamination or adulteration. The SSOPs must include information related to daily preoperational and operational procedures for the plant personnel responsible for monitoring daily plant sanitation activities. Companies must determine whether the SSOPs are effective and, if not, take corrective action. Corrective actions include procedures to ensure the appropriate disposition of contaminated products, restoring sanitary conditions, and preventing the recurrence of direct contamination or product adulteration. SSOPs written for meat and poultry plants must be signed and dated by the individual with overall authority over the establishment to ensure that the company will implement the SSOPs.

Although written SSOPs are not required for all food production establishments, they are the cornerstones of HACCP. SSOPs go beyond specific processes. HACCP plans are developed as a proactive approach to ensure the safety of a product. Interfacing CGMPs and sanitation (SSOPs) permits the development of an effective HACCP program.

HACCP

The development of HACCP programs by companies has become routine since the mid-1990s. In some instances, the FDA and the USDA FSIS have published final rules mandating the development and implementation of HACCP programs to enhance the safety of domestic and imported seafoods, juices, and meat and poultry. In fact, HACCP can be used by all segments of the food industry to effectively and efficiently ensure farm-to-table food safety in the United States (or, for that matter, in any country). HACCP is a proactive prevention-oriented program that addresses food safety through the analysis and control of biological (predominantly pathogenic microorganisms), chemical (including those naturally produced by microorganisms), and physical (metal, wood, and plastic) hazards. The hazards may come from raw material production or the handling, manufacturing, distribution, and consumption of the finished product.

A food processor must identify potential sources of food safety hazards for each product and production line (process). Preventative steps can then be identified to reduce or eliminate potential hazards. Since this is a proactive approach, the manufacturer determines in advance what to do with a product or process should a preventative step fail. This requires continuous monitoring and documentation to verify that the appropriate corrective action was taken. Records can be reviewed to aid in improvement of the plan. HACCP is a continuous system, and HACCP plans require updating as ingredients and processes change and as new technologies emerge.

Background

The HACCP concept was developed nearly 50 years ago by the National Aeronautics and Space Administration and Natick Laboratories as the Failure Mode Effect Analysis system for use in aerospace manufacturing. Through collaborative work by the Pillsbury Company, the National Aeronautics and Space Administration, and Natick Laboratories, the process was applied to foods produced for the U.S. space program. This was done to ensure that the food used would be 100% free of bacterial pathogens. Could you imagine being in space and developing a case of projectile vomiting and severe diarrhea? This initial work led to what is now HACCP.

Although HACCP was introduced to the public (food-manufacturing companies) in the early 1970s the concept was met with limited interest. However, in 1985, a subcommittee of the Food Protection Committee of the National Academy of Sciences made a strong endorsement of HACCP. The National Advisory Committee on Microbiological Criteria for Foods was formed in 1988 and given the mission of developing material to promote the understanding of HACCP. HACCP is now practiced by food manufacturers and food service establishments, and even in the home.

HACCP Basics

The HACCP system is based on seven principles, or steps, which will be discussed in more detail. They are (i) conduct a hazard analysis, (ii) determine the critical control points (CCPs), (iii) establish critical limits, (iv) establish monitoring procedures, (v) establish corrective actions, (vi) establish verification procedures, and (vii) keep records and documentation. There are a number of approaches that can be taken in the development of a HACCP program; however, they all stress a commonsense approach to food safety management. Template HACCP plans can be obtained through the FDA and USDA FSIS for specific products for which HACCP have been mandated.

Before a HACCP plan can be developed, there are a number of activities that must be completed. Obviously, management must support the development and implementation of HACCP. Without such support, efforts to develop and implement HACCP will likely prove futile. A HACCP team should be assembled and should include line workers and personnel involved in sanitation and in production and quality assurance. Including personnel involved in marketing and communication may also be appropriate. To balance the team and bring a fresh perspective, an outside expert may be beneficial. In small companies, it may be difficult to establish an effective HACCP team; this is when an outside expert(s) is especially useful.

The HACCP team must recognize that a generic HACCP plan cannot be developed for all food products produced in a manufacturing plant. A HACCP plan should be specific for each product. However, once a HACCP plan is developed, it may be appropriate to use the plan as a template for the development of subsequent plans. A description of the product should include the name, formulation, method of distribution, and storage requirements. The intended use and anticipated customers for the food should be established. The product may not be intended for use by the general public but rather by infants, the immunocompromised, or some other high-risk group.

The development and verification of a flow diagram describing the production process is an important step that is often overlooked. The flow diagram is essential for hazard analysis and establishing CCPs (Fig. 25.1). Moreover, the diagram becomes a guide and is essential in the verification process. Even though a flow diagram has been constructed, a "walk-through" of the production facility must be done to verify that the diagram is accurate. For example, new equipment, such as metal detectors, may have been added or the route for rework may not have been accounted for. The flow diagram should be modified if and when necessary.

The following narrative discusses the seven principles of HACCP, providing a brief description of each.

Principle 1: hazard analysis.   Through hazard analysis, a list of hazards that are likely to cause injury or illness if not controlled is developed. This should be done systematically, starting with raw materials and following through each step of the food production flow diagram. Hazards that present little or no risk to the consumer should not be considered in an HACCP plan. Indeed, those hazards are generally addressed within GMP programs, a building block for the HACCP system. A hazard as set forth by the National Advisory Committee on Microbiological Criteria for Foods for HACCP is "a biological, chemical, or physical agent that is reasonably likely to cause illness or injury in the absence of its control." Even though a

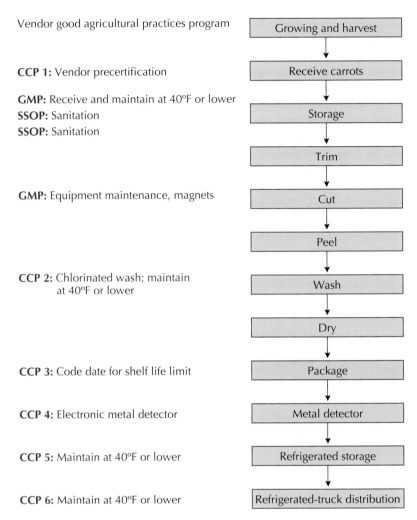

Vendor good agricultural practices program

CCP 1: Vendor precertification

GMP: Receive and maintain at 40°F or lower
SSOP: Sanitation
SSOP: Sanitation

GMP: Equipment maintenance, magnets

CCP 2: Chlorinated wash; maintain
at 40°F or lower

CCP 3: Code date for shelf life limit

CCP 4: Electronic metal detector

CCP 5: Maintain at 40°F or lower

CCP 6: Maintain at 40°F or lower

Growing and harvest → Receive carrots → Storage → Trim → Cut → Peel → Wash → Dry → Package → Metal detector → Refrigerated storage → Refrigerated-truck distribution

Figure 25.1   Representative flow diagram and CCPs for a non-thermally processed product. The flow diagram with CCPs is for peeled refrigerated minicarrots. Note that GAPs, GMPs, and a sanitation program are key elements in the production of a safe, wholesome product.

hazard may not be considered for inclusion in an HACCP plan, this does not mean that food processors can ignore the hazard, especially if the hazard fits the definition of adulteration.

Once the hazards have been identified through the review of ingredients, the entire production process, equipment, storage, and distribution, each hazard must be evaluated. During hazard evaluation, the HACCP team decides which hazards present a significant risk to consumers and must be addressed through the HACCP plan. Generally, this involves determining the severity of the hazard in terms of the illness or injury it could produce and the likelihood of occurrence. For example, the likelihood that an undercooked beef patty might be contaminated with *Escherichia coli* O157:H7 might be low, but the illness that could result is severe and the potential for subsequent illness (e.g., disease involving kidney failure) is high. Therefore, these hazards should be addressed through HACCP. Alternatively, the hazard may be small pieces of cardboard that enter when ingredient packages are opened during mixing. The likelihood of this occurring may be high, but the severity of the hazard is low (the cardboard may not be very tasty, but it presents little if any health hazard). This type of hazard can be addressed through GMPs. The evaluation of the severity of hazards can be based on epidemiological information, scientific research, company records, and information provided by government agencies.

During this process, the HACCP team must identify appropriate control measures. Since not all hazards can be prevented or eliminated, the term "control" is more appropriate. Control measures may vary significantly from one food process to another, even for the same hazard. To control a hazard, a change in processing, such as the addition of metal detectors, or a certificate of analysis from a supplier indicating the absence of a given hazard may be required. Additionally, existing plans, including GMPs and sanitation, can be reviewed to determine whether the hazard is controlled through those plans.

Principle 2: determining CCPs.   Once all significant hazards and control measures have been identified, the CCPs to control each hazard must be identified. A CCP can be any point, step, or procedure at which control can be applied and a food safety hazard (chemical, biological, or physical) can be controlled (prevented, eliminated, or reduced to an acceptable level). A CCP may be a cooler, a metal detector, or a package label indicating storage and cooking instructions. Many processes may be redundant. For example, passing dry materials through multiple sieves to control entry of physical hazards into the final product or the use of multiple metal detectors on a production line. In either case, the last sieve or metal detector is considered the CCP.

The temptation in establishing CCPs in a process is to assign each step as a CCP. This is particularly true for raw products. The number of CCPs for any process should be kept to a minimum to facilitate monitoring and documentation (Fig. 25.1). Otherwise, records are generated that have little meaning with respect to the effectiveness of the HACCP plan. Keep in mind that a CCP may control a hazard that has occurred upstream in the process. For example, the hazard might be enteric pathogens in ground beef being used for cooked beef patties. The cooking process is the CCP, since this step will kill pathogens of concern. This does not mean that the ground beef can be left at room temperature or otherwise abused prior to or after being cooked. Indeed, keeping the ground beef at ≤4°C may also be a CCP.

Principle 3: critical limits.   Critical limits must be set for each CCP. A critical limit is based on predetermined tolerances that must be met at a CCP to prevent, eliminate, or reduce a biological, chemical, or physical hazard to an acceptable level. Critical limits for preventative measures may be set for such factors as time, temperature, pH, water activity, salt concentration, moisture level, and even the presence or concentration of preservatives. Each food company is responsible for ensuring that the critical limits indicated in its HACCP plan will control the identified hazard. Critical limits may be based on in-house experimental studies or derived from regulatory standards and guidelines or review of the scientific literature. Validation (see principle 6 below) that the critical limits are effective and being met is the key to HACCP.

Principle 4: monitoring procedures.   Monitoring is conducted to ensure that a CCP and its limits are effective. The results of monitoring are documented and provide a record for future use in verification. Monitoring may be done automatically (e.g., by computer) or manually. Monitoring activities include measurement of the pH, temperature, and moisture level or water activity. Monitoring must be a planned sequence of observations or measurements to assess whether a CCP is under control. If it is not conducted as a routine, the HACCP plan becomes compromised and contaminated products may reach the consumer. Ideally, monitoring should be continuous and allow advance warning of a problem before violation of a critical limit occurs. However, when continuous monitoring is impractical, a monitoring interval must be established, procedures for the collection of data and equipment to be used must be defined, and the person responsible for monitoring must be identified. The frequency of monitoring will depend in large part on the product and the type of measurement made.

The monitoring procedure(s) for a given CCP should be rapid, which rules out the use of time-consuming analytical tests. Therefore, microbial testing is rarely used for monitoring CCPs. Personnel involved in monitoring must be properly trained, understand the importance of monitoring, and know the proper action to take should deviation from critical limits occur. For example, if ground beef patties were to be cooked for 60 s but the belt increased in speed and the patties were cooked for only 30 s, what should be done? To address this deviation, corrective actions must be established.

Principle 5: corrective actions. HACCP is a proactive system to control hazards that may be associated with a food product. Therefore, corrective actions designed to address deviations from established critical limits should be written into the HACCP plan. Corrective actions should address how the company will fix or correct the cause of deviation to ensure that the process is brought under control and that critical limits are achieved, indicate what is to be done with a product in which deviation from the critical limit occurred, and facilitate the process for determining whether adjustments to the HACCP plan are required.

Corrective action for the beef patties that were not cooked adequately would include correcting the equipment problem, i.e., adjusting the conveyor belt speed. Moreover, the product involved in the deviation must not be released. The product could be sent through the oven again to receive the proper time-temperature treatment, disposed of as waste, or perhaps used in some other product.

Principle 6: verification procedures. Verification within and of the HACCP plan is conducted at several levels. The verification process determines whether the HACCP plan is being followed and whether records of monitoring activities are accurate. The entire HACCP plan should be revalidated through the use of outside auditors. Verification should be done to ensure that the HACCP plan is functioning satisfactorily and in compliance, if required, with government regulations. In the case of cooking ground beef patties, the time-temperature parameters must be verified to kill, for example, *E. coli* O157:H7.

Verification at some level, a given critical limit or the entire HACCP plan, should take place on a regular and predetermined schedule. This may be daily, weekly, monthly, or yearly. The frequency is determined by the operation being verified.

Principle 7: record keeping. Recording keeping is the heart of a HACCP plan. All records generated in association with a HACCP plan must be on file at the food establishment, particularly those relating to CCPs and any action on critical deviations and product disposition. Records provide a means for implementation of changes by the HACCP team to assess and document the safety of products. HACCP plan records should include the names of team members, a product description and intended use, flow diagram with CCPs indicated, types of hazards, critical limits, and a description of monitoring, corrective actions, and types of records.

HACCP Established for Specific Food Industries

In 1996, the USDA FSIS established the Pathogen Reduction and HACCP system program for meat- and poultry-processing plants. The Pathogen Reduction and HACCP rule encompasses more than HACCP. The plan requires all meat and poultry plants to develop and implement HACCP to improve the safety of their products, sets pathogen reduction performance standards for *Salmonella,* requires all meat and poultry plants to develop and implement written SSOPs, and requires animal- and poultry-slaughtering plants to conduct microbial testing for generic *E. coli* to verify the adequacy of their process controls for the prevention of fecal contamination. The plan was precipitated by a large outbreak of foodborne illness caused by the consumption of undercooked hamburgers that were contaminated with *E. coli* O157:H7.

In 1996, the FDA published the *Fish and Fishery Products Hazards and Control Guide* to assist processors in the development of HACCP plans. The plans must address the various hazards to which seafood can be exposed from the fishery to the table, including viruses, bacteria, parasites, natural toxins, and chemical contaminants. Under the regulation, seafood companies must also write SSOPs. The FDA will periodically inspect seafood processors and review HACCP records to determine how well a company is complying over time. The action by the FDA to implement a mandatory HACCP system for the seafood industry is in part linked to the knowledge that more than half of the seafood eaten in the United States is imported from almost 135 countries.

The FDA Center for Food Safety and Applied Nutrition proposed the HACCP fruit and vegetable juice regulation in 1998, which applies to juice products in both interstate and intrastate commerce. As in other HACCP regulations, juice processors are required to evaluate their manufacturing processes to determine whether any chemical, biological, or physical hazards could contaminate their products. The real concern is microbiological hazards. The most notable aspect of this regulation is that processors must implement control measures that achieve a 5-log-unit reduction in the numbers of the most resistant pathogens in their finished products compared with levels that may be present in untreated juice. A combination of methods can be used to achieve this goal. Citrus processors can use surface treatments that achieve the 5-log-unit pathogen reduction. Processors are exempt from the microbiological-hazard requirements of the HACCP regulation if they are making shelf-stable juices or concentrates using a single thermal-processing step. Finally, retail establishments that package and sell juice directly to consumers are not required to comply with the regulation. A driving force behind this HACCP regulation was a large outbreak of foodborne illness linked to the consumption of unpasteurized apple juice that was contaminated with *E. coli* O157:H7.

CONCLUSION

Implementation of GMPs and sanitation and HACCP programs prevents the sale of adulterated food products to the consumer. These measures also have a positive impact by decreasing the number of cases of foodborne illness. In today's global marketplace, measures need to be in place to prevent intentional contamination of food or to prevent food that is potentially contaminated from reaching the consumer. The programs discussed above have mainly concentrated on foods postharvest. On the farm, measures such as good agricultural practices (GAPs) are being used to decrease biological, chemical, and physical hazards that may be associated with raw agricultural products. Indeed, many fruit and vegetable processors are now requiring producers (farmers) to implement GAPs. Reducing the microbiological load of a food at the farm level should aid in keeping the microbiological load of the finished product in an acceptable range.

Summary

- GMPs and sanitation are the building blocks of a HACCP program.
- GMPs are designed to prevent the manufacture of food under unsanitary conditions and the sale of adulterated food.
- Sanitation includes personal and environmental hygiene.
- HACCP is a proactive approach to effectively and efficiently ensure farm-to-table food safety.
- HACCP is mandatory for meat- and poultry-processing plants, fruit and vegetable juice producers, and seafood processors.

Suggested reading

Gould, W.A. (ed.). 1994. *CGMP's/Food Plant Sanitation.* CTI Publications, Inc., Baltimore, Md.

Marriott, N. G. (ed.). 1999. *Principles of Food Sanitation,* 4th ed. Aspen Publishers, Inc., Gaithersburg, Md.

National Advisory Committee on Microbiological Criteria for Foods. 1998. Hazard analysis and critical control point principles and application guidelines. *J. Food Prot.* **61:**762–775.

Questions for critical thought

1. Why is it necessary to have DALs for raw agricultural products?
2. What are the seven HACCP principles? Before a HACCP plan can be developed, what needs to be done?
3. Explain why GMPs and sanitation are considered building blocks (prerequisites) for the development of HACCP.
4. Select a product, such as bagged chopped lettuce or precooked hamburger patties, and develop a HACCP plan for it.
5. When sanitizing agents are used, what factors can influence their efficacy? Explain why.
6. The FDA and USDA FSIS made HACCP mandatory for meat and poultry processors, fruit and vegetable juice producers, and the seafood industry. Write one page for each, detailing events that necessitated mandatory HACCP for those industries.

7. Define adulteration of a food product. Provide a scenario that could lead to a products being considered adulterated.
8. You have been hired by a small company to develop and direct the company's HACCP program. The owner of the company was against hiring you, since he believes that GMPs and sanitation already cover all aspects of HACCP. You must explain to him what an HACCP program does that GMPs and sanitation do not. He requests that you provide him with a two-page report.

Glossary

Acid tolerance response   The response of cells to an initial acid treatment which allows them to survive more severe acid treatments. Abbreviated ATR.

Adenosine triphosphate   A compound that serves as cellular energy currency. Abbreviated ATP.

Aerobe   An organism that requires oxygen to grow.

Aerobic plate count   A count that provides an estimation of the number of microorganisms in a food. Abbreviated APC.

Anaerobe   An organism that cannot grow in the presence of air.

Antigenic   Capable of eliciting an immune (antibody) response. Small molecules may be nonantigenic. Large molecules may have several antigenic sites.

AOAC   *See* Association of Official Analytical Chemists.

APC   *See* Aerobic plate count.

Ascospore   A heat-resistant reproductive fungal spore.

Association of Official Analytical Chemists   An organization involved in validation of testing methods. Abbreviated AOAC.

ATP   *See* Adenosine triphosphate.

ATR   *See* Acid tolerance response.

Attaching and effacing lesions   Lesions that occur when a bacterium (e.g., *Escherichia coli* O157:H7) adheres to the surface of an intestinal epithelial cell, resulting in loss of microvilli (effacement).

Autoxidation   The spontaneous oxidation of a substance.

a_w   *See* Water activity.

Bacteremia   The presence of bacteria in the blood.

CDC   *See* Centers for Disease Control and Prevention.

Centers for Disease Control and Prevention   The agency within the Department of Health and Human Services that tracks foodborne illness and helps solve outbreaks. It has no regulatory authority. Abbreviated CDC.

CFU   *See* Colony-forming unit.

Chelate   To bind ions.

Chlamydospores   Asexually produced resting spores of certain fungi.

-cidal   A suffix indicating the ability to kill.

Colony-forming unit   A colony on an agar plate that in theory arises from a single bacterial cell. Abbreviated CFU.

Commercial sterility   A level of sterility that indicates that an item is free of organisms that can cause illness.

Compatible solutes   "Harmless" compounds which cells accumulate to equilibrate their internal water activity with the water activity of the environment.

Cortex   The spore structure responsible for resistance properties, presumably through dehydration of the core.

Cucurbits   A group of plants that includes squashes, cucumbers, and pumpkins.

Cytokines   Substances or compounds that are produced and secreted by cells of the immune system. These substances play a significant role in the immune response.

Cytotoxic   Lethal to cells.

Differential media   Media that allow specific bacteria to be visualized (e.g., through a color reaction) among a population of other bacteria.

Disulfide linkage   A covalent bond formed between two cysteine molecules in different parts of a protein; serves to stabilize the protein shape.

Emetic toxin   A toxin that causes vomiting.

Endospore   A bacterial spore formed in the body of the mother cell.

Endotoxin   A toxin structurally associated with the cell.

Enterocolitis   Inflammation of the lining of the intestine.

Enterotoxin   A toxin that acts in the gastrointestinal tract.

Environmental Protection Agency   A federal government agency that regulates food-related issues, such as sanitizer efficacy and pesticides. Abbreviated EPA.

EPA   *See* Environmental Protection Agency.

Epidemiology   The study of epidemics; it is used to determine the factors that lead to an outbreak of foodborne illness.

Exotoxin   An excreted toxin.

Extrinsic factor   An external factor, such as temperature or atmosphere, that influences the ability of microbes to grow in a food.

Facultative   Having the ability to do something that is not the preferred mode. For example, a facultative anaerobe can grow in the absence of oxygen but grows better in its presence.

FDA   *See* Food and Drug Administration.

Fecal-oral route   A route of disease transmission from fecal matter to the body via the oral cavity.

Fecal-oral transmission   The transmission of bacteria from one person's feces to another person's mouth, usually due to poor hand washing.

50% lethal dose   The concentration of a substance that will kill 50% of a population. Abbreviated LD_{50}.

Food and Drug Administration   The government agency that has legal authority over all foods except meat, poultry, eggs, and alcohol. It is part of the Department of Health and Human Services. Abbreviated FDA.

Food Safety and Inspection Service   The regulatory arm of the U.S. Department of Agriculture. It inspects all meat and poultry processing plants. Abbreviated FSIS.

Food safety objective   A quantitative goal for the frequency of a particular foodborne illness. Abbreviated FSO.

Frank pathogen   A bacterium that causes illness in healthy people.

FSIS   *See* Food Safety and Inspection Service.

FSO   *See* Food safety objective.

GAPs   *See* Good agricultural practices.

Gastroenteritis   Broadly speaking, a disease or illness that originates in the gut.

Generally recognized as safe   A legal classification of food additives in use before 1958; it includes additives affirmed as safe since that time. Abbreviated GRAS.

Genetic fingerprinting   A nucleic acid-based technique that provides specific identification (a "fingerprint") of a microorganism.

Germinant   A compound that induces spore germination.

Germination   The first irreversible step in the process by which a spore becomes a vegetative cell.

Glyco-   A prefix meaning "containing a sugar."

GMPs   *See* Good manufacturing practices.

Good agricultural practices   Prescribed practices, such as the use of potable water for rinses, prohibition against fertilizing with human manure, and good worker hygiene, that help ensure the microbial safety of food at the farm level. Abbreviated GAPs.

Good manufacturing practices   Prescribed practices, such as rodent control programs, use of hair nets and gloves, and prohibition of jewelry, that help ensure the safety of food in food-processing plants. Abbreviated GMPs.

GRAS   *See* Generally recognized as safe.

HACCP   *See* Hazard Analysis Critical Control Point.

Halotolerant   Able to tolerate high salt concentrations.

Hazard Analysis Critical Control Point   A proactive, prevention-oriented program that addresses food safety through the analysis and control of biological, chemical, and physical hazards. Abbreviated HACCP.

Hemolysin   A compound that causes lysis of red blood cells.

Hemolytic   Able to break open red blood cells.

Hemorrhagic colitis   A disease characterized by bloody diarrhea.

Hermetically sealed   Sealed under a vacuum.

Heterofermentative   Forming lactic acid, acetic acid, and ethanol as fermentation products. Also called heterolactic.

Homeostasis   An attempt by a microorganism to maintain a constant intracellular state, e.g., maintenance of pH.

Homofermentative   Forming only lactic acid as a fermentation product. Also called homolactic.

Host   An organism that provides an environment for a second organism to live in. In the case of foodborne pathogens, the host is the victim.

Immunocompromised   Having an immune system that is unable to combat normal disease processes.

Inoculum (pl., inocula)   The bacteria initially present that initiate growth.

Intrinsic factor   A property, such as pH or water activity, inherent in a food.

Isoelectric point   The pH at which a protein has no net charge.

Isolate   A strain of bacteria obtained ("isolated") from a specific site. Strains of the same species that are environmental or clinical isolates can be quite different.

kGy   *See* Kilogray.

Kilogray   A unit of absorbed radiation equal to 1 joule of energy. Abbreviated kGy.

LD$_{50}$ *See* 50% lethal dose.

Low-acid food   A food with a pH of >4.6 and a water activity of >0.85.

Lyse   To break open.

Lysis   The breakage of cells.

Lysozyme   An enzyme that degrades cell walls.

MAP   *See* Modified-atmosphere packaging.

Meningitis   Inflammation of the tissue (meninges) surrounding the brain and spinal cord.

Mesophile   An organism with an optimal growth range of 20 to 45°C.

4-Methylumbelliferyl-β-D-glucuronide   A substrate used to determine production of β-D-glucuronidase by *Escherichia coli* O157:H7. Abbreviated MUG.

Modified-atmosphere packaging   Storage or packing of foods in increased amounts of CO_2. Abbreviated MAP.

Monoclonal antibody   An antibody that detects a specific cell target.

Most probable number   A statistical method for estimating small populations of bacteria. Abbreviated MPN.

MPN   *See* Most probable number.

MUG   *See* 4-Methylumbelliferyl-β-D-glucuronide.

Mycotoxins   A generic term for the chemical toxins formed by fungi.

Necrotic   Dead.

Neurotoxin   A toxin that acts on nerves.

New-variant Creutzfeldt-Jakob disease   A human neurological disease caused by prions. Abbreviated nvCJD.

nvCJD   *See* New-variant Creutzfeldt-Jakob disease.

Opportunistic pathogen   A bacterium that causes illness only in people with some preexisting medical condition.

Osmotolerant   Tolerant of conditions (i.e., high salt or sugar concentrations) where there is little available water (low water activity).

Outgrowth   The process by which a germinated spore becomes a vegetative cell.

Pandemic   Epidemic over an especially wide geographic area.

Pasteurization   A heat process designed to kill pathogens but not necessarily spoilage organisms.

Plasmid   A piece of circular DNA, separate from the chromosome, that contains genes for traits such as virulence, toxins, bacteriocin production, and various biochemical traits. Plasmids can be transmitted among different species of bacteria.

Plasmolysis   Shrinkage of a cell due to water loss.

Poliomyelitis   A degenerative disease of the muscles that causes paralysis. Also called polio.

Polyclonal antibody   An antibody that detects many cellular targets.

Pomaceous fruits   A group of fruits that includes apples and pears.

Prion   An unusual type of protein which, by changing shape, causes transmissible encephalopathies.

Prostration   An extreme state of exhaustion.

Proteolytic   Producing enzymes that degrade proteins.

Pseudoappendicular syndrome   A syndrome in which, in broad terms, an individual experiences symptoms commonly associated with appendicitis.

Psychrophile   An organism that "loves" to grow in the cold, has an optimum growth temperature of 15°C, and cannot grow at 30°C.

Psychrotroph   An organism that can grow in the cold but has an optimum growth temperature of >20°C and can grow at 30°C.

Ptosis   A drooping effect, generally of the eyelids, in a case of foodborne illness.

Ready-to-eat food   A food that can be eaten without further cooking, such as deli meats.

Reiter's syndrome   An autoimmune disease characterized by arthritis and inflammation around the eye.

rpoS **genes**   Genes that encode DNA-dependent RNA polymerase.

Saprophyte   An organism that survives by living off dead or decaying plant material.

SASP   *See* Small acid-soluble proteins.

SEA   *See* Staphylococcal enterotoxin A.

Selective media   Media that select for the growth of specific bacteria by inhibiting the growth of other bacteria that may be present.

Septicemia   A gross, whole-body infection.

Serological   Able to cause an antibody response.

Serotypes   Varieties of a bacterial species that respond to different antibodies.

Sigma factor   A protein that binds to a DNA-dependent RNA polymerase.

Small acid-soluble proteins   Spore proteins that confer resistance properties. Abbreviated SASP.

Sporadic   Occurring randomly.

Sporulation   The process by which a vegetative cell forms and releases a spore.

Staphylococcal enterotoxin A   One serological type of staphylococcal enterotoxin. Abbreviated SEA.

-static   A suffix indicating the ability to inhibit or stop something.

Temperature abuse   The holding of food at temperatures that permit microbial growth, i.e., 40 to 140°F.

Tenesmus   The sensation of an urgent need to defecate while being unable to do so.

Thermophile   An organism that grows at high temperatures.

Transmissible spongiform encephalopathy   A disease, such as scrapie, kuru, or "mad cow disease," that is caused by prions. Abbreviated TSE.

TSE   *See* Transmissible spongiform encephalopathy.

Turkey "X" disease   A disease of turkeys for which the causal agent was not initially known.

USDA   *See* U.S. Department of Agriculture.

U.S. Department of Agriculture   A cabinet level government department that has legal authority over meat, poultry, and eggs. Abbreviated USDA.

Vehicle   A source or carrier.

Viable but nonculturable   A term applied to cells that cannot be cultured by conventional methods but that still cause illness if ingested. Abbreviated VNC.

Virulent   Causing illness.

VNC   *See* Viable but nonculturable.

Water activity   The measure of water available for microbial growth and chemical reactions, defined as the equilibrium relative humidity of a product. Abbreviated a_w.

Xerotolerant   Capable of tolerating dry conditions.

Crossword Puzzle Answers

Chapter 17

Across
1. bad
4. *anthracis*
5. diarrheal
6. grow
8. tenesmus
10. soil
13. rice
14. thousand
16. eye
17. pasta
18. pie
19. kills

Down
1. *Bacillus*
2. dairy
3. enterotoxin
6. gram
7. emetic
9. spore
11. Otto
12. lots
14. temp
15. drank
16. erops

Chapter 18

Across
1. bacteriophage
5. ATP
6. fermentation
9. nitrite
12. homofermen-
tative
14. raid
15. eyes
18. oxygen
19. seed
20. yeast

Down
2. acidification
3. carbohydrates
4. vinegar
7. cit
8. pentoses
10. proteins
11. *Lactococcus*
13. oxidized
16. sweats
17. lytic

Chapter 24

Across
1. Xerox
3. fee
5. *D* value
7. freeze
8. hydro
9. spore
10. rain
12. ORP
14. pasteurization
17. radiation
18. HUS
19. vent
21. AM
22. torr
23. microwave

Down
1. xero
2. xerotolerant
3. freeze drying
4. energy
5. dehydration
6. add
9. SOS
11. nonthermal
13. pressure
15. activity
16. ziti
20. pico
21. a_w
23. MW

Index

Pressure resistance, spores, 38
Primers, PCR, 60–61
Prions, 289, 294–296
  biology, 290, 296
  history, 294–296
  protein morphing, 295–296
Probiotic bacteria, 7, 316–318
  commercialization, 317
  health benefits and mechanisms, 317
  human GI tract as microbial ecosystem,
    318
Processed products, muscle food spoilage,
  247
Processing and export genes, 315
Produce
  E. coli O157:H7, 120–121
  modified-atmosphere packaging, 263
  spoilage, 258–266
    effect of physiological state, 260–261
    fungal, 259
    mechanism, 260
    microflora, 261–265
    types, 258–260
Propionibacterium freudenreichii, 225–226
Propionibacterium shermanii, 312
Propionic acid, 258, 303
Propyl paraben, 305
Proteins
  primary structure, 296
  protein morphing, 295–296
  secondary structure, 296
  synthesis, 313
Proteolysis
  aroma compounds in cheese, 227
  spoilage of muscle foods, 245–246
Protocatechuic acid, 309
Prusiner, Stanley, 294–296
Pseudoappendicular syndrome, 129, 132–133
Pseudomonas
  Entner-Doudoroff pathway, 24
  preventing growth of, 304, 327
  rapid methods, 59
  spoilage organisms, 242–246, 248–249,
    252–254, 262
Pseudomonas aeruginosa, 262
Pseudomonas fluorescens, 252, 254, 262, 309,
  334
Pseudomonas fragi, 252, 254
Pseudomonas lundensis, 252
Pseudomonas maltophila, 262
Pseudomonas mendocina, 262
Pseudomonas picketti, 262
Pseudomonas putida, 252, 262
"*Pseudomonas putrefaciens*," 70, 262
Pseudomonas stutzeri, 262
Psychrophiles, 20–21
Psychrotrophic clostridia, 35
Psychrotrophic spoilage
  control of product defects, 254–255
  growth of organisms, 253
  lipases in, 253–254
  milk and dairy products, 252–255
  product defects, 253

sources of psychrotrophs in milk, 252
  vegetables, 262
Psychrotrophs, 20–21
  sporeformers, 29
Public Health Security and Bioterrorism Pre-
  paredness and Response Act, 189
Pulsed electric field process, food preserva-
  tion, 341
Pure culture, 6, 12
Putrefactive anaerobes, 35
Pythium, 259

Q
Quality control sanitation programs, 347
Quantitative risk assessment techniques, 69
Quaternary ammonium sanitizers, 351
Quorum sensing, 8, 179–180

R
Radappertization, 339
Radiation lethality, 338
Radiation tolerance, *see* Irradiation tolerance
Radura symbol, 340
Radurization, 339
Rancidity, 245, 254
Rapid methods, 53–63
  based on traditional methods, 56–57
  detection limits, 54
  immunologically based, 57–59
  molecular methods, 60–62
  requirements and validation, 54–56
  sensitivity, 55
  specificity, 55
Ready-to-eat foods, *L. monocytogenes*, 162
Real-time PCR, 60
Redigel system, 58
Redox potential (Eh), 49
Redox potential effect
  C. botulinum, 193–194
  C. perfringens, 206–207
Reductase tests, 50
Refrigeration, 5, 311, 325–327
Regulatory gene, 315
Regulatory status, *L. monocytogenes*, 165–167
Repair, injured microbes, 18
REPFEDs, 330
Replicate organism direct area contact test,
  50
Reservoirs
  B. cereus, 214
  C. botulinum, 195–196
  C. perfringens, 208
  Campylobacter, 102–104
  E. coli O157:H7, 116–118
  S. aureus, 180–181
  Salmonella, 93–94
  Shigella, 144
  V. cholerae, 150
  V. parahaemolyticus, 152
  V. vulnificus, 154
  Y. enterocolitica, 133–134
Respiration, 12
Reveal tests, 59

Reverse-transcriptase PCR, 60, 62
Rhizoctonia, 286
Rhizopus
  preventing growth of, 309
  spoilage organisms, 244, 259–260
  toxicity, 286
Rhizopus chinensis, 237
Rhizopus oligosporus, 237
Rhizopus oryzae, 237
Rhodotorula, 309
Rhodotorula mucilaginosa, 258
Rhodotorula splutinis, 262
Riboprinter, 62
Ribotyping, 62
Risk assessment techniques, 78
  quantitative, 69
Roll tube, 48
Ropiness, bread, 266
Roquefortine, 284
Rosemary, 308
Rotaviruses, 75, 293

S
Saccharomyces
  preventing growth of, 306, 309
  probiotic species, 316
Saccharomyces boulardii, 316
Saccharomyces carlsbergensis, 234
Saccharomyces cerevisiae
  beer making, 234
  bread making, 232
  preventing growth of, 306
  wine making, 235
Saccharomyces ludwigii, 306
SafePath tests, 59
Sage, 308
Sakacins, 313
Salmonella, 85–99
  antibiotic resistance, 95
  biochemical identification, 88
  characteristics of disease
    preventative measures, 94–95
    symptoms and treatment, 94
  characteristics of organism, 88–90
  cytotoxic protein, 98
  discovery, 6
  enterotoxin, 98
  infectious dose, 95–96
  microbiological criteria, 77–79
  modified-atmosphere foods, 93
  outbreaks, 85–87
  pathogenicity
    attachment and invasion, 97
    growth and survival in host cells, 97
    specific and nonspecific human re-
      sponses, 96
  pH tolerance, 19–20, 90–92
  physiology, 90–93
  plasmids, 97–98
  preventing growth of, 308–309
  rapid methods, 56, 58–60
  reservoirs, 93–94
  salt tolerance, 90, 92

Test Yourself

Intermediate Algebra

Joan Van Glabek, Ph.D.
Department of Mathematics
Edison Community College
Naples, FL

Contributing Editors

Carl E. Langenhop, Ph.D.
Emeritus Professor of Mathematics
Southern Illinois University—Carbondale
Carbondale, IL

Charles M. Jones, M.S.
Department of Mathematics
East Texas State University
Commerce, TX

Tony Julianelle, Ph.D.
Department of Mathematics
University of Vermont
Burlington, VT

NTC LEARNINGWORKS
a division of NTC Publishing Group
Lincolnwood, Illinois

Library of Congress Cataloging-in-Publication Data
is available from the Library of Congress.

A *Test Yourself Books, Inc.* Project

6 7 8 9 ML 0 9 8 7 6 5 4 3 2 1

Contents

Preface

These test questions and answers have been written to help you more fully understand the material in an intermediate algebra course. Working through these test questions prior to a test will help you pinpoint areas that need further study. Each answer is followed by a parenthetical topic description that corresponds to a section in your intermediate algebra textbook. Return to your textbook as needed to reread and review areas where you feel unsure about your work. See your professor for additional help, and/or get together with a study group to complete your preparation for each test. Work some problems from the test book *every* day. This will allow you time to properly prepare for your tests.

Practice and a positive attitude (just keep telling yourself how well you are doing) will help you succeed. Good luck with your study of algebra.

Joan Van Glabek, Ph.D.

How to Use this Book

This "Test Yourself" book is part of a unique series designed to help you improve your test scores on almost any type of examination you will face. Too often, you will study for a test—quiz, midterm, or final—and come away with a score that is lower than anticipated. Why? Because there is no way for you to really know how much you understand a topic until you've taken a test. The *purpose* of the test, after all, is to test your complete understanding of the material.

The "Test Yourself" series offers you a way to improve your scores and to actually test your knowledge at the time you use this book. Consider each chapter a diagnostic pretest in a specific topic. Answer the questions, check your answers, and then give yourself a grade. Then, and only then, will you know where your strengths and, more importantly, weaknesses are. Once these areas are identified, you can strategically focus your study on those topics that need additional work.

Each book in this series presents a specific subject in an organized manner, and although each "Test Yourself" chapter may not correspond to exactly the same chapter in your textbook, you should have little difficulty in locating the specific topic you are studying. Written by educators in the field, each book is designed to correspond, as much as possible, to the leading textbooks. This means that you can feel confident in using this book, and that regardless of your textbook, professor, or school, you will be much better prepared for anything you will encounter on your test.

Each chapter has four parts:

 Brief Yourself. All chapters contain a brief overview of the topic that is intended to give you a more thorough understanding of the material with which you need to be familiar. Sometimes this information is presented at the beginning of the chapter, and sometimes it flows throughout the chapter, to review your understanding of various *units* within the chapter.

 Test Yourself. Each chapter covers a specific topic corresponding to one that you will find in your textbook. Answer the questions, either on a separate page or directly in the book, if there is room.

 Check Yourself. Check your answers. Every question is fully answered and explained. These answers will be the key to your increased understanding. If you answered the question incorrectly, read the explanations to *learn* and *understand* the material. You will note that at the end of every answer you will be referred to a specific subtopic within that chapter, so you can focus your studying and prepare more efficiently.

 Grade Yourself. At the end of each chapter is a self-diagnostic key. By indicating on this form the numbers of those questions you answered incorrectly, you will have a clear picture of your weak areas.

There are no secrets to test success. Only good preparation can guarantee higher grades. By utilizing this "Test Yourself" book, you will have a better chance of improving your scores and understanding the subject more fully.

The Real Number System

 Test Yourself

Basic Definitions

One of the steps you may need to solve word problems requires translating statements into symbols. You should recognize the following symbols and understand what each means. More than one phrase can be used to imply an operation. Some of these variations are present in the list.

+	Add, find the sum, find the total, increase
−	Subtract, find the difference, less than
·	Multiply, find the product
÷	Divide, find the quotient
=	Equal, is equal to, any form of "to be"
≠	Not equal
>	Greater than
<	Less than
≥	Greater than or equal to
≤	Less than or equal to

Translate each statement into symbols in 1 through 5.

1. The sum of a number n and 5 is less than or equal to 6.

2. Five less than a number n is equal to 12.

3. The product of 3 and a number x is 15.

4. The quotient of a number x and 8 is greater than 4.

5. The sum of x squared and 2 is 10.

Identify the inequality as true or false in 6 through 8.

6. $-8 < 6$

7. $3 \geq 3$

8. $-5 > -2$

Order of Operations

The following order of operations holds when performing multiple operations:

Simplify inside **parentheses**, working from the innermost set of grouping symbols out.

Simplify any terms involving **exponents**.

Perform **multiplications** and **divisions** in order from left to right.

Perform **additions** and **subtractions** in order from left to right.

It may help you to remember this as **PEMDAS.**

Simplify in 9 through 14.

9. $2 + 7(8 - 3)$

10. $4 - 16 \div 4 + 6^2$

11. $-8 + (-4) \div 2 - 12$

12. $4 + 2[8 - 3(2 - 5)]$

13. $2 \cdot 4^2 - 3 \cdot 2^3$

14. $-4 + 8(2) - 24 \div (-3)$

Sets

The following symbols are used when working with sets:

$A \cup B$ Union. A union B is the set that contains all elements in set A *or* set B.

$A \cap B$ Intersection. A intersection B is the set that contains all elements found in set A *and* set B.

$A \subseteq B$ Subset. A is a subset of B if every element of set A is a member of set B.

$A \not\subset B$ Not a subset. A is not a subset of B if some element of A is not a member of set B.

$a \in A$ Element of. a is an element of set A.

$a \notin B$ Not an element of. a is not an element of B.

Let $A = \{-3, -2, -1, 0, 1\}$, $B = \{-2, 0, 2\}$,
$C = \{-1, 0, 1, 2\}$, and $D = \{3\}$.
Find each set in 15 through 19.

15. $B \cup C$

16. $A \cap B$

17. $\{x \mid x \in A \text{ and } x \geq -1 \}$

18. $\{x \mid x \in A \text{ and } x \notin C\}$

19. $C \cap D$

Sets of Numbers

There are several sets of numbers that are used extensively in algebra. The following symbols describe those sets:

$N = \{x \mid x \text{ is a natural number}\} = \{1, 2, 3, \ldots\}$
$W = \{x \mid x \text{ is a whole number}\} = \{0, 1, 2, 3, \ldots\}$
$I = \{x \mid x \text{ is an integer}\}$
  $= \{\ldots -3, -2, -1, 0, 1, 2, 3, \ldots\}$
$Q = \{x \mid x \text{ is a rational number}\}$
  $= \{a/b \mid b \neq 0, a \text{ and } b \text{ integers}\}$
$H = \{x \mid x \text{ is an irrational number}\}$
$\Re = \{x \mid x \text{ is a real number}\} = Q \cup H$

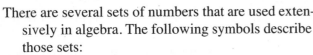

List the numbers in the set $\{-5, -\sqrt{3}, 0, \dfrac{3}{4}, \sqrt{2}, 3.14,$

$\pi, 4\}$ that belong to each of the following sets.

20. natural numbers

21. whole numbers

22. integers

23. rational numbers

24. irrational numbers

25. real numbers

Place $\subseteq$ or $\not\subset$ in each blank to make a true statement in 26 through 36.

26. $\{1, 2, 3\}$ _____ $\Re$

27. $\{\sqrt{2}, \sqrt{5}, \sqrt{7}\}$ _____ Q

28. $\{-\sqrt{2}, 0, 0.555\ldots\}$ _____ H

29. $\{0\}$ _____ N

30. $\{-1, 0, 1\}$ _____ I

31. $\{-\sqrt{5}, \sqrt{3}\}$ _____ $\Re$

32. $\{1/3, 2.\bar{5}\}$ _____ H

33. $\Re$ _____ Q

34. W _____ I

35. W _____ Q

36. Q _____ H

The **additive inverse** of a number a is $-a$.

37. State the additive inverse of 4.

38. State the additive inverse of -8.

39. State the additive inverse of 0.

The **absolute value** of a number is that number's distance from 0; that is,
$|a| = a$ if $a \geq 0$ and $|a| = -a$ if $a < 0$.

---

40. Evaluate: $|-6|$

41. Evaluate: $-|3|$

42. Evaluate: $-|-2|$

43. Evaluate: $|-5| + |-7|$

Graphing Inequalities of the Form $x \,\square\, a$

To graph inequalities on a number line, use shading to show all the numbers that satisfy the inequality.
If the inequality is of the form $x < a$ or $x \leq a$, shade to the left of a on the number line.
If the inequality is of the form $x > a$ or $x \geq a$, shade to the right of a on the number line.

Graphing Inequalities of the Form $x \square a$ (cont.)
If the inequality contains < or >, use a parentheses on a.
If the inequality contains ≤ or ≥, use a bracket on a.

44. Graph $\{x \mid x < 2\}$

45. Graph $\{x \mid x \geq -1\}$

46. Graph $\{x \mid x < -2 \text{ or } x \geq 0\}$

47. Graph $\{x \mid x > -3 \text{ and } x < 2\}$

48. Graph $\{x \mid x < 2 \text{ or } x > -1\}$

49. Graph $\{x \mid x \leq 3 \text{ and } x \geq 5\}$

Properties of Real Numbers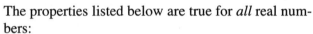

The properties listed below are true for *all* real numbers:

Commutative Property of Addition	$a + b = b + a$
Commutative Property of Multiplication	$ab = ba$
Associative Property of Addition	$a + (b + c) =$ $(a + b) + c$
Associative Property of Multiplication	$a(bc) = (ab)c$
Additive Identity Property	$a + 0 = a = 0 + a$
Multiplicative Identity Property	$a(1) = a = 1(a)$
Additive Inverse Property	$a + (-a) = 0$
Multiplicative Inverse Property	$a\left(\dfrac{1}{a}\right) = 1$ for $a \neq 0$
Distributive Property	$a(b + c) =$ $ab + ac$

State the property that justifies each statement.

50. $2 + (-8) = (-8) + 2$

51. $4(x + 2) = 4x + 8$

52. $2(4xy) = (8)xy$

53. $6 + (-6) = 0$

54. $1x = x$

Use the Distributive Property to simplify:

55. $3(2x + 7)$

56. $\dfrac{1}{3}(9x + 6)$

57. $5(2x + 3y - 8)$

Operations with Real Numbers

To add real numbers:
To add real numbers with the same sign, add their absolute values and use the common sign.
To add two real numbers with different signs, find the absolute value of each number, subtract the smaller absolute value from the larger absolute value and use the sign of the number with the larger absolute value.

To subtract real numbers:
Change the subtraction sign to an addition sign and use the additive inverse of the number being subtracted; that is, $a - b = a + (-b)$. Then follow the rules for adding real numbers.

Find the sum:

58. $-6 + (-8)$

59. $9 + (-3)$

60. $5 + (-11)$

61. $-\dfrac{2}{3} + \left(-\dfrac{1}{4}\right)$

62. $-5.402 + 2.31$

Find the difference:

63. $10 - (-4)$

64. $-7 - 5$

65. $-6 - (-15)$

66. $\dfrac{1}{4} - \dfrac{3}{5}$

67. $-\dfrac{2}{3} - \left(-\dfrac{1}{2}\right)$

68. $-1.4 - 2.58$

To multiply real numbers:
Multiply the absolute values of the numbers and
count the number of negative signs.
If the number of negative signs is an odd number, the
product is negative.
If the number of negative signs is an even number,
the product is positive.
To divide two real numbers:
Divide the absolute values of the numbers.
If the signs of the numbers are the same, the quotient
is positive.
If the signs of the numbers are different, the quotient
is negative.

Perform the indicated operations.

69. $(-8)(-3)$

70. $(-16) \div (-4)$

71. $(-4)(-2)(-5)$

72. $\left(-\dfrac{1}{4}\right) \cdot \left(\dfrac{2}{3}\right)$

73. $\left(\dfrac{5}{7}\right) \div \left(-\dfrac{3}{14}\right)$

Simplify.

74. $(-6)(-3) + 8(-4)$

75. $(-2)^3 - 6^2$

76. $\dfrac{3(-5) + 6(-4)}{3 - 2 \cdot 4}$

77. $4|2 - 5| - 3|6 - 1|$

78. $(-4^2) \div 2 + (-3)(-5)$

Check Yourself

1. $n + 5 \leq 6$ "The sum of n and 5" is written as $n + 5$, and "less than or equal to" is written $\leq$.
    (Basic definitions)

2. $n - 5 = 12$ Be careful with "5 less than"—it means "$- 5$" not "5 $-$". **(Basic definitions)**

3. $3x = 15$ Product means multiply. **(Basic definitions)**

4. $\dfrac{x}{8} > 4$ Quotient means divide. **(Basic definitions)**

5. $x^2 + 2 = 10$. **(Basic definitions)**

6. $-8 < 6$ is true. Since -8 is to the left of 6 on the number line, this is a true statement. **(Basic definitions)**

7. $3 \geq 3$ is true. The symbol $\geq$ means greater than *or* equal to. **(Basic definitions)**

8. $-5 > -2$ is false. Since -5 is to the left of -2 on the number line, $-5 < -2$. **(Basic definitions)**

9. $2 + 7(8 - 3)$

    $= 2 + 7(5)$ Simplify inside parentheses first.

    $= 2 + 35$ Multiplication is done before addition.

    $= 37$ **(Order of operations)**

10. $4 - 16 \div 4 + 6^2$

    $= 4 - 16 \div 4 + 36$ Exponents are simplified first.

$= 4 - 4 + 36$ Division is done before addition and subtraction.

$= 36$ Subtract and add in order from left to right. (**Order of operations**)

11. $-8 + (-4) \div 2 - 12$

$= -8 + (-2) - 12$ Division is done first.

$= -22$ Add and subtract in order from left to right. (**Order of operations**)

12. $4 + 2[8 - 3(2 - 5)]$

$= 4 + 2[8 - 3(-3)]$ Simplify innermost grouping symbols first.

$= 4 + 2[8 + 9]$ Simplify inside brackets.

$= 4 + 2[17]$

$= 4 + 34$ Multiply before adding.

$= 38$ (**Order of operations**)

13. $2 \cdot 4^2 - 3 \cdot 2^3$

$= 2 \cdot 16 - 3 \cdot 8$ Simplify exponents.

$= 32 - 24$ Multiply before subtracting.

$= 8$ (**Order of operations**)

14. $-4 + 8(2) - 24 \div (-3)$

$= -4 + 16 + 8$ Perform multiplications and divisions in order from left to right.

$= 20$ Add from left to right. (**Order of operations**)

15. $B \cup C = \{-2, -1, 0, 1, 2\}$ $\cup$ is the symbol for union and means list all the elements in set *B or C*.

Note that we do not repeat elements, so that even though 2 appears in both *B* and *C*, we only list it once in the union. (**Sets**)

16. $A \cap B = \{-2, 0\}$ $\cap$ is the symbol for intersection and means list all the elements in *A and B*. (**Sets**)

17. $\{x \mid x \in A \text{ and } x \geq -1 \} = \{-1, 0, 1\}$ This set builder notation says to list all elements in set *A* that are greater than of equal to -1. (**Sets**)

18. $\{x \mid x \in A \text{ and } x \notin C\} = \{-3, -2\}$ This set builder notation says to list all elements in set *A* that *are not* in set *C*. (**Sets**)

19. $C \cap D = \{ \} = \emptyset$ There are no elements in set *C* that are also in set *D*, therefore the intersection is the empty set, which can be written as $\{ \}$ or $\emptyset$. (**Sets**)

20.   4   (**Sets of numbers**)

21. $0, 4$ Note that the set of whole numbers includes the set of natural numbers. (**Sets of numbers**)

22. $-5, 0, 4$ Note that the set of integers includes the set of whole numbers. (**Sets of numbers**)

23. $-5, 0, \frac{3}{4}, 3.14, 4$ The rational numbers include all the integers and all terminating and repeating decimals. **(Sets of numbers)**

24. $-\sqrt{3}, \sqrt{2}, \pi$ The irrational numbers include all nonrepeating, nonterminating decimals. **(Sets of numbers)**

25. All the numbers listed are real numbers. Note that the real numbers include rational numbers and irrational numbers. **(Sets of numbers)**

26. $\{1, 2, 3\} \subseteq \Re$ **(Sets of numbers)**

27. $\{\sqrt{2}, \sqrt{5}, \sqrt{7}\} \not\subset Q$ These square roots cannot be written as nonrepeating, nonterminating decimals and therefore are not rational numbers. **(Sets of numbers)**

28. $\{-\sqrt{2}, 0, 0.555\ldots\} \not\subset H$ 0 and $0.555\ldots$ are rational numbers, not irrational numbers. **(Sets of numbers)**

29. $\{0\} \not\subset N$ The natural numbers do not include 0. **(Sets of numbers)**

30. $\{-1, 0, 1\} \subseteq I$ The integers contain both positive and negative whole numbers. **(Sets of numbers)**

31. $\{-\sqrt{5}, \sqrt{3}\} \subseteq \Re$ Both the given numbers are irrational numbers. Since the real numbers contain both rational and irrational numbers, the given set is a subset of the real numbers. **(Sets of numbers)**

32. $\{1/3, 2.\bar{5}\} \not\subset H$ Both 1/3 and $2.\bar{5}$ are rational numbers. **(Sets of numbers)**

33. $\Re \not\subset Q$ The rational numbers are a subset of the real numbers, not vice versa. **(Sets of numbers)**

34. $W \subseteq I$ The whole numbers are contained in the set of integers. **(Sets of numbers)**

35. $W \subseteq Q$ Since every whole number can be written as a fraction (i.e., $5 = \frac{5}{1} = \frac{10}{2}$), the whole numbers are a subset of the rational numbers. **(Sets of numbers)**

36. $Q \not\subset H$ The rational and irrational numbers have no elements in common and therefore neither can be a subset of the other. **(Sets of numbers)**

37. -4 is the additive inverse of 4. **(Additive inverse)**

38. $-(-8) = 8$ so 8 is the additive inverse of -8. **(Additive inverse)**

39. The additive inverse of 0 is 0 since $-0 = 0$. **(Additive inverse)**

40. $|-6| = 6$ Absolute value means distance from 0, without regard to direction. Since -6 is 6 away from 0, the absolute value of -6 is 6; since $-6 < 0$, $|-6| = -(-6) = 6$. **(Absolute value)**

41. $-|3| = -(3) = -3$ Simplify the absolute value first, then take the opposite of the answer. **(Absolute value)**

42. $-|-2| = -(2) = -2$ Simplify absolute value first, then take the opposite of the answer. **(Absolute value)**

43. $|-5| + |-7| = 5 + 7 = 12$ Simplify each absolute value first, then add. **(Absolute value)**

44.   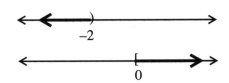   Use) at 2 to indicate that 2 is not included in the solution.

(Graphing inequalities)

45.      Use [at −1 to indicate that −1 is included in the solution.

(Graphing inequalities)

46.   Graph each inequality:

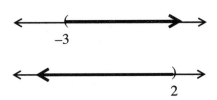

Since "or" means union, we include all numbers already shaded in each of the graphs drawn:

   **(Graphing inequalities)**

47.   Graph each inequality:

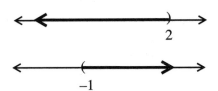

Since "and" means intersection, the answer includes only those numbers that were shaded on both graphs:

(Graphing inequalities)

48.   Graph each inequality:

Since "or" means union, the answer includes all numbers already shaded, which in this case is all the real numbers:

   **(Graphing inequalities)**

49.   Graph each inequality:

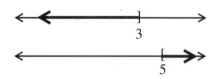

Since "and" means intersection, we look for numbers shaded in both graphs—and there are none. Therefore the solution is Ø. **(Graphing inequalities)**

50. $2 + (-8) = (-8) + 2$ Commutative Property of Addition **(Properties of real numbers)**

51. $4(x + 2) = 4x + 8$ Distributive Property **(Properties of real numbers)**

52. $2(4xy) = (8)xy$ Associative Property of Multiplication **(Properties of real numbers)**

53. $6 + (-6) = 0$ Additive Inverse Property **(Properties of real numbers)**

54. $1x = x$ Multiplicative Identity Property **(Properties of real numbers)**

55. $3(2x + 7) = 3(2x) + 3(7)$ Multiply 3 times each term in parentheses.

$$= 6x + 21 \textbf{ (Properties of real numbers)}$$

56. $\frac{1}{3}(9x + 6) = \frac{1}{3}(9x) + \frac{1}{3}(6)$ Multiply 1/3 times each term in parentheses.

$$= 3x + 2 \textbf{(Properties of real numbers)}$$

57. $5(2x + 3y - 8) = 5(2x) + 5(3y) + 5(-8)$ Multiply 5 times each term in parentheses.

$$= 10x + 15y - 40 \textbf{ (Properties of real numbers)}$$

58. $-6 + (-8) = -14$ Signs are the same, so add their absolute values and use the common sign. **(Operations with real numbers)**

59. $9 + (-3) = 6$ Signs are different, so subtract the absolute values and use the sign of the number with the larger absolute value. **(Operations with real numbers)**

60. $5 + (-11) = -6$ **(Operations with real numbers)**

61. $-\frac{2}{3} + \left(-\frac{1}{4}\right) = -\frac{8}{12} + \left(-\frac{3}{12}\right)$ To add fractions, write each fraction over the least common multiple, 12.

$$= -\frac{11}{12}$$ Signs are the same, so add their absolute values and use the common sign.

(Operations with real numbers)

62. $-5.402 + 2.31 = -3.092$ Signs are different, so subtract the absolute values and use the sign of the number with the larger absolute value. **(Operations with real numbers)**

63. $10 - (-4) = 10 + (+4)$ The additive inverse of –4 is +4.

$= 14$ Signs are the same, add their absolute values and use the common sign.

(Operations with real numbers)

64. $-7 - 5 = -7 + (-5)$ The additive inverse of 5 is –5.

$= -12$

(Operations with real numbers)

65. $-6 - (-15) = -6 + (+15)$ The additive inverse of –15 is 15.

$= 9$ Signs are different, so subtract the absolute values and use the sign of the number with the larger absolute value. **(Operations with real numbers)**

66. $\dfrac{1}{4} - \dfrac{3}{5} = \dfrac{1}{4} + \left(-\dfrac{3}{5}\right)$ The additive inverse of $\dfrac{3}{5}$ is $-\dfrac{3}{5}$.

$= \dfrac{5}{20} + \left(-\dfrac{12}{20}\right)$ Write each fraction as an equivalent fraction over 20.

$= -\dfrac{7}{20}$ Signs are different, so subtract the absolute values and use the sign of the number with the larger absolute value. **(Operations with real numbers)**

67. $-\dfrac{2}{3} - \left(-\dfrac{1}{2}\right) = -\dfrac{2}{3} + \left(\dfrac{1}{2}\right)$ The additive inverse of $-\dfrac{1}{2}$ is $\dfrac{1}{2}$.

$= -\dfrac{4}{6} + \dfrac{3}{6}$ Write each fraction as an equivalent fraction over 6.

$= -\dfrac{1}{6}$

(Operations with real numbers)

68. $-1.4 - 2.58 = -1.4 + (-2.58)$ The additive inverse of 2.58 is –2.58.

$= -3.98$

(Operations with real numbers)

69. $(-8)(-3) = 24$ $8 \cdot 3 = 24$ and the number of negative signs is even, so the product is positive. **(Operations with real numbers)**

70. $(-16) \div (-4) = 4$ $16 \div 4 = 4$ and the number of negative signs is even, so the quotient is positive. **(Operations with real numbers)**

71. $(-4)(-2)(-5) = -40$ The number of negative signs is odd, so the product is negative. **(Operations with real numbers)**

72. $\left(-\dfrac{1}{4}\right) \cdot \left(\dfrac{2}{3}\right) = \dfrac{-1 \cdot 2}{4 \cdot 3}$ Multiply the numerators; multiply the denominators.

$\qquad\qquad = -\dfrac{1}{6}$ Reduce the answer to lowest terms. **(Operations with real numbers)**

73. $\left(\dfrac{5}{7}\right) \div \left(-\dfrac{3}{14}\right) = \left(\dfrac{5}{7}\right) \cdot \left(-\dfrac{14}{3}\right)$ To divide, invert and multiply.

$\qquad\qquad = \dfrac{5 \cdot -14}{7 \cdot 3}$ Multiply the numerators; multiply the denominators.

$\qquad\qquad = -\dfrac{10}{3}$ Reduce the answer to lowest terms. **(Operations with real numbers)**

74. $(-6)(-3) + 8(-4) = +18 + (-32)$ Perform multiplications in order from left to right.

$\qquad\qquad = -14$ Signs are different so subtract the absolute values and use the sign of the number with the larger absolute value. **(Operations with real numbers)**

75. $(-2)^3 - 6^2 = -8 - 36$ Simplify exponents.

$\qquad\qquad = -8 + (-36)$ The additive inverse of 36 is –36.

$\qquad\qquad = -44$ The signs are the same; add the absolute values and use the common sign.

(Operations with real numbers)

76. $\dfrac{3(-5) + 6(-4)}{3 - 2 \cdot 4} = \dfrac{-15 + (-24)}{3 - 8}$ Simplify the numerator; simplify the denominator.

$\qquad\qquad = \dfrac{-39}{-5} = \dfrac{39}{5}$ Numbers have the same sign, so the quotient is positive.

(Operations with real numbers)

77. $4|2 - 5| - 3|6 - 1| = 4|-3| - 3|5|$ Simplify inside absolute value bars.

$\qquad\qquad = 4(3) - 3(5)$ Evaluate the absolute values.

$\qquad\qquad = 12 - 15$ Multiply before subtracting.

$\qquad\qquad = -3$

(Operations with real numbers)

78. $(-4^2) \div 2 + (-3)(-5) = -16 \div 2 + (-3)(-5)$ Evaluate exponents first.

$\qquad\qquad = -8 + (+15)$ Perform multiplications and divisions in order from left to right.

$\qquad\qquad = 7$ Signs are different, so subtract the absolute values and use the sign of the number with the larger absolute value. **(Operations with real numbers)**

Grade Yourself

Circle the question numbers that you had incorrect. Then indicate the number of questions you missed. If you answered more than three questions incorrectly, you need to focus on that topic. (If a topic has less than three questions and you had at least one wrong, we suggest you study that topic also. Read your textbook, a review book, or ask your teacher for help.)

Subject: The Real Number System

Topic	Question Numbers	Number Incorrect
Basic definitions	1, 2, 3, 4, 5, 6, 7, 8	
Order of operations	9, 10, 11, 12, 13, 14	
Sets	15, 16, 17, 18, 19	
Sets of numbers	20, 21, 22, 23, 24, 25, 26, 27, 28, 29, 30, 31, 32, 33, 34, 35, 36	
Additive inverse	37, 38, 39	
Absolute value	40, 41, 42, 43	
Graphing inequalities	44, 45, 46, 47, 48, 49	
Properties of real numbers	50, 51, 52, 53, 54, 55, 56, 57	
Operations with real numbers	58, 59, 60, 61, 62, 63, 64, 65, 66, 71, 72, 73, 74, 75, 76, 77, 78	

Solving Linear Equations and Inequalities

Test Yourself

Solving Linear Equations in One Variable

To simplify an expression, combine similar terms (that is, terms with exactly the same variables and exponents). You may need to use the rules for order of operations and the distributive property to remove parentheses before you can combine all the similar terms.

Simplify:

1. $4x - 8x - 7 - 3$

2. $5y - 3(2y - 1)$

3. $4(6x - 3) - 8x + 4$

Addition and Multiplication Property of Equality

To help you isolate the variable, remember that you can add the same quantity to both sides of an equation, and you can multiply both sides of an equation by the same nonzero quantity

Solve:

4. $3x - 5 = 4$

5. $\frac{1}{2}x + 4 = 1$

6. $3x - 2 = 8 - x$

Solving Linear Equations

1. If the equation contains fractions, you may clear all the fractions by multiplying both sides of the equation by the least common multiple of all the denominators.
2. If the equation contains decimals, you may clear all the decimals by multiplying both sides of the equation by a power of 10.
3. Simplify the left side of the equation
4. Simplify the right side of the equation
5. Use addition (subtraction) to get all the terms with variables on one side of the equation and all the terms with constants on the other side of the equation.
6. Divide both sides of the equation by the coefficient of the variable.

Solve

7. $5 + 2(12 - x) = 2x - 7$

8. $4(2x - 5) + 3x = 3 - 2(x + 4)$

9. $\frac{y}{3} - \frac{2y}{5} = 2$

10. $0.2x - 0.6 = -0.4(x - 1) + 0.2$

Types of Solution Sets

If the process of solving a linear equation yields an equation with no variables and a true statement, (such as $0 = 0$), the solution set is the set of all real numbers.

If the process of solving a linear equation yields an equation with no variables and a false statement, (such as $0 = 1$), the solution set is the empty set, $\varnothing$.

Solve:

11. $3(x-1) = 6 + 3x$

12. $\dfrac{1}{2}(4x + 2) = 4x - 2(x-3) - 5$

Formulas

Evaluating Formulas

Formulas can be evaluated if you are given values for all but one of the variables in the formula. Always substitute given values into the formula using parentheses to help you prevent sign errors and order of operation errors.

13. Find A in $A = Prt$ if $P = 2000$, $r = 0.08$, and $t = 10$

14. Find y in $y = mx + b$ if $m = \dfrac{2}{3}$, $x = 6$, and $b = -4$

15. Find h in $A = \dfrac{1}{2}h(b_1 + b_2)$ if $A = 120$,

$b_1 = 12$, and $b_2 = 16$

Solving Formulas for a Variable

To solve a formula for a particular variable, use the same rules for solving any linear equation:

1. If the equation contains fractions, you may clear all the fractions by multiplying both sides of the equation by the least common multiple of all the denominators.
2. If the equation contains decimals, you may clear all the decimals by multiplying both sides of the equation by a power of 10.
3. Simplify the left side of the equation.
4. Simplify the right side of the equation.
5. Use addition (subtraction) to get all the terms with the desired variable on one side of the equation and all other terms on the other side of the equation.
6. Divide both sides of the equation by the coefficient of the variable. Note that the coefficient may not be a number in these problems.

16. Solve $A = Prt$ for P

17. Solve $y = mx + b$ for x

18. Solve $A = \dfrac{1}{2}h(b_1 + b_2)$ for h

Using Formulas to Solve Word Problems

You may have a list of formulas to memorize, or you may have formulas stored in your calculator. Be sure you know what you are responsible for memorizing before the test. Often, formulas are written using the first letter of the quantity as the variable in the formula. In $D = rt$, D represents distance, r represents rate, and t represents time.

19. Ron traveled 224 miles in $3\dfrac{1}{2}$ hours. Find his average speed.

20. If $6,000 is invested in an account that earns simple interest at a rate of 5%, find the interest earned after 8 years.

21. The diameter of a circle is 22 cm. Find the circumference of the circle. Round your answer to four decimal places.

Solving Linear Inequalities in One Variable

To solve a linear inequality in one variable, use the same rules as for solving a linear equation in one variable, *except*:

If you multiply or divide by a negative number, you reverse the inequality symbol.

Solutions are often accompanied by a number line graph. Remember that brackets are used to include an endpoint, and parentheses are used to exclude an endpoint.

Solve and graph:

22. $4x - 8 < 4$

23. $-6m \le 3$

24. $3y + 5 > y - 3$

25. $4(2x + 3) - 6 \ge 2(4 - x)$

Solving Compound Inequalities
When two inequalities are joined by the word "or," the solution contains all numbers that solve either inequality.
When two inequalities are joined by the word "and," the solution contains only numbers that solve both inequalities.
A compound inequality of the form $a < x < b$ represents $x > a$ *and* $x < b$.

Solve and graph.

26. $2x - 1 > 5$ or $2x - 1 < -5$

27. $-9 \le 3x + 1 \le 9$

28. $5x + 4 > -6$ and $5x + 4 < 6$

Solving Absolute Value Equations and Inequalities

To solve an absolute value equation or inequality, you usually rewrite the equation or inequality without the absolute value bars. Memorize the forms below:

With Absolute Value Bars	Without Absolute Value Bars
$\|x\| = a$, $a > 0$	$x = a$ or $x = -a$
$\|x\| = \|y\|$	$x = y$ or $x = -y$
$\|x\| < a$	$-a < x < a$
$\|x\| > a$	$x > a$ or $x < -a$

Note that > can be replaced by $\ge$ and < can be replaced by $\le$ above.

Solve:

29. $\|7x + 2\| = 16$

30. $\|3x - 4\| + 2 = 8$

31. $\|2y + 5\| - 4 = 6$

32. $\|6x - 3\| = \|2x + 4\|$

33. $\|x - 4\| = \|x - 6\|$

34. $\|2x - 1\| > 5$

35. $\|3x + 1\| \le 9$

36. $\|5x + 4\| < 6$

37. $\|2x - 8\| + 3 \ge 4$

Types of Solution Sets
The following absolute value equation and inequality forms lead to solutions sets that are empty or contain all real numbers:

Form	Solution Set
$|x| = $ a negative number	$\varnothing$
$|x| < $ a negative number or 0	$\varnothing$
$|x| > $ a negative number	$\Re$

Note that < can be replaced by $\le$ and > can be replaced by $\ge$.

Solve:

38. $\|5x - 12\| = -6$

39. $\|4(m + 3)\| < -3$

40. $\|2p + 6\| > -1$

41. $\|3y - 1\| - 4 > -2$

Word Problems

You should recognize more phrases that indicate the basic operations of addition, subtraction, multiplication, and division as well as phrases that indicate equality and inequalities. Some are listed below.
Addition: plus, increased by, sum, added to, total, more than
Subtraction: difference, decreased by, less than, minus
Multiplication: product, times, twice, double, triple
Division: quotient, divided by, reciprocal, per
Equality: is, is the same as, equal to
Less than or equal to: at most
Greater than or equal to: at least

42. If your test scores are 98, 75, and 85, what must you score on the fourth test to have an average of at least 90? Assume a maximum score of 100 on each test.

43. John's employer does not allow him to work over 40 hours per week. If John has already worked 8 hours, 5 hours, 7 hours, and 6 hours in one week, what is the most he can work to complete the week?

Consecutive Integers

Consecutive integers are represented using $x, x + 1$, $x + 2 \ldots$

Consecutive even integers are represented using x, $x + 2, x + 4 \ldots$

Consecutive odd integers are represented using x, $x + 2, x + 4 \ldots$

44. The sum of the first and twice the second of three consecutive integers is 4 less than 3 times the third. Find the integers.

45. Twice the sum of two consecutive even integers is 100. Find the integers.

46. The sum of the first and third consecutive odd integers is 21 more than five times the second consecutive odd integer. Find the integers.

Geometry Problems

Some formulas you should know include:

Rectangle:
$$A = L \cdot W$$
$$P = 2L + 2W$$

Circle:
$$A = \pi r^2$$
$$C = 2\pi r = \pi d$$

Angles:
Right angle = 90°
Straight angle = 180°

Complementary angles sum to 90°
Supplementary angles sum to 180°

Whenever possible, draw a picture and label it with the given information. Assign a variable to one of the unknown quantities and express the other unknowns in terms of that variable.

47. The length of a rectangle is 2 inches more than 3 times the width. The perimeter is 40 inches. Find the length and width.

48. The sum of two complementary angles is 90°. If one angle is 10° more than 3 times the other, find the angles.

Interest Problems

The formula for simple interest is $I = Prt$, where I is the interest earned, P is the principal (the money invested), r is the rate written as a decimal, and t is the time in years. You may need to convert interest from a per cent to a decimal (divide by 100), or convert time into years.

49. Pat has $36,000 to invest. She invests part of the money in an account that pays 6% interest, and the remainder in an account that pays 8% interest. If she earns $2640 in one year, how much was invested in each account?

50. Hillary invested $8,000 in two accounts, one that pays 6.5% interest and one that pays 8% interest. If she earned $610 in one year, how much was invested in each account?

 # Check Yourself

1. $4x - 8x - 7 - 3 = -4x - 10$ **(Solving linear equations in one variable)**

2. $5y - 3(2y - 1) =$

    $5y - 6y + 3$ Use the distributive property.

    $-y + 3$ Combine similar terms. **(Solving linear equations in one variable)**

3. $4(6x - 3) - 8x + 4$

    $= 24x - 12 - 8x + 4$ Use the distributive property.

    $= 16x - 8$ Combine similar terms. **(Solving linear equations in one variable)**

4. $3x - 5 = 4$ Add 5 to both sides of the equation to get $3x - 5 + 5 = 4 + 5$ or $3x = 9$.

$\dfrac{3x}{3} = \dfrac{9}{3}$ Divide both sides of the equation by 3.

$x = 3$ Simplify the right side. (**Addition and multiplication property of equality**)

5. $\dfrac{1}{2}x + 4 = 1$

$2\left(\dfrac{1}{2}x + 4\right) = 2\,(1)$ Multiply both sides of the equation by 2.

$x + 8 = 2$ Use the distributive property.

$x + 8 - 8 = 2 - 8$ Subtract 8 from both sides of the equation.

$x = -6$ Simplify the right side. (**Addition and multiplication property of equality**)

6. $3x - 2 = 8 - x$

$3x - 2 + x + 2 = 8 - x + x + 2$ Add $x + 2$ to both sides of the equation.

$4x = 10$ Combine similar terms.

$\dfrac{4x}{4} = \dfrac{10}{4}$ Divide both sides of the equation by 4.

$x = \dfrac{10}{4} = \dfrac{5}{2}$ Reduce. (**Addition and multiplication property of equality**)

7. $5 + 2\,(12 - x) = 2x - 7$

$5 + 24 - 2x = 2x - 7$ Use the distributive property.

$29 - 2x = 2x - 7$ Combine similar terms.

$29 - 2x + 2x + 7 = 2x - 7 + 2x + 7$ Add $2x$ and 7 to both sides of the equation.

$36 = 4x$ Combine similar terms.

$\dfrac{36}{4} = \dfrac{4x}{4}$ Divide both sides of the equation by 4.

$9 = x$ Simplify the left side of the equation. (**Solving linear equations**)

8. $4\,(2x - 5) + 3x = 3 - 2\,(x + 4)$

$8x - 20 + 3x = 3 - 2x - 8$ Use the distributive property.

$11x - 20 = -2x - 5$ Combine similar terms on the left side, then on the right side.

$11x - 20 + 2x + 20 = -2x - 5 + 2x + 20$ Add $2x$ and 20 to both sides.

$13x = 15$ Combine similar terms.

$$\frac{13x}{13} = \frac{15}{13} \quad \text{Divide both sides by 13.}$$

$$x = \frac{15}{13} \quad \textbf{(Solving linear equations)}$$

9. $$\frac{y}{3} - \frac{2y}{5} = 2$$

$$15\left(\frac{y}{3} - \frac{2y}{5}\right) = 15\,(2) \quad \text{Multiply both sides by 15.}$$

$$15 \cdot \frac{y}{3} - 15 \cdot \frac{2y}{5} = 30 \quad \text{Use the distributive property.}$$

$$5y - 6y = 30 \quad \text{Multiply.}$$

$$-y = 30 \quad \text{Combine similar terms on the right side.}$$

$$\frac{-y}{-1} = \frac{30}{-1} \quad \text{Divide both sides by } -1.$$

$$y = -30 \quad \textbf{(Solving linear equations)}$$

10. $$0.2x - 0.6 = -0.4\,(x - 1) + 0.2$$

$$10\,(0.2x - 0.6) = 10\,[-0.4\,(x - 1) + 0.2] \quad \text{Multiply both sides by 10 to clear decimals.}$$

$$2x - 6 = -4\,(x - 1) + 2 \quad \text{Use the distributive property.}$$

$$2x - 6 = -4x + 4 + 2$$

$$2x - 6 + 4x + 6 = -4x + 6 + 4x + 6 \quad \text{Add } 4x \text{ and 6 to both sides.}$$

$$6x = 12 \quad \text{Combine similar terms.}$$

$$\frac{6x}{6} = \frac{12}{6} \quad \text{Divide both sides by 6.}$$

$$x = 2 \quad \text{Simplify the right side.} \ \textbf{(Solving linear equations)}$$

11. $$3\,(x - 1) = 6 + 3x$$

$$3x - 3 = 6 + 3x \quad \text{Use the distributive property.}$$

$$3x - 3 + -3x + 3 = 6 + 3x + -3x + 3 \quad \text{Add } -3x \text{ and 3 to both sides.}$$

$$0 = 9 \quad \text{False statement}$$

Since there is no variable and the resulting statement is false, there is no solution. The solution set is the empty set, ø. **(Solution sets)**

12. $$\frac{1}{2}(4x+2) = 4x-2(x-3)-5$$

$$2x+1 = 4x-2x+6-5 \quad \text{Use the distributive property.}$$

$$2x+1 = 2x+1 \quad \text{Combine similar terms.}$$

$$2x+1-2x-1 = 2x+1-2x-1 \quad \text{Add } -2x \text{ and } -1 \text{ to both sides.}$$

$$0 = 0 \quad \text{True statement}$$

Since there is no variable and the resulting statement is true, the solution set is all real numbers.

(Solution sets)

13. $A = Prt$

$A = (2000)(0.08)(10) \quad$ Substitute the given values into the formula using parentheses.

$A = 1600$ **(Formulas)**

14. $y = mx + b$

$y = \left(\frac{2}{3}\right)(6) + (-4) \quad$ Substitute the given values into the formula using parentheses.

$y = 4 + (-4) \quad$ Multiplication is done before addition.

$y = 0$ **(Formulas)**

15. $$A = \frac{1}{2}h(b_1 + b_2)$$

$$(120) = \frac{1}{2}h((12)+(16)) \quad \text{Substitute the given values into the formula using parentheses.}$$

$$120 = \frac{1}{2}h(28) \quad \text{Simplify inside the parentheses.}$$

$$120 = 14h \quad \text{Multiply.}$$

$$\frac{120}{14} = \frac{14h}{14} \quad \text{Divide both sides by 14.}$$

$$\frac{60}{7} = h \quad \text{Simplify. } \textbf{(Formulas)}$$

16. $\dfrac{A}{rt} = \dfrac{Prt}{rt} \quad$ Divide both sides by rt.

$\dfrac{A}{rt} = P \quad$ Simplify. **(Formulas)**

17. $y = mx + b$

$y - b = mx + b - b \quad$ Subtract b from both sides.

$$\frac{y-b}{m} = \frac{mx}{m} \quad \text{Divide both sides by } m.$$

$$\frac{y-b}{m} = x \quad \text{Simplify. (Formulas)}$$

18. $A = \frac{1}{2}h\,(b_1 + b_2)$

$$2A = 2\left(\frac{1}{2}h\,(b_1 + b_2)\right) = h\,(b_1 + b_2) \quad \text{Multiply both sides by 2 to eliminate fractions.}$$

$$\frac{2A}{b_1 + b_2} = \frac{h\,(b_1 + b_2)}{b_1 + b_2} \quad \text{Divide both sides by } b_1 + b_2.$$

$$\frac{2A}{b_1 + b_2} = h \quad \text{Simplify. (Formulas)}$$

19. $D = rt \quad$ Choose the appropriate formula.

$224 = r\,(3.5) \quad$ 224 miles is distance; $3\frac{1}{2}$ hours is time.

$$\frac{224}{3.5} = r \quad \text{Divide both sides by 3.5.}$$

$64 \text{ mph} = r \quad$ Use the appropriate units: miles/hour = mph. **(Formulas)**

20. $I = Prt \quad$ Choose the appropriate formula.

$I = 6000\,(0.05)\,(8) \quad$ Covert the rate of 5% to a decimal.

$I = \$3840 \quad$ Multiply. **(Formulas)**

21. $C = \pi d \quad$ Choose the appropriate formula.

$C = \pi\,(22) \quad$ Substitute $d = 22$.

$C \approx 69.1150 \text{ cm} \quad$ Use a calculator to multiply. **(Formulas)**

22. $4x - 8 < 4$

$4x - 8 + 8 < 4 + 8 \quad$ Add 8 to both sides.

$$\frac{4x}{4} < \frac{12}{4} \quad \text{Divide both sides by 4.}$$

$x < 3 \quad$ Simplify.

(Solving linear inequalities in one variable)

23. $\dfrac{-6m}{-6} \geq \dfrac{3}{-6}$ Divide both sides by –6. Be careful to reverse the inequality symbol.

$m \geq -\dfrac{1}{2}$

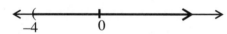   Use a bracket at –1/2 to show that it is included in the solution.

(Solving linear inequalities in one variable)

24. $3y + 5 > y - 3$

$3y + 5 - y - 5 > y + (-3) - y - 5$ Add –y and – 5 to both sides of the inequality.

$2y > -8$ Combine similar terms.

$\dfrac{2y}{2} > \dfrac{-8}{2}$ Divide both sides by 2.

$y > -4$

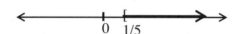   Use a parenthesis at –4 to show that it is *not* included

in the solution. **(Solving linear inequalities in one variable)**

25. $4(2x + 3) - 6 \geq 2(4 - x)$

$8x + 12 - 6 \geq 8 - 2x$ Use the distributive property to remove parentheses.

$8x + 6 \geq 8 - 2x$ Combine similar terms on each side of the inequality.

$8x + 6 + 2x - 6 \geq 8 - 2x + 2x - 6$ Add 2x and – 6 to both sides of the inequality.

$10x \geq 2$ Combine similar terms.

$\dfrac{10x}{10} \geq \dfrac{2}{10}$ Divide both sides of the inequality by 10.

$x \geq \dfrac{1}{5}$

   Use a bracket at 1/5 to show that it is included in the solution.

(Solving linear inequalities in one variable)

26.   Solve each inequality:

$2x - 1 > 5$ or $2x - 1 < -5$

$2x > 6$ $2x < -4$ Add 1 to both sides of the inequalities.

$x > 3$ $\qquad\qquad$ $x < -2$ $\qquad$ Divide both sides of the inequalities by 2.

The solution set contains numbers that are greater than 3 or less than –2:

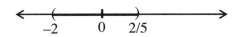 **(Solving compound inequalities)**

27. This can be solved by working on all three parts simultaneously:

$-9 \le 3x + 1 \le 9$

$-10 \le 3x \le 8$ $\quad$ Subtract 1 from all three parts.

$\dfrac{-10}{3} \le \dfrac{3x}{3} \le \dfrac{8}{3}$ $\quad$ Divide each part by 3.

$\dfrac{-10}{3} \le x \le \dfrac{8}{3}$ $\quad$ Simplify.

(Solving compound inequalities)

28. $5x + 4 > -6$ $\quad$ and $\quad$ $5x + 4 < 6$

$\qquad 5x > -10$ $\quad$ and $\qquad 5x < 2$ $\qquad$ Subtract 4 from both sides of the inequalities.

$\qquad\quad x > -2$ $\quad$ and $\qquad\quad x < \dfrac{2}{5}$ $\qquad$ Divide each side by 5.

(Solving compound inequalities)

29. $|7x + 2| = 16$

$7x + 2 = 16$ $\quad$ or $\quad$ $7x + 2 = -16$ $\quad$ Rewrite the absolute value equation without absolute value bars.

$\quad 7x = 14$ $\quad$ or $\qquad 7x = -18$ $\quad$ Subtract 2 from both sides.

$\quad\; x = 2$ $\quad$ or $\qquad\; x = -\dfrac{18}{7}$ $\quad$ Divide both sides by 7.

The solution set is $\{-18/7, 2\}$. **(Solving absolute value equations and inequalities)**

30. $|3x - 4| + 2 = 8$

$|3x - 4| = 6$ $\quad$ Isolate the absolute value bars by subtracting 2 from both sides.

$3x - 4 = 6$ $\quad$ or $\quad$ $3x - 4 = -6$ $\quad$ Rewrite the absolute value equation without absolute value bars.

$\quad 3x = 10$ $\quad$ or $\qquad 3x = -2$ $\quad$ Add 4 to both sides.

$$x = \frac{10}{3} \quad \text{or} \quad x = -\frac{2}{3} \quad \text{Divide both sides by 3.}$$

The solution set is {–2/3, 10/3}. (**Solving absolute value equations and inequalities**)

31. $|2y + 5| - 4 = 6$

$|2y + 5| = 10$ Isolate the absolute value bars by adding 4 to both sides.

$2y + 5 = 10 \quad \text{or} \quad 2y + 5 = -10$ Rewrite the absolute value equation without absolute value bars.

$2y = 5 \quad \text{or} \quad 2y = -15$ Subtract 5 from both sides.

$y = \frac{5}{2} \quad \text{or} \quad y = \frac{-15}{2}$ Divide both sides by 2.

The solution set is {–15/2, 5/2}. (**Solving absolute value equations and inequalities**)

32. $|6x - 3| = |2x + 4|$

$6x - 3 = 2x + 4 \quad \text{or} \quad 6x - 3 = -(2x + 4)$ Rewrite without absolute value bars.

$4x = 7 \quad \text{or} \quad 8x = -1$ Solve each equation.

$x = \frac{7}{4} \quad \text{or} \quad x = -\frac{1}{8}$

The solution set is {–1/8, 7/4}. (**Solving absolute value equations and inequalities**)

33. $|x - 4| = |x - 6|$

$x - 4 = x - 6 \quad\quad\quad \text{or} \quad x - 4 = -(x - 6)$ Rewrite without absolute value bars.

$-4 = -6 \ \text{(A false statement)} \quad \text{or} \quad 2x = 10$

In this case, only one of the equations led to a possible solution. If $2x = 10, x = 5$. Therefore, the solution set is {5}. (**Solving absolute value equations and inequalities**)

34. $|2x - 1| > 5$

$2x - 1 > 5 \quad \text{or} \quad 2x - 1 < -5$ Rewrite without the absolute value bars.

$2x > 6 \quad \text{or} \quad 2x < -4$ Add 1 to both sides of the inequality.

$x > 3 \quad \text{or} \quad x < -2$ Divide both sides by 2.

In interval notation, the solution set would be written as $(\infty, -2) \cup (3, \infty)$. (**Solving absolute value equations and inequalities**)

35. $|3x + 1| \le 9$

$-9 \le 3x + 1 \le 9$ Rewrite without the absolute value bars.

This was already solved in #27. The solution is $\frac{-10}{3} \le x \le \frac{8}{3}$ or $\left[-\frac{10}{3}, \frac{8}{3}\right]$. (**Solving absolute value equations and inequalities**)

36. $|5x + 4| < 6$

    $-6 < 5x + 4 < 6$ Rewrite without the absolute value bars.

    $-10 < 5x < 2$

    $-2 < x < \dfrac{2}{5}$

    Note that the inequality could have been solved in two parts (see #28) or by working on all three parts simultaneously as was shown here. The solution set using interval notation is $(-2, 2/5)$. (**Solving absolute value equations and inequalities**)

37. $|2x - 8| + 3 \geq 4$

    $|2x - 8| \geq 1$ Isolate the absolute value bars by subtracting 3 from both sides.

    $2x - 8 \geq 1$ or $2x - 8 \leq -1$ Rewrite without the absolute value bars.

    $x \geq \dfrac{9}{2}$ or $x \leq \dfrac{7}{2}$ Solve each inequality.

    The solution set is $(-\infty, 7/2] \cup [9/2, \infty)$. (**Solving absolute value equations and inequalities**)

38. $|5x - 12| = -6$ fits the form of an absolute value equal to a negative number which has a solution set that is empty, or $\emptyset$. (**Solving absolute value equations and inequalities**)

39. $|4(m + 3)| < -3$ fits the form of an absolute value less than a negative number which has a solution set that is empty, or $\emptyset$. (**Solving absolute value equations and inequalities**)

40. $|2p + 6| > -1$ fits the form of an absolute value greater than a negative number which has a solution set of all real numbers, $\Re$. (**Solving absolute value equations and inequalities**)

41. $|3y - 1| - 4 > -2$ must first be written in a form with the absolute value expression isolated on one side of the inequality:

    $|3y - 1| > 2$ This does not fit one of the special types of solutions, so apply the rules to write the inequality without the absolute value bars:

    $3y - 1 > 2$ or $3y - 1 < -2$

    $y > 1$ or $y < -\dfrac{1}{3}$ Solve each inequality.

    The solution set can be written as $(-\infty, -1/3) \cup (1, \infty)$. (**Solving absolute value equations and inequalities**)

42. Let x = score on the fourth test.

    $\dfrac{98 + 75 + 85 + x}{4} \geq 90$ "At least" translates to $\geq$.

    $4 \cdot \left(\dfrac{98 + 75 + 85 + x}{4} \right) \geq 4 \cdot 90$ Multiply both sides by 4 to clear the fraction.

$258 + x \geq 360$ Simplify both sides.

$x \geq 102$ Subtract 258 from both sides.

If we assume that test scores cannot be greater than 100, it is impossible to have an average of at least 90. **(Word problems)**

43.   Let x = number of hours of work to complete the week.

$8 + 5 + 7 + 6 + x \leq 40$ Total number of hours must be less than or equal to 40.

$26 + x \leq 40$ Combine similar terms.

$x \leq 14$ Subtract 26 from both sides.

Therefore John can work a maximum of 14 more hours during that week. **(Word problems)**

44.   Let x = first integer. Then $x + 1$ = second consecutive integer and $x + 2$ = third consecutive integer.

$x + 2(x + 1) = 3(x + 2) - 4$ "Four less than" means $- 4$.

$x + 2x + 2 = 3x + 6 - 4$ Use the distributive property to remove parentheses.

$3x + 2 = 3x + 2$ Combine similar terms on each side.

$0 = 0$

The variable has been eliminated and the resulting statement is true. Therefore, x can be any integer. **(Word problems)**

45.   Let x = first even integer. Then $x + 2$ = next even integer.

$2(x + x + 2) = 100$

$2(2x + 2) = 100$ Combine similar terms inside the parentheses.

$4x + 4 = 100$ Use the distributive property to remove parentheses.

$4x = 96$ Subtract 4 from both sides of the equation.

$x = \dfrac{96}{4}$ Divide both sides by 4.

$x = 24$ Simplify the fraction.

Therefore, the integers are 24 and 24+2 = 26. **(Word problems)**

46.   Let x = first odd integer. Then $x + 2$ = second consecutive odd integer and $x + 4$ = third consecutive odd integer.

$x + x + 4 = 21 + 5(x + 2)$ Translate into symbols.

$2x + 4 = 21 + 5x + 10$ Combine similar terms on the left side; use the distributive property on the right.

$2x + 4 = 31 + 5x$ Combine similar terms on the right.

$-3x = 27$ Subtract $5x$ from both sides; subtract 4 from both sides.

$x = -9$ Divide both sides by -3.

Therefore the integers are $-9, -7,$ and -5. **(Word problems)**

47.   Let $w =$ width since the length is easily expressed in terms of the width. Draw a picture and label it:

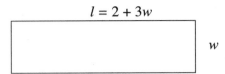

$P = 2l + 2w$ Use the formula for the perimeter of a rectangle.

$40 = 2(2 + 3w) + 2(w)$ Substitute 40 for the perimeter, $2 + 3w$ for length and w for width.

$40 = 4 + 6w + 2w$ Use the distributive property to remove parentheses.

$40 = 4 + 8w$ Combine similar terms.

$36 = 8w$ Subtract 4 from both sides.

$\dfrac{36}{8} = w$ Divide both sides by 8.

$4.5 = w$ Write the fraction as a decimal.

$l = 2 + 3w = 2 + 3(4.5) = 2 + 13.5 = 15.5$ Solve for l.

Therefore, the length is 15.5 inches and the width is 4.5 inches. **(Word problems)**

48.   Let $x =$ one angle. Then $90 - x =$ other angle.

$x = 10 + 3(90 - x)$ Translate into symbols.

$x = 10 + 270 - 3x$ Use the distributive property to remove parentheses.

$4x = 280$ Combine similar terms; add $3x$ to both sides.

$x = 70$ Divide both sides by 4.

$90 - x = 90 - 70 = 20$ Solve for the other angle.

Therefore, the two angles are $70°$ and $20°$. **(Word problems)**

49.   Let $x =$ the amount invested at 6% interest. Then $36000 - x =$ the amount invested at 8% interest.

$0.06x + 0.08(36000 - x) = 2640$ Use $I = Prt$ with $t = 1$. Find the amount of interest by multiplying the interest rate times the amount of money invested.

$6x + 8(36000 - x) = 264000$ Multiply by 100 to clear the decimals.

$6x + 288000 - 8x = 264000$ Use the distributive property to remove parentheses.

$-2x = -24000$ Combine similar terms; subtract 288000 from both sides.

$x = 12000$ Divide both sides by -2.

$36000 - x = 36000 - 12000 = 24000$ Solve for the other amount.

Therefore, \$12,000 was invested at 6% and \$24,000 was invested at 8%. **(Word problems)**

50.   Let x = the amount invested at 6.5%. Then $8000 - x$ = the amount invested at 8%.

$0.065x + 0.08(8000 - x) \ = \ 610$ Multiply the rate times the amount invested to get interest.

$65x + 80(8000 - x) \ = \ 610000$ Multiply by 1000 to clear decimals.

$65x + 640000 - 80x \ = \ 610000$ Use the distributive property to remove parentheses.

$-15x \ = \ -30000$ Combine similar terms on the left; subtract 640000 from both sides.

$x \ = \ 2000$ Divide both sides by –15.

$8000 - x \ = \ 8000 - 2000 \ = \ 6000$ Solve for the other amount.

Therefore, Hillary invested \$2000 at 6.5% and \$6000 at 8%. **(Word problems)**

Grade Yourself

Circle the question numbers that you had incorrect. Then indicate the number of questions you missed. If you answered more than three questions incorrectly, you need to focus on that topic. (If a topic has less than three questions and you had at least one wrong, we suggest you study that topic also. Read your textbook, a review book, or ask your teacher for help.)

Subject: Solving Linear Equations and Inequalities

Topic	Question Numbers	Number Incorrect
Solving linear equations in one variable	1, 2, 3	
Addition and multiplication property of equality	4, 5, 6	
Solving linear equations	7, 8, 9, 10	
Types of solution sets	11, 12	
Formulas	13, 14, 15, 16, 17, 18, 19, 20, 21	
Solving linear inequalities in one variable	22, 23, 24, 25	
Solving compound inequalities	26, 27, 28	
Solving absolute value equations and inequalities	29, 30, 31, 32, 33, 34, 35, 36, 37, 38, 39, 40, 41	
Word problems	42, 43, 44, 45, 46, 47, 48, 49, 50	

Exponents and Polynomials

 Test Yourself

Zero and Negative Exponents

Definition: $a^0 = 1$, $a \neq 0$

$$a^{-n} = \frac{1}{a^n}, \ a \neq 0$$

Previously the exponent laws were used for positive integers. These laws also apply to zero and negative exponents.

Simplify.

1. 6^0

2. -3^0

3. $4x^0$, $x \neq 0$

4. 5^{-2}

5. -4^{-2}

6. $(-2)^{-3}$

7. $8x^{-1}$

8. $(2x)^{-2}$

Scientific Notation

Writing Numbers in Scientific Notation
1. Move the decimal point to the right of the first nonzero digit. Count the number of places you moved the decimal point.

Writing Numbers in Scientific Notation (continued)
2. Multiply the number found in Step 1 times 10 raised to the number of places you moved the decimal point. If the original number was less than 1, the exponent on 10 is negative. If the original number was greater than 1, the exponent on 10 is positive.

Write each number in scientific notation.

9. 42,300

10. 0.0016

Writing a Number in Expanded Form
1. If the exponent on 10 is positive, move the decimal point to the right the same number of places as the exponent. You may need to add zeros.
2. If the exponent on 10 is negative, move the decimal point to the left the same number of places as the exponent. You may need to add zeros.

---

11. Write in expanded form: 2.32×10^4

12. Write in expanded form: 1.28×10^{-3}

13. Write in expanded form: 4.16×10^{-1}

Properties of Exponents

You should now be able to use the exponents laws with integer exponents:

$$a^x \cdot a^y = a^{x+y}$$

$$(a^x)^y = a^{xy}$$

$$(ab)^x = a^x b^x$$

$$\left(\frac{a}{b}\right)^x = \frac{a^x}{b^x}, b \neq 0$$

$$\frac{a^x}{a^y} = a^{x-y}, a \neq 0$$

Simplify. Write your answers with positive exponents.

14. $5^2 \cdot 5^{-4}$

15. $x^2 \cdot x^{-4} \cdot x^{-3}$

16. $(2^3)^{-2}$

17. $(4x)^{-2}$

18. $\left(\frac{3}{2}\right)^{-2}$

19. $\dfrac{x^{-6}}{x^{-2}}$

20. $\dfrac{4^{-2} a^{-3} b^2}{4^{-1} a^{-5} b^{-1}}$

21. Use scientific notation to compute $(0.0004)^2$

22. Use scientific notation to compute
     $(4,600)(0.0002)$

23. Use scientific notation to compute
     $\dfrac{(0.00026)(0.0001)}{0.00013}$

Classifying Polynomials

Polynomials are classified by the number of terms:
   A polynomial with one term is a *monomial*.
   A polynomial with two terms is a *binomial*.
   A polynomial with three terms is a *trinomial*.

The **degree** of a term is the sum of the exponents on the variables in the term.

The **degree** of a polynomial is the highest degree of any term in the polynomial.

Classify each polynomial as monomial, binomial, or trinomial and state the degree.

24. $4x - 8$

25. $x^2 - x - 6$

26. $3xy^2z^3$

Adding and Subtracting Polynomials

To add polynomials:
Group like (or similar) terms and then add numerical
   coefficients of like terms.
To subtract polynomials:
Change the sign of each term being subtracted, and
   then add.
Note that like (or similar) terms have exactly the
   same variables raised to exactly the same powers.

Perform the indicated operations.

27. $(2x^2 - 5x + 8) + (6x^2 - 7x - 12)$

28. $(x^5 + 3x^2 - 5x - 10) + (-2x^5 + 4x^3 - 6x^2 + 2)$

29. $(x^2 - 2x - 5) - (3x^2 - 7x + 6)$

30. $(4y^5 - 2y^2 - 3) - (6y^5 + 2y^4 - 7y^2 + 8y + 7)$

Evaluating Polynomials

To evaluate a polynomial, substitute the given value
   of the variable(s) into the polynomial.
If $P(x) = x^2 - 4x + 2$, find $P(3)$ means substitute
   3 in for x.

31. Find $P(3)$ for $P(x) = x^2 - 4x + 2$

32. Find $f(-1)$ for $f(x) = -x^2 + 5x - 7$

Multiplying Polynomials

In general, the distributive law is used to multiply
   polynomials. Any similar terms produced by the
   multiplication should be combined.

Perform the indicated operations.

33. $(2x - 5)(3x + 1)$

34. $(3y - 5)(y^2 + 3y - 8)$

35. $(m^2 - 2m + 1)(2m^2 + 3m - 4)$

Multiplication Shortcuts

To shorten the multiplication of certain polynomials, you must recognize when the shortcuts apply.

FOIL can be used to multiply two binomials. FOIL means: multiply the **F**irst terms of each binomial, multiply the **O**uter two terms of each binomial, multiply the **I**nner two terms of each binomial, and multiply the **L**ast two terms of each binomial. Often the terms created when multiplying the outer two terms and the inner two terms are similar terms and can be combined. In symbols:

$(a + b)(c + d) = ac + ad + bc + bd$

When squaring a binomial, square the first term, double the product of the first term times the second term, and square the last term. In symbols:

$(a + b)^2 = a^2 + 2ab + b^2$

When multiplying the sum and difference of two like terms, square the first term then subtract the square of the last term. In symbols:

$(a + b)(a - b) = a^2 - b^2$

Perform the indicated operations.

36. $(4x - 3)(2x + 1)$

37. $(3x + 2y)(5x + 4y)$

38. $(3x - 1)(2x - 5)$

39. $(b - 4)^2$

40. $(m + 2)^2$

41. $(3x - 5y)^2$

42. $(2r + 3s)^2$

43. $(x - y)(x + y)$

44. $(2a - 3b)(2a + 3b)$

45. $(3m + 5n)(3m - 5n)$

Factoring

Factoring is the reverse process of multiplying. A factoring problem can be checked by multiplying. However, be careful to factor completely.

In any factoring problem, the first step is factoring out the greatest common factor (GCF)—that is, the largest number that divides evenly into each term with common variables raised to the lowest exponent appearing in any of the terms.

Factor.

46. $4x^2 + 10x^3y$

47. $36a^4b^4 - 72a^2b^2 + 24a^3b$

48. $2x^2 + 5$

49. $5(x + y) + 6m(x + y)$

Factoring by Grouping

When a polynomial contains four or more terms, try to factor by grouping the terms. One technique is to group the first two terms, the last two terms, and factor each pair. If the resulting expressions in parentheses are equal, factor out that expression as the greatest common factor. If the expressions in parentheses are not equal, try changing the order of the terms and begin again.

Factor.

50. $3x + bx + 3y + by$

51. $x^2y^2 + 4x^2 + 4y^2 + 16$

52. $a^2x^2 + 4b^2 + 4a^2 + b^2x^2$

Factoring Trinomials

Several techniques exist for factoring trinomials. Regardless of which method you use, remember to:

1. Begin every factoring problem by factoring out the greatest common factor.
2. Check your answer by multiplying the factors in your answer.

Factor.

53. $x^2 - x - 6$

54. $y^2 - 12y + 32$

55. $x^2 + xy - 6y^2$

56. $6x^2 - 13x - 5$

57. $8x^2 - 26x + 15$

58. $15x^2 + 22x + 8$

59. $2x^2 - 14x + 24$

60. $3x^3y - 11x^2y - 20xy$

61. $30x^5 + 132x^4 + 48x^3$

Factoring Special Polynomials

Some polynomials fit special patterns that can be factored immediately by the following formulas:

1. $x^2 - y^2 = (x - y)(x + y)$
   Difference of Two Squares

2. $x^2 + y^2$ does not factor
   Sum of Two Squares

3. $x^2 \pm 2xy + y^2 = (x \pm y)^2$
   Perfect Square Trinomial

4. $x^3 + y^3 = (x + y)(x^2 - xy + y^2)$
   Sum of Two Cubes

5. $x^3 - y^3 = (x - y)(x^2 + xy + y^2)$
   Difference of Two Cubes

You should memorize these formulas.

Factor.

62. $9x^2 - 25$

63. $16 - x^2y^2$

64. $49x^2 + 4$

65. $4x^2 + 20xy + 25y^2$

66. $36x^2 - 84x + 49$

67. $125x^3 - 1$

68. $8a^3b^3 + 27c^3$

69. $12ax^2 - 27ay^2$

70. $16a^3p + 54p$

71. $5x^4 - 5$

Factoring By Grouping

Factoring by grouping may be needed for polynomials with four terms where three of the terms can be grouped and factored as a perfect square trinomial.

Factor.

72. $x^2 + 4xy + 4y^2 - 9$

73. $16 - x^2 + 6xy - 9y^2$

Solving Equations by Factoring

The Zero-Factor Property can be used to solve a factorable equation:

$$\text{If } a \cdot b = 0, \text{ then } a = 0 \text{ or } b = 0$$

While many of the equations at this level are quadratic equations, the Zero-Factor Property can be used regardless of the number of factors. Remember to get 0 on one side of the equation and to factor the other side completely.

Solve each equation.

74. $(3x - 1)(5x + 4) = 0$

75. $2x(x - 1)(2x + 3) = 0$

76. $4x^2 + 16x = 0$

77. $\frac{1}{5}x^2 - \frac{3}{5}x - 2 = 0$

78. $8x^2 = 10x + 3$

79. $4x^2 = 16$

80. $y(y - 2) = 8$

Check Yourself

1. $6^0 = 1$ (**Zero and negative exponents**)

2. $-3^0 = -1$ Remember the exponent only applies to the 3, and the negative in front is really -1 times 3^0. (**Zero and negative exponents**)

3. $4x^0 = 4(1) = 4$ (**Zero and negative exponents**)

4. $5^{-2} = \dfrac{1}{5^2} = \dfrac{1}{25}$ (**Zero and negative exponents**)

5. $-4^{-2} = -\dfrac{1}{4^2} = -\dfrac{1}{16}$ The exponent applies only to 4, not -4. (**Zero and negative exponents**)

6. $(-2)^{-3} = \dfrac{1}{(-2)^3} = \dfrac{1}{-8} = -\dfrac{1}{8}$ Here the exponent -3 applies to the expression in parentheses. (**Zero and negative exponents**)

7. $8x^{-1} = 8\left(\dfrac{1}{x}\right) = \dfrac{8}{x}$ (**Zero and negative exponents**)

8. $(2x)^{-2} = \dfrac{1}{(2x)^2} = \dfrac{1}{4x^2}$ Note that the expression in parentheses is raised to the -2 power. (**Zero and negative exponents**)

9. $42{,}300 = 4.23 \times 10^4$ The decimal point moves 4 places and the original number is greater than 1, so the exponent on 10 is $+4$, which is usually written as 4. (**Scientific notation**)

10. $0.0016 = 1.6 \times 10^{-3}$ The decimal point moves 3 places and the original number is less than 1, so the exponent on 10 is -3. (**Scientific notation**)

11. $2.32 \times 10^4 = 23200$ To expand, move the decimal point to the right if the exponent on 10 is positive and to the left if the exponent on 10 is negative. Here the exponent of $+4$ means to move the decimal point 4 places to the right. Add zeros as needed. (**Scientific notation**)

12. $1.28 \times 10^{-3} = 0.00128$ The -3 exponent on 10 means the decimal point must be moved 3 places to the left. Zeros are added as needed. (**Scientific notation**)

13. $4.16 \times 10^{-1} = 0.416$ (**Scientific notation**)

14. $5^2 \cdot 5^{-4} = 5^{2 + (-4)} = 5^{-2} = \dfrac{1}{5^2} = \dfrac{1}{25}$ When the bases are the same, we add the exponents. Since the directions indicate that answers must be written with positive exponents, use the definition of a negative exponent to write 5^{-2} as $\dfrac{1}{5^2}$. (**Properties of exponents**)

15. $x^2 \cdot x^{-4} \cdot x^{-3} = x^{2 + (-4) + (-3)} = x^{-5} = \dfrac{1}{x^5}$ When bases are the same, add the exponents. Be careful to write the final answer with a positive exponent. (**Properties of exponents**)

16. $(2^3)^{-2} = 2^{3 \cdot (-2)} = 2^{-6} = \dfrac{1}{2^6} = \dfrac{1}{64}$ An exponent outside a parentheses multiplies times the exponent inside the parentheses. (**Properties of exponents**)

17. $(4x)^{-2} = \dfrac{1}{(4x)^2} = \dfrac{1}{4^2 x^2} = \dfrac{1}{16x^2}$ Use the definition of a negative exponent to write $(4x)^{-2}$ as $\dfrac{1}{(4x)^2}$.
    Then use the property that the exponent outside the parentheses multiplies times each exponent inside the parentheses. (**Properties of exponents**)

18. $\left(\dfrac{3}{2}\right)^{-2} = \dfrac{3^{-2}}{2^{-2}} = \dfrac{\dfrac{1}{3^2}}{\dfrac{1}{2^2}} = \dfrac{1}{3^2} \cdot \dfrac{2^2}{1} = \dfrac{4}{9}$ Using this process uses the property $\left(\dfrac{a}{b}\right)^x = \dfrac{a^x}{b^x}$ and the definition of a

    negative exponent to simplify the expression. However, there is a shortcut: $\left(\dfrac{a}{b}\right)^{-x} = \left(\dfrac{b}{a}\right)^x$. This would

    allow you to invert the fraction in the parentheses and make the exponent positive:

    $\left(\dfrac{3}{2}\right)^{-2} = \left(\dfrac{2}{3}\right)^2 = \dfrac{2^2}{3^2} = \dfrac{4}{9}$ Of course, you will get the same answer using either technique. Use the method

    you have the most confidence in. (**Properties of exponents**)

19. $\dfrac{x^{-6}}{x^{-2}} = x^{-6 - (-2)} = x^{-6 + 2} = x^{-4} = \dfrac{1}{x^4}$ When dividing and the bases are equal, subtract the exponents.

    (**Properties of exponents**)

20. $\dfrac{4^{-2} a^{-3} b^2}{4^{-1} a^{-5} b^{-1}} = 4^{-2 - (-1)} a^{-3 - (-5)} b^{2 - (-1)} = 4^{-1} a^2 b^3 = \dfrac{a^2 b^3}{4}$ Subtract exponents for each base. Then sim-

    plify 4^{-1} to $\dfrac{1}{4}$. (**Properties of exponents**)

21. Write 0.0004 in scientific notation as 4×10^{-4}. Then use exponent laws to raise 4 and 10^{-4} to the second power:

    $(4 \times 10^{-4})^2 = 4^2 \cdot (10^{-4})^2 = 16 \times 10^{-8}$ Now write the answer in scientific notation:

    $16 \times 10^{-8} = 1.6 \times 10^{-7}$ (**Scientific notation**)

22. $(4,600)(0.0002) = (4.6 \times 10^3)(2 \times 10^{-4})$ Write each number in scientific notation.

    $= 4.6 \times 2 \times 10^3 \times 10^{-4} = 9.2 \times 10^{-1}$ Remember to add exponents when multiplying equal bases of 10.
    (**Scientific notation**)

23. $\dfrac{(0.00026)\,(0.0001)}{0.00013} = \dfrac{(2.6\times10^{-4})\,(1\times10^{-4})}{1.3\times10^{-4}}$ Write each number in scientific notation.

  $= \dfrac{2.6\times1}{1.3}\times10^{-4-4-(-4)}$

  $= 2\times10^{-4}$ **(Scientific notation)**

24. $4x-8$ is a binomial of degree one. **(Classifying polynomials)**

25. x^2-x-6 is a trinomial of degree two. **(Classifying polynomials)**

26. $3xy^2z^3$ is a monomial of degree six. **(Classifying polynomials)**

27. $(2x^2-5x+8)+(6x^2-7x-12) = 8x^2-12x-4$. Combine the like terms. The remaining terms cannot be combined further because there are no like terms. **(Adding and subtracting polynomials)**

28. $(x^5+3x^2-5x-10)+(-2x^5+4x^3-6x^2+2) = -x^5+4x^3-3x^2-5x-8$ **(Adding and subtracting polynomials)**

29. $(x^2-2x-5)-(3x^2-7x+6) = x^2-2x-5-3x^2+7x-6$ Change the sign of each term in the polynomial being subtracted.

  $= -2x^2+5x-11$ Combine similar terms. **(Adding and subtracting polynomials)**

30. $(4y^5-2y^2-3)-(6y^5+2y^4-7y^2+8y+7) = 4y^5-2y^2-3-6y^5-2y^4+7y^2-8y-7$ Change the sign of each term in the polynomial being subtracted.

  $= -2y^5-2y^4+5y^2-8y-10$ Combine similar terms. **(Adding and subtracting polynomials)**

31. Find $P(3)$ for $P(x) = x^2-4x+2$ means replace x with 3 in P. Thus,
  $$P(3) = (3)^2-4(3)+2 = 9-12+2 = -1.\ \textbf{(Evaluating polynomials)}$$

32. Find $f(-1)$ for $f(x) = -x^2+5x-7$ means replace x with -1 in f. Thus,
  $$f(-1) = -(-1)^2+5(-1)-7 = -1-5-7 = -13.\ \textbf{(Evaluating polynomials)}$$

33. $(2x-5)(3x+1) = 2x(3x+1)-5(3x+1) = 6x^2+2x-15x-5 = 6x^2-13x-5$
  (Multiplying polynomials)

34. $(3y-5)(y^2+3y-8) = 3y(y^2+3y-8)-5(y^2+3y-8) =$
  $3y^3+9y^2-24y-5y^2-15y+40 = 3y^3+4y^2-39y+40$ **(Multiplying polynomials)**

35. $(m^2-2m+1)(2m^2+3m-4) = 2m^4+3m^3-4m^2-4m^3-6m^2+8m+2m^2+3m-4$

  $= 2m^4-m^3-8m^2+11m-4$ **(Multiplying polynomials)**

36. $(4x-3)(2x+1) = 8x^2+4x-6x-3 = 8x^2-2x-3$ Note that the Outer and Inner products are similar terms and can be added to simplify the answer. **(Multiplying polynomials)**

37. $(3x + 2y)(5x + 4y) = 15x^2 + 12xy + 10xy + 8y^2 = 15x^2 + 22xy + 8y^2$ (**Multiplying polynomials**)

38. $(3x - 1)(2x - 5) = 6x^2 - 15x - 2x + 5 = 6x^2 - 17x + 5$ (**Multiplying polynomials**)

39. $(b - 4)^2 = b^2 + 2(b)(-4) + (-4)^2 = b^2 - 8b + 16$ To square a binomial, square the first term, double the product of the first term times the second term, and square the last term. (**Multiplying polynomials**)

40. $(m + 2)^2 = m^2 + 2(m)(+2) + 2^2 = m^2 + 4m + 4$ (**Multiplying polynomials**)

41. $(3x - 5y)^2 = (3x)^2 + 2(3x)(-5y) + (-5y)^2 = 9x^2 - 30xy + 25y^2$ (**Multiplying polynomials**)

42. $(2r + 3s)^2 = (2r)^2 + 2(2r)(3s) + (3s)^2 = 4r^2 + 12rs + 9s^2$ (**Multiplying polynomials**)

43. $(x - y)(x + y) = x^2 - y^2$ When multiplying the sum and difference of two like terms, square the first term then subtract the square of the last term. (**Multiplying polynomials**)

44. $(2a - 3b)(2a + 3b) = (2a)^2 - (3b)^2 = 4a^2 - 9b^2$ (**Multiplying polynomials**)

45. $(3m + 5n)(3m - 5n) = (3m)^2 - (5n)^2 = 9m^2 - 25n^2$ (**Multiplying polynomials**)

46. $4x^2 + 10x^3y = 2x^2(2 + 5xy)$ Factor out the greatest common factor of $2x^2$. Note that 2 and $5xy$ have no common factors. (**Factoring**)

47. $36a^4b^4 - 72a^2b^2 + 24a^3b = 12a^2b(3a^2b^3 - 6b + 2a)$ 12 is the largest number that divides evenly into $36, 72$, and 24. 2 is the largest power of a in any term. 1 is the largest power of b in any term. Therefore, factor out $12a^2b$ as the greatest common factor. (**Factoring**)

48. $2x^2 + 5$ has no common variables and 1 is the only number that divides evenly into 2 and 5. We say that $2x^2 + 5$ does not factor or is prime. (**Factoring**)

49. $5(x + y) + 6m(x + y) = (x + y)(5 + 6m)$ Notice that each term contains the expression $(x + y)$. This is the greatest common factor. (**Factoring**)

50. $3x + bx + 3y + by = (3x + bx) + (3y + by)$ Group the first two terms and the last two terms.

    $= x(3 + b) + y(3 + b)$ Factor each binomial.

    $= (3 + b)(x + y)$ Each term contains the expression $(3 + b)$, so that can be factored out.

    (**Factoring**)

51. $x^2y^2 + 4x^2 + 4y^2 + 16 = (x^2y^2 + 4x^2) + (4y^2 + 16)$ Group the first two terms and the last two terms.

    $= x^2(y^2 + 4) + 4(y^2 + 4)$ Factor each binomial.

    $= (y^2 + 4)(x^2 + 4)$ Factor out $y^2 + 4$. (**Factoring**)

52. $a^2x^2 + 4b^2 + 4a^2 + b^2x^2 = (a^2x^2 + 4b^2) + (4a^2 + b^2x^2)$ Grouping the first two terms and the last two terms does not lead to any possible common factoring. Try rearranging the terms:

$$a^2 x^2 + 4a^2 + 4b^2 + b^2 x^2 = (a^2 x^2 + 4a^2) + (4b^2 + b^2 x^2) \quad \text{Group the rearranged terms.}$$

$$= a^2 (x^2 + 4) + b^2 (4 + x^2) \quad \text{Note that although the terms in the parenthetical}$$

expressions are in different orders, the expressions are equal, and so $(x^2 + 4)$ can be factored out:

$$= (x^2 + 4)(a^2 + b^2)$$

(Factoring)

53. $x^2 - x - 6 = (x - 3)(x + 2)$ **(Factoring)**

54. $y^2 - 12y + 32 = (y - 4)(y - 8)$ **(Factoring)**

55. $x^2 + xy - 6y^2 = (x + 3y)(x - 2y)$ **(Factoring)**

56. $6x^2 - 13x - 5 = (3x + 1)(2x - 5)$ **(Factoring)**

57. $8x^2 - 26x + 15 = (4x - 3)(2x - 5)$ **(Factoring)**

58. $15x^2 + 22x + 8 = (5x + 4)(3x + 2)$ **(Factoring)**

59. $2x^2 - 14x + 24 = 2(x^2 - 7x + 12) \quad$ Factor out the common factor 2 first.

$$= 2(x - 3)(x - 4) \quad \textbf{(Factoring)}$$

60. $3x^3 y - 11x^2 y - 20xy = xy(3x^2 - 11x - 20) \quad$ Factor out the greatest common factor first.

$$= xy(3x + 4)(x - 5) \quad \textbf{(Factoring)}$$

61. $30x^5 + 132x^4 + 48x^3 = 6x^3(5x^2 + 22x + 8) \quad$ Factor out the greatest common factor first.

$$= 6x^3(5x + 2)(x + 4) \quad \textbf{(Factoring)}$$

62. $9x^2 - 25 = (3x)^2 - 5^2 = (3x - 5)(3x + 5) \quad$ Use the formula for the difference of two squares.
    (Factoring)

63. $16 - x^2 y^2 = (4 - xy)(4 + xy) \quad$ Use the formula for the difference of two squares. **(Factoring)**

64. $49x^2 + 4 \quad$ This is the sum of two squares and does not factor. **(Factoring)**

65. $4x^2 + 20xy + 25y^2 = (2x + 5y)^2 \quad$ Use the formula for a perfect square trinomial. Note that the middle
    term, $20xy$, is $2(2x)(5y)$ which is double the square root of the first term times the square root of the
    last term. This is a requirement for a trinomial to be a perfect square trinomial. **(Factoring)**

66. $36x^2 - 84x + 49 = (6x - 7)^2 \quad$ Use the formula for a perfect square trinomial. **(Factoring)**

67. $125x^3 - 1 = (5x)^3 - (1)^3 \quad$ Write each term as a perfect cube.

$$= (5x - 1)(25x^2 + 5x + 1) \quad \text{Use the formula for the difference of two cubes. } \textbf{(Factoring)}$$

68. $8a^3b^3 + 27c^3 = (2ab)^3 + (3c)^3$ Write each term as a perfect cube.

    $= (2ab + 3c)(4a^2b^2 - 6abc + 9c^2)$ Use the formula for the sum of two cubes. **(Factoring)**

69. $12ax^2 - 27ay^2 = 3a(4x^2 - 9y^2)$ Remember to factor out the greatest common factor first.

    $= 3a(2x - 3y)(2x + 3y)$ Factor the binomial using the formula for the difference of two squares. **(Factoring)**

70. $16a^3p + 54p = 2p(8a^3 + 27)$ Factor out the greatest common factor first.

    $= 2p((2a)^3 + (3)^3)$ Write each term as a perfect cube.

    $= 2p(2a + 3)(4a^2 - 6a + 9)$ Use the formula for the sum of two cubes. **(Factoring)**

71. $5x^4 - 5 = 5(x^4 - 1)$ Factor out the greatest common factor first.

    $= 5(x^2 - 1)(x^2 + 1)$ Factor the binomial as a difference of two squares. Note that you must continue factoring since one of the new factors $(x^2 - 1)$ can be factored:

    $= 5(x - 1)(x + 1)(x^2 + 1)$ The factor $x^2 + 1$ is a sum of two squares and does not factor further. **(Factoring)**

72. $x^2 + 4xy + 4y^2 - 9 = (x^2 + 4xy + 4y^2) - 9$ Group the first three terms. Note that they can be factored as a perfect square trinomial:

    $= (x + 2y)^2 - 9$ Now this binomial is a difference of two squares and can be factored:

    $= ((x + 2y) - 3)((x + 2y) + 3) = (x + 2y - 3)(x + 2y + 3)$ **(Factoring)**

73. $16 - x^2 + 6xy - 9y^2 = 16 - (x^2 - 6xy + 9y^2)$ Group the last three terms together and factor out -1.

    $= 16 - (x - 3y)^2$ Factor the trinomial as a perfect square trinomial. Note that the binomial is a difference of two squares and can be factored further:

    $= (4 - (x - 3y))(4 + (x - 3y))$ Be careful to use the parentheses around the term $x - 3y$.

    $= (4 - x + 3y)(4 + x - 3y)$ Remove the inner parentheses by distributing the signs properly. **(Factoring)**

74. $(3x - 1)(5x + 4) = 0$ The expression on the left side is already factored and equal to 0, so set each factor equal to zero and solve:

    $3x - 1 = 0$ or $5x + 4 = 0$

    $3x = 1$ $5x = -4$

    $x = \dfrac{1}{3}$ or $x = -\dfrac{4}{5}$

    (Solving equations by factoring)

75. $2x(x-1)(2x+3) = 0$ The expression on the left side is already factored and equal to 0, so set each factor equal to zero and solve:

$2x = 0$	or	$x - 1 = 0$	or	$2x + 3 = 0$
$x = 0$		$x = 1$		$2x = -3$
				$x = -\dfrac{3}{2}$

(Solving equations by factoring)

76. $4x^2 + 16x = 0$ 0 is already on the right side of the equation, but the left side must be factored:

$4x(x+4) = 0$ Now set each factor equal to 0 and solve:

$4x = 0$	$x + 4 = 0$
$x = 0$	$x = -4$

(Solving equations by factoring)

77. $\dfrac{1}{5}x^2 - \dfrac{3}{5}x - 2 = 0$ Since the equation contains fractions, it will be easier if we eliminate the fractions by multiplying by the least common denominator 5:

$$5\left(\frac{1}{5}x^2 - \frac{3}{5}x - 2\right) = 5(0)$$

$$x^2 - 3x - 10 = 0$$

Now factor the left side of the equation:

$$(x-5)(x+2) = 0$$

Set each factor equal to 0 and solve:

$x - 5 = 0$	$x + 2 = 0$
$x = 5$	$x = -2$

(Solving equations by factoring)

78. $8x^2 = 10x + 3$ Get 0 on the right side first.

$$8x^2 - 10x - 3 = 0$$

$(4x+1)(2x-3) = 0$ Factor the left side of the equation.

$4x + 1 = 0$	$2x - 3 = 0$
$4x = -1$	$2x = 3$
$x = -\dfrac{1}{4}$	$x = \dfrac{3}{2}$

(Solving equations by factoring)

79. $4x^2 = 16$

   $4x^2 - 16 = 0$ Get 0 on the right side first.

   $4(x^2 - 4) = 0$ Factor the left side. Note that after the common factor 4 is factored out, the binomial should be factored as a difference of two squares:

   $4(x-2)(x+2) = 0$

   $x - 2 = 0 \qquad x + 2 = 0$
   $x = 2 \qquad x = -2$

   Note that when setting the factors equal to 0, we cannot set 4 equal to 0 since $4 \neq 0$. (**Solving equations by factoring**)

80. $y(y-2) = 8$ It is very tempting to set each factor equal to 8 and solve, but the Zero-Factor property only applies when 0 is on one side of the equation.

   $y(y-2) - 8 = 0$

   $y^2 - 2y - 8 = 0$

   $(y-4)(y+2) = 0$

   $y - 4 = 0 \qquad y + 2 = 0$
   $y = 4 \qquad y = -2$

   (**Solving equations by factoring**)

Grade Yourself

Circle the question numbers that you had incorrect. Then indicate the number of questions you missed. If you answered more than three questions incorrectly, you need to focus on that topic. (If a topic has less than three questions and you had at least one wrong, we suggest you study that topic also. Read your textbook, a review book, or ask your teacher for help.)

Subject: Exponents and Polynomials

Topic	Question Numbers	Number Incorrect
Zero and negative exponents	1, 2, 3, 4, 5, 6, 7, 8	
Scientific notation	9, 10, 11, 12, 13, 21, 22, 23	
Properties of exponents	14, 15, 16, 17, 18, 19, 20	
Classifying polynomials	24, 25, 26	
Adding and subtracting polynomials	27, 28, 29, 30	
Evaluating polynomials	31, 32	
Multiplying polynomials	33, 34, 35, 36, 37, 38, 39, 40, 41, 42, 43, 44, 45	
Factoring	46, 47, 48, 49, 50, 51, 52, 53, 54, 55, 56, 57, 58, 59, 60, 61, 62, 63, 64, 65, 66, 67, 68, 69, 70, 71, 72, 73	
Solving equations by factoring	74, 75, 76, 77, 78, 79, 80	

Rational Expressions

4

 ## Test Yourself

Reducing Rational Expressions

A rational expression is in reduced form when there are no common factors in the numerator and denominator. To reduce a rational expression:
1. Factor numerator and denominator.
2. Divide out common factors.
3. Note factors that reduce to –1; e.g., $\dfrac{x-3}{3-x} = -1$ since $3 - x = (-1)(x-3)$.

Reduce to lowest terms.

1. $\dfrac{x^2 - 16}{x + 4}$

2. $\dfrac{2x^2 - 11x - 6}{x - 6}$

3. $\dfrac{8x^3 - y^3}{4x^2 - y^2}$

4. $\dfrac{12a + 8ab - 3b - 2b^2}{20a - 4ab - 5b + b^2}$

5. $\dfrac{x - 5y}{5y - x}$

6. $\dfrac{7xy - 1}{1 + 7xy}$

Multiplying Rational Expressions

To multiply rational expressions, use $\dfrac{A}{B} \cdot \dfrac{C}{D} = \dfrac{AC}{BD}$, $B, D \neq 0$, for A, B, C, D polynomials.
Factor each numerator and denominator and divide out common factors. All answers should be in reduced form.

Multiply.

7. $\dfrac{6a^2}{b^2} \cdot \dfrac{4b^8}{3a^3}$

8. $\dfrac{x^2 - 9}{x^2 - 7x + 12} \cdot \dfrac{x^2 - 16}{x^3 + 6x^2 + 9x}$

9. $\dfrac{r^3 - 8s^3}{r^2 - 4rs + 4s^2} \cdot \dfrac{r^2 - 4s^2}{r^2 + 4rs + 4s^2}$

10. $\dfrac{6x^2 - xy - 2y^2}{2x^3y - 9x^2y^2 - 5xy^3} \cdot \dfrac{2x^4 - 11x^3y + 5x^2y^2}{2y^2 - 3xy}$

Dividing Rational Expressions

To divide rational expressions, use
$\dfrac{A}{B} \div \dfrac{C}{D} = \dfrac{A}{B} \cdot \dfrac{D}{C} = \dfrac{AD}{BC}$, $B, C, D \neq 0$, for A, B, C, D polynomials. Factor each numerator and denominator, invert the divisor, and divide out common factors. All answers should be reduced.

Divide and reduce to lowest terms.

11. $\dfrac{6c^2d^3}{14c} \div \dfrac{12cd^2}{7d}$

12. $\dfrac{x^2 + 2x - 8}{x^2 - 2x} \div \dfrac{x^2 - 2x - 24}{x^2y^3}$

13. $\dfrac{12x^2 - 5x - 2}{16x^2 - 1} \div \dfrac{6x^2 - x - 2}{64x^3 - 1}$

14. $\dfrac{a + b}{a^2 - b^2} \div \dfrac{4ab}{a^2 - 2ab + b^2}$

Perform the indicated operations and reduce to lowest terms.

15. $\dfrac{4rs}{5t} \cdot \dfrac{10st^2}{12r} \div \dfrac{s^3}{6r^3}$

16. $\dfrac{3x^2y^3}{4z} \div 6xy^2$

17. $\dfrac{x^2 - x - 6}{x + 1} \div \left(\dfrac{x^2 - 9}{x^2 - 1} \cdot \dfrac{x - 1}{x + 3} \right)$

Adding and Subtracting Rational Expressions with Common Denominators

To add or subtract rational expression with a common denominator use:

$\dfrac{A}{C} + \dfrac{B}{C} = \dfrac{A + B}{C}$

$\dfrac{A}{C} - \dfrac{B}{C} = \dfrac{A - B}{C}$, for A, B, C polynomials and $C \neq 0$

All answers should be in reduced form.

Combine into a single fraction and reduce.

18. $\dfrac{6x}{3x - 1} - \dfrac{2}{3x - 1}$

19. $\dfrac{3}{5a^2} - \dfrac{2a - 1}{5a^2}$

20. $\dfrac{2x - 5}{x + 1} - \dfrac{6x - 2}{x + 1}$

21. $\dfrac{3x^2 - 10x}{(3x + 1)\,(4x - 3)} + \dfrac{3x^2 - 3x - 5}{(3x + 1)\,(4x - 3)}$

Adding and Subtracting Rational Expressions

Rational expressions must have the same denominator before they can be added or subtracted. The common denominator should contain each different factor raised to the highest exponent found in any denominator. To write a given rational expression as an equivalent rational expression with the common denominator, multiply the numerator *and* denominator by factors that are missing from the original denominator.

Combine into a single fraction and simplify.

22. $\dfrac{5}{8x} - \dfrac{3}{2x}$

23. $\dfrac{3}{y - 1} + \dfrac{2}{y}$

24. $3x - 2 + \dfrac{x}{x - 4}$

25. $\dfrac{4}{3x + 9} - \dfrac{7}{4x + 12}$

26. $\dfrac{6}{8x^2 - 14x - 15} - \dfrac{3}{2x - 5}$

27. $\dfrac{3}{8x^2y^3} - \dfrac{2}{6xy} + \dfrac{1}{4y^3}$

28. $\dfrac{1}{p^2}\left(\dfrac{p}{p - 4} + \dfrac{p}{p + 4} \right)$

29. $\dfrac{2x - y}{3x + 12y}\left(\dfrac{x}{2x - y} - \dfrac{4y}{y - 2x} \right)$

Complex Fractions

There are two methods used to simplify complex fractions.

Method 1:
1. Multiply every term by the least common denominator of all the denominators
2. Simplify and reduce if possible.

Method 2:
1. Simplify the numerator of the complex fraction to a single fraction.
2. Simplify the denominator of the complex fraction to a single fraction.
3. Invert and multiply by the divisor of the complex fraction.
4. Reduce if possible.

Simplify. Reduce all answers to lowest terms.

30. $\dfrac{\frac{5}{8}}{\frac{7}{12}}$

31. $\dfrac{\frac{x^2}{4x^3y}}{\frac{6xy^2}{2x^2y^3}}$

32. $\dfrac{\frac{x^2-1}{x-1}}{\frac{x-1}{x^2-1}}$

33. $\dfrac{5+\frac{2}{y}}{3-\frac{1}{y}}$

34. $\dfrac{6-\frac{1}{x}-\frac{15}{x^2}}{9-\frac{27}{x}+\frac{20}{x^2}}$

35. $\dfrac{\frac{6}{x^2}-\frac{6}{y^2}}{x-y}$

Dividing a Polynomial by a Monomial

When the divisor is a monomial, divide each term of the numerator by the denominator. Remember that exponents on the same base in a division are subtracted.

Divide.

36. $\dfrac{28y^8 - 14y^6 + 7y^2}{7y^2}$

37. $\dfrac{4a^2b^2 + 8a^3b - 6ab^3}{4ab^2}$

Dividing a Polynomial by a Polynomial

When the divisor is a polynomial with more than one term, the division follows rules similar to rules for long division with numbers.

Divide.

38. $\dfrac{3x^2 - 2x - 5}{x + 1}$

39. $(9x^3 - 3x^2 - 3x + 4) \div (3x + 2)$

40. $(x^4 - 2x + 5) \div (x - 2)$

Solving Equations Containing Rational Terms

To solve an equation containing rational terms:
1. Multiply both sides of the equation by the least common denominator.
2. Divide out common factors. Note: No denominators should remain after this step, if you have multiplied by the least common denominator.
3. Solve.
4. Eliminate any solutions that make any denominator in the original equation equal 0.

Solve.

41. $\dfrac{2}{x} - \dfrac{3}{4} = \dfrac{13}{4x} - 1$

42. $\dfrac{3}{2x + 1} = \dfrac{6}{x - 2}$

43. $1 + \dfrac{3}{x} = \dfrac{x^2 + 2x - 4}{x^2 - x}$

44. $\dfrac{2x}{x + 2} = \dfrac{x}{x + 3} - \dfrac{3}{x^2 + 5x + 6}$

Applications

If the equation used to solve a word problem contains fractions, use the technique of multiplying by the least common denominator to clear fractions.

45. If a person's heart beats 72 beats per minute, how often does it beat in 25 seconds?

46. If the ratio of the width of a rectangle to its length is $\frac{5}{9}$ and its length is 27 meters, find its width.

47. There are 2.54 centimeters per inch. How many centimeters are there in 1 foot?

48. The numerator of a fraction is 4 less than the denominator. If $\frac{1}{2}$ is added to the fraction, the result is $\frac{1}{14}$ more than twice the original fraction. Find the fraction.

✔ Check Yourself

1. $\dfrac{x^2 - 16}{x + 4} = \dfrac{(x + 4)(x - 4)}{x + 4}$ Factor the numerator.

    $= x - 4$ Divide out common factors. (**Reducing rational expressions**)

2. $\dfrac{2x^2 - 11x - 6}{x - 6} = \dfrac{(x - 6)(2x + 1)}{x - 6}$ Factor the numerator.

    $= 2x + 1$ Divide out common factors. (**Reducing rational expressions**)

3. $\dfrac{8x^3 - y^3}{4x^2 - y^2} = \dfrac{(2x - y)(4x^2 + 2xy + y^2)}{(2x - y)(2x + y)} = \dfrac{4x^2 + 2xy + y^2}{2x + y}$ Factor and divide out common factors.
    (**Reducing rational expressions**)

4. $\dfrac{12a + 8ab - 3b - 2b^2}{20a - 4ab - 5b + b^2} = \dfrac{(4a - b)(3 + 2b)}{(4a - b)(5 - b)} = \dfrac{3 + 2b}{5 - b}$ Factor by grouping; divide out common factors.
    (**Reducing rational expressions**)

5. $\dfrac{x - 5y}{5y - x} = -1$ Note that the terms on opposite sides of the subtraction signs are equal. This reduces to -1
    since $x - 5y = -(5y - x)$. (**Reducing rational expressions**)

6. $\dfrac{7xy - 1}{1 + 7xy}$ This does not reduce. (**Reducing rational expressions**)

7. $\dfrac{6a^2}{b^2} \cdot \dfrac{4b^8}{3a^3} = \dfrac{\overset{2}{\cancel{6a^2}}}{\cancel{b^2}} \cdot \dfrac{4b^{\overset{6}{\cancel{8}}}}{\underset{a}{\cancel{3a^3}}} = \dfrac{8b^6}{a}$ (**Multiplying rational expressions**)

8. $\dfrac{x^2-9}{x^2-7x+12} \cdot \dfrac{x^2-16}{x^3+6x^2+9x} = \dfrac{(x-3)\,(x+3)}{(x-3)\,(x-4)} \cdot \dfrac{(x-4)\,(x+4)}{x\,(x+3)\,(x+3)}$ Factor.

$= \dfrac{(x+4)}{x\,(x+3)}$ Divide out common factors. (**Multiplying rational expressions**)

9. $\dfrac{r^3-8s^3}{r^2-4rs+4s^2} \cdot \dfrac{r^2-4s^2}{r^2+4rs+4s^2} = \dfrac{(r-2s)\,(r^2+2rs+4s^2)}{(r-2s)\,(r-2s)} \cdot \dfrac{(r-2s)\,(r+2s)}{(r+2s)\,(r+2s)}$ Factor

$= \dfrac{r^2+2rs+4s^2}{r+2s}$ Divide out common factors $x-3, x-4,$ and $x+3.$

(Multiplying rational expressions)

10. $\dfrac{6x^2-xy-2y^2}{2x^3y-9x^2y^2-5xy^3} \cdot \dfrac{2x^4-11x^3y+5x^2y^2}{2y^2-3xy} = \dfrac{(3x-2y)\,(2x+y)}{xy\,(2x+y)\,(x-5y)} \cdot \dfrac{x^2\,(x-5y)\,(2x-y)}{y\,(2y-3x)}$ Factor.

$= \dfrac{-x\,(2x-y)}{y^2}$ Divide out common factors.

(Multiplying rational expressions)

11. $\dfrac{6c^2d^3}{14c} \div \dfrac{12cd^2}{7d} = \dfrac{6c^2d^3}{14c} \cdot \dfrac{7d}{12cd^2}$ Invert divisor and multiply.

$= \dfrac{d^2}{4}$ Divide out common factors. (**Dividing rational expressions**)

12. $\dfrac{x^2+2x-8}{x^2-2x} \div \dfrac{x^2-2x-24}{x^2y^3} = \dfrac{(x-2)\,(x+4)}{x\,(x-2)} \cdot \dfrac{x^2y^3}{(x+4)\,(x-6)}$ Invert divisor and multiply. Factor.

$= \dfrac{xy^3}{x-6}$ Divide out common factors. (**Dividing rational expressions**)

13. $\dfrac{12x^2-5x-2}{16x^2-1} \div \dfrac{6x^2-x-2}{64x^3-1} = \dfrac{(3x-2)\,(4x+1)}{(4x+1)\,(4x-1)} \cdot \dfrac{(4x-1)\,(16x^2+4x+1)}{(2x+1)\,(3x-2)}$ Invert and multiply. Factor.

$= \dfrac{16x^2+4x+1}{2x+1}$ Divide out common factors. (**Dividing rational expressions**)

14. $\dfrac{a+b}{a^2-b^2} \div \dfrac{4ab}{a^2-2ab+b^2} = \dfrac{a+b}{(a+b)\,(a-b)} \cdot \dfrac{(a-b)\,(a-b)}{4ab}$ Invert divisor and multiply. Factor.

$= \dfrac{a-b}{4ab}$ Divide out common factors. (**Dividing rational expressions**)

15. $\dfrac{4rs}{5t} \cdot \dfrac{10st^2}{12r} \div \dfrac{s^3}{6r^3} = \dfrac{4rs}{5t} \cdot \dfrac{10st^2}{12r} \cdot \dfrac{6r^3}{s^3}$ Invert divisor and multiply.

$$= \dfrac{4tr^3}{s} \text{ (Dividing rational expressions)}$$

16. $\dfrac{3x^2y^3}{4z} \div 6xy^2 = \dfrac{3x^2y^3}{4z} \cdot \dfrac{1}{6xy^2} = \dfrac{xy}{8z}$ Invert divisor and multiply. Divide out common factors.

   (Dividing rational expressions)

17. $\dfrac{x^2 - x - 6}{x + 1} \div \left(\dfrac{x^2 - 9}{x^2 - 1} \cdot \dfrac{x - 1}{x + 3} \right) = \dfrac{x^2 - x - 6}{x + 1} \div \left(\dfrac{(x - 3)\,(x + 3)}{(x - 1)\,(x + 1)} \cdot \dfrac{x - 1}{x + 3} \right)$ Simplify inside the parentheses by factoring and dividing out common factors.

   $= \dfrac{x^2 - x - 6}{x + 1} \div \left(\dfrac{x - 3}{x + 1} \right) = \dfrac{(x - 3)\,(x + 2)}{x + 1} \cdot \dfrac{x + 1}{x - 3}$ Invert and multiply.

   $= x + 2$ Divide out common factors. **(Dividing rational expressions)**

18. $\dfrac{6x}{3x - 1} - \dfrac{2}{3x - 1} = \dfrac{6x - 2}{3x - 1} = \dfrac{2\,(3x - 1)}{3x - 1} = 2$ Combine the numerators. Reduce by factoring and dividing out common factors. **(Adding and subtracting rational expressions)**

19. $\dfrac{3}{5a^2} - \dfrac{2a - 1}{5a^2} = \dfrac{3 - (2a - 1)}{5a^2}$ Be careful to distribute the subtraction across the whole numerator of the subtrahend.

   $= \dfrac{3 - 2a + 1}{5a^2} = \dfrac{-2a + 4}{5a^2}$ **(Adding and subtracting rational expressions)**

20. $\dfrac{2x - 5}{x + 1} - \dfrac{6x - 2}{x + 1} = \dfrac{2x - 5 - (6x - 2)}{x + 1} = \dfrac{2x - 5 - 6x + 2}{x + 1} = \dfrac{-4x - 3}{x + 1} = -\dfrac{4x + 3}{x + 1}$ Note that the negative can be factored out in front of the fraction by changing the signs of the numerator. **(Adding and subtracting rational expressions)**

21. $\dfrac{3x^2 - 10x}{(3x + 1)\,(4x - 3)} + \dfrac{3x^2 - 3x - 5}{(3x + 1)\,(4x - 3)} = \dfrac{6x^2 - 13x - 5}{(3x + 1)\,(4x - 3)} = \dfrac{(3x + 1)\,(2x - 5)}{(3x + 1)\,(4x - 3)} = \dfrac{2x - 5}{4x - 3}$

   (Adding and subtracting rational expressions)

22. $\dfrac{5}{8x} - \dfrac{3}{2x} = \dfrac{5}{8x} - \dfrac{3}{2x} \cdot \dfrac{4}{4} = \dfrac{5}{8x} - \dfrac{12}{8x}$ The least common denominator is $8x$. Multiply the second fraction by 4/4 to get an equivalent fraction with a denominator of $8x$.

   $= \dfrac{-7}{8x} = -\dfrac{7}{8x}$ Combine the numerators. **(Adding and subtracting rational expressions)**

23. $\dfrac{3}{y-1} + \dfrac{2}{y} = \dfrac{y}{y} \cdot \dfrac{3}{y-1} + \dfrac{y-1}{y-1} \cdot \dfrac{2}{y}$ The least common denominator is $y(y-1)$.

$= \dfrac{3y + 2y - 2}{y(y-1)} = \dfrac{5y - 2}{y(y-1)}$ Combine the numerators and simplify.

(Adding and subtracting rational expressions)

24. $3x - 2 + \dfrac{x}{x-4} = \dfrac{x-4}{x-4} \cdot \dfrac{3x-2}{1} + \dfrac{x}{x-4}$ The least common denominator is $x-4$.

$= \dfrac{3x^2 - 2x - 12x + 8}{x-4} + \dfrac{x}{x-4}$ Multiply the numerator and denominator of the first fraction.

$= \dfrac{3x^2 - 13x + 8}{x-4}$ Combine the numerators, keep the common denominator.

(Adding and subtracting rational expressions)

25. $\dfrac{4}{3x+9} - \dfrac{7}{4x+12} = \dfrac{4}{3(x+3)} - \dfrac{7}{4(x+3)}$ Factor each denominator. The LCD is $3 \cdot 4(x+3)$.

$= \dfrac{4}{4} \cdot \dfrac{4}{3(x+3)} - \dfrac{3}{3} \cdot \dfrac{7}{4(x+3)}$ Multiply each fraction by a form of 1 to get the LCD.

$= \dfrac{16 - 21}{12(x+3)}$ Multiply and combine the numerators.

$= \dfrac{-5}{12(x+3)}$ or $- \dfrac{5}{12(x+3)}$ The answer can be written with the negative in the

numerator or in front of the entire fraction. **(Adding and subtracting rational expressions)**

26. $\dfrac{6}{8x^2 - 14x - 15} - \dfrac{3}{2x-5} = \dfrac{6}{(2x-5)(4x+3)} - \dfrac{3}{2x-5}$ Factor the denominator of the first fraction.

$= \dfrac{6}{(2x-5)(4x+3)} - \dfrac{(4x+3)}{(4x+3)} \cdot \dfrac{3}{(2x-5)}$ Multiply by a form of 1.

$= \dfrac{6}{(2x-5)(4x+3)} - \dfrac{12x+9}{(4x+3)(2x-5)}$ Multiply.

$= \dfrac{6 - 12x - 9}{(2x-5)(4x+3)}$ Combine the numerators. Be careful to distribute the

subtraction across the whole numerator of the subtrahend.

$= \dfrac{-12x - 3}{(2x-5)(4x+3)}$ or $- \dfrac{12x+3}{(2x-5)(4x+3)}$ or $- \dfrac{3(4x+1)}{(2x-3)(4x+3)}$ Note: No com-

mon factors. **(Adding and subtracting rational expressions)**

27. $\dfrac{3}{8x^2y^3} - \dfrac{2}{6xy} + \dfrac{1}{4y^3} = \dfrac{3}{3} \cdot \dfrac{3}{8x^2y^3} - \dfrac{4xy^2}{4xy^2} \cdot \dfrac{2}{6xy} + \dfrac{6x^2}{6x^2} \cdot \dfrac{1}{4y^3}$ Multiply each fraction by a form of 1.

$$= \dfrac{9 - 8xy^2 + 6x^2}{24x^2y^3}$$ Combine the numerators.

(Adding and subtracting rational expressions)

28. $\dfrac{1}{p^2}\left(\dfrac{p}{p-4} + \dfrac{p}{p+4}\right) = \dfrac{1}{p^2}\left(\dfrac{p+4}{p+4} \cdot \dfrac{p}{p-4} + \dfrac{p-4}{p-4} \cdot \dfrac{p}{p+4}\right)$ Work inside the parenthesis to get the LCD.

$$= \dfrac{1}{p^2}\left(\dfrac{p^2 + 4p + p^2 - 4p}{(p+4)\,(p-4)}\right)$$ Multiply the numerators and combine.

$$= \dfrac{1}{p^2}\left(\dfrac{2p^2}{(p+4)\,(p-4)}\right)$$ Combine similar terms.

$$= \dfrac{2}{(p+4)\,(p-4)}$$ Divide out common factor of p^2.

(Adding and subtracting rational expressions)

29. $\dfrac{2x-y}{3x+12y}\left(\dfrac{x}{2x-y} - \dfrac{4y}{y-2x}\right) = \dfrac{2x-y}{3x+12y}\left(\dfrac{x}{2x-y} + \dfrac{4y}{2x-y}\right)$ Write $y - 2x = -2x + y = -(2x-y)$

$$= \dfrac{2x-y}{3x+12y} \cdot \dfrac{x+4y}{2x-y}$$ The fractions inside the parentheses have a common denominator, so combine the numerators.

$$= \dfrac{2x-y}{3\,(x+4y)} \cdot \dfrac{x+4y}{2x-y}$$ Factor.

$$= \dfrac{1}{3}$$ Divide out common factors.

(Adding and subtracting rational expressions)

30. $\dfrac{\frac{5}{8}}{\frac{7}{12}} = \dfrac{5}{8} \cdot \dfrac{12}{7} = \dfrac{15}{14}$ Use Method 2; invert and multiply by the divisor $\dfrac{7}{12}$. **(Complex fractions)**

31. $\dfrac{\frac{x^2}{4x^3y}}{\frac{6xy^2}{2x^2y^3}} = \dfrac{x^2}{4x^3y} \cdot \dfrac{2x^2y^3}{6xy^2} = \dfrac{1}{12}$ Use Method 2; invert and multiply by the divisor.

(Complex fractions)

32. $\dfrac{\dfrac{x^2-1}{x-1}}{\dfrac{x-1}{x^2-1}} = \dfrac{x^2-1}{x-1}\cdot\dfrac{x^2-1}{x-1}$ Use Method 2; invert and multiply.

$\qquad = \dfrac{(x-1)(x+1)}{x-1}\cdot\dfrac{(x-1)(x+1)}{x-1} = (x+1)^2$ Factor and divide out common factors.

(Complex fractions)

33. $\dfrac{5+\dfrac{2}{y}}{3-\dfrac{1}{y}} = \dfrac{y\left(5+\dfrac{2}{y}\right)}{y\left(3-\dfrac{1}{y}\right)}$ Use Method 1; multiply by the least common denominator, y.

$\qquad = \dfrac{5y+2}{3y-1}$ Use the distributive property to multiply. Divide out common factors.

(Complex fractions)

34. $\dfrac{6-\dfrac{1}{x}-\dfrac{15}{x^2}}{9-\dfrac{27}{x}+\dfrac{20}{x^2}} = \dfrac{x^2\left(6-\dfrac{1}{x}-\dfrac{15}{x^2}\right)}{x^2\left(9-\dfrac{27}{x}+\dfrac{20}{x^2}\right)}$ Use Method 1; multiply by the least common denominator, x^2.

$\qquad = \dfrac{6x^2-x-15}{9x^2-27x+20}$ Use the distributive property to multiply. Divide out common factors.

$\qquad = \dfrac{(3x-5)(2x+3)}{(3x-5)(3x-4)} = \dfrac{2x+3}{3x-4}$ Factor and divide out common factors.

(Complex fractions)

35. $\dfrac{\dfrac{6}{x^2}-\dfrac{6}{y^2}}{x-y} = \dfrac{x^2y^2\left(\dfrac{6}{x^2}-\dfrac{6}{y^2}\right)}{x^2y^2(x-y)}$ Use Method 1; multiply by the least common denominator, x^2y^2.

$\qquad = \dfrac{6y^2-6x^2}{x^2y^2(x-y)} = \dfrac{6(y^2-x^2)}{x^2y^2(x-y)} = \dfrac{6(y-x)(y+x)}{x^2y^2(x-y)}$ Multiply. Then factor the numerator.

$\qquad = \dfrac{-6(y+x)}{x^2y^2}$ Use $\dfrac{y-x}{x-y}=-1$

This problem can be easily done with Method 2 also:

$\dfrac{\dfrac{6}{x^2}-\dfrac{6}{y^2}}{x-y} = \dfrac{\dfrac{y^2}{y^2}\cdot\dfrac{6}{x^2}-\dfrac{x^2}{x^2}\cdot\dfrac{6}{y^2}}{x-y}$ Multiply in the numerator to get a single fraction.

$$= \frac{\dfrac{6y^2 - 6x^2}{x^2 y^2}}{\dfrac{x-y}{1}} = \frac{6y^2 - 6x^2}{x^2 y^2} \cdot \frac{1}{x-y}$$ Invert and multiply. From this point, the solution is the same as for

Method 1. (**Complex fractions**)

36. $\dfrac{28y^8 - 14y^6 + 7y^2}{7y^2} = \dfrac{28y^8}{7y^2} - \dfrac{14y^6}{7y^2} + \dfrac{7y^2}{7y^2}$ Divide each term of the numerator by the denominator.

$$= 4y^6 - 2y^4 + 1$$ Reduce each fraction. (**Dividing polynomials**)

37. $\dfrac{4a^2b^2 + 8a^3b - 6ab^3}{4ab^2} = \dfrac{4a^2b^2}{4ab^2} + \dfrac{8a^3b}{4ab^2} - \dfrac{6ab^3}{4ab^2}$ Divide each term of the numerator by the denominator.

$$= a + \frac{2a^2}{b} - \frac{3b}{2}$$ Reduce each fraction. (**Dividing polynomials**)

38. $\dfrac{3x^2 - 2x - 5}{x+1}$:

$$\begin{array}{r} 3x - 5 \\ x+1 \overline{) 3x^2 - 2x - 5} \\ \underline{3x^2 + 3x} \\ -5x - 5 \\ \underline{-5x - 5} \\ 0 \end{array}$$

The quotient is $3x - 5$, and the remainder is 0.

(**Dividing polynomials**)

39. $(9x^3 - 3x^2 - 3x + 4) \div (3x + 2)$:

$$\begin{array}{r} 3x^2 - 3x + 1 \\ 3x+2 \overline{) 9x^3 - 3x^2 - 3x + 4} \\ \underline{9x^3 + 6x^2} \\ -9x^2 - 3x \\ \underline{-9x^2 - 6x} \\ +3x + 4 \\ \underline{3x + 2} \\ 2 \end{array}$$

The quotient is $3x^2 - 3x + 1$, and the remainder is 2. (**Dividing polynomials**)

40. $(x^4 - 2x + 5) \div (x - 2)$:

$$\begin{array}{r} x^3 + 2x^2 + 4x + 6 \\ x - 2 \overline{\smash{\big)}\, x^4 + 0x^3 + 0x^2 - 2x + 5} \\ \underline{x^4 - 2x^3} \\ 2x^3 + 0x^2 \\ \underline{2x^3 - 4x^2} \\ 4x^2 - 2x \\ \underline{4x^2 - 8x} \\ 6x + 5 \\ \underline{6x - 12} \\ 17 \end{array}$$

Insert $0x^n$ for missing powers in dividend.

The quotient is $x^3 + 2x^2 + 4x + 6$ and the remainder is 17. The answer can also be written as

$x^3 + 2x^2 + 4x + 6 + \dfrac{17}{x - 2}$. **(Dividing polynomials)**

41. $\dfrac{2}{x} - \dfrac{3}{4} = \dfrac{13}{4x} - 1$ The LCD is $4x$, so multiply each side of the equation by $4x$.

$4x\left(\dfrac{2}{x} - \dfrac{3}{4}\right) = 4x\left(\dfrac{13}{4x} - 1\right)$

$4x \cdot \dfrac{2}{x} - 4x \cdot \dfrac{3}{4} = 4x \cdot \dfrac{13}{4x} - 4x \cdot 1$ Use the distributive property.

$8 - 3x = 13 - 4x$ Divide out common factors.

$x = 5$ Solve for x. **(Solving equations containing rational terms)**

42. $\dfrac{3}{2x + 1} = \dfrac{6}{x - 2}$ The LCD is $(2x + 1)(x - 2)$.

$(2x + 1)(x - 2)\left(\dfrac{3}{2x + 1}\right) = (2x + 1)(x - 2)\left(\dfrac{6}{x - 2}\right)$ Multiply each side by the LCD.

$(x - 2)\,3 = (2x + 1)\,6$ Divide out common factors.

$3x - 6 = 12x + 6$ Multiply.

$-9x = 12$

$x = -\dfrac{12}{9} = -\dfrac{4}{3}$ **(Solving equations containing rational terms)**

43. $1 + \dfrac{3}{x} = \dfrac{x^2 + 2x - 4}{x^2 - x}$ The LCD is $x(x - 1)$.

$x(x - 1)\left(1 + \dfrac{3}{x}\right) = x(x - 1)\left(\dfrac{x^2 + 2x - 4}{x^2 - x}\right)$ Multiply each side by the LCD.

$x(x-1) + 3(x-1) = x^2 + 2x - 4$ Divide out common factors.

$x^2 - x + 3x - 3 = x^2 + 2x - 4$ Multiply.

$x^2 + 2x - 3 = x^2 + 2x - 4$ Combine similar terms.

$1 = 0$

Since $1 \neq 0$, there is no solution to the equation, often written as $\varnothing$.
   (Solving equations containing rational terms)

44. $\dfrac{2x}{x+2} = \dfrac{x}{x+3} - \dfrac{3}{x^2+5x+6}$ Factor $x^2 + 5x + 6 = (x+2)(x+3)$. The LCD is $(x+2)(x+3)$

   $(x+2)(x+3)\left(\dfrac{2x}{x+2}\right) = (x+2)(x+3)\left(\dfrac{x}{x+3}\right) - (x+2)(x+3)\left(\dfrac{3}{(x+2)(x+3)}\right)$

   $(x+3)\,2x = (x+2)\,x - 3$

   $2x^2 + 6x = x^2 + 2x - 3$ Multiply.

   $x^2 + 4x + 3 = 0$

   $(x+3)(x+1) = 0$ Factor.

   $x+3 = 0 \quad x+1 = 0$ Set each factor equal to 0.

   $x = -3 \qquad x = -1$

   Since $x = -3$ would make one of the denominators in the original equation equal to 0, -3 cannot be a solution. The only solution is $x = -1$ (check in the original equation to verify).
   (Solving equations containing rational terms)

45. $\dfrac{72}{1 \text{ min}} = \dfrac{x}{25 \text{ sec}}$ Use a proportion.

   $\dfrac{72}{60 \text{ sec}} = \dfrac{x}{25 \text{ sec}}$ Make the units match by converting 1 minute to 60 seconds.

   $\dfrac{72(25)}{60} = x$ Multiply both sides by 25.

   $30 = x$

   Therefore, the person's heart would beat 30 times in 25 seconds. **(Applications)**

46. $\dfrac{5}{9} = \dfrac{x}{27}$ Use a proportion to compare width to length.

   $\dfrac{27 \cdot 5}{9} = x$ Multiply both sides of the equation by 27.

$15 = x$

Therefore, the width of the rectangle is 15 meters. (**Applications**)

47. $\dfrac{2.54 \text{ cm}}{1 \text{ in}} = \dfrac{x \text{ cm}}{1 \text{ ft}}$ Set up a proportion to compare centimeters to inches.

$\dfrac{2.54 \text{ cm}}{1 \text{ in}} = \dfrac{x \text{ cm}}{12 \text{ in}}$ Convert 1 foot into 12 inches.

$12\,(2.54) = x$ Multiply both sides of the equation by 12.

$30.48 = x$

Therefore, there are 30.48 centimeters in 1 foot. (**Applications**)

48. Let $x =$ the denominator of the fraction. Then $x - 4$ is the numerator of the fraction.

$\dfrac{x-4}{x} + \dfrac{1}{2} = \dfrac{1}{14} + 2\left(\dfrac{x-4}{x}\right)$ Write an equation to represent the given information.

$14x\left(\dfrac{x-4}{x} + \dfrac{1}{2}\right) = 14x\left(\dfrac{1}{14} + 2\left(\dfrac{x-4}{x}\right)\right)$ Multiply both sides by the LCD, $14x$.

$14\,(x-4) + 7x = x + 28\,(x-4)$ Distribute $14x$ and divide out common factors.

$14x - 56 + 7x = x + 28x - 112$ Multiply.

$21x - 56 = 29x - 112$ Combine similar terms.

$-8x = -56$

$x = 7$

Therefore, the numerator is $7 - 4 = 3$, so the fraction is $\dfrac{3}{7}$. (**Applications**)

Grade Yourself

Circle the question numbers that you had incorrect. Then indicate the number of questions you missed. If you answered more than three questions incorrectly, you need to focus on that topic. (If a topic has less than three questions and you had at least one wrong, we suggest you study that topic also. Read your textbook, a review book, or ask your teacher for help.)

Subject: *Rational Expressions*

Topic	Question Numbers	Number Incorrect
Reducing rational expressions	1, 2, 3, 4, 5, 6	
Multiplying rational expressions	7, 8, 9, 10	
Dividing rational expressions	11, 12, 13, 14, 15, 16, 17	
Adding and subtracting rational expressions	18, 19, 20, 21, 22, 23, 24, 25, 26, 27, 28, 29	
Complex fractions	30, 31, 32, 33, 34, 35	
Dividing polynomials	36, 37, 38, 39, 40	
Solving equations containing rational terms	41, 42, 43, 44	
Applications	45, 46, 47, 48	

Linear Equations and Inequalities in Two Variables

 Test Yourself

Graphing Points

When graphing an ordered pair (x, y) remember that the first number given is the x-coordinate (on the horizontal axis) and the second number is the y-coordinate (on the vertical axis).

1. Plot the points $(2, 3), (3, 2), (3, 0), (-1, 3)$, $(-2, -3)$, and $(4, -1)$ on the same set of axes.

Graphing Lines Using a Table of Values

One method used for graphing lines is to find three points on the line (two points *must* be found, the third is a check point) and draw a line through the points. It is often convenient to complete the table letting $x = 1, 2$, and 3, but any numbers can be substituted for x (or y) to find the other coordinate.

Graph.

2. $x - 2y = 6$

3. $y = 4x$

4. $3x + 2y = 6$

Graphing Linear Equations Using the *x*- and *y*-intercepts

The table of values is often limited to the x- and y-intercepts. When $x = 0$, the y value is called the y-intercept. When $y = 0$, the x value is called the x-intercept.

Graph.

5. $3x + 4y = 12$

6. $2x - y = 4$

7. $y = -2x$

Graphing Horizontal and Vertical Lines

The graph of $x = a$ is a vertical line through the point $(a, 0)$.

The graph of $y = b$ is a horizontal line through the point $(b, 0)$.

Graph.

8. $x = 2$

9. $y = -4$

10. $x - 3 = 0$

11. $y = 0$

The Distance Formula

The distance between two points (x_1, y_1) and (x_2, y_2) is $d = \sqrt{(x_2 - x_1)^2 + (y_2 - y_1)^2}$. Note that when given two points you may label either point as (x_1, y_1) and the other as (x_2, y_2).

Find the distance between each of the following pairs of points.

12. $(-2, 6)$ and $(-5, -1)$

13. $(4, 3)$ and $(0, 0)$

The Midpoint Formula

If (x_1, y_1) and (x_2, y_2) are the endpoints of a line segment, the midpoint of the line segment is

$$M = \left(\frac{x_1 + x_2}{2}, \frac{y_1 + y_2}{2} \right)$$

Be careful to use addition in this formula, and to write each answer as an ordered pair. One way to remember this formula is to note that the x-coordinate is the average of the x values and the y-coordinate is the average of the y values.

Find the midpoint of the line segment whose endpoints are given.

14. $(-2, 6)$ and $(-5, -1)$

15. $(4, 3)$ and $(0, 0)$

Finding the Slope of a Line Given Two Points

Given two points (x_1, y_1) and (x_2, y_2) on a line, the slope of the line is

$$m = \frac{y_2 - y_1}{x_2 - x_1}$$

Be careful to put the difference of the y-coordinates in the numerator and the difference of the x-coordinates in the denominator. However, it does *not* matter which point is labeled as (x_1, y_1) with the other as (x_2, y_2).

Find the slope of the line through the given pair of points.

16. $(2, 5)$ and $(4, 10)$.

17. $(3, 5)$ and $(6, -2)$.

18. $(2, 6)$ and $(2, -4)$.

19. $(-3, 4)$ and $(2, 4)$.

Graphing Lines Using the Slope-Intercept Formula

To graph a line using $y = mx + b$:
1. Solve the given equation for y.
2. Put a point on the y axis at b (b is the y-intercept).
3. Use slope $= m = \dfrac{\text{rise}}{\text{run}}$ to locate another point on the line.
4. Draw a line through the two points.

Graph.

20. $y = \dfrac{2}{3}x - 1$

21. $y = -\dfrac{1}{3}x + 2$

22. $3x + y = 4$

Slope of Parallel and Perpendicular Lines

Parallel line have equal slopes.
Perpendicular lines have slopes that are negative reciprocals of each other (for example,

$y = 2x + 3$ and $y = -\dfrac{1}{2}x + 3$ have slopes of 2

and $-1/2$, respectively).

---

23. Find the slope of a line parallel to the line $y = 3x + 5$

24. Find the slope of a line parallel to the line containing the points $(-4, 1)$ and $(2, -6)$.

25. Find the slope of a line perpendicular to the line $y = 4x - 3$

26. Find the slope of a line perpendicular to the line containing the points $(-3, 5)$ and $(-2, 3)$.

Finding the Equation of a Line

Given various information about a line, you should be able to find the equation of the line. You should memorize the following formulas.

$y = mx + b$ slope-intercept form of a line

$y - y_1 = m(x - x_1)$ point-slope form of a line

The equation of a horizontal line through (p, q) is

  $y = q$.

The equation of a vertical line through (p, q) is

  $x = p$.

$ax + by = c$ standard form of a line

$ax + by + c = 0$ general form of a line

27. Write an equation of the line with slope $\frac{2}{3}$ and y-intercept 2.

28. Write an equation of the line with $m = -\frac{1}{2}$ that contains the point $(-1, 3)$. Write your answer in standard form.

29. Write an equation of the line containing the points $(-3, 4)$ and $(2, -1)$. Write your answer in slope-intercept form.

30. Write an equation of the horizontal line through $(2, -3)$.

31. Write an equation of the vertical line through $(1, 6)$.

32. Write an equation of the line through $(-2, -3)$ and parallel to $2x + 3y = 6$. Write the answer in standard form.

33. Write an equation of the line through $(1, -4)$ and perpendicular to $y = 2x + 4$. Write the answer in slope-intercept form.

34. Write an equation of the line through $(2, 5)$ and parallel to $x = 4$.

35. Write an equation of the line through $(-4, -2)$ and perpendicular to $x = 2$.

Graphing Linear Inequalities

Graphing inequalities generally means there will be shading in the answer. Linear inequalities have either a solid or dotted line as a boundary, and shading on one side of the line.

The graph of a linear inequality containing $\leq$ or $\geq$ has a solid line as the boundary. The graph of a linear inequality containing $<$ or $>$ has a dotted line as the boundary.

To determine where shading occurs, choose a point not on the boundary. Substitute the point's coordinates into the given inequality. If the resulting inequality is true, shade the side of the line containing the test point. If the resulting inequality is false, shade the side of the line *not* containing the test point.

Graph.

36. $y \leq 2x - 3$

37. $2x - 4y < 8$

38. $x > -1$

39. $y \leq 3$

Graphing Systems of Inequalities

To graph a system of inequalities, graph and shade each inequality. If possible, use colored pencils to help you distinguish the shading.

If the intersection of the inequalities is required, the solution is where the shading overlaps.

If the union of the inequalities is required, the solution contains *all* the shaded areas.

40. $2x + 3y \leq 4$ and $x > -2$.

41. $-1 \leq y < 4$

42. $y < \frac{1}{2}x - 3$ or $x \leq 1$.

Variation

To translate word problems involving variation, you will need three basic formulas.

Direct variation: y varies directly as x or y is proportional to x: use $y = kx$

Indirect variation: y varies inversely as x or y varies indirectly as x: use $y = \frac{k}{x}$

Joint variation: y varies jointly as x and z: use $y = kxz$

43. y varies directly as x. If x is 20 when y is 90, find y when x is 40.

44. Hooke's Law states that the force needed to stretch a spring is proportional to the distance it is

stretched. If a force of 5 pounds stretches a spring a distance of 2 cm, find the force needed to stretch the same spring 12 cm.

45. Boyle's Law states that the pressure of a compressed gas is inversely proportional to the volume of the gas. If there is a pressure of 28 pounds per square inch when the volume of gas is 50 cubic inches, find the pressure when the gas has a volume of 200 cubic inches.

46. r varies jointly as s and t^3. If r is 36 when s is 4 and t is 2, find r when s is 3 and t is 4.

✓ Check Yourself

1.

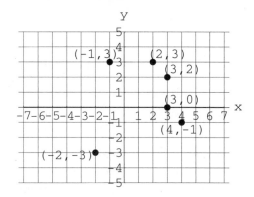

(Graphing lines)

2. $x - 2y = 6$ Use a table of values:

x	y
1	−5/2
2	−2
3	−3/2

(Graphing lines)

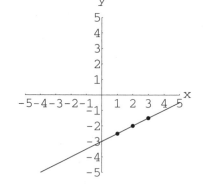

3. $y = 4x$ Use a table of values:

x	y
1	4
2	8
3	12

(Graphing lines)

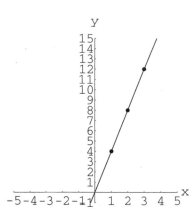

4. $3x + 2y = 6$ Use a table of values:

x	y
1	3/2
2	0
3	−3/2

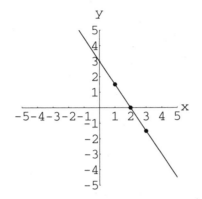

(Graphing lines)

5. $3x + 4y = 12$ Find the x- and y-intercepts:

x	y
0	3
4	0

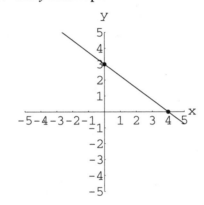

(Graphing lines)

6. $2x - y = 4$ Find the x- and y-intercepts:

x	y
0	−4
2	0

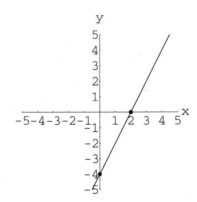

(Graphing lines)

7. $y = -2x$ Find the x- and y-intercepts:

x	y
0	0
0	0

Note that the *x* and *y*-intercept are the same. Since one point is not enough to determine a line, you must find another point. For example, let $x = 1$. Then $y = -2(1) = -2$:

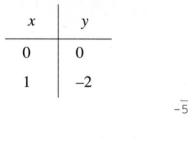

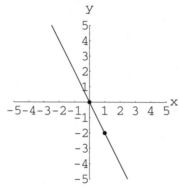

(Graphing lines)

8. $x = 2$ The graph will be a vertical line:

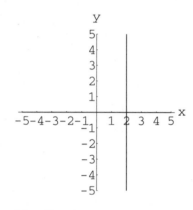

(Graphing lines)

9. $y = -4$ The graph will be a horizontal line:

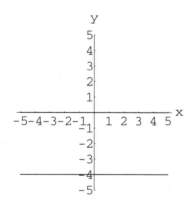

(Graphing lines)

10. $x - 3 = 0$ First write this as $x = 3$ by adding 3 to both sides of the equation. The graph is a vertical line:

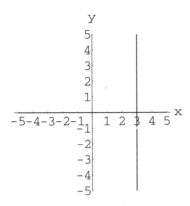

(Graphing lines)

11. $y = 0$ The graph is a horizontal line through the origin:

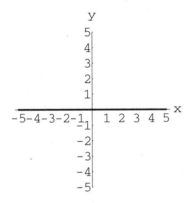

(Graphing lines)

12. Label $(-2, 6)$ as (x_1, y_1) and label $(-5, -1)$ as (x_2, y_2). Then substitute into the formula:

$$d = \sqrt{(-5 - (-2))^2 + (-1 - 6)^2}$$
$$= \sqrt{(-5 + 2)^2 + (-7)^2}$$
$$= \sqrt{9 + 49}$$
$$= \sqrt{58}$$

(The distance and midpoint formula)

13. Label $(4, 3)$ as (x_2, y_2) and $(0, 0)$ as (x_1, y_1).

$$d = \sqrt{(4 - 0)^2 + (3 - 0)^2}$$
$$= \sqrt{4^2 + 3^2}$$
$$= \sqrt{16 + 9}$$
$$= \sqrt{25}$$
$$= 5$$

(The distance and midpoint formula)

14. Label $(-2, 6)$ as (x_1, y_1) and $(-5, -1)$ as (x_2, y_2). Then substitute into the formula:

$$M = \left(\frac{-2 + (-5)}{2}, \frac{6 + (-1)}{2} \right)$$

$$= \left(-\frac{7}{2}, \frac{5}{2} \right)$$

(The distance and midpoint formula)

15. Label $(4, 3)$ as (x_1, y_1) and $(0, 0)$ as (x_2, y_2).

$$M = \left(\frac{4 + 0}{2}, \frac{3 + 0}{2} \right)$$

$$= \left(2, \frac{3}{2} \right)$$

(The distance and midpoint formula)

16. Let $(2, 5)$ be (x_1, y_1) and $(4, 10)$ be (x_2, y_2). Then substitute into the formula:

$$m = \frac{y_2 - y_1}{x_2 - x_1}$$

$$= \frac{10 - 5}{4 - 2}$$

$$= \frac{5}{2}$$

(Slope)

17. Let $(3, 5) = (x_1, y_1)$ and $(6, -2) = (x_2, y_2)$.

$$m = \frac{-2 - 5}{6 - 3} = -\frac{7}{3} \text{ (Slope)}$$

18. Let $(2, 6) = (x_1, y_1)$ and $(2, -4) = (x_2, y_2)$.

$$m = \frac{-4 - 6}{2 - 2} = -\frac{10}{0}$$ which is undefined (division by 0). Note that the given points are on a vertical line (they have the same x-coordinate). The slope of any vertical line is undefined. **(Slope)**

19. Let $(-3, 4) = (x_1, y_1)$ and $(2, 4) = (x_2, y_2)$.

$$m = \frac{4 - 4}{2 - (-3)} = \frac{0}{5} = 0$$. Note that the given points are on a horizontal line (they have the same y-coordinate). The slope of any horizontal line is 0. **(Slope)**

20. $y = \frac{2}{3}x - 1$ The y-intercept is $(0, -1)$. The slope is $\frac{2}{3}$ so rise 2 units and run 3 units from $(0, -1)$ to get the point $(3, 1)$.

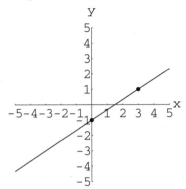

(Graphing lines)

21. $y = -\frac{1}{3}x + 2$ The y-intercept is $(0, 2)$. The slope is $-\frac{1}{3}$ so fall 1 unit and run 3 units from $(0, 2)$ to get the point $(3, 1)$.

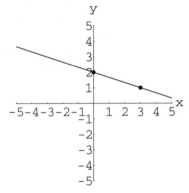

(Graphing lines)

22. $3x + y = 4$ Solve the equation for y to get the slope-intercept form: $y = -3x + 4$. The y-intercept is 4 and the slope is $-3 = -\frac{3}{1}$.

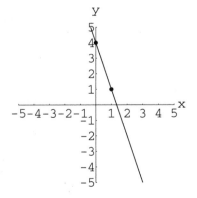

(Graphing lines)

23. Any line parallel to the line $y = 3x + 5$ has the same slope. Since this equation is already in slope-intercept form with slope 3, the slope of any line parallel to the line has slope 3. **(Slope)**

24. To find the slope of a line parallel to the line containing the points (–4, 1) and (2, –6), use the slope formula:

$$m = \frac{-6-1}{2-(-4)} = \frac{-7}{6} = -\frac{7}{6}.$$ Any line parallel to this line has the same slope, $-\frac{7}{6}$. **(Slope)**

25. The slope of the line $y = 4x - 3$ is 4. The slope of any line perpendicular to this line is the negative reciprocal of 4, which is $-\frac{1}{4}$. **(Slope)**

26. The slope of the line containing the points (–3, 5) and (–2, 3) is:

$$m = \frac{3-5}{-2-(-3)} = \frac{-2}{1} = -2$$

The slope of any line perpendicular to this has slope $\frac{1}{2}$. **(Slope)**

27. Use the slope-intercept form $y = mx + b$: An equation of the line with slope $\frac{2}{3}$ and y-intercept 2 is

$$y = \frac{2}{3}x + 2.$$ **(Finding the equation of a line)**

28. Since you are given a point and the slope, use the point-slope formula: $y - y_1 = m(x - x_1)$:

$$y - 3 = -\frac{1}{2}(x - (-1)).$$

To write the answer in standard form, the x and y term must be on the left side of the equation:

$$y - 3 = -\frac{1}{2}(x + 1)$$

$$2(y - 3) = 2\left(-\frac{1}{2}(x + 1)\right) \quad \text{Multiply both sides of the equation by 2.}$$

$$2y - 6 = -x - 1$$

$$x + 2y = 5 \quad \text{Answer is now in standard form } ax + by = c. \textbf{ (Finding the equation of a line)}$$

29. When you are given two points, first find the slope using $m = \frac{y_2 - y_1}{x_2 - x_1}$:

$$m = \frac{-1-4}{2-(-3)} = \frac{-5}{5} = -1.$$ Now use the point-slope formula.

$$y - 4 = -1(x - (-3))$$

To write the answer in slope-intercept form, isolate y:

$$y - 4 = -x - 3$$

$$y = -x + 1 \textbf{ (Finding the equation of a line)}$$

30. $y = -3$ **(Finding the equation of a line)**

31. $x = 1$ **(Finding the equation of a line)**

32. The required line must have the same slope as $2x + 3y = 6$ since parallel lines have equal slopes. Write $2x + 3y = 6$ in slope-intercept form to find its slope:

$$3y = -2x + 6$$

$$y = -\frac{2}{3}x + 2$$

Now use the point-slope formula with point $(-2, -3)$ and slope $-\frac{2}{3}$:

$$y - (-3) = -\frac{2}{3}(x - (-2))$$

To write the answer in standard form, clear fractions and get the x and y term on the left side of the equation:

$$3(y + 3) = 3\left(-\frac{2}{3}(x + 2)\right)$$

$$3y + 9 = -2x - 4$$

$$2x + 3y = -13 \text{ (Finding the equation of a line)}$$

33. To be perpendicular to $y = 2x + 4$, the required line must have slope $-\frac{1}{2}$. Now use the point-slope formula:

$$y - (-4) = -\frac{1}{2}(x - 1)$$

To write the answer in slope-intercept form, isolate y:

$$y + 4 = -\frac{1}{2}x + \frac{1}{2}$$

$$y = -\frac{1}{2}x - \frac{7}{2} \text{ (Finding the equation of a line)}$$

34. Since $x = 4$ is a vertical line, any line parallel to it must also be a vertical line. An equation of the vertical line through $(2, 5)$ is $x = 2$. **(Finding the equation of a line)**

35. Since $x = 2$ is a vertical line, any line perpendicular to it must be a horizontal line. An equation of the horizontal line through $(-4, -2)$ is $y = -2$. **(Finding the equation of a line)**

36. The boundary line should be solid because the inequality contains $\leq$. The boundary line is $y = 2x - 3$. Check $(0, 0)$ in the inequality $y \leq 2x - 3$:

$$0 \leq 2(0) - 3$$
$$0 \leq -3$$

Since 0 is not less than -3, this is a false statement. Shade the side of the line that does not contain the test point $(0, 0)$.

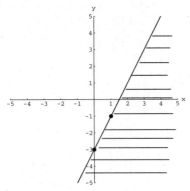

(Graphing linear inequalities)

37. The boundary line should be dotted because the inequality contains <. The boundary line is $2x - 4y = 8$. Check $(0, 0)$ in the inequality:

$$2x - 4y < 8$$
$$2(0) - 4(0) < 8 \quad \text{Substitute } (0, 0) \text{ for } (x, y)$$
$$0 < 8$$

Since 0 is less than 8, this is a true statement. Shade the side of the line that contains the test point $(0, 0)$.

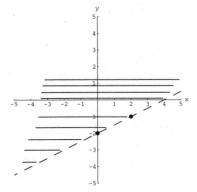

(Graphing linear inequalities)

38. The boundary line is a vertical dotted line. The boundary line is $x = -1$. Check $(0, 0)$ in the inequality.

$$x > -1$$
$$0 > -1$$

Since 0 is greater than −1, this is a true statement. Shade the side of the line that contains the test point $(0, 0)$.

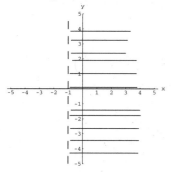

(Graphing linear inequalities)

39. The boundary line is a solid horizontal line. The boundary line is $y = 3$. Check $(0, 0)$ in the inequality.

$y \leq 3$

$0 \leq 3$

Since $0 \leq 3$ is a true statement, shade the side of the line that contains $(0, 0)$.

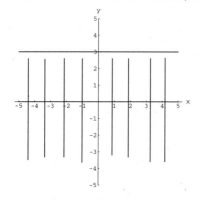

(Graphing linear inequalities)

40. The boundary line for $2x + 3y \leq 4$ is a solid line, and the boundary line for $x > -2$ is a dotted line. The word "and" joining the two inequalities indicates that the intersection of the graphs is required. Shade each graph and use only the part of the shading that overlaps:

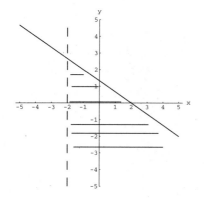

(Graphing systems of inequalities)

41. The inequality means $y \geq -1$ and $y < 4$. The horizontal boundary line $y = -1$ is a solid line, and the horizontal boundary line $y = 4$ is a dotted line.

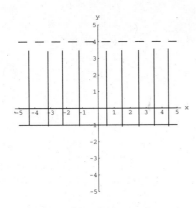

(Graphing systems of inequalities)

42. The word "or" means to include any shading in the answer.

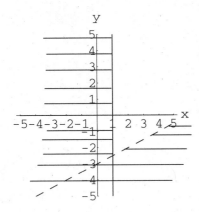

(Graphing systems of inequalities)

43. Since y varies directly as x, use $y = kx$. Use the value of $x = 20$ when $y = 90$ to find the value of k:

$$y = kx$$
$$90 = k(20)$$
$$\frac{90}{20} = k$$
$$\frac{9}{2} = k$$

Now find y when $x = 40$:

$$y = \frac{9}{2}(40)$$
$$= 180$$

(Variation)

44. Let F be the force and x be the distance the spring is stretched. Then $F = kx$. Find the value of k:

$F = kx$

$5 = k(2)$

$\dfrac{5}{2} = k$

Now find F when $x = 12$:

$F = \dfrac{5}{2}(12) = 30$ pounds. **(Variation)**

45. Let P = pressure and V = volume. Then Boyle's Law is

$P = \dfrac{k}{V}$.

Find k:

$28 = \dfrac{k}{50}$

$28(50) = k$

$1400 = k$

Now use the value of k to find P when $V = 200$:

$P = \dfrac{1400}{200} = 7$

Therefore, the pressure is 7 pounds per square inch. **(Variation)**

46. Use the formula for joint variation to write: $r = kst^3$

Find k when $r = 36$ and $s = 4$ and $t = 2$:

$36 = k(4)(2)^3$

$36 = k(32)$

$\dfrac{36}{32} = k$

$\dfrac{9}{8} = k$

Now find r when $s = 3$ and $t = 4$:

$r = \dfrac{9}{8}(3)(4)^3$

$r = \dfrac{9}{8}(3)(64)$

$r = 216$

(Variation)

Grade Yourself

Circle the question numbers that you had incorrect. Then indicate the number of questions you missed. If you answered more than three questions incorrectly, you need to focus on that topic. (If a topic has less than three questions and you had at least one wrong, we suggest you study that topic also. Read your textbook, a review book, or ask your teacher for help.)

Subject: Linear Equations and Inequalities in Two Variables

Topic	Question Numbers	Number Incorrect
Graphing lines	1, 2, 3, 4, 5, 6, 7, 8, 9, 10, 11, 20, 21, 22	
The distance and midpoint formula	12, 13, 14, 15	
Slope	16, 17, 18, 19, 23, 24, 25, 26	
Finding the equation of a line	27, 28, 29, 30, 31, 32, 33, 34, 35	
Graphing linear inequalities	36, 37. 38, 39	
Graphing systems of inequalities	40, 41, 42	
Variation	43, 44, 45, 46	

Exponents and Radicals

6

 Test Yourself

Rational Exponents

The exponent laws used previously for integer exponents can be expanded to include rational exponents by defining:

$x^{1/n} = \sqrt[n]{x}$ for $x \geq 0$ and n a positive integer

$x^{1/n} = \sqrt[n]{x}$ for $x < 0$ and n an odd positive integer

$x^{m/n} = (x^{1/n})^m$ provided $x^{1/n}$ is a real number.

Note that these definitions imply that the even root of a negative number is not a real number.

Simplify.

1. $81^{1/2}$

2. $125^{1/3}$

3. $(-64)^{1/4}$

4. $(-32)^{1/5}$

5. $27^{2/3}$

6. $16^{3/4}$

Exponent Laws

The laws previously used for integers also apply when a and b in the following are rational:

$$x^a \cdot x^b = x^{a+b} \qquad (x^a)^b = x^{ab}$$

$$\frac{x^a}{x^b} = x^{a-b} \qquad \left(\frac{x}{y}\right)^a = \frac{x^a}{y^a}$$

$$(xy)^a = x^a y^a \qquad x^{-a} = \frac{1}{x^a}$$

Note: Division by zero is not allowed.

Simplify. Write all answers with positive exponents.

7. $5^{1/3} \cdot 5^{1/4}$

8. $\dfrac{6^{2/3}}{6^{1/4}}$

9. $(64x^3 y^{1/4})^{1/3}$

10. $\left(\dfrac{27}{125}\right)^{4/3}$

11. $25^{-1/2}$

12. $y^{2/3} \cdot y^{-1/5}$

13. $\dfrac{(a^{2/3})^2}{(a^{1/4})^3}$

14. $(25m^3 n^4)^{-1/2}$

15. $\left(\dfrac{x^{-2/3}}{y^{-1/5}}\right)^4 (x^{1/3} y^{-2/5})^{-2}$

71

Radicals

Rational exponents are often written using radical notation. From the definition we have

$$x^{m/n} = \left(\sqrt[n]{x}\right)^m = \sqrt[n]{x^m}$$

Write each of the following in radical form.

16. $11^{1/2}$

17. $5^{2/3}$

18. $4x^{2/3}$

19. $(3x)^{1/4}$

Write each of the following in exponent form.

20. $\sqrt{21}$

21. $\sqrt[3]{x^2}$

22. $\left(\sqrt[4]{y}\right)^3$

Simplifying Radicals

A radical is in simplified form when:
1. The index of the radical is greater than any exponent in the radicand.
2. There are no fractions under the radical.
3. There are no radicals in the denominator of a fraction.

Note that in all of the exercises, it is assumed that all variables represent nonnegative real numbers.

Simplify.

23. $\sqrt{56x^4y^5}$

24. $\sqrt[3]{135a^4b^6z}$

25. $\sqrt{\dfrac{25}{49}}$

26. $\sqrt{\dfrac{3}{100}}$

27. $\sqrt[3]{\dfrac{1}{8}}$

28. $\sqrt{\dfrac{y}{4}}$

Multiplying Monomials

When the indices on two radicals are equal, the radicals can be multiplying by:

$$\left(a\sqrt[n]{x}\right)\left(b\sqrt[n]{y}\right) = ab\sqrt[n]{xy}$$

That is, multiply the coefficients, multiply the radicands, and keep the index the same.

Simplify.

29. $\left(2\sqrt{3xy}\right)\left(3\sqrt{6x}\right)$

30. $\left(\sqrt[3]{5x}\right)\left(3\sqrt[3]{75x^2y}\right)$

Rationalizing a Monomial Denominator

To rationalize a denominator that is a monomial, multiply the numerator and denominator by a radical that will make the denominator rational. Often, you can multiply by the denominator.

31. $\dfrac{4}{\sqrt{3}}$

32. $\dfrac{2}{\sqrt{27}}$

33. $\dfrac{2\sqrt{3x}}{\sqrt{5y}}$

34. $\sqrt[3]{\dfrac{3x}{4y}}$

Adding and Subtracting Radical Expressions

Similar radicals (same index, same radicand) can be combined into a single term. To add (or subtract) similar radicals, add (or subtract) the expressions in front of the radicals and keep the index and radicand the same. You may have to simplify some of the terms in an expression before it can be simplified.

Simplify.

35. $4\sqrt{27} + 2\sqrt{75}$

36. $2\sqrt{125} - 4\sqrt{20} + 6\sqrt{45}$

37. $4x\sqrt{6x^4y^3} - 7y\sqrt{24x^6y}$

38. $x\sqrt[3]{54x^4y^5} - 8x\sqrt[3]{2x^4y^5}$

Multiplying Radical Expressions

Combine the distributive property and the rule

$(a\sqrt[n]{x})(b\sqrt[n]{y}) = ab\sqrt[n]{xy}$ to multiply radical expressions involving more than monomials.

Multiply.

39. $3\sqrt{2}(5\sqrt{10} + 4\sqrt{3})$

40. $(2 + 5\sqrt{3})(3 - 4\sqrt{3})$

41. $(4\sqrt{3} - 2)(4\sqrt{3} + 2)$

42. $(\sqrt{x} + 3\sqrt{y})(2\sqrt{x} - \sqrt{y})$

43. $(\sqrt{10} - 9)^2$

Rationalizing a Binomial Denominator

When the denominator contains a binomial with one or more radicals, multiply the numerator and denominator by the conjugate of the denominator (that is, same terms with the opposite sign between terms). Use the fact that

$(a - b)(a + b) = a^2 - b^2$

to shorten your work.

Rationalize the denominator and simplify.

44. $\dfrac{6}{3 - \sqrt{5}}$

45. $\dfrac{3 + 2\sqrt{5}}{4\sqrt{2} + 3\sqrt{5}}$

46. $\dfrac{3 + \sqrt{x}}{4 - \sqrt{x}}$

Solving Radical Equations

To solve an equation containing one or more square roots, isolate one of the radicals and square both sides of the equation. If there is still a square root in the equation, isolate it and square both sides. When no radicals containing variables remain, solve the resulting equation.

Be careful to check *all* solutions. This process of squaring both sides of an equation can produce answers that do not check.

Note: If the radicals in the equation are not square roots, squaring both sides will not eliminate them. In general, raise both sides of the equation to the power that is the index of the given radical.

Solve.

47. $\sqrt{2x - 4} = 4$

48. $\sqrt{5y + 3} + 2 = 0$

49. $\sqrt[3]{3t + 6} - 3 = 0$

50. $\sqrt{x + 1} = 2 - \sqrt{x}$

51. $\sqrt{3x + 10} - 2 = \sqrt{2x - 1}$

52. $\sqrt{x - 1} - 3 = \sqrt{x + 2}$

Imaginary Numbers

The complex number system uses the imaginary number i defined by:

$$\sqrt{-1} = i \text{ and } i^2 = -1$$

Square roots containing negative numbers can be simplified by first factoring out $\sqrt{-1}$ as i.

Simplify.

53. $\sqrt{-25}$

54. $\sqrt{-72}$

Adding and Subtracting Complex Numbers

Although i is a constant, we can add or subtract complex numbers by treating i like a variable.

Perform the indicated operations.

55. $(4 + 7i) + (2 - 3i)$

56. $(2 - 8i) - (6 + i)$

57. $6 - (3 - 2i)$

Multiplying Complex Numbers

Use the distributive property to multiply complex numbers. Remember to simplify any powers of i ($i^2 = -1$, $i^3 = -i$, $i^4 = 1$, etc.).

Perform the indicated operations.

58. $2i(3 - 7i)$

59. $(2 + 5i)(3 + 6i)$

60. $(4 - 3i)(4 + 3i)$

Dividing Complex Numbers

A complex number is not simplified if the denominator contains i. The following steps are used to simplify quotients:
1. If the denominator is a monomial containing i, multiply by $\dfrac{i}{i}$ and simplify.

2. If the denominator is a binomial containing i, multiply the numerator and denominator by the conjugate of the denominator.

Writing Complex Numbers in Standard Form

The standard form of a complex number is $a + bi$, where a and b are real numbers. If the directions on a test indicate that all answers must be in standard form, be careful to write answers in that form. Fractions such as $\dfrac{2 + 5i}{6}$ should be written as $\dfrac{1}{3} + \dfrac{5}{6}i$.

Find each quotient. Write each answer in standard form.

61. $\dfrac{6 - 3i}{i}$

62. $\dfrac{3 + 2i}{4i}$

63. $\dfrac{2 + 5i}{4 - 3i}$

Check Yourself

1. $81^{1/2} = \sqrt{81} = 9$ **(Rational exponents)**

2. $125^{1/3} = \sqrt[3]{125} = 5$ **(Rational exponents)**

3. $(-64)^{1/4} = \sqrt[4]{-64}$, which is not a real number. **(Rational exponents)**

4. $(-32)^{1/5} = \sqrt[5]{-32} = -2$. Note that the odd root of a negative number is negative, while the even root of a negative number cannot be simplified using the real number system. **(Rational exponents)**

5. $27^{2/3} = (\sqrt[3]{27})^2 = 3^2 = 9$. Or by factoring 27 as 3^3, we can write:

$$27^{2/3} = (3^3)^{2/3} = 3^{3 \cdot \frac{2}{3}} = 3^2 = 9 \text{ (Rational exponents)}$$

6. $16^{3/4} = \left(\sqrt[4]{16}\right)^3 = 2^3 = 8$ **(Rational exponents)**

7. $5^{1/3} \cdot 5^{1/4} = 5^{\frac{1}{3}+\frac{1}{4}} = 5^{\frac{4+3}{12}} = 5^{7/12}$ **(Rational exponents)**

8. $\dfrac{6^{2/3}}{6^{1/4}} = 6^{\frac{2}{3}-\frac{1}{4}} = 6^{\frac{8-3}{12}} = 6^{5/12}$ **(Rational exponents)**

9. $\left(64x^3y^{1/4}\right)^{1/3} = 64^{1\cdot\frac{1}{3}} x^{3\cdot\frac{1}{3}} y^{\frac{1}{4}\cdot\frac{1}{3}}$

$$= 64^{1/3} x^1 y^{1/12} = 4xy^{1/12} \textbf{ (Rational exponents)}$$

10. $\left(\dfrac{27}{125}\right)^{4/3} = \dfrac{27^{4/3}}{125^{4/3}} = \dfrac{3^4}{5^4} = \dfrac{81}{625}$ **(Rational exponents)**

11. $25^{-1/2} = \dfrac{1}{25^{1/2}} = \dfrac{1}{\sqrt{25}} = \dfrac{1}{5}$ **(Rational exponents)**

12. $y^{2/3} \cdot y^{-1/5} = y^{\frac{2}{3}+\left(-\frac{1}{5}\right)} = y^{\frac{10-3}{15}} = y^{7/15}$ **(Rational exponents)**

13. $\dfrac{\left(a^{2/3}\right)^2}{\left(a^{1/4}\right)^3} = \dfrac{a^{\frac{2}{3}\cdot 2}}{a^{\frac{1}{4}\cdot 3}} = \dfrac{a^{4/3}}{a^{3/4}} = a^{\frac{4}{3}-\frac{3}{4}} = a^{\frac{16-9}{12}} = a^{7/12}$ **(Rational exponents)**

14. $\left(25m^3n^4\right)^{-1/2} = \dfrac{1}{\left(25m^3n^4\right)^{1/2}}$ Use the definition of negative exponent.

$$= \dfrac{1}{25^{1/2}m^{3/2}n^{4/2}} = \dfrac{1}{5m^{3/2}n^2} \textbf{ (Rational exponents)}$$

15. $\left(\dfrac{x^{-2/3}}{y^{-1/5}}\right)^4 \left(x^{1/3}y^{-2/5}\right)^{-2} = \left(\dfrac{x^{-8/3}}{y^{-4/5}}\right)\left(x^{-2/3}y^{4/5}\right)$

$$= x^{-\frac{8}{3}+\left(-\frac{2}{3}\right)} y^{\frac{4}{5}-\left(-\frac{4}{5}\right)}$$

$$= x^{-10/3}y^{8/5}$$

$$= \dfrac{y^{8/5}}{x^{10/3}} \quad \text{Write the answer with positive exponents. \textbf{(Rational exponents)}}$$

16. $11^{1/2} = \sqrt{11}$ Note that an index of 2 is not written. **(Radicals)**

17. $5^{2/3} = \sqrt[3]{5^2} = \sqrt[3]{25}$ **(Radicals)**

18. $4x^{2/3} = 4\sqrt[3]{x^2}$ Note that the exponent of 2/3 applies only to x by the order of operations, not 4. **(Radicals)**

19. $(3x)^{1/4} = \sqrt[4]{3x}$ Note that the exponent applies to the whole expression in parentheses. **(Radicals)**

20. $\sqrt{21} = 21^{1/2}$ Note that the understood index is 2. **(Radicals)**

21. $\sqrt[3]{x^2} = x^{2/3}$ **(Radicals)**

22. $\left(\sqrt[4]{y}\right)^3 = y^{3/4}$ **(Radicals)**

23. $\sqrt{56x^4y^5} = \sqrt{2^3 \cdot 7x^4y^5} = 2x^2y^2\sqrt{14y}$ **(Simplifying radicals)**

24. $\sqrt[3]{135a^4b^6z} = \sqrt[3]{3^3 \cdot 5a^4b^6z} = 3ab^2\sqrt[3]{5az}$ **(Simplifying radicals)**

25. $\sqrt{\dfrac{25}{49}} = \dfrac{5}{7}$ **(Simplifying radicals)**

26. $\sqrt{\dfrac{3}{100}} = \dfrac{\sqrt{3}}{\sqrt{100}} = \dfrac{\sqrt{3}}{10}$ **(Simplifying radicals)**

27. $\sqrt[3]{\dfrac{1}{8}} = \dfrac{1}{2}$ **(Simplifying radicals)**

28. $\sqrt{\dfrac{y}{4}} = \dfrac{\sqrt{y}}{\sqrt{4}} = \dfrac{\sqrt{y}}{2}$ **(Simplifying radicals)**

29. $(2\sqrt{3xy})(3\sqrt{6x}) = 2 \cdot 3\sqrt{(3xy)(6x)} = 6\sqrt{18x^2y} = 6\sqrt{3^2 \cdot 2x^2y} = 18x\sqrt{2y}$ **(Multiplying and dividing radicals)**

30. $(\sqrt[3]{5x})\left(3\sqrt[3]{75x^2y}\right) = 1 \cdot 3 \ \sqrt[3]{(5x)(75x^2y)} = 3\sqrt[3]{375x^3y} = 3\sqrt[3]{5^3 \cdot 3x^3y} = 15x \ \sqrt[3]{3y}$ **(Multiplying and dividing radicals)**

31. $\dfrac{4}{\sqrt{3}} = \dfrac{4}{\sqrt{3}} \cdot \dfrac{\sqrt{3}}{\sqrt{3}}$ Multiply numerator and denominator by $\sqrt{3}$.

$\qquad = \dfrac{4\sqrt{3}}{\sqrt{9}} = \dfrac{4\sqrt{3}}{3}$ **(Multiplying and dividing radicals)**

32. $\dfrac{2}{\sqrt{27}} = \dfrac{2}{\sqrt{3^3}}$

$\qquad = \dfrac{2}{3\sqrt{3}}$

$\qquad = \dfrac{2}{3\sqrt{3}} \cdot \dfrac{\sqrt{3}}{\sqrt{3}}$ Note that you need only multiply by $\dfrac{\sqrt{3}}{\sqrt{3}}$, not $\dfrac{3\sqrt{3}}{3\sqrt{3}}$.

$\qquad = \dfrac{2\sqrt{3}}{3\sqrt{9}}$

$$= \frac{2\sqrt{3}}{3 \cdot 3} \quad \text{Simplify the radical.}$$

$$= \frac{2\sqrt{3}}{9} \quad \textbf{(Multiplying and dividing radicals)}$$

33. $\dfrac{2\sqrt{3x}}{\sqrt{5y}} = \dfrac{2\sqrt{3x}}{\sqrt{5y}} \cdot \dfrac{\sqrt{5y}}{\sqrt{5y}}$

$$= \frac{2\sqrt{15xy}}{\left(\sqrt{5y}\right)^2}$$

$$= \frac{2\sqrt{15xy}}{5y} \quad \textbf{(Multiplying and dividing radicals)}$$

34. $\sqrt[3]{\dfrac{3x}{4y}} = \dfrac{\sqrt[3]{3x}}{\sqrt[3]{2^2 y}}$

$$= \frac{\sqrt[3]{3x}}{\sqrt[3]{2^2 y}} \cdot \frac{\sqrt[3]{2y^2}}{\sqrt[3]{2y^2}}$$

$$= \frac{\sqrt[3]{6xy^2}}{\sqrt[3]{2^3 y^3}}$$

$$= \frac{\sqrt[3]{6xy^2}}{2y} \quad \textbf{(Multiplying and dividing radicals)}$$

35. $4\sqrt{27} + 2\sqrt{75} = 4\sqrt{9 \cdot 3} + 2\sqrt{25 \cdot 3} \quad$ Factor each radicand.

$$= 4 \cdot 3\sqrt{3} + 2 \cdot 5\sqrt{3} \quad \text{Factor out perfect squares.}$$

$$= 12\sqrt{3} + 10\sqrt{3}$$

$$= (12 + 10)\sqrt{3} \quad \text{Use the distributive property.}$$

$$= 22\sqrt{3} \quad \textbf{(Multiplying and dividing radicals)}$$

36. $2\sqrt{125} - 4\sqrt{20} + 6\sqrt{45} = 2\sqrt{25 \cdot 5} - 4\sqrt{4 \cdot 5} + 6\sqrt{9 \cdot 5} \quad$ Factor each radicand.

$$= 2 \cdot 5\sqrt{5} - 4 \cdot 2\sqrt{5} + 6 \cdot 3\sqrt{5} \quad \text{Factor out perfect squares.}$$

$$= 10\sqrt{5} - 8\sqrt{5} + 18\sqrt{5}$$

$$= 20\sqrt{5} \quad \textbf{(Multiplying and dividing radicals)}$$

37. $4x\sqrt{6x^4y^3} - 7y\sqrt{24x^6y} = 4x \cdot x^2 \cdot y\sqrt{6y} - 7y \cdot 2x^3\sqrt{6y}$ Simplify each radical.

$$= 4x^3y\sqrt{6y} - 14x^3y\sqrt{6y}$$

$$= -10x^3y\sqrt{6y} \text{ (\textbf{Multiplying and dividing radicals})}$$

38. $x\sqrt[3]{54x^4y^5} - 8x\sqrt[3]{2x^4y^5} = x \cdot 3 \cdot x \cdot y\sqrt[3]{2xy^2} - 8x \cdot x \cdot y\sqrt[3]{2xy^2}$ Simplify each radical.

$$= 3x^2y\sqrt[3]{2xy^2} - 8x^2y\sqrt[3]{2xy^2}$$

$$= -5x^2y\sqrt[3]{2xy^2} \text{ (\textbf{Multiplying and dividing radicals})}$$

39. $3\sqrt{2}\,(5\sqrt{10} + 4\sqrt{3}) = 15\sqrt{20} + 12\sqrt{6}$ Use the distributive property.

$$= 15\sqrt{4 \cdot 5} + 12\sqrt{6}$$

$$= 30\sqrt{5} + 12\sqrt{6} \quad \text{Note that this expressions cannot be simplified because the}$$
$$\text{radicands are not the same. (\textbf{Multiplying and dividing radicals})}$$

40. $(2 + 5\sqrt{3})\,(3 - 4\sqrt{3}) = 6 - 8\sqrt{3} + 15\sqrt{3} - 20\sqrt{9}$

$$= 6 + 7\sqrt{3} - 60 \quad \text{Combine similar radicals and simplify } \sqrt{9} = 3.$$

$$= -54 + 7\sqrt{3} \quad \text{Combine similar terms. (\textbf{Multiplying and dividing radicals})}$$

41. $(4\sqrt{3} - 2)\,(4\sqrt{3} + 2) = (4\sqrt{3})^2 - 2^2$ Use $(a - b)\,(a + b) = a^2 - b^2$.

$$= 16 \cdot 3 - 4$$

$$= 48 - 4$$

$$= 44 \text{ (\textbf{Multiplying and dividing radicals})}$$

42. $(\sqrt{x} + 3\sqrt{y})\,(2\sqrt{x} - \sqrt{y}) = 2\sqrt{x^2} - \sqrt{xy} + 6\sqrt{xy} - 3\sqrt{y^2}$ Use the distributive property.

$$= 2x + 5\sqrt{xy} - 3y \quad \text{Simplify radicals and combine similar radicals.}$$

(Multiplying and dividing radicals)

43. $(\sqrt{10} - 9)^2 = (\sqrt{10} - 9)\,(\sqrt{10} - 9)$

$$= \sqrt{100} - 9\sqrt{10} - 9\sqrt{10} + 81 \quad \text{Use the distributive property.}$$

$$= 10 - 18\sqrt{10} + 81 \quad \text{Simplify radicals and combine similar radicals.}$$

$$= 91 - 18\sqrt{10}$$

or use $(a - b)^2 = a^2 - 2ab + b^2$ to write

$$(\sqrt{10}-9)^2 = (\sqrt{10})^2 - 2\sqrt{10}\,(9) + 9^2 = 10 - 18\sqrt{10} + 81 = 91 - 18\sqrt{10}$$
(Multiplying and dividing radicals)

44. $\dfrac{6}{3-\sqrt{5}} = \dfrac{6}{3-\sqrt{5}} \cdot \dfrac{3+\sqrt{5}}{3+\sqrt{5}}$ Multiply numerator and denominator by the conjugate of the denominator.

$$= \frac{6(3+\sqrt{5})}{3^2 - (\sqrt{5})^2}$$ Use $(a-b)(a+b) = a^2 - b^2$

$$= \frac{6(3+\sqrt{5})}{9-5}$$

$$= \frac{6(3+\sqrt{5})}{4}$$

$$= \frac{\overset{3}{\cancel{6}}(3+\sqrt{5})}{\underset{2}{\cancel{4}}}$$ Simplify the denominator; reduce the fraction.

$$= \frac{9+3\sqrt{5}}{2}$$ **(Multiplying and dividing radicals)**

45. $\dfrac{3+2\sqrt{5}}{4\sqrt{2}+3\sqrt{5}} = \dfrac{3+2\sqrt{5}}{4\sqrt{2}+3\sqrt{5}} \cdot \dfrac{4\sqrt{2}-3\sqrt{5}}{4\sqrt{2}-3\sqrt{5}}$ Multiply numerator and denominator by the

conjugate of the denominator.

$$= \frac{12\sqrt{2} - 9\sqrt{5} + 8\sqrt{10} - 6\sqrt{25}}{(4\sqrt{2})^2 - (3\sqrt{5})^2}$$

$$= \frac{12\sqrt{2} - 9\sqrt{5} + 8\sqrt{10} - 30}{32 - 45}$$

$$= \frac{12\sqrt{2} - 9\sqrt{5} + 8\sqrt{10} - 30}{-13} = -\frac{12\sqrt{2} - 9\sqrt{5} + 8\sqrt{10} - 30}{13}$$

Note that the answer may also be written as $\dfrac{9\sqrt{5} + 30 - 12\sqrt{2} - 8\sqrt{10}}{13}$ where the −1 has been distributed

through the numerator and the order of the terms has been changed.
(Multiplying and dividing radicals)

46. $\dfrac{3+\sqrt{x}}{4-\sqrt{x}} = \dfrac{3+\sqrt{x}}{4-\sqrt{x}} \cdot \dfrac{4+\sqrt{x}}{4+\sqrt{x}}$

$$= \frac{12 + 7\sqrt{x} + x}{16 - x}$$ **(Multiplying and dividing radicals)**

47. $\sqrt{2x-4} = 4$

$2x - 4 = 16$ Square both sides of the equation.

$$2x = 20$$
$$x = 10$$ Solve for x.

Check: $\sqrt{2(10) - 4} \overset{?}{=} 4$

$\sqrt{16} = 4$ So the solution checks. **(Solving radical equations)**

48. $\sqrt{5y + 3} + 2 = 0$

$\sqrt{5y + 3} = -2$ Isolate the radical.

$5y + 3 = 4$ Square both sides of the equation.

$5y = 1$
$y = \dfrac{1}{5}$ Solve for y.

Check: $\sqrt{5\left(\dfrac{1}{5}\right) + 3} + 2 \overset{?}{=} 0$

$\sqrt{4} + 2 \neq 0$ The solution does not check. Therefore there is no solution.

(Solving radical equations)

49. $\sqrt[3]{3t + 6} - 3 = 0$

$\sqrt[3]{3t + 6} = 3$ Isolate the radical.

$3t + 6 = 27$ Cube both sides of the equation.

$3t = 21$
$t = 7$ Solve for t.

The solution does check. **(Solving radical equations)**

50. $\sqrt{x + 1} = 2 - \sqrt{x}$

$x + 1 = 2^2 + 2(2)(-\sqrt{x}) + (-\sqrt{x})^2$ Square both sides of the equation.

$x + 1 = 4 - 4\sqrt{x} + x$ Simplify the right side of the equation.

$-3 = -4\sqrt{x}$ Isolate the radical.

$9 = 16x$ Square both sides of the equation.

$\dfrac{9}{16} = x$ Solve for x.

Check: $\sqrt{\dfrac{9}{16} + 1} \overset{?}{=} 2 - \sqrt{\dfrac{9}{16}}$

$$\sqrt{\frac{9}{16} + 1} = \sqrt{\frac{25}{16}} = \frac{5}{4} \quad \text{Simplify the left side.}$$

$$2 - \sqrt{\frac{9}{16}} = 2 - \frac{3}{4} = \frac{5}{4} \quad \text{Simplify the right side.}$$

Since both sides equal 5/4, the solution checks. (**Solving radical equations**)

51. $\sqrt{3x + 10} - 2 = \sqrt{2x - 1}$

   $(\sqrt{3x + 10})^2 + 2(\sqrt{3x + 10})(-2) + (-2)^2 = 2x - 1 \quad$ Square both sides of the equation.

   $3x + 10 - 4\sqrt{3x + 10} + 4 = 2x - 1 \quad$ Simplify the left side of the equation.

   $x + 15 = 4\sqrt{3x + 10} \quad$ Isolate the radical.

   $x^2 + 30x + 225 = 16(3x + 10) \quad$ Square both sides of the equation.

   $x^2 - 18x + 65 = 0 \quad$ This is a quadratic equation that can be solved by factoring.

   $(x - 5)(x - 13) = 0$

   $x = 5 \qquad x = 13$

   Both solutions check. Therefore the solution set is $\{5, 13\}$. (**Solving radical equations**)

52. $\sqrt{x - 1} - 3 = \sqrt{x + 2}$

   $x - 1 + 2(\sqrt{x - 1})(-3) + (-3)^2 = x + 2 \quad$ Square both sides of the equation.

   $x - 1 - 6\sqrt{x - 1} + 9 = x + 2 \quad$ Simplify the left side of the equation.

   $-6\sqrt{x - 1} = -6 \quad$ Isolate the radical.

   $36(x - 1) = 36 \quad$ Square both sides of the equation.

   $36x - 36 = 36$

   $\qquad 36x = 72 \quad$ Solve for x.

   $\qquad\quad x = 2$

   Check: $\sqrt{2 - 1} - 3 \overset{?}{=} \sqrt{2 + 2}$

   $\qquad\qquad 1 - 3 \ne 2 \quad$ The solution does not check. Therefore there is no solution, written $\{\ \}$ or $\varnothing$.

   (**Solving radical equations**)

53. $\sqrt{-25} = \sqrt{-1 \cdot 25} = \sqrt{-1} \cdot \sqrt{25} = i5$ or $5i$ (**Imaginary numbers**)

54. $\sqrt{-72} = \sqrt{-1 \cdot 72} = i\sqrt{36 \cdot 2} = 6i\sqrt{2}$ (**Imaginary numbers**)

55. $(4 + 7i) + (2 - 3i) = (4 + 2) + (7i - 3i) = 6 + 4i$ Combine similar terms. (**Adding and subtracting complex numbers**)

56. $(2 - 8i) - (6 + i) = 2 - 8i - 6 - i = -4 - 9i$ Distribute the subtraction sign; combine similar terms.

(**Adding and subtracting complex numbers**)

57. $6 - (3 - 2i) = 6 - 3 + 2i = 3 + 2i$ (**Adding and subtracting complex numbers**)

58. $2i (3 - 7i) = 6i - 14i^2$

$= 6i - 14(-1)$ Use $i^2 = -1$.

$= 6i + 14$ (**Multiplying and dividing complex numbers**)

59. $(2 + 5i)(3 + 6i) = 6 + 12i + 15i + 30i^2$ Use the distributive property.

$= 6 + 27i + 30(-1)$ Combine similar terms; use $i^2 = -1$.

$= -24 + 27i$ (**Multiplying and dividing complex numbers**)

60. $(4 - 3i)(4 + 3i) = 4^2 - (3i)^2$ Use $(a - b)(a + b) = a^2 - b^2$.

$= 16 - 9i^2$

$= 16 - 9(-1)$ Use $i^2 = -1$.

$= 25$ (**Multiplying and dividing complex numbers**)

61. $\dfrac{6 - 3i}{i} = \dfrac{6 - 3i}{i} \cdot \dfrac{i}{i}$ Multiply numerator and denominator by i.

$= \dfrac{6i - 3i^2}{i^2}$ Multiply.

$= \dfrac{6i - 3(-1)}{-1}$ Use $i^2 = -1$.

$= \dfrac{6i + 3}{-1} = -3 - 6i$ Write the answer in standard form $a + bi$.

(**Multiplying and dividing complex numbers**)

62. $\dfrac{3 + 2i}{4i} = \dfrac{3 + 2i}{4i} \cdot \dfrac{i}{i}$ Multiply numerator and denominator by i.

$= \dfrac{3i + 2i^2}{4i^2}$ Multiply.

$= \dfrac{3i - 2}{-4}$ Use $i^2 = -1$.

$$= -\frac{3}{4}i + \frac{1}{2} = \frac{1}{2} - \frac{3}{4}i \qquad \text{Write the answer in standard form } a + bi.$$

(Multiplying and dividing complex numbers)

63. $\dfrac{2 + 5i}{4 - 3i} = \dfrac{2 + 5i}{4 - 3i} \cdot \dfrac{4 + 3i}{4 + 3i}$ Multiply numerator and denominator by the conjugate of the denominator.

$$= \frac{8 + 6i + 20i + 15i^2}{16 - 9i^2} \qquad \text{Multiply.}$$

$$= \frac{8 + 26i - 15}{16 + 9} \qquad \text{Use } i^2 = -1.$$

$$= \frac{-7 + 26i}{25} = -\frac{7}{25} + \frac{26}{25}i \qquad \text{Write the answer in standard form.}$$

(Multiplying and dividing complex numbers)

Grade Yourself

Circle the question numbers that you had incorrect. Then indicate the number of questions you missed. If you answered more than three questions incorrectly, you need to focus on that topic. (If a topic has less than three questions and you had at least one wrong, we suggest you study that topic also. Read your textbook, a review book, or ask your teacher for help.)

Subject: Exponents and Radicals

Topic	Question Numbers	Number Incorrect
Rational exponents	1, 2, 3, 4, 5, 6, 7, 8, 9, 10, 11, 12, 13, 14, 15	
Radicals	16, 17, 18, 19, 20, 21, 22	
Simplifying radicals	23, 24, 25, 26, 27, 28	
Multiplying and dividing radicals	29, 30, 31, 32, 33, 34, 39, 40, 41, 42, 43, 44, 45, 46	
Adding and subtracting radical expressions	35, 36, 37, 38	
Solving radical equations	47, 48, 49, 50, 51, 52	
Imaginary numbers	53, 54	
Adding and subtracting complex numbers	55, 56, 57, 58	
Multiplying and dividing complex numbers	59, 60, 61, 62, 63	

Quadratic Equations and Inequalities

Test Yourself

The Square Root Property

If a quadratic equation can be written in the form $x^2 = a$, the Square Root Property can be used to solve the equation:

If $x^2 = a$, then $x = \pm\sqrt{a}$.

Solve.

1. $x^2 = 16$

2. $x^2 - 9 = 0$

3. $(3a - 5)^2 = 36$

4. $p^2 + 25 = 0$

5. $(2y + 5)^2 = 11$

Solving Quadratic Equations by Completing the Square

To solve a quadratic equation by completing the square:

1. Write the equation in the form $ax^2 + bx = c$.

2. Divide each term by a to get $x^2 + \dfrac{b}{a}x = \dfrac{c}{a}$.

3. Add $\left(\dfrac{1}{2}\text{times the coefficient of } x\right)^2$ to both sides of the equation.

4. Factor the left side of the equation as a perfect square trinomial.

5. Simplify the right side of the equation.
6. Use the Square Root Property to solve the equation.

Solve by completing the square.

6. $m^2 + 6m - 7 = 0$

7. $2x^2 - 5x + 1 = 0$

8. $3x^2 + 4x + 3 = 0$

Solving Quadratic Equations with the Quadratic Formula

You should memorize the Quadratic Formula:

The solutions to $ax^2 + bx + c = 0$, $a \neq 0$, are

$$x = \frac{-b \pm \sqrt{b^2 - 4ac}}{2a}.$$

Be careful to write the equation in standard form $ax^2 + bx + c = 0$ before determining the values of a, b, and c.

To help you organize your work, write out what a, b, and c equal before substituting into the formula.

Solve.

9. $5x^2 - 2x - 7 = 0$

10. $x^2 - 5x + 8 = 0$

11. $4x(x - 5) = 3x - 2$

12. $\dfrac{x^2}{2} - 1 = \dfrac{x}{3}$

13. $\dfrac{3x}{x-2} - \dfrac{6}{x+3} = \dfrac{8}{(x-2)(x+3)}$

Applications

Be careful to check that the solutions to word problems are reasonable. Eliminate answers that are impossible, such as negative time.

14. If an object is thrown downward with an initial velocity of 4 feet per second, the distance s that it travels in time t is given by $s = 16t^2 + 4t$. How long does it take the object to fall 60 feet?

15. Joan and Ros are preparing a mathematics project. If they work together, it will take 5 hours to complete the project. If each worked alone, Joan could complete the project in 2 hours more than Ros. How long would it take Ros to complete the project alone?

Determining Types of Solutions from the Discriminant

By computing the discriminant, you can determine the number and nature of the solutions (or roots) of a quadratic equation with real number coefficients. Assuming a, b, and c are rational numbers,

1. If $b^2 - 4ac > 0$ and is a perfect square, the equation has two rational solutions.

2. If $b^2 - 4ac > 0$ and is not a perfect square, the equation has two irrational solutions.

3. If $b^2 - 4ac = 0$, the equation has one rational solution.

4. If $b^2 - 4ac < 0$, the equation has two complex number solutions.

Use the discriminant of the quadratic formula to determine the number and type of solutions for each equation.

16. $2x^2 + 2x - 1 = 0$

17. $10y(y+4) = 15y - 15$

18. $2x^2 + 5x + 4 = 0$

19. $1 + \dfrac{6}{x} + \dfrac{9}{x^2} = 0$

Finding Quadratic Equations Using Solutions

Given solutions a and b to a quadratic equation, multiply out $(x-a)(x-b) = 0$ to find an equation with the given solutions.

20. Find a quadratic equation with solutions 3 and –5.

21. Find a quadratic equation with solutions $4\sqrt{5}$ and $-2\sqrt{5}$.

22. Find a quadratic equation with solutions $y = 5i$ and $y = -5i$.

Solving Equations Using Substitution

Some equations that may not appear to be quadratic can be put in standard form by an appropriate substitution. You may need to replace an expression in parentheses with a single variable or replace a variable to a power other than 1 with a variable raised to the first power. When making substitutions, do not reuse a variable that is in the original equation. Remember to complete the problem by solving for the original variable.

Solve.

23. $(4x-1)^2 - 3(4x-1) - 10 = 0$

24. $2x^{2/3} - x^{1/3} - 3 = 0$

25. $x - 10\sqrt{x} + 24 = 0$

26. $4y^4 + 3 = 13y^2$

Solving Quadratic Inequalities

To solve a factorable quadratic inequality:
1. Isolate 0 on the right side of the inequality.
2. Factor the left side of the inequality.
3. Find the split points (also called critical points) by setting each factor equal to 0 and solving.
4. Draw a number line and use the split points to divide the number line into regions.

5. Write the factors on the left edge of the number line.
6. Pick a number in each region created by the split points, substitute it into each factor, and record the resulting sign (+ or –) on the number line.
7. Use the sign rules for products to determine regions to be shaded for the solution. Include the split points in the solution if the inequality contains ≤ or ≥. Do not include the split points in the solution if the inequality contains < or >.

Solve.

27. $x^2 - x - 6 \le 0$

28. $2x^2 > 5x + 3$

Higher Degree Inequalities

Use a process similar to the process for quadratic inequalities to solve higher degree inequalities, adding extra split points for each additional factor.

Solve.

29. $(x + 1)(x - 2)(x + 4) \ge 0$

30. $x(x - 2)^2(x + 1) < 0$

Rational Inequalities

To solve inequalities containing rational expressions:
1. Isolate 0 on the right side of the inequality.
2. Write the left side as one fraction (you may need to find the least common denominator and combine fractions).
3. Split points are found by setting the numerator and denominator equal to 0 and solving.
4. Use a number line and regions to find the sign of the numerator and denominator and the region(s) needed for the solution.
5. If the inequality contains ≤ or ≥, do *not* include split point(s) where the denominator equals 0.

Solve.

31. $\dfrac{x - 3}{x + 5} \ge 0$

32. $\dfrac{4}{x - 1} \le 2$

✓ Check Yourself

1. $x^2 = 16$

$\sqrt{x^2} = \pm\sqrt{16}$ Take the square root of both sides of the equation. Be sure to use ± on the right side.

$x = \pm4$ **(The square root property)**

2. $x^2 - 9 = 0$

$x^2 = 9$ Isolate x^2.

$\sqrt{x^2} = \pm\sqrt{9}$ Take the square root of both sides of the equation.

$x = \pm3$ **(The square root property)**

3. $(3a-5)^2 = 36$

$\sqrt{(3a-5)^2} = \pm\sqrt{36}$ Take the square root of both sides of the equation. Use ± on one side only.

$3a-5 = \pm 6$ To solve for a, set up two equations and solve separately.

$\begin{array}{ll} 3a-5 = 6 & 3a-5 = -6 \\ \quad 3a = 11 & \quad 3a = -1 \\ \quad\quad a = \dfrac{11}{3} & \quad\quad a = -\dfrac{1}{3} \end{array}$ Solve each equation for a. (**The square root property**)

4. $p^2 + 25 = 0$

$p^2 = -25$ Isolate p^2.

$\sqrt{p^2} = \pm\sqrt{-25}$ Take the square root of both sides of the equation.

$p = \pm 5i$ Use $\sqrt{-1} = i$ to simplify the right side. (**The square root property**)

5. $(2y+5)^2 = 11$

$\sqrt{(2y+5)^2} = \pm\sqrt{11}$ Take the square root of both sides of the equation.

$\begin{array}{l} 2y+5 = \pm\sqrt{11} \\ \quad 2y = -5 \pm \sqrt{11} \\ \quad\quad y = \dfrac{-5 \pm \sqrt{11}}{2} \end{array}$ Solve for y.

(The square root property)

6. $m^2 + 6m - 7 = 0$

$m^2 + 6m = 7$ Write in the form $ax^2 + bx = c$.

Since the coefficient of m^2 is 1, dividing by 1 does not change the equation.

$m^2 + 6m + 9 = 7 + 9$ Add $\left(\dfrac{1}{2} \cdot 6\right)^2 = 3^2 = 9$ to both sides of the equation.

$(m+3)^2 = 16$ Factor the left side of the equation; simplify the right side of the equation.

$\sqrt{(m+3)^2} = \pm\sqrt{16}$ Use the Square Root Property to solve.

$m+3 = \pm 4$

$\begin{array}{ll} m+3 = 4 & m+3 = -4 \\ \quad m = 1 & \quad m = -7 \end{array}$ Set up two equations and solve each for m. (**Solving quadratic equations by completing the square**)

7. $2x^2 - 5x + 1 = 0$

$2x^2 - 5x = -1$ Write in the form $ax^2 + bx = c$.

$x^2 - \dfrac{5}{2}x = -\dfrac{1}{2}$ Divide each term by 2, the coefficient of x^2.

$x^2 - \dfrac{5}{2}x + \dfrac{25}{16} = -\dfrac{1}{2} + \dfrac{25}{16}$ Add $\left(\dfrac{1}{2}\cdot -\dfrac{5}{2}\right)^2 = \dfrac{25}{16}$ to both sides of the equation.

$\left(x - \dfrac{5}{4}\right)^2 = \dfrac{17}{16}$ Factor the left side of the equation; simplify the right side of the equation.

$x - \dfrac{5}{4} = \pm\sqrt{\dfrac{17}{16}}$ Use the Square Root Property to solve.

$x = \dfrac{5}{4} \pm \dfrac{\sqrt{17}}{4}$ **(Solving quadratic equations by completing the square)**

8. $3x^2 + 4x + 3 = 0$

$3x^2 + 4x = -3$ Write in the form $ax^2 + bx = c$.

$x^2 + \dfrac{4}{3}x = -1$ Divide each term by 3, the coefficient of x^2.

$x^2 + \dfrac{4}{3}x + \dfrac{4}{9} = -1 + \dfrac{4}{9}$ Add $\left(\dfrac{1}{2}\cdot\dfrac{4}{3}\right)^2 = \left(\dfrac{2}{3}\right)^2 = \dfrac{4}{9}$ to both sides of the equation.

$\left(x + \dfrac{2}{3}\right)^2 = -\dfrac{5}{9}$ Factor the left side of the equation; simplify the right side of the equation.

$x + \dfrac{2}{3} = \pm\sqrt{-\dfrac{5}{9}}$ Use the Square Root Property to solve.

$x = -\dfrac{2}{3} \pm \dfrac{i\sqrt{5}}{3}$ **(Solving quadratic equations by completing the square)**

9. $5x^2 - 2x - 7 = 0$

Identify a, b, c: $a = 5$, $b = -2$, $c = -7$

$x = \dfrac{-(-2) \pm \sqrt{(-2)^2 - 4(5)(-7)}}{2(5)}$ Substitute into the Quadratic Formula.

$x = \dfrac{2 \pm \sqrt{4 + 140}}{10}$ Simplify.

$x = \dfrac{2 \pm \sqrt{144}}{10}$

$x = \dfrac{2 \pm 12}{10}$ Since the radical contained a perfect square, the solutions can be simplified further:

$x = \dfrac{2 + 12}{10}$ $x = \dfrac{2 - 12}{10}$

$x = \dfrac{14}{10}$ $x = \dfrac{-10}{10}$ Solve each equation for x.

$x = \dfrac{7}{5}$ $x = -1$

Note that the original equation could have been solved by factoring:

$(5x - 7)(x + 1) = 0$

$5x - 7 = 0$

$\qquad 5x = 7 \qquad\qquad x + 1 = 0$

$\qquad\quad x = \dfrac{7}{5} \qquad\qquad x = -1$

If the test directions allow you to choose which method you may use, use the method easiest for you. The solutions will be the same. However, if a quadratic equation does not factor, you must use either completing the square or the quadratic formula to solve. (**Solving quadratic equations with the quadratic formula**)

10. $x^2 - 5x + 8 = 0$

$a = 1, b = -5, c = 8$

$x = \dfrac{-(-5) \pm \sqrt{(-5)^2 - 4(1)(8)}}{2(1)}$

$= \dfrac{5 \pm \sqrt{25 - 32}}{2}$

$= \dfrac{5 \pm \sqrt{-7}}{2}$

$= \dfrac{5 \pm i\sqrt{7}}{2}$ (**Solving quadratic equations with the quadratic formula**)

11. $4x(x - 5) = 3x - 2$

$4x^2 - 20x = 3x - 2$ Write the equation in standard form.

$4x^2 - 23x + 2 = 0$

Identify a, b, c: $a = 4, b = -23, c = 2$.

$x = \dfrac{-(-23) \pm \sqrt{(-23)^2 - 4(4)(2)}}{2(4)}$

$$= \frac{23 \pm \sqrt{529-32}}{8}$$

$$= \frac{23 \pm \sqrt{497}}{8}$$

(Solving quadratic equations with the quadratic formula)

12. $\dfrac{x^2}{2} - 1 = \dfrac{x}{3}$

$6\left(\dfrac{x^2}{2} - 1 = \dfrac{x}{3}\right)$ Multiply both sides of the equation by 6 to eliminate the fractions.

$3x^2 - 2x - 6 = 0$ Write the equation in standard form.

$a = 3, b = -2, c = -6$

$x = \dfrac{-(-2) \pm \sqrt{(-2)^2 - 4(3)(-6)}}{2(3)}$

$= \dfrac{2 \pm \sqrt{4+72}}{6}$

$= \dfrac{2 \pm \sqrt{76}}{6}$

$= \dfrac{2 \pm 2\sqrt{19}}{6}$

$= \dfrac{1 \pm \sqrt{19}}{3}$

(Solving quadratic equations with the quadratic formula)

13. $\dfrac{3x}{x-2} - \dfrac{6}{x+3} = \dfrac{8}{(x-2)(x+3)}$

$(x-2)(x+3) \cdot \left(\dfrac{3x}{x-2} - \dfrac{6}{x+3}\right) = (x-2)(x+3) \cdot \dfrac{8}{(x-2)(x+3)}$ Multiply both sides of the equation

by $(x-2)(x+3)$ to clear fractions.

$3x(x+3) - 6(x-2) = 8$

$3x^2 + 9x - 6x + 12 = 8$

$3x^2 + 3x + 4 = 0$ Write the equation in standard form.

$a = 3, b = 3, c = 4$

$x = \dfrac{-(3) \pm \sqrt{3^2 - 4(3)(4)}}{2(3)}$

$$= \frac{-3 \pm \sqrt{9 - 48}}{6}$$

$$= \frac{-3 \pm \sqrt{-39}}{6}$$

$$= \frac{-3 \pm i\sqrt{39}}{6}$$

(Solving quadratic equations with the quadratic formula)

14. Since the given distance is 60 feet, $s = 60$. Substitute this value into the given equation:

$60 = 16t^2 + 4t$

$16t^2 + 4t - 60 = 0$ Write the equation in standard form.

$4t^2 + t - 15 = 0$ Divide each term of the equation by 4.

$a = 4$, $b = 1$, $c = -15$

$$x = \frac{-(1) \pm \sqrt{1^2 - 4(4)(-15)}}{2(4)}$$

$$= \frac{-1 \pm \sqrt{241}}{8}$$

Use $\sqrt{241} \approx 15.5$ to complete the computation:

$$x = \frac{-1 + 15.5}{8} = 1.8 \text{ seconds}$$

The other solution, $x = \dfrac{-1 - 15.5}{8}$ must be eliminated since time cannot be negative in this problem.

(Applications)

15. Let x = the number of hours for Ros to complete the project alone.

Then $x + 2$ = the number of hours for Joan to complete the project alone.

$\dfrac{1}{x}$ = the amount of work completed by Ros in 1 hour.

$\dfrac{1}{x + 2}$ = the amount of work completed by Joan in 1 hour.

$\dfrac{1}{x} + \dfrac{1}{x + 2} = \dfrac{1}{5}$ Together, in 1 hour they complete $\dfrac{1}{5}$ of the project.

$5x(x + 2)\left(\dfrac{1}{x} + \dfrac{1}{x + 2}\right) = 5x(x + 2)\left(\dfrac{1}{5}\right)$ Multiply both sides of the equation by $5x(x + 2)$.

$5(x + 2) + 5x = x(x + 2)$

$5x + 10 + 5x = x^2 + 2x$

$x^2 - 8x - 10 = 0$ Write the equation in standard form.

$a = 1, b = -8, c = -10$

$x = \dfrac{-(-8) \pm \sqrt{(-8)^2 - 4(1)(-10)}}{2(1)}$

$ = \dfrac{8 \pm \sqrt{104}}{2}$

If $x = \dfrac{8 + \sqrt{104}}{2} \approx 9.1$. If $x = \dfrac{8 - \sqrt{104}}{2} \approx -1.1$.

Since time cannot be negative in this problem, eliminate the -1.1 solution. If $x \approx 9.1$, it would take Ros approximately 9.1 hours to complete the project alone. **(Applications)**

16. $2x^2 + 2x - 1 = 0$

$a = 2, b = 2, c = -1$

$b^2 - 4ac = 2^2 - 4(2)(-1) = 4 + 8 = 12$

Since $12 > 0$ but is not a perfect square, the equation has 2 irrational solutions. **(The discriminant)**

17. $10y(y + 4) = 15y - 15$

$10y^2 + 40y = 15y - 15$

$10y^2 + 25y + 15 = 0$ Write the equation in standard form.

$a = 10, b = 25, c = 15$

$b^2 - 4ac = 25^2 - 4(10)(15) = 625 - 600 = 25$

Since $25 > 0$ and is a perfect square, the equation has 2 rational solutions. **(The discriminant)**

18. $2x^2 + 5x + 4 = 0$

$a = 2, b = 5, c = 4$

$b^2 - 4ac = 5^2 - 4(2)(4) = 25 - 32 = -7$

Since $-7 < 0$, the equation has 2 complex number solutions. **(The discriminant)**

19. $1 + \dfrac{6}{x} + \dfrac{9}{x^2} = 0$

$x^2\left(1 + \dfrac{6}{x} + \dfrac{9}{x^2} = 0\right)$ Multiply both sides of the equation by x^2 to clear fractions.

$$x^2 + 6x + 9 = 0$$

$$a = 1, b = 6, c = 9$$

$$b^2 - 4ac = 6^2 - 4(1)(9) = 36 - 36 = 0$$

Since the discriminant equals 0, the equation has one rational solution. **(The discriminant)**

20. $x = 3$ implies that $x - 3 = 0$. $x = -5$ implies that $x - (-5) = 0$. Thus,

$$(x - 3)(x - (-5)) = 0$$

$$(x - 3)(x + 5) = 0$$

$$x^2 + 2x - 15 = 0 \text{ \textbf{(Finding quadratic equations)}}$$

21. $(x - 4\sqrt{5})(x - (-2\sqrt{5})) = 0$

$$(x - 4\sqrt{5})(x + 2\sqrt{5}) = 0$$

$$x^2 - 2\sqrt{5}x - 40 = 0 \text{ \textbf{(Finding quadratic equations)}}$$

22. $(y - 5i)(y - (-5i)) = 0$

$$(y - 5i)(y + 5i) = 0$$

$$y^2 - 25i^2 = 0$$

$$y^2 + 25 = 0 \quad \text{Use } i^2 = -1. \text{ \textbf{(Finding quadratic equations)}}$$

23. $(4x - 1)^2 - 3(4x - 1) - 10 = 0$

Let $u = 4x - 1$ and write the equation as:

$$u^2 - 3u - 10 = 0 \quad \text{This equation can be solved by factoring.}$$

$$(u - 5)(u + 2) = 0$$

$$u - 5 = 0 \qquad u + 2 = 0$$
$$u = 5 \qquad u = -2$$

Now solve for x by using $u = 4x - 1$:

$$4x - 1 = 5 \qquad 4x - 1 = -2$$
$$4x = 6 \qquad 4x = -1$$
$$x = \frac{6}{4} \qquad x = -\frac{1}{4}$$
$$x = \frac{3}{2}$$

(Solving equations using substitution)

24. $2x^{2/3} - x^{1/3} - 3 = 0$ Notice that $x^{2/3}$ is $(x^{1/3})^2$.

Let $u = x^{1/3}$. Then $u^2 = x^{2/3}$. Now substitute.

$2u^2 - u - 3 = 0$ This equation can be solved by factoring.

$(2u - 3)(u + 1) = 0$

$2u - 3 = 0$

$\quad 2u = 3 \qquad\qquad u + 1 = 0$

$\qquad u = \dfrac{3}{2} \qquad\qquad u = -1$

Now solve for x by using $u = x^{1/3}$:

$x^{1/3} = \dfrac{3}{2}$

$(x^{1/3})^3 = \left(\dfrac{3}{2}\right)^3 \qquad x^{1/3} = -1$

$\qquad x = \dfrac{27}{8} \qquad (x^{1/3})^3 = (-1)^3$

$\qquad\qquad\qquad\qquad x = -1$

(Solving equations using substitution)

25. $x - 10\sqrt{x} + 24 = 0$ Notice that x^1 is $(\sqrt{x})^2$.

Let $u = \sqrt{x}$. Then $u^2 = x^1$.

$u^2 - 10u + 24 = 0$

$(u - 6)(u - 4) = 0$

$\quad u = 6 \qquad\qquad u = 4$

$\quad \sqrt{x} = 6 \qquad\qquad \sqrt{x} = 4$

$(\sqrt{x})^2 = 6^2 \qquad (\sqrt{x})^2 = 4^2$

$\quad x = 36 \qquad\qquad x = 16$

Note that both of these solutions check in the original equation. On a test, check all solutions.

(Solving equations using substitution)

26. $4y^4 + 3 = 13y^2$

$4y^4 - 13y^2 + 3 = 0$

Let $u = y^2$ so $u^2 = y^4$.

$4u^2 - 13u + 3 = 0$

$$(4u - 1)(u - 3) = 0$$

$$u = \frac{1}{4} \qquad u = 3$$

$$y^2 = \frac{1}{4}$$

$$\sqrt{y^2} = \pm\sqrt{\frac{1}{4}}$$

$$y = \pm\frac{1}{2}$$

$$y^2 = 3$$

$$\sqrt{y^2} = \pm\sqrt{3}$$

$$y = \pm\sqrt{3}$$

(Solving equations using substitution)

27. $x^2 - x - 6 \le 0$

$(x - 3)(x + 2) \le 0$

The split points are 3 and –2. Draw a number line with the indicated split points.

The solution is [–2, 3] which can also be written as $-2 \le x \le 3$. **(Solving higher order inequalities)**

28. $2x^2 > 5x + 3$

$2x^2 - 5x - 3 > 0$

$(2x + 1)(x - 3) > 0$

The split points are –1/2 and 3. Draw a number line with the indicated split points.

The solution is $\left(-\infty, -\frac{1}{2}\right) \cup (3, \infty)$ which can also be written as $x < -\frac{1}{2}$ or $x > 3$.

(Solving higher order inequalities)

29. $(x + 1)(x - 2)(x + 4) \ge 0$

The split points are –1, 2, and –4. Draw a number line with the indicated split points.

$$
\begin{array}{llll}
x+1 & - & - & + & + \\
x-2 & - & - & - & + \\
x+4 & - & + & + & +
\end{array}
$$

$(x+1)\,(x-2)\,(x+4)$ $-$ -4 $+$ -1 $-$ 2 $+$

The solution is $[-4,-1] \cup [2, \infty)$ which can also be written as $-4 \le x \le -1$ or $x \ge 2$.

(Solving higher order inequalities)

30. $x\,(x-2)^2\,(x+1) < 0$

The split points are $0, 2$ and -1.

$$
\begin{array}{llll}
x & - & - & + & + \\
(x-2)^2 & + & + & + & + \\
x+1 & - & + & + & +
\end{array}
$$

$x\,(x-2)^2\,(x+1)$ $+$ -1 $-$ 0 $+$ 2 $+$

The solution is $(-1, 0)$ which can also be written as $-1 < x < 0$. **(Solving higher order inequalities)**

31. $\dfrac{x-3}{x+5} \ge 0$

The split points are 3 and -5, found by setting the numerator and denominator equal to 0 and solving.

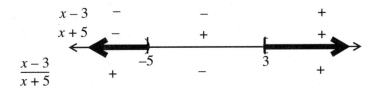

$$
\begin{array}{lll}
x-3 & - & - & + \\
x+5 & - & + & +
\end{array}
$$

$\dfrac{x-3}{x+5}$ $+$ -5 $-$ 3 $+$

The solution is $(-\infty, -5) \cup [3, \infty]$ which can also be written $x < -5$ or $x \ge 3$. Notice that $x = -5$ is *not* included in the solution because it would make the denominator equal 0. **(Solving rational inequalities)**

32. $\dfrac{4}{x-1} \le 2$

$\dfrac{4}{x-1} - 2 \le 0$ Get 0 on the right side of the inequality.

$\dfrac{4}{x-1} - \dfrac{2\,(x-1)}{x-1} \le 0$ Get a common denominator on the left side.

$\dfrac{4 - 2x + 2}{x-1} \le 0$

$\dfrac{-2x + 6}{x-1} \le 0$

The split points are 3 and 1.

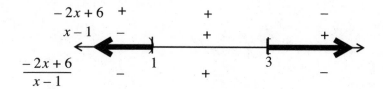

The solution is $(-\infty, 1) \cup [3, \infty]$, which can also be written as $x < 1$ or $x \geq 3$.

Note that $x = 1$ is *not* included in the solution because it makes the denominator equal 0.

(Solving rational inequalities)

Grade Yourself

Circle the question numbers that you had incorrect. Then indicate the number of questions you missed. If you answered more than three questions incorrectly, you need to focus on that topic. (If a topic has less than three questions and you had at least one wrong, we suggest you study that topic also. Read your textbook, a review book, or ask your teacher for help.)

Subject: *Quadratic Equations and Inequalities*

Topic	Question Numbers	Number Incorrect
The square root property	1, 2, 3, 4, 5	
Solving quadratic equations by completing the square	6, 7, 8	
Solving quadratic equations with the quadratic formula	9, 10, 11, 12, 13	
Applications	14, 15	
The discriminant	16, 17, 18, 19	
Finding quadratic equations	20, 21, 22	
Solving equations using substitution	23, 24, 25, 26	
Solving higher order inequalities	27, 28, 29	
Solving rational inequalities	31, 32	

Systems of Linear Equations

8

 ## Test Yourself

Solving a System of Linear Equations by Graphing

To solve a system of linear equations by graphing, graph each line using the same set of axes. Three possibilities exist:

1. The lines intersect in exactly 1 point (called a consistent system of independent equations).
2. The lines do not intersect so there is no solution (called an inconsistent system of independent equations).
3. The lines coincide to produce an infinite number of solutions (called a consistent system of dependent equations).

Solve each system by graphing.

1. $x + y = 3$
   $x - y = -1$

2. $y = x + 3$
   $y = -\dfrac{1}{2}x$

3. Determine whether the given equations are dependent or independent.

   $3x + y = 6$
   $6x + 2y = 12$

Solving a System of Linear Equations by Addition

The addition method (also called the elimination method) involves adding the equations together in a way that will eliminate one of the variables.

To solve a system of equations by addition:

1. Write each equation in the form $ax + by = c$.
2. If the coefficients in the two equations of either x or y are additive inverses, add the equations. If the coefficients are not additive inverses, eliminate x's by multiplying each equation by the coefficient of x from the other equation. If the signs are the same, multiply one of the equations by -1.
3. Add the equations together.
4. Solve for y.
5. Substitute the value of y into either of the original equations to solve for x. The answer should be an ordered pair, (x, y).
6. Check your answer.

Note: You may eliminate y in the same manner.

Solve each system.

4. $2x - y = 9$
   $4x + 3y = 23$

5. $2x + 4y = 8$
   $3x - 5y = -21$

6. $\frac{1}{4}x - \frac{1}{2}y = -\frac{5}{4}$

   $\frac{1}{3}x + \frac{1}{4}y = -\frac{3}{4}$

7. $6x - 10y = -2$

   $9x + 15y = 9$

8. $7x - 9y = 2$

   $-28x + 36y = -8$

9. $x - 5y = 8$

   $4x - 20y = 10$

Solving a System of Linear Equations by Substitution

Substitution is often used when one equation is already solved for one variable or when one equation can be easily solved for one variable.

To solve by substitution:

1. Solve one of the equations for x or y, if necessary.
2. Substitute the expression from Step 1 into the other equation.
3. Solve the resulting equation.
4. Use the value from Step 3 to substitute into the equation from Step 1 to solve for the other variable.
5. Check your solution.

Solve.

10. $y = x + 4$

    $6x - 3y = -15$

11. $x + 4y = 10$

    $2x + 5y = 17$

12. $y = 4x - 5$

    $8x - 2y = 10$

13. $x = 3y + 1$

    $3x - 9y = 6$

Solving a System of Linear Equations in Three Variables

To solve a system of three equations in three variables, use the addition method to eliminate one variable from two equations at a time.

Then solve the resulting system of two equations in two variables. The solution should be written as an ordered triple (x, y, z).

Solve.

14. $2x + 3y - z = -8$

    $x - 3y + 2z = 7$

    $5x + 60y + 2z = 20$

15. $2x + y + 3z = 4$

    $2x + 3y - z = -4$

    $3x - 2y - 5z = -13$

16. $x - 3y = -14$

    $2x + y + 5z = 0$

    $-2y + 7z = -8$

Applications

Some applications contain information that produces more than one equation. When a system of equations is needed to solve such a problem, be sure to decide what each variable represents. You may need to draw and label a diagram, or use a table to help you organize the given information.

17. Roses sell for $3 each and carnations for $1.50 each. If a mixed bouquet of 28 blossoms costs $54, how many of each type of flower is in the bouquet?

18. Caitlyn invests money in two accounts, one paying 6% interest per year and one paying 8% interest per year. If she invested a total of $12,000 and received $800 interest for the year, how much did she invest in each account?

19. Felix's coin collection consists of 155 coins with a total value of $15.50. If the coins are nickels, dimes, and quarters, and twice the number of dimes is 70 more than the sum of the number of nickels and quarters, how many of each coin are in his collection?

20. The business office supply store sells three sizes of labels (small medium, large) costing $.75, $1.00, and $1.50 per package. One day they sold 39 packages of labels. The number of packages of medium size labels was twice the total number of packages of small and large labels. The receipts

for the labels was $40.25. How many packages of each type of label were sold?

Determinants

The determinant of the 2×2 matrix $\begin{bmatrix} a & b \\ c & d \end{bmatrix}$ is

$$\begin{vmatrix} a & b \\ c & d \end{vmatrix} = ad - bc$$

The determinant of $\begin{bmatrix} a & b & c \\ d & e & f \\ g & h & i \end{bmatrix}$ can be found by:

1. Writing the first two columns adjacent to the original three columns.
2. Multiplying the elements down the diagonals containing three entries and then adding.
3. Multiplying the elements up the diagonals containing three entries and then adding.
4. Subtracting the answer in step 3 from the answer in step 2.

Note: The method stated above is valid only for 3×3 matrices.

Find the value of each determinant.

21. $\begin{vmatrix} 2 & 5 \\ 1 & 6 \end{vmatrix}$

22. $\begin{vmatrix} 4 & 2 \\ 3 & 0 \end{vmatrix}$

23. $\begin{vmatrix} 1 & 0 \\ 0 & 1 \end{vmatrix}$

24. $\begin{vmatrix} 3 & -6 \\ 2 & 4 \end{vmatrix}$

25. $\begin{vmatrix} 3 & 2 & 2 \\ -1 & -1 & 1 \\ 0 & 1 & 2 \end{vmatrix}$

26. $\begin{vmatrix} -1 & 6 & 3 \\ 2 & -4 & 1 \\ -3 & -2 & 5 \end{vmatrix}$

Solving a System of Linear Equations Using Cramer's Rule

Cramer's Rule uses determinants to solve systems of equations. The rule is stated below for solving two equations in two variables and three equations in three variables.

Note: The denominator for x and y (and for x, y, and z in the three equation case) is the same. Thus, once you have found that value, you can use it for each denominator. Also, Cramer's Rule is *not* valid if the denominator equals 0.

The solution to $\begin{aligned} a_1 x + b_1 y &= c \\ a_2 x + b_2 y &= d \end{aligned}$ is

$$x = \frac{\begin{vmatrix} c & b_1 \\ d & b_2 \end{vmatrix}}{\begin{vmatrix} a_1 & b_1 \\ a_2 & b_2 \end{vmatrix}} \quad \text{and} \quad y = \frac{\begin{vmatrix} a_1 & c \\ a_2 & d \end{vmatrix}}{\begin{vmatrix} a_1 & b_1 \\ a_2 & b_2 \end{vmatrix}}$$

The solution to $\begin{aligned} a_1 x + b_1 y + c_1 z &= d \\ a_2 x + b_2 y + c_2 z &= e \\ a_3 x + b_3 y + c_3 z &= f \end{aligned}$ is

$$x = \frac{\begin{vmatrix} d & b_1 & c_1 \\ e & b_2 & c_2 \\ f & b_3 & c_3 \end{vmatrix}}{\begin{vmatrix} a_1 & b_1 & c_1 \\ a_2 & b_2 & c_2 \\ a_3 & b_3 & c_3 \end{vmatrix}} \quad y = \frac{\begin{vmatrix} a_1 & d & c_1 \\ a_2 & e & c_2 \\ a_3 & f & c_3 \end{vmatrix}}{\begin{vmatrix} a_1 & b_1 & c_1 \\ a_2 & b_2 & c_2 \\ a_3 & b_3 & c_3 \end{vmatrix}}$$

$$z = \frac{\begin{vmatrix} a_1 & b_1 & d \\ a_2 & b_2 & e \\ a_3 & b_3 & f \end{vmatrix}}{\begin{vmatrix} a_1 & b_1 & c_1 \\ a_2 & b_2 & c_2 \\ a_3 & b_3 & c_3 \end{vmatrix}}$$

Solve, using Cramer's Rule.

27. $2x + 3y = 5$
     $-3x - 5y = -9$

28. $4x + y = 7$
     $-2x + 3y = 14$

29. Solve for x using Cramer's Rule. Set up the determinants needed to solve for y and z.
    $$2x + y + 3z = 4$$
    $$2x + 3y - z = -4$$
    $$3x - 2y - 5z = -13$$

 ✓ # Check Yourself

1. Graph each line. The lines appear to intersect at $(1, 2)$. The solution to the system is in fact $(1, 2)$, as can be verified by substituting 1 for x and 2 for y in each equation.

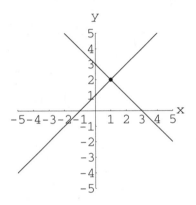

(Solving a system of linear equations by graphing)

2. Graph each line. The lines appear to intersect at $(-2, 1)$. The solution to the system is $(-2, 1)$, as can be verified by substitution.

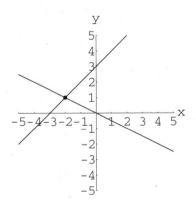

(Solving a system of linear equations by graphing)

3. Graph each line. Since the graphs are the same line (the *x*- and *y*-intercepts are the same for each line), the system is dependent.

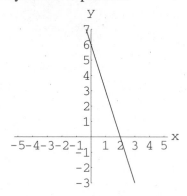

(Solving a system of linear equations by graphing)

4. To eliminate *y*, multiply the first equation by 3:

$$3(2x - y) = 3(9) \qquad \Rightarrow \qquad 6x - 3y = 27$$

Now add the equations:

$$
\begin{array}{r}
6x - 3y = 27 \\
\underline{4x + 3y = 23} \\
10x \qquad = 50 \\
x = 5
\end{array}
$$

Now substitute *x* = 5 into either of the original equations:

$$
\begin{aligned}
2(5) - y &= 9 \\
-y &= -1 \\
y &= 1
\end{aligned}
$$

Therefore, the solution to the system is (5, 1). **(Solving a system of linear equations by addition)**

5. To eliminate *x*, multiply the first equation by 3 and the second equation by –2. Then add the results.

$$
\begin{array}{l}
3(2x + 4y) = 3(8) \\
-2(3x - 5y) = -2(-21)
\end{array}
\quad \Rightarrow \quad
\begin{array}{r}
6x + 12y = 24 \\
\underline{-6x + 10y = 42} \\
22y = 66 \\
y = 3
\end{array}
$$

Now solve for *x* by substituting *y* = 3 into either of the original equations:

$$
\begin{aligned}
2x + 4(3) &= 8 \\
2x &= -4 \\
x &= -2
\end{aligned}
$$

(Solving a system of linear equations by addition)

6. First eliminate the fractions in each equation. Multiply the first equation by 4 and the second equation by 12:

$$4\left(\frac{1}{4}x - \frac{1}{2}y\right) = 4\left(-\frac{5}{4}\right)$$
$$12\left(\frac{1}{3}x + \frac{1}{4}y\right) = 12\left(-\frac{3}{4}\right)$$

$\Rightarrow$

$$x - 2y = -5$$
$$4x + 3y = -9$$

To eliminate x, multiply the first equation by -4 and then add the result to the second equation.

$$-4(x - 2y) = -4(-5) \qquad \Rightarrow$$

$$\begin{array}{r} -4x + 8y = 20 \\ \underline{4x + 3y = -9} \\ 11y = 11 \\ y = 1 \end{array}$$

Now solve for x by substituting $y = 1$ into either of the original equations:

$$\frac{1}{4}x - \frac{1}{2}(1) = -\frac{5}{4}$$
$$\frac{1}{4}x = -\frac{3}{4}$$
$$x = -3$$

Therefore, the solution to the system is $(-3, 1)$. **(Solving a system of linear equations by addition)**

7. To eliminate y, multiply the first equation by 3 and the second equation by 2 and then add the resulting equations. Note that you could multiply the first equation by 15 and the second equation by 10, but multiplying to produce the least common multiple of 10 and 15 which is 30 is a little simpler.

$$3(6x - 10y) = 3(-2)$$
$$2(9x + 15y) = 2(9)$$

$\Rightarrow$

$$\begin{array}{r} 18x - 30y = -6 \\ \underline{18x + 30y = 18} \\ 36x = 12 \\ x = \frac{12}{36} \\ x = \frac{1}{3} \end{array}$$

Now solve for y:

$$6\left(\frac{1}{3}\right) - 10y = -2$$
$$-10y = -4$$
$$y = \frac{4}{10}$$
$$y = \frac{2}{5}$$

Therefore, the solution is $\left(\frac{1}{3}, \frac{2}{5}\right)$. **(Solving a system of linear equations by addition)**

8. Multiply the first equation by 4 and then add the result to the second equation.

$$4\,(7x - 9y) \;=\; 4\,(2) \qquad \Rightarrow \qquad \begin{aligned} 28x - 36y &= 8 \\ -28x + 36y &= -8 \\ \hline 0 &= 0 \end{aligned}$$

When both variables are eliminated and the resulting equation is true, the equations are dependent (same line) and there are an infinite number of solutions. The solution may be written as $\{(x, y)|\ 7x - 9y = 2\,\}$. **(Solving a system of linear equations by addition)**

9. Multiply the first equation by –4 and then add the result to the second equation.

$$-4\,(x - 5y) \;=\; -4\,(8) \qquad \Rightarrow \qquad \begin{aligned} -4x + 20y &= -32 \\ 4x - 20y &= 10 \\ \hline 0 &= -22 \end{aligned}$$

When both variables are eliminated and the resulting equation is false, the system is inconsistent (parallel lines) and there is no solution to the system. **(Solving a system of linear equations by addition)**

10. $y = x + 4$ \qquad Substitute $x + 4$ for y in the second equation:
$6x - 3y = -15$

$$\begin{aligned} 6x - 3\,(x + 4) &= -15 \\ 6x - 3x - 12 &= -15 \\ 3x &= -3 \\ x &= -1 \end{aligned}$$

Now find y by substituting into $y = x + 4$:

$$y = (-1) + 4 = 3$$

The solution to the system is $(-1, 3)$. **(Solving a system of linear equations by substitution)**

11. $\quad x + 4y = 10$
$\quad 2x + 5y = 17$

Solve for x in the first equation:

$$x = -4y + 10$$

Now substitute for x in the second equation:

$$\begin{aligned} 2\,(-4y + 10) + 5y &= 17 \\ -8y + 20 + 5y &= 17 \\ -3y &= -3 \\ y &= 1 \end{aligned}$$

Then $x = -4y + 10 = -4\,(1) + 10 = 6$

The solution to the system is $(6, 1)$. **(Solving a system of linear equations by substitution)**

12. $y = 4x - 5$

    $8x - 2y = 10$

Substitute $4x - 5$ for y in the second equation:

$8x - 2(4x - 5) = 10$

    $8x - 8x + 10 = 10$

              $0 = 0$

When both variables are eliminated and the resulting equation is true, the equations are dependent (same line) and there are an infinite number of solutions. (**Solving a system of linear equations by substitution**)

13. $x = 3y + 1$

    $3x - 9y = 6$

$3(3y + 1) - 9y = 6$

    $9y + 3 - 9y = 6$

             $0 = 3$

When both variables are eliminated and the resulting equation is false, the system is inconsistent (parallel lines) and there is no solution to the system. (**Solving a system of linear equations by substitution**)

14. $2x + 3y - z = -8$

        $x - 3y + 2z = 7$

    $5x + 60y + 2z = 20$

Eliminate z from the equations. Use equations 1 and 2:

$$2(2x + 3y - z) = 2(-8) \quad \Rightarrow \quad \begin{array}{r} 4x + 6y - 2z = -16 \\ x - 3y + 2z = 7 \\ \hline 5x + 3y = -9 \end{array}$$

Use equations 2 and 3:

$$-1(x - 3y + 2z) = -1(7) \quad \Rightarrow \quad \begin{array}{r} -x + 3y - 2z = -7 \\ 5x + 60y + 2z = 20 \\ \hline 4x + 63y = 13 \end{array}$$

Now solve the new system in x and y:

$$\begin{aligned} -4(5x + 3y) &= -4(-9) \\ 5(4x + 63y) &= 5(13) \end{aligned} \quad \Rightarrow \quad \begin{array}{r} -20x - 12y = 36 \\ 20x + 315y = 65 \\ \hline 303y = 101 \\ y = \dfrac{1}{3} \end{array}$$

If $y = \dfrac{1}{3}$, then $5x + 3\left(\dfrac{1}{3}\right) = -9$

                              $5x = -10$

                                $x = -2$

Now use the x and y values to find z using one of the original equations:

$$2x + 3y - z = -8$$

$$2(-2) + 3\left(\frac{1}{3}\right) - z = -8$$

$$-4 + 1 - z = -8$$

$$z = 5$$

Therefore, the solution to the system is $\left(-2, \frac{1}{3}, 5\right)$. **(Solving a system of linear equations in three variables)**

15. $2x + y + 3z = 4$
     $2x + 3y - z = -4$
     $3x - 2y - 5z = -13$

Eliminate x. Use equations 1 and 2:

$$-1(2x + y + 3z) = -1(4) \qquad \Rightarrow \qquad \begin{array}{r} -2x - y - 3z = -4 \\ 2x + 3y - z = -4 \\ \hline 2y - 4z = -8 \end{array}$$

Use equations 2 and 3:

$$\begin{array}{l} -3(2x + 3y - z) = -3(-4) \\ 2(3x - 2y - 5z) = 2(-13) \end{array} \qquad \Rightarrow \qquad \begin{array}{r} -6x - 9y + 3z = 12 \\ 6x - 4y - 10z = -26 \\ \hline -13y - 7z = -14 \end{array}$$

Now solve the new system in y and z:

$$\begin{array}{l} -7(2y - 4z) = -7(-8) \\ 4(-13y - 7z) = 4(-14) \end{array} \qquad \Rightarrow \qquad \begin{array}{r} -14y + 28z = 56 \\ -52y - 28z = -56 \\ \hline -66y \qquad = 0 \\ y = 0 \end{array}$$

$$2y - 4z = -8$$

$$2(0) - 4z = -8$$

$$z = 2$$

Now use the y and z values to find x:

$$2x + y + 3z = 4$$

$$2x + 0 + 3(2) = 4$$

$$2x = -2$$

$$x = -1$$

The solution to the system is $(-1, 0, 2)$. **(Solving a system of linear equations in three variables)**

16. $x - 3y = -14$ Since equation 3 is already missing x, eliminate x from equations 1 and 2:

$2x + y + 5z = 0$

$-2y + 7z = -8$

$-2(x - 3y) = -2(-14)$ $\Rightarrow$ $\begin{aligned} -2x + 6y &= 28 \\ 2x + y + 5z &= 0 \\ \hline 7y + 5z &= 28 \end{aligned}$

The new system consists of the equation that resulted from combining equations 1 and 2, and original equation 3.

$\begin{aligned} 7(-2y + 7z) &= 7(-8) \\ 2(7y + 5z) &= 2(28) \end{aligned}$ $\Rightarrow$ $\begin{aligned} -14y + 49z &= -56 \\ 14y + 10z &= 56 \\ \hline 59z &= 0 \\ z &= 0 \end{aligned}$

$-2y + 7z = -8$

$-2y + 7(0) = -8$

$y = 4$

$x - 3y = -14$

$x - 3(4) = -14$

$x = -2$

The solution to the system is $(-2, 4, 0)$. (**Solving a system of linear equations in three variables**)

17. Let $R =$ the number of roses in the bouquet and let $C =$ the number of carnations in the bouquet.

$R + C = 28$ Set up a system of equations using the information given in the problem statement.

$3R + 1.5C = 54$

$-3(R + C) = -3(28)$ $\Rightarrow$ $\begin{aligned} -3R - 3C &= -84 \\ 3R + 1.5C &= 54 \\ \hline -1.5C &= -30 \\ C &= 20 \end{aligned}$

$R + C = 28$

$R + 20 = 28$

$R = 8$

Therefore, there are 8 roses and 20 carnations in the bouquet. (**Applications**)

18. Let $x =$ amount invested at 6% and let $y =$ the amount invested at 8%.

$x + y = 12000$ Write a system of equations. Eliminate the decimals in equation 2:

$0.06x + 0.08y = 800$

$100(0.06x + 0.08y) = 100(800)$ $\Rightarrow$ $6x + 8y = 80000$

Eliminate x:

$$-6(x+y) = -6(12000) \quad \Rightarrow \quad \begin{aligned} -6x - 6y &= -72000 \\ \underline{6x + 8y = 80000} \\ 2y &= 8000 \\ y &= 4000 \end{aligned}$$

$$\begin{aligned} x + y &= 12000 \\ x + 4000 &= 12000 \\ x &= 8000 \end{aligned}$$

Therefore, she invested $8000 at 6% interest and $4000 at 8% interest. **(Applications)**

19. Let x = the number of nickels, y = the number of dimes and z = the number of quarters.

$x + y + z = 155$ There are 155 coins in the collection.

$0.05x + 0.10y + 0.25z = 15.50$ The value of the collection is $15.50.

$2y = 70 + x + z$ Twice the number of dimes is 70 more than the number of nickels and quarters.

The system becomes:

$$\begin{aligned} x + y + z &= 155 \\ 5x + 10y + 25z &= 1550 \\ -x + 2y - z &= 70 \end{aligned}$$

Eliminate x. Use equations 1 and 2:

$$-5(x+y+z) = -5(155) \quad \Rightarrow \quad \begin{aligned} -5x - 5y - 5z &= -775 \\ \underline{5x + 10y + 25z = 1550} \\ 5y + 20z &= 775 \end{aligned}$$

Use equations 1 and 3:

$$\begin{aligned} x + y + z &= 155 \\ \underline{-x + 2y - z = 70} \\ 3y &= 225 \\ y &= 75 \end{aligned}$$

$$\begin{aligned} 5y + 20z &= 775 \\ 5(75) + 20z &= 775 \\ 20z &= 400 \\ z &= 20 \end{aligned}$$

$$\begin{aligned} x + y + z &= 155 \\ x + 75 + 20 &= 155 \\ x &= 60 \end{aligned}$$

Therefore, his collection contained 60 nickels, 75 dimes and 20 quarters. **(Applications)**

20. Let x = the number of packages of small labels sold, y = the number of packages of medium labels sold, and z = the number of packages of large labels sold.

$$x + y + z = 39$$
$$0.75x + 1y + 1.5z = 40.25$$
$$y = 2(x+z)$$

Set up equations to represent the given information.

The system becomes:

$$x + y + z = 39$$
$$75x + 100y + 150z = 4025$$
$$-2x + y - 2z = 0$$

Eliminate z. Use equations 1 and 2:

$$-150(x+y+z) = 150(39) \quad \Rightarrow \quad \begin{array}{r} -150x - 150y - 150z = -5850 \\ 75x + 100y + 150z = 4025 \\ \hline -75x - 50y = -1825 \end{array}$$

Use equations 1 and 3:

$$2(x+y+z) = 2(39) \quad \Rightarrow \quad \begin{array}{r} 2x + 2y + 2z = 78 \\ -2x + y - 2z = 0 \\ \hline 3y = 78 \\ y = 26 \end{array}$$

$$-75x - 50y = -1825$$
$$-75x - 50(26) = -1825$$
$$-75x - 1300 = -1825$$
$$-75x = -525$$
$$x = 7$$

$$x + y + z = 39$$
$$7 + 26 + z = 39$$
$$z = 6$$

Therefore, 7 packages of small labels were sold, 26 packages of medium labels were sold, and 6 packages of large labels were sold. (**Applications**)

21. $\begin{vmatrix} 2 & 5 \\ 1 & 6 \end{vmatrix} = 2 \cdot 6 - (5 \cdot 1) = 12 - 5 = 7$ (**Determinants**)

22. $\begin{vmatrix} 4 & 2 \\ 3 & 0 \end{vmatrix} = 4 \cdot 0 - (2 \cdot 3) = 0 - 6 = -6$ (**Determinants**)

23. $\begin{vmatrix} 1 & 0 \\ 0 & 1 \end{vmatrix} = 1 \cdot 1 - (0 \cdot 0) = 1 - 0 = 1$ (**Determinants**)

24. $\begin{vmatrix} 3 & -6 \\ 2 & 4 \end{vmatrix} = 3 \cdot 4 - (-6 \cdot 2) = 12 - (-12) = 12 + 12 = 24$ (**Determinants**)

25.

$$0 \qquad 3 \qquad -4$$
$$\begin{array}{cccccc} 3 & 2 & 2 & 3 & 2 \\ -1 & -1 & 1 & -1 & -1 \\ 0 & 1 & 2 & 0 & 1 \end{array}$$
$$-6 \qquad 0 \qquad -2$$

$-6 + 0 + (-2) = -8$ Find the sum of the products going down the diagonals.

$0 + 3 + (-4) = -1$ Find the sum of the products going up the diagonals.

$-8 - (-1) = -7$ Downward products minus upward products gives the determinant of -7.
(Determinants)

26.

$$36 \qquad 2 \qquad 60$$
$$\begin{array}{cccccc} -1 & 6 & 3 & -1 & 6 \\ 2 & -4 & 1 & 2 & -4 \\ -3 & -2 & 5 & -3 & -2 \end{array}$$
$$20 \qquad -18 \qquad -12$$

$(20 - 18 - 12) - (36 + 2 + 60) = -108$

The value of the determinant is -108. **(Determinants)**

27. $x = \dfrac{\begin{vmatrix} 5 & 3 \\ -9 & -5 \end{vmatrix}}{\begin{vmatrix} 2 & 3 \\ -3 & -5 \end{vmatrix}} = \dfrac{5\,(-5) - 3\,(-9)}{2\,(-5) - 3\,(-3)} = \dfrac{-25 + 27}{-10 + 9} = \dfrac{2}{-1} = -2$

$y = \dfrac{\begin{vmatrix} 2 & 5 \\ -3 & -9 \end{vmatrix}}{\begin{vmatrix} 2 & 3 \\ -3 & -5 \end{vmatrix}} = \dfrac{2\,(-9) - 5\,(-3)}{-1} = \dfrac{-3}{-1} = 3$

Therefore, the solution to the system is $(-2, 3)$. **(Solving a system of linear equations using Cramer's Rule)**

28. $x = \dfrac{\begin{vmatrix} 7 & 1 \\ 14 & 3 \end{vmatrix}}{\begin{vmatrix} 4 & 1 \\ -2 & 3 \end{vmatrix}} = \dfrac{7\,(3) - 1\,(14)}{4\,(3) - 1\,(-2)} = \dfrac{21 - 14}{12 + 2} = \dfrac{7}{14} = \dfrac{1}{2}$

$y = \dfrac{\begin{vmatrix} 4 & 7 \\ -2 & 14 \end{vmatrix}}{\begin{vmatrix} 4 & 1 \\ -2 & 3 \end{vmatrix}} = \dfrac{4\,(14) - 7\,(-2)}{14} = \dfrac{70}{14} = 5$

Therefore the solution to the system is $\left(\dfrac{1}{2}, 5\right)$ **(Solving a system of linear equations using Cramer's Rule)**

29. $x = \dfrac{\begin{vmatrix} 4 & 1 & 3 \\ -4 & 3 & -1 \\ -13 & -2 & -5 \end{vmatrix}}{\begin{vmatrix} 2 & 1 & 3 \\ 2 & 3 & -1 \\ 3 & -2 & -5 \end{vmatrix}} = \dfrac{66}{-66} = -1$

$y = \dfrac{\begin{vmatrix} 2 & 4 & 3 \\ 2 & -4 & -1 \\ 3 & -13 & -5 \end{vmatrix}}{\begin{vmatrix} 2 & 1 & 3 \\ 2 & 3 & -1 \\ 3 & -2 & -5 \end{vmatrix}}$

$z = \dfrac{\begin{vmatrix} 2 & 1 & 4 \\ 2 & 3 & -4 \\ 3 & -2 & -13 \end{vmatrix}}{\begin{vmatrix} 2 & 1 & 3 \\ 2 & 3 & -1 \\ 3 & -2 & -5 \end{vmatrix}}$

(Solving a system of linear equations using Cramer's Rule)

Grade Yourself

Circle the question numbers that you had incorrect. Then indicate the number of questions you missed. If you answered more than three questions incorrectly, you need to focus on that topic. (If a topic has less than three questions and you had at least one wrong, we suggest you study that topic also. Read your textbook, a review book, or ask your teacher for help.)

Subject: Systems of Linear Equations

Topic	Question Numbers	Number Incorrect
Solving a system of linear equations by graphing	1, 2, 3	
Solving a system of linear equations by addition	4, 5, 6, 7, 8, 9	
Solving a system of linear equations by substitution	10, 11, 12, 13	
Solving a system of linear equations in three variables	14, 15, 16	
Applications	17, 18, 19, 20	
Determinants	21, 22, 23, 24, 25, 26	
Solving a system of linear equations using Cramer's Rule	27, 28, 29	

Functions

9

 Test Yourself

Functions

A function is a set of ordered pairs in which no *x*-coordinate is repeated. There are various ways to identify functions, depending on the type of information you are given.

1. Given a set of ordered pairs, check the *x*-coordinates. If no *x*-coordinate is repeated, the set of ordered pairs describes a function.
2. Given a graph, draw vertical lines through the graph. If any vertical line intersects more than 1 point on the graph, the graph does not describe a function. (This is called the vertical line test.)
3. Given a mapping diagram, each element in the first set must map to only one element in the second set.
4. Given an equation, graph the equation and use the vertical line test to determine whether the equation represents a function.

Identify which of the following are functions.

1. $\{(2, 5), (1, 3), (2, -5)\}$

2. $\{(0, 1), (2, 1), (3, 4)\}$

3.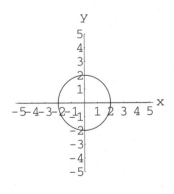

4.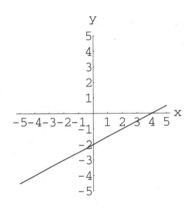

5.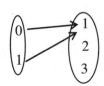

5.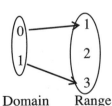

6.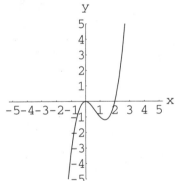

Domain   Range

7.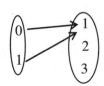

Domain   Range

8.

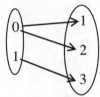

Domain   Range

9. $y = \dfrac{1}{2}x - 3$

10. $y = \sqrt{x - 4}$

Finding the Domain of a Function

The domain of a function is the set of all permissible *x*-values.

Given a set of ordered pairs, the domain is the set of *x*-coordinates.

Given an equation:

1. Often the domain is the set of all real numbers.
2. If there is a variable in the denominator, set the denominator equal to 0 and solve. The domain consists of all real numbers except the value(s) that makes the denominator 0.
3. If there is a variable in a square root (or an even indexed root), set the radicand greater than or equal to 0 and use that solution as the domain.

Find the domain.

11. $\{(2, 4), (1, 3), (3, 5)\}$

12. $y = 4x - 1$

13. $y = \dfrac{6}{3x - 1}$

14. $y = \sqrt{x + 2}$

15. $y = \dfrac{\sqrt{x + 4}}{x}$

Function Notation

Function notation consists of a function name, usually $f, g,$ or h, followed in parentheses by the variable that represents the domain elements: $f(x)$, $g(x)$, $h(x)$ read f of x, g of x and h of x.

Note that $f(x)$ does *not* mean f times x.

Finding $f(3)$ means finding the *y*-value paired with the *x*-value 3.

For example, given $f(x) = 2x + 1$, then $f(3)$ is obtained by replacing x with 3 in $2x + 1$.

16. Find $f(3)$ if $f(x) = 2x + 1$

17. Find $g(-2)$ if $g(x) = x^2 - 2x + 1$

18. Find $f(a)$ if $f(x) = 4x - 6$

19. Find $f(a + h)$ if $f(x) = x^2 + 4x$

Operations with Functions

We add, subtract, multiply and divide functions using the following definitions.

$$(f + g)(x) = f(x) + g(x)$$
$$(f - g)(x) = f(x) - g(x)$$
$$(f \cdot g)(x) = f(x) \cdot g(x)$$
$$\left(\dfrac{f}{g}\right)(x) = \dfrac{f(x)}{g(x)}, g(x) \neq 0$$

If $f(x) = 3x - 6$ and $g(x) = 2x^2 + x - 1$, find:

20. $(f + g)(2)$

21. $(f - g)(-4)$

22. $(f \cdot g)(x)$

23. $\left(\dfrac{f}{g}\right)(1)$

Composition of Functions

The composition $f \circ g$ of functions f and g is defined as

$$(f \circ g)(x) = f(g(x)).$$

That is, first find $g(x)$ and then use that result to find f at $g(x)$.

24. If $f(x) = 2x - 5$ and $g(x) = x^2$, find $(f \circ g)(3)$

25. If $f(x) = 2x - 5$ and $g(x) = x^2$, find $(g \circ f)(3)$

26. If $f(x) = 2x - 5$ and $g(x) = x^2$, find $(f \circ g)(x)$

27. If $f(x) = 2x - 5$ and $g(x) = x^2$, find
    $(g \circ f)(x)$

28. If $f(x) = 4x + 2$ and $g(x) = 3x^2 - 4x$, find
    $(f \circ g)(x)$

Graphing Functions

To graph a function, prepare a table of values. The $f(x)$ can be thought of as y. Although you need only two points to graph a line, to graph other functions, you will probably need more than two points. Functions may be identified by names:

Linear function: $f(x) = ax + b$, graph is a nonhorizontal line if $a \neq 0$

Constant function: $f(x) = c$, graph is a horizontal line

Quadratic function: $f(x) = ax^2 + bx + c$, $a \neq 0$, graph is a parabola

Polynomial function: $f(x) =$ polynomial in x, graph is a smooth curve that *may* contain peaks and valleys

Square Root function: $f(x) = \sqrt{ax + b}$, $a \neq 0$, graph is one branch of a parabola

---

Graph and identify the type of function.

29. $f(x) = \dfrac{1}{2}x - 3$

30. $f(x) = x^2 + 2$

31. $f(x) = -4$

32. $f(x) = \sqrt{x + 2}$

33. $f(x) = x^3 - 1$

34. $f(x) = -\dfrac{3}{2}x + 3$

One-to-One Functions

In a function, each x-coordinate is paired with a unique y-coordinate. In a one-to-one function, each y coordinate is paired with a unique x coordinate.

A one-to-one function passes the horizontal line test—that is, any horizontal line passes through the graph of a one-to-one function at most once.

35. Is the graph below the graph of a one-to-one function?

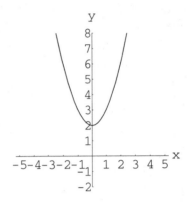

36. Is the graph below the graph of a one-to-one function?

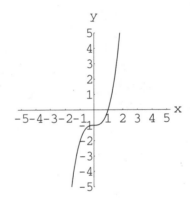

37. Is the graph below the graph of a one-to-one function?

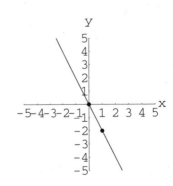

Inverse Functions

The inverse of a function *f*, written $f^{-1}(x)$, can be found by any one of the following methods:
1. Reversing the coordinates in a set of ordered pairs.
2. Interchanging *x* and *y* and solving for the new *y* if you are given an equation.
3. Reflecting a given graph across the line $y = x$.

Note: Only one-to-one functions have inverses that are functions.

Find the inverse.

38. $\{(2,5),(1,3),(4,-6)\}$

39. $y = 3x + 2$

40. $f(x) = \frac{2}{3}x - 4$

41. $f(x) = x^3 + 1$

42.

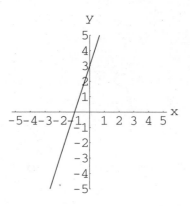

43.

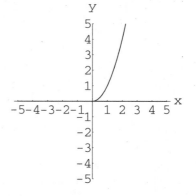

Check Yourself

1. Not a function. Note the repeated *x*-value of 2. **(Functions)**

2. This is a function. There are no repeated *x*-values. **(Functions)**

3. Not a function. The graph does not pass the vertical line test. **(Functions)**

4. This is a function. The graph passes the vertical line test. **(Functions)**

5. This is a function. The graph passes the vertical line test. **(Functions)**

6. This is a function. No element in the domain is used more than once. **(Functions)**

7. This is a function. The set of ordered pairs is $\{(0,1),(1,3)\}$. **(Functions)**

8. Not a function. The set of ordered pairs is $\{(0,1),(0,2),(1,3)\}$. The element 0 is used twice. **(Functions)**

9. Draw the graph. Note that it passes the vertical line test. Yes, this is a function.

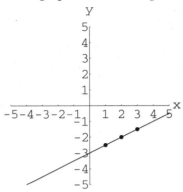

(Functions)

10.

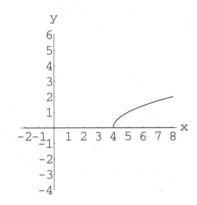

Draw the graph. Note that is passes the vertical line test. Yes, this is a function. Also note that for each $x \geq 4$, only one y-value is specified. **(Functions)**

11. The domain is $\{2, 1, 3\}$ **(Finding the domain of a function)**

12. The domain is all real numbers, written $(-\infty, \infty)$. **(Finding the domain of a function)**

13. $3x - 1 = 0$ Set the denominator equal to 0 and solve.

$$3x = 1$$

$$x = \frac{1}{3}$$

The domain is all real numbers except $\frac{1}{3}$. **(Finding the domain of a function)**

14. $x + 2 \geq 0$ Set the radicand ≥ 0 and solve.

$$x \geq -2$$

The domain is $x \geq -2$. **(Finding the domain of a function)**

15. $x + 4 \geq 0$ *and* $x \neq 0$ (since this would make the denominator undefined)

$$x \geq -4$$

The domain is all real numbers greater than or equal to -4 except 0. **(Finding the domain of a function)**

16. $f(3) = 2(3) + 1 = 7$ **(Function notation)**

17. $g(-2) = (-2)^2 - 2(-2) + 1$
    $\qquad\quad = 4 + 4 + 1$
    $\qquad\quad = 9$

    (Function notation)

18. $f(a) = 4(a) - 6 = 4a - 6$ **(Function notation)**

19. $f(a + h) = (a + h)^2 + 4(a + h)$ Replace each occurrence of x with $(a + h)$.

    $\qquad\quad\;\; = a^2 + 2ah + h^2 + 4a + 4h$

    (Function notation)

20. $(f + g)(2) = f(2) + g(2)$
    $\qquad\qquad\; = (3(2) - 6) + (2(2)^2 + 2 - 1)$
    $\qquad\qquad\; = 6 - 6 + 8 + 2 - 1$
    $\qquad\qquad\; = 9$

    Or, first add $f + g$:

    $f(x) + g(x) = (3x - 6) + (2x^2 + x - 1) = 2x^2 + 4x - 7$

    Now replace x with 2:

    $(f + g)(2) = 2(2^2) + 4(2) - 7 = 9$ **(Operations with functions)**

21. $(f - g)(-4) = f(-4) - g(-4)$
    $\qquad\qquad\quad = (3(-4) - 6) - (2(-4)^2 + (-4) - 1)$
    $\qquad\qquad\quad = (-18) - (27)$
    $\qquad\qquad\quad = -45$

    (Operations with functions)

22. $(f \cdot g)(x) = f(x) \cdot g(x)$
    $\qquad\qquad\; = (3x - 6)(2x^2 + x - 1)$
    $\qquad\qquad\; = 6x^3 + 3x^2 - 3x - 12x^2 - 6x + 6$
    $\qquad\qquad\; = 6x^3 - 9x^2 - 9x + 6$

    (Operations with functions)

23. $\left(\dfrac{f}{g}\right)(1) = \dfrac{f(1)}{g(1)}$

    $\qquad\qquad = \dfrac{3(1) - 6}{2(1)^2 + 1 - 1}$

    $\qquad\qquad = \dfrac{-3}{2}$

    (Operations with functions)

24. $(f \circ g)(3) = f(g(3)) = f(3^2) = f(9) = 2(9) - 5 = 13$ **(Composition of functions)**

25. $(g \circ f)(3) = g(f(3)) = g(2(3) - 5) = g(1) = 1^2 = 1$

Note from #24 and #25 that in general $f(g(x)) \neq g(f(x))$; thus usually $f \circ g \neq g \circ f$. **(Composition of functions)**

26. $(f \circ g)(x) = f(g(x)) = f(x^2) = 2(x^2) - 5 = 2x^2 - 5$ **(Composition of functions)**

27. $(g \circ f)(x) = g(f(x)) = g(2x - 5)$
$$= (2x - 5)^2$$
$$= 4x^2 - 20x + 25$$

(Composition of functions)

28. $f(g(x)) = f(3x^2 - 4x)$
$$= 4(3x^2 - 4x) + 2$$
$$= 12x^2 - 16x + 2$$

(Composition of functions)

29. 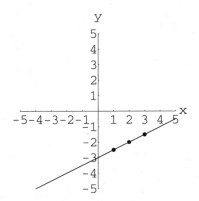 linear function **(Graphing functions)**

30. 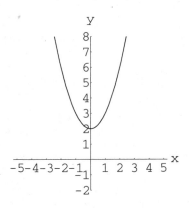 quadratic function **(Graphing functions)**

31. 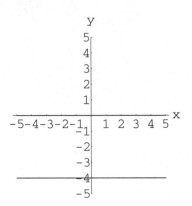   constant function **(Graphing functions)**

32. 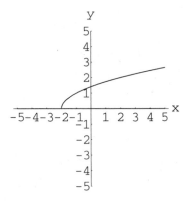   square root function **(Graphing functions)**

33. 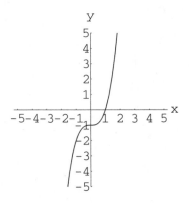   polynomial function **(Graphing functions)**

34. 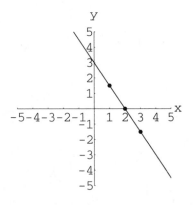   linear function **(Graphing functions)**

35. No. Draw several horizontal lines and observe that some of them intersect more than one point on the graph. **(One-to-one functions)**

36. Yes. The graph passes the horizontal line test. **(One-to-one functions)**

37. Yes **(One-to-one functions)**

38. $\{(5, 2), (3, 1), (-6, 4)\}$ Interchange the x and y coordinates. **(Inverse functions)**

39. $y = 3x + 2$

   $x = 3y + 2$ Interchange x and y.

   $\dfrac{x - 2}{3} = y$ Solve for y.

   $f^{-1}(x) = \dfrac{x - 2}{3}$ Use the notation for the inverse function. **(Inverse functions)**

40. $f(x) = \dfrac{2}{3}x - 4$ can be written $y = \dfrac{2}{3}x - 4$.

   $x = \dfrac{2}{3}y - 4$ Interchange x and y.

   $x + 4 = \dfrac{2}{3}y$

   $\dfrac{3}{2}(x + 4) = y$ Solve for y.

   $f^{-1}(x) = \dfrac{3}{2}(x + 4)$ Use the notation for the inverse function. **(Inverse functions)**

41. $y = x^3 + 1$ Set $y = f(x)$.

   $x = y^3 + 1$ Interchange x and y.

   $x - 1 = y^3$

   $\sqrt[3]{x - 1} = y$ Solve for y.

   $f^{-1}(x) = \sqrt[3]{x - 1}$ Use the notation for the inverse function. **(Inverse functions)**

42.

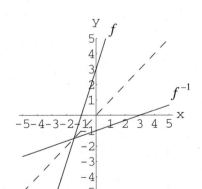

Reflect the given graph across the line $y = x$.

(Inverse functions)

43.

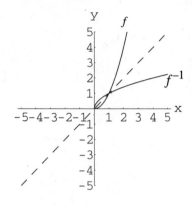

(Inverse functions)

Grade Yourself

Circle the question numbers that you had incorrect. Then indicate the number of questions you missed. If you answered more than three questions incorrectly, you need to focus on that topic. (If a topic has less than three questions and you had at least one wrong, we suggest you study that topic also. Read your textbook, a review book, or ask your teacher for help.)

Subject: Functions

Topic	Question Numbers	Number Incorrect
Functions	1, 2, 3, 4, 5, 6, 7, 8, 9, 10	
Finding the domain of a function	11, 12, 13, 14, 15	
Function notation	16, 17, 18, 19	
Operations with functions	20, 21, 22, 23	
Composition of functions	24, 25, 26, 27 28	
Graphing functions	29, 30, 31, 32, 33, 34	
One-to-one functions	35, 36, 37	
Inverse functions	38, 39, 40, 41, 42, 43	

Conic Sections

 Test Yourself

Parabolas

Parabolas of the form $y = ax^2 + bx + c, a \neq 0$, have a **vertex**. The vertex is the lowest or highest point of the parabola, depending on whether the parabola opens up or down. The **axis** of a parabola is a vertical line through the vertex. The parabola is **symmetric** about its axis, that is, if folded along its axis the two sides of the parabola would coincide.

If $a > 0$, the parabola opens up and if $a < 0$ the parabola opens down.

Graphing Parabolas

To graph a parabola of the form $y = ax^2 + bx + c$
1. Find the x-intercept(s) by setting $y = 0$. (You may need to use the quadratic formula if
   $ax^2 + bx + c$ doesn't factor.)
2. Find the y-intercept by setting $x = 0$.
3. Find the vertex using the formula $x = -\dfrac{b}{2a}$ for the
   x-coordinate of the vertex.
4. Find point(s) to the left and right of the vertex.
5. Connect the points with a smooth curve.

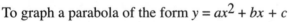

Graph each parabola. State the vertex and x-intercept(s).

1. $y = x^2 + 4x$

2. $y = 2x^2 - 8x + 8$

3. $y = 3x^2 - 6x + 1$

Finding the Vertex by Completing the Square

The formula for finding the x-coordinate of the vertex of a parabola can be derived by completing the square on the equation $y = ax^2 + bx + c, a \neq 0$. Since we have already used completing the square to solve quadratic equations, you may find it easier to complete the square than to remember $x = -\dfrac{b}{2a}$ as a formula for the vertex.

To find the vertex of $y = ax^2 + bx + c$ by completing the square:
1. Move the constant term to the left side of the equation.
2. Factor out a from $ax^2 + bx$ if $a \neq 1$.
3. Add $(\frac{1}{2}$ the coefficient of $x)^2$ inside the parentheses. Add a times $(\frac{1}{2}$ the coefficient of $x)^2$ to the left side of the equation.
4. Factor the right side as a perfect square trinomial.
5. Solve for y.
6. The equation should now be in the form
   $y = a(x - h)^2 + k$ where (h, k) is the vertex of the parabola.

Find the vertex of each parabola by completing the square.

4. $y = x^2 - 4x + 8$

126

5. $y = 2x^2 - 16x + 8$

6. $y = -x^2 - 2x - 3$

7. $y = x^2 + x + 1$

Applications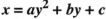

We can use either the formula or completing the square to find the vertex of a parabola. Finding the highest or lowest point on a parabola can be used to find the maximum or minimum in an application that involves a quadratic function.

8. A company selling a home computer version of a video game finds that they will make a weekly profit of P dollars where $P = -x^2 + 150x - 4425$ where x is the number of copies sold per week. Find the number of copies they should sell to make the maximum profit. What is the maximum profit?

Parabolas of the Form
$x = ay^2 + by + c$

Features of parabolas of the form $x = ay^2 + by + c$
1. The parabola opens left if $a < 0$ and opens right if $a > 0$.
2. If you complete the square on y and write the equation as $x = a(y - k)^2 + h$, the vertex is (h, k).
3. The axis is a horizontal line through the vertex.

9. Graph $x = -2(y - 1)^2 + 3$.

10. Graph $x = 3y^2 + 6y + 5$.

Circles and Ellipses

Equation of a Circle. The equation of a circle with center (h, k) and radius r is
$(x - h)^2 + (y - k)^2 = r^2$.
In the equation of a circle, notice that the coordinates of the center are *subtracted* from x and y in the parentheses.

11. Find an equation for the circle with center at $(2, 3)$ and radius 4.

12. Find an equation for the circle with center at $(-2, 6)$ and radius 3.

13. Find an equation for the circle with center at the origin and radius 2.

Finding the Center and Radius
When an equation of a circle is written in the form $(x - h)^2 + (y - k)^2 = r^2$, we know the vertex is (h, k) and the radius is r. If the equation is *not* in that form, we must complete the square on the x's and y's to write it in that form.

Find the center and radius of each circle.

14. $(x - 2)^2 + (y - 1)^2 = 25$

15. $(x + 3)^2 + (y - 4)^2 = 49$

16. $x^2 + (y - 2)^2 = 16$

17. $x^2 + y^2 = 9$

Graph each equation.

18. $x^2 + y^2 - 4x + 2y - 4 = 0$

19. $x^2 + y^2 + 6y + 5 = 0$

Ellipses Centered at the Origin

Equation of an ellipse centered at $(0, 0)$
The equation of an ellipse centered at $(0, 0)$ with x-intercepts at a and $-a$ and y-intercepts at b and $-b$ is:

$$\frac{x^2}{a^2} + \frac{y^2}{b^2} = 1 \qquad \text{Standard Form}$$

20. Sketch the graph of $\frac{x^2}{16} + \frac{y^2}{9} = 1$.

21. Sketch the graph of $9x^2 + 4y^2 = 36$.

Ellipses with Center (h, k)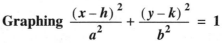

Graphing $\frac{(x - h)^2}{a^2} + \frac{(y - k)^2}{b^2} = 1$

1. Locate the center at (h, k).
2. Move a units to the left of (h, k) and place a point there. Move a units to the right of (h, k) and place a point there.
3. Move b units up from (h, k) and place a point there. Move b units down from (h, k) and place a point there.
4. Connect the four points with a smooth curve.

22. Sketch the graph of $\dfrac{(x-2)^2}{9} + \dfrac{(y-1)^2}{4} = 1$.

Hyperbolas Centered at (0, 0)

Hyperbolas centered at the origin
A hyperbola has two vertices, a center, and may open left and right or up and down.
The standard form of a hyperbola with center $(0, 0)$

that opens left and right is: $\dfrac{x^2}{a^2} - \dfrac{y^2}{b^2} = 1$.

The vertices are $(-a, 0)$ and $(a, 0)$.
The standard form of a hyperbola with center $(0, 0)$

that opens up and down is: $\dfrac{y^2}{b^2} - \dfrac{x^2}{a^2} = 1$.

The vertices are $(0, b)$ and $(0, -b)$
Note that when the equation is written in standard form and the x^2 term is positive, the hyperbola opens left and right. When the equation is written in standard form and the y^2 term is positive, the hyperbola opens up and down. In addition to using the vertices and the direction a hyperbola opens, we also need the **asymptotes** to determine the width of the openings. An asymptote is a line that the curve approaches, but never crosses. These asymptotes are the extended diagonals of the **fundamental rectangle**; that is, the rectangle with corners at $(-a, b), (a, b), (a, -b), (-a, -b)$.

---

23. Sketch the graph of $\dfrac{x^2}{9} - \dfrac{y^2}{16} = 1$.

24. Sketch the graph of $\dfrac{y^2}{25} - \dfrac{x^2}{4} = 1$.

25. Sketch the graph of $4x^2 - 49y^2 = 196$.

Nonlinear Systems of Equations and Second-Degree Inequalities

Systems of nonlinear equations are generally solved using substitution or addition.
When both of the equations contain squared x and y terms, it may be easier to use the addition method.

---

26. Solve the system:

$x^2 + y^2 = 1$

$2y = x + 1$

27. Solve the system:

$x^2 + y^2 = 16$

$y = x^2 - 4$

28. Solve the system:

$4x^2 + 3y^2 = 256$

$x^2 + y^2 = 80$

29. Solve the system:

$\dfrac{y^2}{4} - \dfrac{x^2}{16} = 1$

$x^2 + y^2 = 9$

Second-Degree Inequalities

When graphing inequalities:
Use solid curves for boundaries when the inequality contains $\leq$ or $\geq$.
Use dotted curves for boundaries when the inequality contains $>$ or $<$.
Choose a point in each region formed to determine whether the region should be shaded. If the chosen point makes the inequality true, shade that region.

---

30. Graph $x^2 + y^2 < 9$.

31. Graph $y \leq (x-3)^2 - 2$.

32. Graph $4x^2 < 9y^2 + 36$.

Systems of Inequalities

The solution to a system of inequalities is the intersection of the solutions of each inequality. You may wish to use a different colored pencil for each inequality. The solution to the system is where the colors overlap.

33. Graph.

$x^2 + y^2 < 9$

$y \le 2x$

34. Graph.

$y \le (x - 3)^2 - 2$

$2x^2 + y^2 \le 16$

 Check Yourself

1. $y = x^2 + 4x$

    Step 1.　　Find the x-intercepts by setting $y = 0$.

    $0 = x^2 + 4x$

    $0 = x(x + 4)$　Factor.

    $x = 0$　or　$x + 4 = 0$　Set each factor equal to 0.

    $\qquad\qquad\qquad x = -4$

    Step 2.　Find the y-intercept by setting $x = 0$.

    $y = 0^2 + 4(0) = 0$ when $x = 0$

    Step 3.　　Find the vertex.

    $a = 1, b = 4$　　Identify a and b.

    $x = -\dfrac{b}{2a}$　Use the formula for the x-coordinate of the vertex.

    $x = -\dfrac{4}{2\,(1)}$　Substitute $a = 1, b = 4$.

    $y = (-2)^2 + 4(-2) = 4 - 8 = -4$ 　　　　　　　　　Find the y-coordinate of the vertex.

    Step 4.

    Let's organize the points in a table:

x	y
-4	0
-2	-4
0	0

    If $x = -3, f(-3) = (-3)^2 + 4(-3) = 9 - 12 = -3$ 　　　Find a point to the left of the vertex.

    If $x = -1, f(-1) = (-1)^2 + 4(-1) = 1 - 4 = -3$ 　　　Find a point to the right of the vertex.

The table becomes

x	y
-3	-3
-1	-3

and the graph is

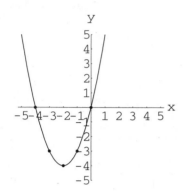

Step 5.

The vertex is $(-2, -4)$ and the x-intercepts are -4 and 0. **(Parabolas)**

2. $y = 2x^2 - 8x + 8$

Step 1.   Find the x-intercepts by setting $y = 0$.

$0 = 2x^2 - 8x + 8$

$0 = 2(x^2 - 4x + 4)$ Factor out 2.

$0 = 2(x - 2)(x - 2)$ Factor.

$x - 2 = 0$ Set the factor equal to 0.

$x = 2$ There is only one x-intercept.

Step 2.   Find the y-intercept by setting $x = 0$.

$y = 2(0)^2 - 8(0) + 8 = 8$ when $x = 0$.

Step 3.   Find the vertex.

$a = 2, b = -8$ Identify a and b.

$x = -\dfrac{b}{2a}$ Use the formula for the x-coordinate of the vertex.

$x = -\dfrac{(-8)}{2\,(2)} = 2$ Substitute $a = 2, b = -8$.

$y = 2(2)^2 - 8(2) + 8 = 0$ when $x = 2$.   Find the y-coordinate of the vertex.

Step 4.   Organize the data in a table.

Find y for $x = 1$ and $x = 3$.

x	y
0	8
1	2
2	0
3	2

$y = 2(1)^2 - 8(1) + 8 = 2 - 8 + 8 = 2$ when $x = 1$

$y = 2(3)^2 - 8(3) + 8 = 2(9) - 24 + 8 = 2$ when $x = 3$

Step 5.

Graph the points and connect with a smooth curve.

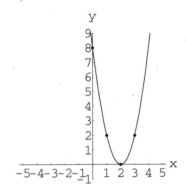

The vertex is $(2, 0)$ and the x-intercept is $(2, 0)$. **(Parabolas)**

3. $y = 3x^2 - 6x + 1$

Step 1.   Find the x-intercepts by setting $y = 0$.

$0 = 3x^2 - 6x + 1$

$a = 3, b = -6, c = 1$ The equation does not factor, so we use the quadratic formula.

$x = \dfrac{-b \pm \sqrt{b^2 - 4ac}}{2a}$ Write the quadratic formula.

$x = \dfrac{-(-6) \pm \sqrt{(-6)^2 - 4(3)(1)}}{2(3)}$ Substitute a, b, and c.

$x = \dfrac{6 \pm \sqrt{36 - 12}}{6}$ Simplify the radicand.

$x = \dfrac{6 \pm \sqrt{24}}{6}$

$$x = \frac{6 \pm 2\sqrt{6}}{6} \quad \text{Simplify the radical.} \quad \sqrt{24} = \sqrt{4}\sqrt{6} = 2\sqrt{6}$$

$$x = \frac{3 \pm \sqrt{6}}{3} \quad \text{Reduce.}$$

$\frac{3 + \sqrt{6}}{3}$ and $\frac{3 - \sqrt{6}}{3}$ are the exact x-intercepts

Approximations from a calculator give $x \approx 1.8$ and $x \approx 0.2$.

Step 2.   Find the y-intercept by setting $x = 0$.

$y = 3(0)^2 - 6(0) + 1 = 1$ when $x = 0$

Step 3.   Find the vertex.

$a = 3, b = -6$ Identify a and b.

$$x = -\frac{b}{2a} \quad \text{Use the formula for the } x\text{-coordinate of the vertex.}$$

$$x = -\frac{(-6)}{2\,(3)} = 1 \quad \text{Substitute } a = 3, b = -6.$$

$y = 3(1)^2 - 6(1) + 1 = -2$ Find the y-coordinate of the vertex.

Step 4.

Organizing our data in a table and finding (x) for $x = -1$ and $x = 2$ we have

x	y
0	1
$\frac{3 - \sqrt{6}}{3} \approx 0.2$	0
1	-2
$\frac{3 + \sqrt{6}}{3} \approx 1.8$	0
2	1

$y = 3(2)^2 - 6(2) + 1 = 12 - 12 + 1 = 1$ when $x = 2$

Step 5.

Graph the points and connect with a smooth curve.

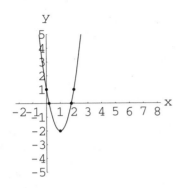

(Parabolas)

4. $y = x^2 - 4x + 8$

$y - 8 = x^2 - 4x$ Subtract 8 from both sides.

$y - 8 + 4 = (x^2 - 4x + 4)$ Add $(\frac{1}{2} \cdot -4)^2 = (-2)^2 = 4$ to both sides.

$y - 4 = (x - 2)^2$ Factor the right side.

$y = (x - 2)^2 + 4$ Solve for y.

The vertex is $(2, 4)$. **(Finding the vertex by completing the square)**

5. $y = 2x^2 - 16x + 8$

$y - 8 = 2x^2 - 16x$ Subtract 8 from both sides.

$y - 8 = 2(x^2 - 8x)$ Factor out 2.

$y - 8 + 32 = 2(x^2 - 8x + 16)$ Add$(\frac{1}{2} \cdot -8)^2 = (-4)^2 = 16$ inside the parentheses. Add $2 \cdot 16 = 32$ to the left side.

$y + 24 = 2(x - 4)^2$ Factor the right side.

$y = 2(x - 2)^2 - 24$ Solve for y.

The vertex is $(4, -24)$. **(Finding the vertex by completing the square)**

6. $y = -x^2 - 2x - 3$

$y + 3 = -x^2 - 2x$ Add 3 to both sides.

$y + 3 = -(x^2 + 2x)$ Factor out -1.

$y + 3 - 1 = -(x^2 + 2x + 1)$ Add $(\frac{1}{2} \cdot 2)^2 = 1$ inside the parentheses. Add $-1 \cdot 1$ to the left side.

$y + 2 = -(x + 1)^2$ Factor the right side.

$y = -(x + 1)^2 - 2$ Solve for y.

Note: $(x + 1)^2 = (x - (-1))^2 = (x - h)^2$, so $h = -1$.

The vertex is $(-1, -2)$. **(Finding the vertex by completing the square)**

7. $y = x^2 + x + 1$

$y - 1 = x^2 + x$ Subtract 1 from both sides.

$y - 1 + \dfrac{1}{4} = x^2 + x + \dfrac{1}{4}$ Add $(\dfrac{1}{2} \cdot 1)^2 = \dfrac{1}{4}$ to both sides.

$y - \dfrac{3}{4} = (x + \dfrac{1}{2})^2$ Factor the right side.

$y = (x + \dfrac{1}{2})^2 + \dfrac{3}{4}$ Solve for y.

The vertex is $\left(-\dfrac{1}{2}, \dfrac{3}{4} \right)$. **(Finding the vertex by completing the square)**

8. Since the profit equation is quadratic, its graph will be a parabola. Since $a < 0$, the parabola opens down. The vertex will be at the highest point on the curve, and so will represent the maximum profit. Let's find the vertex using $x = -\dfrac{b}{2a}$:

$a = -1, b = 150$ Identify a and b.

$x = -\dfrac{b}{2a} = -\dfrac{150}{2(-1)} = 75$ Find the x-coordinate of the vertex.

When 75 copies of the program are sold,

$P = -(75)^2 + 150(75) - 4425$ Find the profit.

$P = -5625 + 11250 - 4425$ Simplify.

$P = 1200$ Simplify.

The maximum profit of $1200 is made when 75 copies per week are sold. **(Applications)**

9. Since $a < 0$, the parabola opens left. The vertex is $(3, 1)$. When $y = 0$,

$x = -2(0 - 1)^2 + 3 = -2(1) + 3 = 1$. By symmetry (or by substitution) the point $(1, 2)$ is also on the graph.

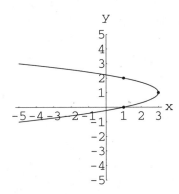

(Parabolas)

10. The parabola opens to the right since $a = 3$. Let's complete the square on y:

$x - 5 = 3y^2 + 6y$ Subtract 5 from both sides.

$x - 5 = 3(y^2 + 2y)$ Factor out 3

$x - 5 + 3 = 3(y^2 + 2y + 1)$ Add $(\frac{1}{2} \cdot 2)^2 = 1$ inside the parentheses. Add $3 \cdot 1 = 3$ on the left.

$x - 2 = 3(y + 1)^2$ Factor the right side.

$x = 3(y + 1)^2 + 2$ Solve for x.

The vertex is $(2, -1)$. When $y = 0, x = 5$. By symmetry (or by substitution) the point $(5, -2)$ is also on the graph.

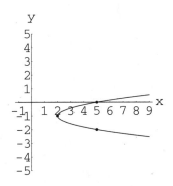

(Parabolas)

11. $(h, k) = (2, 3)$ $r = 4$ Identify h, k, and r.

$(x - h)^2 + (y - k)^2 = r^2$ Write the equation of a circle.

$(x - 2)^2 + (y - 3)^2 = (4)^2$ Substitute for h, k, and r.

$(x - 2)^2 + (y - 3)^2 = 16$ **(Circles and ellipses)**

12. $(h, k) = (-2, 6)$ $r = 3$ Identify h, k, and r.

$(x - h)^2 + (y - k)^2 = r^2$ Write the equation of a circle.

$(x - (-2))^2 + (y - 6)^2 = (3)^2$ Substitute for h, k, and r.

$(x + 2)^2 + (y - 6)^2 = 9$ Simplify inside the parentheses. **(Circles and ellipses)**

13. $(h, k) = (0, 0)$ $r = 2$ Center at the origin means $(h, k) = (0, 0)$.

    $(x - h)^2 + (y - k)^2 = r^2$ Write the equation of a circle.

    $(x - 0)^2 + (y - 0)^2 = (2)^2$ Substitute for h, k, and r.

    $x^2 + y^2 = 4$ Simplify. **(Circles and ellipses)**

14. $(x - 2)^2 + (y - 1)^2 = 25$

    $(x - 2)^2 + (y - 1)^2 = (5)^2$ Write in the form $(x - h)^2 + (y - k)^2 = r^2$.

    $h = 2, k = 1, r = 5$ Identify h, k, and r.

    The center is at (2, 1) and the radius is 5. **(Circles and ellipses)**

15. $(x + 3)^2 + (y - 4)^2 = 49$

    $(x - (-3))^2 + (y - 4)^2 = (7)^2$ Write in the form $(x - h)^2 + (y - k)^2 = r^2$.

    $h = -3, k = 4, r = 7$ Identify h, k, and r.

    The center is at (−3, 4) and the radius is 7. **(Circles and ellipses)**

16. $x^2 + (y - 2)^2 = 16$

    $(x - 0)^2 + (y - 2)^2 = (4)^2$ Write in the form $(x - h)^2 + (y - k)^2 = r^2$.

    $h = 0, k = 2, r = 4$ Identify h, k, and r.

    The center is at (0, 2) and the radius is 4. **(Circles and ellipses)**

17. $x^2 + y^2 = 9$

    $(x - 0)^2 + (y - 0)^2 = 3^2$ Write in the form $(x - h)^2 + (y - k)^2 = r^2$.

    $h = 0, k = 0, r = 3$ Identify h, k, and r.

    The center is at (0, 0) and the radius is 3. **(Circles and ellipses)**

18. $x^2 + y^2 - 4x + 2y - 4 = 0$

    First rewrite the equation in the form $(x - h)^2 + (y - k)^2 = r^2$ to find the center and radius.

    $x^2 - 4x + y^2 + 2y = 4$ Complete the square on x and on y.

    $x^2 - 4x + 4 + y^2 + 2y + 1 = 4 + 4 + 1$ $(\frac{1}{2} \cdot 4)^2 = 4$ $(\frac{1}{2} \cdot 2)^2 = 1$

    $(x - 2)^2 + (y + 1)^2 = 9$

    $(x - 2)^2 + (y - (-1))^2 = 3^2$ Write in the form $(x - h)^2 + (y - k)^2 = r^2$.

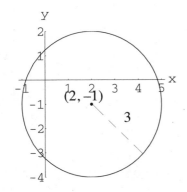

The center is at $(2, -1)$ and the radius is 3. **(Circles and ellipses)**

19. $x^2 + y^2 + 6y + 5 = 0$

$x^2 + y^2 + 6y = -5$ We only need to complete the square on y.

$x^2 + (y^2 + 6y + 9) = -5 + 9$ $\qquad$ $(\frac{1}{2} \cdot 6)^2 = 9$

$x^2 + (y + 3)^2 = 4$

$(x + 0)^2 + (y - (-3))^2 = (2)^2$

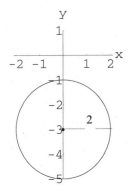

The center is at $(0, -3)$ and the radius is 2. **(Circles and ellipses)**

20. $a^2 = 16$ so the x-intercepts are $+4$ and -4. $b^2 = 9$ so the y-intercepts are $+3$ and -3. Plot the intercepts and connect with a smooth curve:

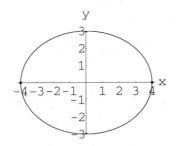

(Circles and ellipses)

21. First rewrite the equation in standard form:

$$\frac{9x^2}{36} + \frac{4y^2}{36} = \frac{36}{36} \quad \text{Divide both sides by 36.}$$

$$\frac{x^2}{4} + \frac{y^2}{9} = 1 \quad \text{Standard form}$$

$a^2 = 4$ so the x-intercepts are $+2$ and -2

$b^2 = 9$ so the y-intercepts are $+3$ and -3

Plot the intercepts and connect with a smooth curve:

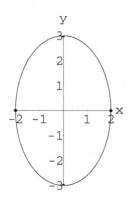

(Circles and ellipses)

22. The center is at $(2, 1)$. Since $a^2 = 9$, move 3 units to the left and right of $(2, 1)$ to get $(-1, 1)$ and $(5, 1)$ and place points there. Since $b^2 = 4$, move 2 units up and down from $(2, 1)$ to get $(2, 3)$ and $(2, -1)$ and place points there. Connect the four points with a smooth curve.

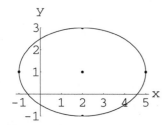

(Circles and ellipses)

23. $\dfrac{x^2}{9} - \dfrac{y^2}{16} = 1$

The hyperbola opens left and right since the x^2 term is positive. Here $a^2 = 9$, $b^2 = 16$ so $a = 3$ and $b = 4$. The vertices are $(3, 0)$ and $(-3, 0)$. The corners of the fundamental rectangle are $(-3, 4), (3, 4), (3, -4)$ and $(-3, -4)$.

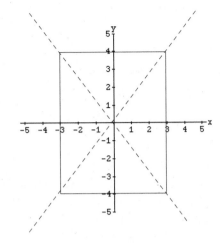

Lightly draw in the fundamental rectangle. Use dotted lines for the asymptotes. Sketch the hyperbola, using a smooth curve for each branch that approaches the asymptotes and passes through the vertex.

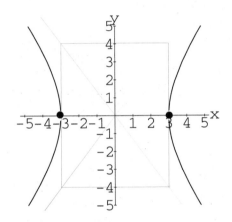

(Hyperbolas)

24. The hyperbola opens up and down since the y^2 term is positive. In this case, $a^2 = 4$, $b^2 = 25$, so $a = 2$ and $b = 5$. The vertices are $(0, 5)$ and $(0, -5)$. The corners of the fundamental rectangle are $(-2, 5)$, $(2, 5)$, $(2, -5)$, $(-2, -5)$. Sketch the fundamental rectangle. Sketch the asymptotes. Sketch the hyperbola.

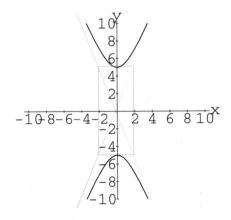

(Hyperbolas)

25. First rewrite the equation in standard form:

$$\frac{4x^2}{196} - \frac{49y^2}{196} = \frac{196}{196} \quad \text{Divide by 196}$$

$$\frac{x^2}{49} - \frac{y^2}{4} = 1$$

The hyperbola opens left and right since the x^2 term is positive. Also $a^2 = 49$, $b^2 = 4$ so $a = 7$ and $b = 2$. The vertices are $(7,0)$ and $(-7,0)$. The corners of the fundamental rectangle are $(-7,2), (7,2), (7,-2), (-7,-2)$. Note that although the fundamental rectangle and asymptotes are guides to help you sketch the graph, they are not actually part of the graph. The fundamental rectangle and asymptotes have been erased to show the actual graph of the hyperbola.

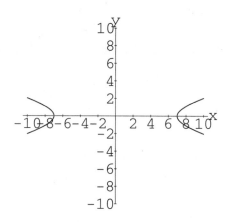

(Hyperbolas)

26. The substitution method can be used.

$2y = x + 1$ Solve for x.

$2y - 1 = x$

$(2y - 1)^2 + y^2 = 1$ Substitute $x = 2y - 1$ into the first equation.

$4y^2 - 4y + 1 + y^2 = 1$ Simplify.

$5y^2 - 4y + 1 = 1$ Add similar terms.

$5y^2 - 4y = 0$ Subtract 1 from each side.

$y(5y - 4) = 0$ Factor.

$y = 0 \quad$ or $\quad 5y - 4 = 0$ so $y = \dfrac{4}{5}$ Set each factor equal to 0 and solve.

Now find the x-coordinates.

If $y = 0$, then $x = 2(0) - 1 = -1$ giving the point $(-1, 0)$.

If $y = \dfrac{4}{5}$, then $x = 2\left(\dfrac{4}{5}\right) - 1 = \dfrac{3}{5}$ giving the point $\left(\dfrac{3}{5}, \dfrac{4}{5}\right)$.

The graphs intersect at $(-1, 0)$ and $\left(\dfrac{3}{5}, \dfrac{4}{5}\right)$. Though not required to graph the system, it may help you to do so:

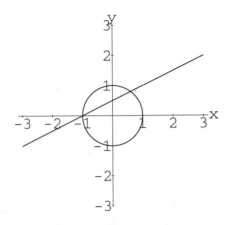

(Nonlinear systems of equations and second-degree inequalities)

27. Substitution works well here since the second equation is already solved for y.

$x^2 + (x^2 - 4)^2 = 16$ Substitute $y = x^2 - 4$ into the first equation.

$x^2 + x^4 - 8x^2 + 16 = 16$ Use FOIL.

$x^4 - 7x^2 + 16 = 16$ Combine similar terms.

$x^4 - 7x^2 = 0$ Subtract 16 from both sides.

$x^2(x^2 - 7) = 0$ Factor.

$x^2 = 0$ or $x^2 - 7 = 0$ so

$x = \pm\sqrt{7}$

Now find the y-coordinates.

If $x = 0$, $y = (0)^2 - 4 = 0 - 4 = -4$ giving the point $(0, -4)$.

If $x = \sqrt{7}$, $y = (\sqrt{7})^2 - 4 = 7 - 4 = 3$ giving the point $(\sqrt{7}, 3)$.

If $x = -\sqrt{7}$, $y = (-\sqrt{7})^2 - 4 = 7 - 4 = 3$ giving the point $(-\sqrt{7}, 3)$.

The graphs intersect at $(0, -4)$, $(\sqrt{7}, 3)$, and $(-\sqrt{7}, 3)$.

The graph of the system is:

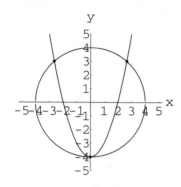

(Nonlinear systems of equations and second-degree inequalities)

28. Let's eliminate the y^2 terms:

$$4x^2 + 3y^2 = 256$$

$$-3x^2 - 3y^2 = -240 \qquad \text{Multiply the second equation by } -3.$$

$$x^2 \qquad = \quad 16 \qquad\qquad \text{Add the equations.}$$

$$x = \pm\sqrt{16} \quad \text{Use the square root property.}$$

$$x = \pm 4$$

We can find the y-coordinates using either equation.

If $x = 4$ If $x = -4$

$$(4)^2 + y^2 = 80 \qquad\qquad (-4)^2 + y^2 = 80$$

$$16 + y^2 = 80 \qquad\qquad\quad 16 + y^2 = 80$$

$$y^2 = 64 \qquad\qquad\qquad\quad y^2 = 64$$

$$y = \pm 8 \qquad\qquad\qquad\quad y = \pm 8$$

The solutions are $(4, 8), (4, -8), (-4, 8),$ and $(-4, -8)$.

The graph of the system is:

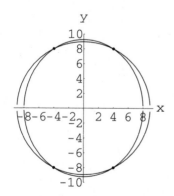

(Nonlinear systems of equations and second-degree inequalities)

29. We can solve this system by addition. First eliminate the fractions in the first equation.

$$16\left(\frac{y^2}{4} - \frac{x^2}{16}\right) = 16\,(1)$$ Multiply both sides by the LCD 16.

$4y^2 - x^2 = 16$ Rewrite the second equation so the variables line up.

$$\underline{y^2 + x^2 = 9}$$

$5y^2 \quad\;\; = 25$ Add the equations.

$y^2 \quad\;\; = 5$ Divide by 5.

$y = \pm\sqrt{5}$ Use the square root property.

Now find the x-coordinates.

If $y = \sqrt{5}$	If $y = -\sqrt{5}$	
$x^2 + (\sqrt{5})^2 = 9$	$x^2 + (-\sqrt{5})^2 = 9$	Substitute into $x^2 + y^2 = 9$.
$x^2 + 5 = 9$	$x^2 + 5 = 9$	
$x^2 = 4$	$x^2 = 4$	
$x = \pm 2$	$x = \pm 2$	

The solutions are $(2, \sqrt{5}), (-2, \sqrt{5}), (2, -\sqrt{5})$, and $(-2, -\sqrt{5})$.

The graph is:

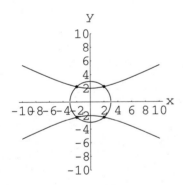

(Nonlinear systems of equations and second-degree inequalities)

30. The boundary of the graph is the circle $x^2 + y^2 = 9$. The circle has center $(0, 0)$ and radius 3. Because the inequality symbol is $<$, the circle will be dashed.

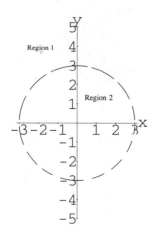

The boundary has separated the plane into two regions. To decide where to shade, select a point in each region, substitute the coordinates into the inequality, and shade regions where the inequality is true.

Region 1   Region 2

Use $(4, 0)$ Use $(0, 0)$ Select a point in each region.

$4^2 + 0^2 \overset{?}{<} 9$ $0^2 + 0^2 \overset{?}{<} 9$ Substitute into the inequality.

$\qquad 16 < 9 \qquad\qquad\qquad\qquad 0 < 9$

$\qquad$ False $\qquad\qquad\qquad\qquad$ True

Since the point in region 2 produced a true statement, shade region 2:

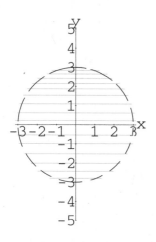

(Nonlinear systems of equations and second-degree inequalities)

31. The boundary of the graph is the parabola $y = (x - 3)^2 - 2$. The parabola opens up and has vertex $(3, -2)$. Because the inequality is $\leq$, the parabola is solid. The parabola separates the plane into two regions.

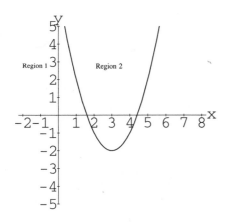

Region 1	Region 2	
$(0,0)$	$(3,0)$	Select a point in each region.
$0 \overset{?}{\le} (0-3)^2 - 2$	$0 \overset{?}{\le} (3-3)^2 - 2$	Substitute into the inequality.
$0 \overset{?}{\le} 9-2$	$0 \overset{?}{\le} 0-2$	
$0 \le 7$ (True)	$0 \le -2$ (False)	

Since the point in region 1 produced a true statement, shade region 1:

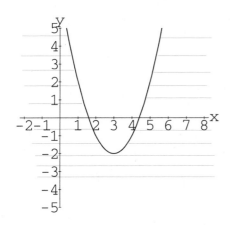

(Nonlinear systems of equations and second-degree inequalities)

32. Rewrite $4x^2 < 9y^2 + 36$.

   $4x^2 - 9y^2 < 36$ Subtract $9y^2$ from both sides of the inequality.

   $$\frac{4x^2}{36} - \frac{9y^2}{36} < 1 \qquad \text{Divide each term by 36.}$$

   $$\frac{x^2}{9} - \frac{y^2}{4} < 1 \qquad \text{Simplify.}$$

   The boundary of the graph is the hyperbola $\dfrac{x^2}{9} - \dfrac{y^2}{4} = 1$. The hyperbola opens left and right and has vertices at $(3,0)$ and $(-3,0)$. Because the inequality symbol is $<$, the hyperbola will be dashed.

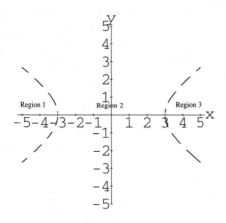

The hyperbola separates the plane into three regions.

Select a point in each region.

Region 1	Region 2	Region 3
$(-4, 0)$	$(0, 0)$	$(4, 0)$

Select a point in each region.

Region 1

$$4(-4)^2 \overset{?}{<} 9(0)^2 + 36$$

$$64 \overset{?}{<} 0 + 36$$

$$64 < 36$$

False

Region 2

$$4(0)^2 \overset{?}{<} 9(0)^2 + 36$$

$$0 \overset{?}{<} 0 + 36$$

$$0 < 36$$

True

Region 3

$$4(4)^2 \overset{?}{<} 9(0)^2 + 36$$

$$64 \overset{?}{<} 0 + 36$$

$$64 < 36$$

False

Shade region 2:

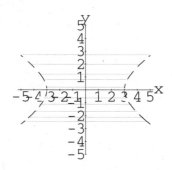

(Nonlinear systems of equations and second-degree inequalities)

33. We graphed the solution to $x^2 + y^2 < 9$ in Example 21. Now graph the line $y = 2x$, which has slope $\frac{2}{1}$ and y-intercept 0. The line is solid. Shade below the line:

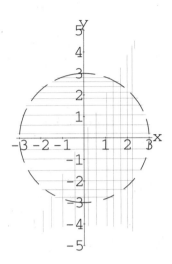

The solution is the region with crosshatched shading. **(Systems of inequalities)**

34. We graphed the solution to $y \le (x-3)^2 - 2$ in Example 22.

Now graph $2x^2 + y^2 \le 16$.

$$\frac{2x^2}{16} + \frac{y^2}{16} \le \frac{16}{16} \quad \text{Rewrite the inequality.}$$

$$\frac{x^2}{8} + \frac{y^2}{16} \le 1$$

The boundary of $\frac{x^2}{8} + \frac{y^2}{16} \le 1$ is the ellipse $\frac{x^2}{8} + \frac{y^2}{16} = 1$. The ellipse has x-intercepts at $2\sqrt{2}$ and $-2\sqrt{2}$ and y-intercepts at 4 and –4. Because the inequality is $\le$, the ellipse is solid. Shade inside the ellipse:

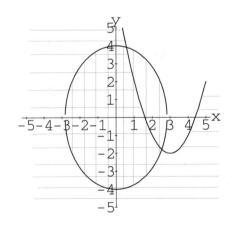

The solution is outside the parabola and inside the ellipse.

(Systems of inequalities)

Grade Yourself

Circle the question numbers that you had incorrect. Then indicate the number of questions you missed. If you answered more than three questions incorrectly, you need to focus on that topic. (If a topic has less than three questions and you had at least one wrong, we suggest you study that topic also. Read your textbook, a review book, or ask your teacher for help.)

Subject: Conic Sections

Topic	Question Numbers	Number Incorrect
Parabolas	1, 2, 3, 9, 10	
Finding the vertex by completing the square	4, 5, 6, 7	
Applications	8	
Circles and ellipses	11, 12, 13, 14, 15, 16, 17, 18, 19, 20, 21, 22	
Hyperbolas	23, 24, 25	
Nonlinear systems of equations and second degree inequalities	26, 27, 28, 29, 30, 31, 32	
Systems of inequalities	33, 34	

Exponential and Logarithmic Functions

Test Yourself

Exponential Functions

The function $f(x) = a^x$ where $a > 0$ and $a \neq 1$ is called an exponential function. The number a is called the base.

We graph exponential functions by finding and plotting several ordered pairs that satisfy the equation and connecting the points with a smooth curve.

Graph.

1. $y = 3^x$

2. $y = \left(\frac{1}{3}\right)^x$

3. $y = 5^x$

Solving Exponential Equations

To solve exponential equations (equations where the variable appears as the exponent), use the following property:

If $a^x = a^y$, then $x = y$ for $a > 0, a \neq 1$.

Note that the bases must be equal to make use of this property.

Solve each equation.

4. $8^x = 4$

5. $3^x = \frac{1}{81}$

6. $9^{x+2} = 27$

7. The population of a small Florida island is given by the equation

$$y = 100(2^{2x})$$

where x is the time in years from 1990. Find the population in the year 1995.

Logarithms

A logarithm is an exponent. Logarithmic form provides an alternate way to write exponential expressions such as $2^3 = 8$.

Definition. $\log_a x = y$ means $a^y = x$ for $a > 0, a \neq 1$.

$\log_a x$ is read "the log of x to the base a." A visual aid using a small loop may help you translate from logarithmic form to exponential form:

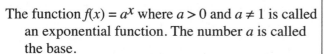

$\log_2 8 = 3$ means $2^3 = 8$

$\log_4 16 = 2$ means $4^2 = 16$

$\log_{10} 1000 = 3$ means $10^3 = 1000$

Rewrite each expression in exponential form.

8. $\log_6 36 = 2$

9. $\log_{10} 0.01 = -2$

10. $\log_5 1 = 0$

11. $\log_8 4 = \frac{2}{3}$

Rewrite each expression in logarithmic form.

12. $10^2 = 100$

13. $4^{-3} = \frac{1}{64}$

14. $16^{3/4} = 8$

Graphing Logarithmic Functions

It is usually easier to rewrite a logarithmic function in exponential form before setting up a table of values to graph a logarithmic function. The steps we'll use to graph logarithmic functions are:
1. Rewrite the logarithmic equation in exponential form.
2. Make a table of values, choosing values for the exponent and finding x-values.
3. Plot the points listed in the table and connect with a smooth curve.

Graph.

15. $y = \log_2 x$

16. $\log_{1/2} x$

Solving Logarithmic Equations

We solve logarithmic equations by first rewriting them in exponential form. Get the bases equal and use the property:
$\quad$ If $a^x = a^y$, then $x = y$, $a > 0$, $a \neq 1$.

Solve.

17. $\log_2 x = 4$

18. $\log_x 9 = 2$

19. $\log_{27} 3 = x$

Simplifying Logarithmic Expressions

To simplify a logarithmic expression, set it equal to x, and solve for x.

Simplify.

20. $\log_6 36$

21. $\log_8 1$

22. $\log_{\sqrt{3}} \sqrt{3}$

23. $\log_2 (\log_4 4)$

Logarithmic Applications

The Richter scale is used to measure the magnitude of an earthquake. The formula for the magnitude, M, is
$$M = \log_{10} I$$
where the intensity, I, is the number of times greater the earthquake is than the smallest measurable activity on the seismograph.

24. If an earthquake has magnitude 5 on the Richter scale, how many times more intense is it than the smallest measurable activity?

Properties of Logarithms

There are three properties of logarithms used to convert multiplication into addition, division into subtraction, and power problems into multiplication problems.
1. The log of a product is the sum of the logs:
$\log_a(xy) = \log_a x + \log_a y$
2. The log of a quotient is the difference of the logs:
$\log_a\left(\frac{x}{y}\right) = \log_a x - \log_a y$
3. The log of a number raised to a power is the product of that power and the log:
$\log_a x^b = b(\log_a x)$

Rewrite as a sum or difference, using the properties of logarithms.

25. $\log_2 5x$

26. $\log_{10} \frac{3m}{n}$

27. $\log_5 \sqrt{x}$

28. $\log_3 \dfrac{4xy}{pq}$

29. $\log_5 x^2 y^3$

30. $\log_2 \dfrac{x \sqrt[3]{y}}{z^5}$

Writing Expressions as Single Logarithms

To write an expression as a single logarithm (shrinking the expression), rewrite the sum of two logs as the log of a product, or the difference of two logs as the log of a quotient, and rewrite coefficients of log terms as exponents.

Write each expression as a single logarithm.

31. $\log_4 x + \log_4 y$

32. $\log_5 x - \log_5 z$

33. $4\log_2 x$

34. $2\log_5 x + 6\log_5 y - \dfrac{1}{2}\log_5 z$

35. $\dfrac{1}{3}\log_7 x - \log_7 y - \dfrac{1}{2}\log_7 z$

Common Logarithms

Logarithms with a base of 10 are called **common logarithms**. Because the base of 10 occurs so often with logarithms, it does not have to be written.

$$\log_{10} x = \log x$$

Since $\log 10 = 1$ ($10^1 = 10$), we can easily simplify common logarithms of powers of 10:

$\log 1000 = \log 10^3 = 3(\log 10) = 3(1) = 3$
$\log 100 = \log 10^2 = 2(\log 10) = 2(1) = 3$
$\log 10 = \log 10^1 = 1(\log 10) = 1(1) = 1$
$\log 1 = \log 10^0 = 0(\log 10) = 0(1) = 0$
$\log 0.1 = \log 10^{-1} = -1(\log 10) = -1(1) = -1$
$\log 0.01 = \log 10^{-2} = -2(\log 10) = -2(1) = -2$
$\log 0.001 = \log 10^{-3} = -3(\log 10) = -3(1) = -3$

In $\log 1640 = 3.2148$, the 3 is called the *characteristic* and .2148 is called the *mantissa*.

Finding Common Logs with a Calculator

We use the log key on a calculator to find common logarithms of numbers that are not powers of 10. Use your owner's manual to find out how your calculator works with logarithms.

Use a calculator to find each logarithm. Round each answer to four decimal places.

36. log 1640

37. log 0.015

38. log 4,550,000

39. log 0.000026

Finding the Antilog

If we need x in the equation $\log x = y$, we are looking for the **antilog** of y. Use your owner's manual to find out how your calculator works with logarithms.

Find x.

40. $\log x = 2.3711$

41. $\log x = -1.9208$

Find each antilog.

42. antilog 4.2058

43. anitlog -1.3046

Solving Exponential Equations

When it is not obvious how to make the bases equal, we take the logarithm of each side and solve the resulting equation. If an *exact* solution is required, leave the logarithm(s) in your answer. Otherwise, use your calculator to find the logarithms and complete the solution.

Solve.

44. $5^x = 12$

45. $4^{x+1} = 6$

Logarithmic Equations

To solve a logarithmic equation, we use the definition of a logarithm to write the equation in exponential form. The following steps may be used.

1. Use the properties of logarithms to write the equation with a single logarithm or in the form $\log_b N = \log_b M$ where $N = M$.
2. Write $\log_b x = y$ as $b^y = x$.
3. Solve the exponential equation.
4. Use only solutions for which the logarithms are defined.

Solve.

46. $\log_4 x + \log_4 3 = 2$

47. $\log_6 x + \log_6 (x-1) = 1$

48. $\log 2 + \log(5x - 2) = \log(8x + 2)$

Natural Logarithms and Change of Base

The number e is an irrational number approximately equal to 2.718281828. The natural logarithm of x, written $\ln x$, means $\log_e x$. Thus

$$\ln x = y \text{ means } e^y = x$$

Simplify.

49. $\ln e$

50. $\ln e^2$

51. $\ln e^3$

Use the properties of logarithms to rewrite each expression as a sum or difference.

52. $\ln x^2 y^3$

53. $\ln \dfrac{\sqrt{x}}{y}$

Write each expression as a single logarithm.

54. $3\ln x - \dfrac{2}{3}\ln y$

55. $\dfrac{1}{2}\ln x + \ln y - 3\ln z$

Natural Logarithms With a Calculator

Scientific calculators usually have a natural logarithm key labeled $\boxed{\ln}$. Consult your owner's manual to see how your calculator works with natural logarithms.

For example,

$$\ln 2.59 \approx 0.9517$$
$$\ln 0.37 \approx -0.9943$$

Change of Base Formula

Scientific calculators have common logarithm (base 10) and natural logarithm (base e) keys. We can find logarithms with a different base by using the change of base formula.

$$\log_a x = \frac{\log x}{\log a} = \frac{\ln x}{\ln a} \text{ for } a > 0, a \neq 1 \text{ and } x > 0.$$

Note that the two forms given in the change of base formula use base 10 and base e. Any base *could* be used, but the convenience of a calculator makes the use of these bases preferable.

Find each logarithm.

56. $\log_5 8$

57. $\log_7 56$

58. $\log_2 12$

Check Yourself

1. We'll complete the table of values, choosing *x*-values and finding y:

x	y
−3	1/27
−2	1/9
−1	1/3
0	1
1	3
2	9

$$3^{-3} = \frac{1}{3^3} = \frac{1}{27} \quad \text{Remember } x^{-n} = \frac{1}{x^n}.$$

$$3^{-2} = \frac{1}{3^2} = \frac{1}{9}$$

$$3^{-1} = \frac{1}{3^1} = \frac{1}{3}$$

$$3^0 = 1 \quad \text{Remember } x^0 = 1 \text{ for } x \neq 0.$$

Plot the points and connect with a smooth curve:

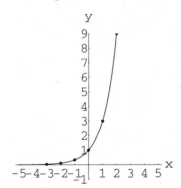

(Graphing exponential functions)

2. Use a table of values, choosing x-values and finding $f(x)$:

x	y
-3	27
-2	9
-1	3
0	1
1	1/3
2	1/9
3	1/27

$$\left(\frac{1}{3}\right)^{-3} = \left(\frac{3}{1}\right)^{3} = 27$$

$$\left(\frac{1}{3}\right)^{2} = \left(\frac{3}{1}\right)^{2} = 9$$

$$\left(\frac{1}{3}\right)^{-1} = \left(\frac{3}{1}\right)^{1} = 3$$

$$\left(\frac{1}{3}\right)^{0} = 1$$

Plot the points and connect with a smooth curve:

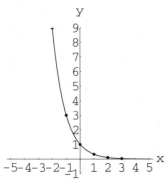

(Graphing exponential functions)

3. Use a table of values:

x	y
-3	1/125
-2	1/25
-1	1/5

x	y
0	1
1	5
2	25
3	125

Plot the points and connect with a smooth curve:

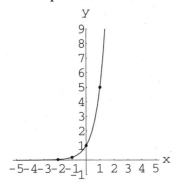

(Graphing exponential functions)

4. $8^x = 4$

$(2^3)^x = 2^2$ Rewrite each side using 2 as the base.

$2^{3x} = 2^2$ Use $(x^a)^b = x^{ab}$.

$3x = 2$ Since the bases are equal, the exponents must be equal.

$x = \dfrac{2}{3}$ Divide by 3.   **(Solving exponential equations)**

5. $3^x = \dfrac{1}{81}$

$3^x = \dfrac{1}{3^4}$ Write 81 with a base of 3.

$3^x = 3^{-4}$ Use $\dfrac{1}{x^n} = x^{-n}$.

$x = -4$ Since the bases are equal, the exponents must be equal.   **(Solving exponential equations)**

6. $9^{x+2} = 27$

$(3^2)^{x+2} = 3^3$ Write each side with a base of 3.

$3^{2x+4} = 3^3$ Use $(x^a)^b = x^{ab}$.

$2x + 4 = 3$ Since the bases are equal, the exponents must be equal.

$2x = -1$ Solve for x.

$x = -\dfrac{1}{2}$ **(Solving exponential equations)**

7. We are considering a population change for 5 years' time $(1995 - 1990 = 5)$, so $x = 5$.

$y = 100(2^{2x})$ Given population equation.

$y = 100(2^{2(5)})$ Substitute $x = 5$.

$y = 100(2^{10})$ Multiply the exponents.

$y = 100(1024)$

$y = 102,400$

The population will be about 102,400 in the year 1995.   **(Applications)**

8. $\log_6 36 = 2$ means $6^2 = 36$ **(Logarithmic functions)**

9. $\log_{10} 0.01 = -2$ means $10^{-2} = 0.01$ **(Logarithmic functions)**

10. $\log_5 1 = 0$ means $5^0 = 1$ **(Logarithmic functions)**

11. $\log_8 4 = \dfrac{2}{3}$ means $8^{2/3} = 4$ **(Logarithmic functions)**

12. $10^2 = 100$ means $\log_{10} 100 = 2$ A logarithm is an exponent, so 2 is the logarithm.
        (Logarithmic functions)

13. $4^{-3} = \dfrac{1}{64}$ means $\log_4 \dfrac{1}{64} = -3$ **(Logarithmic functions)**

14. $16^{3/4} = 8$ means $\log_{16} 8 = \dfrac{3}{4}$ **(Logarithmic functions)**

15. $y = \log_2 x$ means $2^y = x$ Rewrite in exponential form.

$2^{-2} = \dfrac{1}{2^2} = \dfrac{1}{4}$ Substitute y-values into $2^y = x$.

$2^{-1} = \dfrac{1}{2^1} = \dfrac{1}{2}$

$2^0 = 1$

$2^1 = 2$

$2^2 = 4$

The completed table is:

x	y
1/4	−2
1/2	−1
1	0
2	1
4	2

Plot the points and connect with a smooth curve:

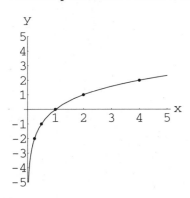

(Logarithmic functions)

16. $y = \log_{1/2}x$ means $\left(\dfrac{1}{2}\right)^y = x$ Rewrite in exponential form.

x	y
4	−2
2	−1
1	0
1/2	1
1/4	2

$$\left(\frac{1}{2}\right)^{-2} = 2^2 = 4$$

$$\left(\frac{1}{2}\right)^{-1} = 2^1 = 2$$

Plot the points and connect with a smooth curve:

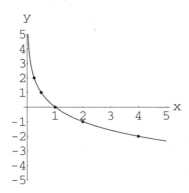

(Logarithmic functions)

17. $\log_2 x = 4$ means $2^4 = x$ Rewrite in exponential form.

     $x = 16$ Simplify 2^4 **(Solving logarithmic equations)**

18. $\log_x 9 = 2$ means $x^2 = 9$ Rewrite in exponential form.

     $x = \pm\sqrt{9}$ Use the square root property.

     $x = \pm 3$ Simplify the radical.

The solution is $x = 3$ because logarithms are only defined for bases greater than 0 and not equal to 1.
     (Solving logarithmic equations)

19. $\log_{27} 3 = x$ means $27^x = 3$ Rewrite in exponential form.

     $(3^3)^x = 3$ $27 = 3^3$.

     $3^{3x} = 3^1$ Use $(x^a)^b = x^{ab}$.

     $3x = 1$ If $x^a = x^b$, then $a = b$.

     $x = \dfrac{1}{3}$ **(Solving logarithmic equations)**

20. $\log_6 36 = x$ Set the expression equal to x.

     $6^x = 36$ Rewrite in exponential form.

     $6^x = 6^2$ Get the bases equal.

     $x = 2$ If $a^x = a^y$, $x = y$.   **(Solving logarithmic equations)**

21. $\log_8 1 = x$ Set the expression equal to x.

     $8^x = 1$ Rewrite in exponential form.

     $8^x = 8^0$ Use the definition of 0 as an exponent to write $1 = 8^0$.

     $x = 0$ If $a^x = a^y$, $x = y$.   **(Solving logarithmic equations)**

22. $\log_{\sqrt{3}}\sqrt{3} = x$ Set the expression equal to x.

    $(\sqrt{3})^x = \sqrt{3}$ Rewrite in exponential form.

    $(\sqrt{3})^x = (\sqrt{3})^1$ $\sqrt{3} = \sqrt{3}^1$.

    $x = 1$ If $a^x = a^y$, $x = y$.   **(Solving logarithmic equations)**

23. $\log_2(\log_4 4)$

Begin by simplifying the logarithm in parentheses.

    $\log_4 4 = x$ Set the expression equal to x.

    $4^x = 4$ Rewrite in exponential form.

    $4^x = 4^1$ $4 = 4^1$.

    $x = 1$ If $a^x = a^y$, $x = y$.

Now replace the expression in parentheses with 1:

    $\log_2 1 = x$ Set the expression equal to x.

    $2^x = 1$ Rewrite in exponential form.

    $2^x = 2^0$ Rewrite $1 = 2^0$.

    $x = 0$ If $a^x = a^y$, $x = y$.   **(Solving logarithmic equations)**

24. $M = \log_{10} I$ Write the formula.

    $5 = \log_{10} I$ The magnitude is 5.

    $10^5 = I$ Rewrite in exponential form.

    $100,000 = I$ Simplify 10^5.

A magnitude of 5 on the Richter scale means the earthquake is 100,000 times greater than the smallest measurable activity.   **(Applications)**

25. $\log_2 5x = \log_2 5 + \log_2 x$ Since $5x$ is a product, use property 1.   **(Properties of logarithms)**

26. $\log_{10}\dfrac{3m}{n} = \log_{10} 3m - \log_{10} n$ Use property 2 to rewrite the quotient.

    $= \log_{10} 3 + \log_{10} m - \log_{10} n$ Use property 1 to rewrite the log of the product $3m$.
(Properties of logarithms)

27. $\log_5\sqrt{x} = \log_5 x^{1/2}$ Rewrite $\sqrt{x} = x^{1/2}$.

         $= \dfrac{1}{2}\log_5 x$ Use property 3 to write the exponent in front of the log.
       (Properties of logarithms)

28. $\log_3\dfrac{4xy}{pq} = \log_3 4xy - \log_3 pq$ Use property 2 to rewrite the log of a quotient.

      $= \log_3 4 + \log_3 x + \log_3 y - (\log_3 p + \log_3 q)$ Use property 1 to rewrite the log of each product.

      $= \log_3 4 + \log_3 x + \log_3 y - \log_3 p - \log_3 q$ Distribute the negative to remove the parentheses.
       (Properties of logarithms)

29. $\log_5 x^2 y^3 = \log_5 x^2 + \log_5 y^3$ Use property 1 to rewrite the log of a product.

     $= 2\log_5 x + 3\log_5 y$ Use property 3 to write the exponent in front of the log.
       (Properties of logarithms)

30. $\log_2\dfrac{x\sqrt[3]{y}}{z^5} = \log_2 x\sqrt[3]{y} - \log_2 z^5$ Use property 2 to rewrite the log of a quotient.

         $= \log_2 xy^{1/3} - \log_2 z^5$ Rewrite $\sqrt[3]{y} = y^{1/3}$.

         $= \log_2 x + \log_2 y^{1/3} - \log_2 z^5$ Use property 1 to rewrite the log of a product.

         $= \log_2 x + \dfrac{1}{3}\log_2 y - 5\log_2 z$ Use property 3 to write the exponent in front of the log.
       (Properties of logarithms)

31. $\log_4 x + \log_4 y = \log_4 xy$ Use property 1 to write a sum of logs as the log of a product.
       (Properties of logarithms)

32. $\log_5 x - \log_5 z = \log_5\left(\dfrac{x}{z}\right)$ Use property 2 to write a difference of logs as the log of a quotient.

       (Properties of logarithms)

33. $4\log_2 x = \log_2 x^4$ Use property 3 to write 4 as the exponent.   **(Properties of logarithms)**

34. $2\log_5 x + 6\log_5 y - \dfrac{1}{2}\log_5 z = \log_5 x^2 + \log_5 y^6 - \log_5 z^{1/2}$ Use property 3.

         $= \log_5 x^2 y^6 - \log_5 z^{1/2}$ Use property 1.

         $= \log_5\dfrac{x^2 y^6}{z^{1/2}}$ Use property 2.

         $= \log_5\dfrac{x^2 y^6}{\sqrt{z}}$ Write $z^{1/2}$ as $\sqrt{z}$.   **(Properties of logarithms)**

35. $\frac{1}{3}\log_7 x - \log_7 y - \frac{1}{2}\log_7 z = \log_7 x^{1/3} - \log_7 y - \log_7 z^{1/2}$ Use property 3.

$$= \log_7 \frac{x^{1/3}}{y} - \log_7 z^{1/2}$$ Use property 2.

$$= \log_7 (\frac{x^{1/3}}{y} \div z^{1/2})$$ Use property 2.

$$= \log_7 \frac{x^{1/3}}{yz^{1/2}} \qquad\qquad \frac{x^{1/3}}{y} \div z^{1/2} = \frac{x^{1/3}}{y} \cdot \frac{1}{z^{1/2}}.$$

$$= \log_7 \frac{\sqrt[3]{x}}{y\sqrt{z}}$$ Use $x^{1/b} = \sqrt[b]{x}$.   **(Properties of logarithms)**

36. $\log 1640 = 3.2148$ **(Evaluating logarithms and antilogarithms)**

37. $\log 0.015 = -1.8239$ **(Evaluating logarithms and antilogarithms)**

38. $\log 4{,}550{,}000 = 6.6580$ **(Evaluating logarithms and antilogarithms)**

39. $\log 0.000026 = -4.5850$ **(Evaluating logarithms and antilogarithms)**

40. $\log x = 2.3711$

    $x = 235$ **(Evaluating logarithms and antilogarithms)**

41. $\log x = -1.9208$

    $x = 0.012$ **(Evaluating logarithms and antilogarithms)**

42. antilog 4.2058

    antilog $4.2058 = 16062$ **(Evaluating logarithms and antilogarithms)**

43. antilog -1.3046

    antilog $-1.3046 = 0.04959$ **(Evaluating logarithms and antilogarithms)**

44. $5^x = 12$ Since the bases 5 and 12 cannot easily be made equal, take the log of both sides.

    $\log 5^x = \log 12$ Take the log of both sides.

    $x\log 5 = \log 12$ Use property 3 to write the exponent in front of the log

    $\dfrac{x\log 5}{\log 5} = \dfrac{\log 12}{\log 5}$ Divide by log 5.

    $x = \dfrac{\log 12}{\log 5}$ This is the exact answer.

    $x \approx \dfrac{1.0792}{0.6990}$ Use a calculator to find the logs.

    $x \approx 1.544$ This is an approximation.   **(Solving exponential equations)**

45. $4^{x+1} = 6$

$\log 4^{x+1} = \log 6$ Take the log of both sides.

$(x+1)\log 4 = \log 6$ Use property 3.

$\dfrac{(x+1)\log 4}{\log 4} = \dfrac{\log 6}{\log 4}$ Divide by log 4.

$x + 1 = \dfrac{\log 6}{\log 4}$

$x + 1 - 1 = \dfrac{\log 6}{\log 4} - 1$ Subtract 1 from both sides.

$x = \dfrac{\log 6}{\log 4} - 1$ This is the exact answer.

$x \approx \dfrac{0.7782}{0.6021} - 1$ Use a calculator to find the logs.

$x \approx 0.292$ This is an approximation.   **(Solving exponential equations)**

46. $\log_4 x + \log_4 3 = 2$

$\log_4 3x = 2$ Use property 1 to write the sum of logs as the log of a product.

$4^2 = 3x$ Write in exponential form.

$16 = 3x$ Simplify.

$\dfrac{16}{3} = x$ Divide by 3.

Check: Since $\log_4 \dfrac{16}{3}$ is defined, our solution set is $\{\dfrac{16}{3}\}$.   **(Solving logarithmic equations)**

47. $\log_6 x + \log_6 (x-1) = 1$

$\log_6 x(x-1) = 1$ Use property 1 to write the sum of logs as the log of a product.

$6^1 = x(x-1)$ Write in exponential form.

$6 = x^2 - x$ Use the distributive property.

$x^2 - x - 6 = 0$ Solve the quadratic equation.

$(x-3)(x+2) = 0$ Factor.

$x - 3 = 0$ or $x + 2 = 0$ Set each factor equal to 0.

$x = 3$ or $x = -2$ Solve each equation.

We cannot include –2 in the solution set because $\log_6(-2)$ does not exist (there is no exponent we can raise 6 to that equals a negative number). Note that $\log_6 3$ and $\log_6(3-1)$ are both defined. Therefore, the solution set is {3}.   **(Solving logarithmic equations)**

48. $\log 2 + \log(5x - 2) = \log(8x + 2)$

$\log 2(5x - 2) = \log(8x + 2)$ Use property 1 to write the sum of logs as the log of a product.

$\log 2(5x - 2) - \log(8x + 2) = 0$ Subtract $\log(8x + 2)$ from both sides.

$\log\dfrac{2(5x - 2)}{8x + 2} = 0$ Use property 2 to write the difference of logs as the log of a quotient.

$10^0 = \dfrac{2(5x - 2)}{8x + 2}$ Write in exponential form. The understood base is 10.

$1 = \dfrac{2(5x - 2)}{8x + 2}$ $10^0 = 1$

$(8x + 2)1 = (8x + 2)\cdot\dfrac{2(5x - 2)}{8x + 2}$ Multiply by the LCD to clear fractions.

$8x + 2 = 2(5x - 2)$

$8x + 2 = 10x - 4$ Use the distributive property.

$8x + 2 - 10x = 10x - 4 - 10x$ Subtract $10x$ from both sides.

$-2x + 2 = -4$ Combine similar terms.

$-2x + 2 - 2 = -4 - 2$ Subtract 2 from both sides.

$-2x = -6$ Combine similar terms.

$x = 3$ Divide both sides by –2.

Since $\log(5\cdot 3 - 2) = \log 13$ and $\log(8\cdot 3 + 2) = \log 25$ are both defined, the solution set is {3}.
    (Solving logarithmic equations)

49. $\ln e = y$

$e^y = e$ Write the natural log in exponential form.

$e^y = e^1$ Write $e = e^1$.

$y = 1$ Since the bases are equal, the exponents are equal.

$\ln e = 1$ **(Natural logarithms)**

50. $\ln e^2 = 2\ln e$ Use property 3.

$= 2(1)$ Use $\ln e = 1$.

$= 2$ **(Natural logarithms)**

51. $\ln e^3 = 3\ln e$ Use property 3.

    $= 3(1)$ Use $\ln e = 1$.

    $= 3$ **(Natural logarithms)**

52. $\ln x^2 y^3 = \ln x^2 + \ln y^3$ Use property 1 to write the ln of a product as the sum of natural logarithms.

    $= 2\ln x + 3\ln y$ Use property 3. **(Natural logarithms)**

53. $\ln \dfrac{\sqrt{x}}{y} = \ln\sqrt{x} - \ln y$ Use property 2 to write the natural log of a quotient as the difference of natural logs.

    $= \ln x^{1/2} - \ln y$ Write $\sqrt{x} = x^{1/2}$.

    $= \dfrac{1}{2}\ln x - \ln y$ Use property 3.

    (Natural logarithms)

54. $3\ln x - \dfrac{2}{3}\ln y = \ln x^3 - \ln y^{2/3}$ Use property 3.

    $= \ln \dfrac{x^3}{y^{2/3}}$ Use property 2.

    $= \ln \dfrac{x^3}{\sqrt[3]{y^2}}$ **(Natural logarithms)**

55. $\dfrac{1}{2}\ln x + \ln y - 3\ln z$

    $= \ln x^{1/2} + \ln y - \ln z^3$ Use property 3.

    $= \ln x^{1/2} y - \ln z^3$ Use property 1.

    $= \ln \dfrac{x^{1/2} y}{z^3}$ Use property 2.

    $= \ln \dfrac{\sqrt{x}\, y}{z^3}$ Use $x^{1/2} = \sqrt{x}$.

    (Natural logarithms)

56. $\log_5 8 = \dfrac{\log 8}{\log 5}$ Use the change of base formula.

    $= \dfrac{0.9031}{0.6990}$ Use a calculator find each log.

    $= 1.29$ **(Change of base formula)**

57. $\log_7 56 = \dfrac{\log 56}{\log 7}$ Use the change of base formula.

    $= \dfrac{1.7482}{0.8451}$ Use a calculator find each log.

    $= 2.07$

    (Change of base formula)

58. $\log_2 12 = \dfrac{\log 12}{\log 2}$ Use the change of base formula.

    $= \dfrac{1.0792}{0.3010}$ Use a calculator find each log.

    $= 3.59$

    (Change of base formula)

Grade Yourself

Circle the question numbers that you had incorrect. Then indicate the number of questions you missed. If you answered more than three questions incorrectly, you need to focus on that topic. (If a topic has less than three questions and you had at least one wrong, we suggest you study that topic also. Read your textbook, a review book, or ask your teacher for help.)

Subject: Exponential and Logarithmic Functions

Topic	Question Numbers	Number Incorrect
Graphing exponential functions	1, 2, 3	
Solving exponential equations	4, 5, 6	
Applications	7, 24	
Logarithmic functions	8, 9, 10, 11, 12, 13, 14, 15, 16	
Solving logarithmic equations	17, 18, 19, 20, 21, 22, 23	
Properties of logarithms	25, 26, 27, 28, 29, 30, 31, 32, 33, 34, 35	
Evaluating logarithms and antilogarithms	36, 37, 38, 39, 40, 41, 42, 43	
Solving exponential equations	44, 45	
Solving logarithmic equations	46, 47, 48	
Natural logarithms	49, 50, 51, 52, 53, 54, 55	
Change of base formula	56, 57, 58	

Sequences and Series

Sequences

A **sequence** is an ordered list of numbers. Each number in the list is called a **term** of the sequence. We use subscripts to identify specific terms in a sequence.

Using a simple continuation rule, write the next three terms of each sequence.

1. $3, 6, 9, 12 \ldots$

2. $\dfrac{1}{3}, \dfrac{1}{5}, \dfrac{1}{7}, \dfrac{1}{9} \ldots$

3. $1, -\dfrac{1}{2}, \dfrac{1}{4}, -\dfrac{1}{8}, \dfrac{1}{16} \ldots$

Find the first five terms of the sequence whose n-th term is:

4. $a_n = 2n - 1$

5. $a_n = \dfrac{n+1}{n}$

6. $a_n = n^2$

7. $a_n = (-1)^{n+1}$

Recursion Formulas

A recursion formula is a formula for finding the n-th term of a sequence using one or more earlier terms of the sequence. The sequence $3, 6, 9, 12 \ldots$ has terms:

$a_1 = 3$
$a_2 = 6$
$a_3 = 9$
$a_4 = 12$

This sequence could be defined recursively as:
$a_1 = 3$
$a_n = a_{n-1} + 3 \quad$ for $n > 1$

To find a_2, n would equal 2 so
$a_2 = a_{2-1} + 3 \quad$ Substitute $n = 2$.
$\quad = a_1 + 3 \quad$ Simplify.
$\quad = 3 + 3 = 6 \quad$ Replace a_1 with 3.

To find a_3, n would equal 3, so
$a_3 = a_{3-1} + 3 \quad$ Substitute $n = 3$.
$\quad = a_2 + 3 \quad$ Simplify.
$\quad = 6 + 3 = 9 \quad$ Replace a_2 with 6.

Note that each new term is found by adding the previous term and 3.

Write the first five terms of the sequences defined by the recursion formulas.

8. $a_1 = 5 \quad a_n = -2a_{n-1} \qquad n > 1$

9. $a_1 = 1 \quad a_n = 3a_{n-1} + 1 \quad n > 1$

10. $a_1 = 2 \quad a_n = 2a_{n-1} + n \quad n > 1$

Series

A **series** is a sum of the terms of a sequence. Special summation notation is used to write series.

$$\sum_{i=1}^{4} (3i + 1)$$ means replace i with integers from 1 to 4 in the formula $3i + 1$ and sum the terms.

The **index of the summation**, i, is used to indicate the beginning and ending integer. In the formulas for the terms, i may be used as a term, factor, or exponent. Do not confuse the index i with the imaginary number i. The series we deal with do not involve imaginary numbers.

Evaluate.

11. $\sum_{i=1}^{5} (2i - 3)$

12. $\sum_{i=2}^{6} i^2$

13. $\sum_{i=1}^{4} (i-1)(i+4)$

14. $\sum_{i=3}^{5} \dfrac{i^2 - 3}{2i + 4}$

Variables in Formulas

The formula in a series may contain variables other than the index. These variables do *not* change as the index changes.

Write out the terms in each series.

15. $\sum_{i=1}^{4} (2x - i)$

16. $\sum_{i=2}^{4} x^i$

17. $\sum_{i=1}^{3} \dfrac{i^2}{x^2}$

Writing Sums Using Summation Notation

Summation notation can be used to write a sum of terms if we can find a formula for the terms. This usually requires a trial and error process, and note that the answers are not unique.

Write in summation notation.

18. $9 + 16 + 25 + 36 + 49$

19. $1 + 8 + 27 + 64$

20. $\dfrac{1}{2} + \dfrac{2}{3} + \dfrac{3}{4} + \dfrac{4}{5} + \dfrac{5}{6}$

Arithmetic Sequences

An **arithmetic sequence** or **arithmetic progression** consists of terms that differ by a **common difference**, d.
If the terms of an arithmetic sequence are listed, the common difference d can be found by subtracting any term from the term that follows it.

Find the common difference, d for these arithmetic sequences.

21. $2, 4, 6, 8, 10 \ldots$

22. $6, 11, 16, 21, 2 \ldots$

23. $4, 1, -2, -5, -8 \ldots$

24. $1, \dfrac{4}{3}, \dfrac{5}{3}, 2, \dfrac{7}{3}, \dfrac{8}{3}, 3 \ldots$

Writing Terms Given a_1 and d

Since the terms of an arithmetic sequence differ by d, the common difference, each term can be found by adding d to the previous term.

Write the first five terms for each arithmetic sequence with the given first term a_1 and common difference d.

25. $a_1 = 5 \quad d = 4$

26. $a_1 = 7 \quad d = -2$

Finding the General Term

The general term of an arithmetic sequence with first
term a_1 and common difference d is
$$a_n = a_1 + (n-1)d$$

Find the indicated term of the arithmetic sequence.

27. $a_1 = 6 \; d = -4$ Find a_{23}.

28. $4, -2, -8, -14 \ldots$ Find a_{100}.

Finding a_n Given Two Terms

If we know two terms of an arithmetic sequence, we
can find a_1 and d by solving a system of two equations in two unknowns.

29. Find a_{24} for the arithmetic sequence where $a_{11} = 35$ and $a_{15} = 47$.

Arithmetic Series

Sum of n Terms of an Arithmetic Sequence
The sum S_n of the first n terms of any arithmetic
sequence with first term a_1, n-th term a_n, and
common difference d is
$$S_n = \frac{n(a_1 + a_n)}{2}$$
or
$$S_n = \frac{n}{2}[2a_1 + (n-1)d]$$

We use the first formula when both the first term and
n-th term are given or can be easily found.

30. Find the sum of the first 10 positive integers.

31. Find S_9 for the arithmetic sequence with $a_1 = 3$ and $d = 8$.

Geometric Sequences

A **geometric sequence** is a sequence where each
term after the first term is a constant multiple of
the preceding term. The constant multiple is
called the **common ratio, r**.
If the terms of a geometric sequence are listed, the
common ratio r can be found by dividing any term
by the term preceding it.

We find terms of a geometric sequence by multiplying by r; $a_n = a_{n-1} r$, $n > 1$.

Find the common ratio, r for these geometric
sequences.

32. $3, 6, 12, 24, 48 \ldots$

33. $2, -6, 18, -54, 162 \ldots$

34. $4, 1, \dfrac{1}{4}, \dfrac{1}{16}, \dfrac{1}{64} \ldots$

Find the first five terms for each geometric sequence
with the given first term a_1 and common ratio r.

35. $a_1 = 5, r = -2$

36. $a_1 = 1, r = \dfrac{1}{2}$

Finding the General Term

The general term of a geometric sequence with first
term a_1 and common ratio r is
$$a_n = a_1 r^{n-1}$$
We can substitute directly into the formula if we
know a_1 and r.

Find the indicated term of the geometric sequence.

37. $a_1 = 5$ $r = -3$ Find a_4.

38. $1, 3, 9, 27, 81 \ldots$ Find a_{10}.

Find a formula for the general term of each geometric
sequence.

39. $3, 6, 12, 24 \ldots$

40. $\dfrac{1}{4}, \dfrac{3}{4}, \dfrac{9}{4}, \dfrac{27}{4} \ldots$

41. Find the general term of a geometric sequence with $a_2 = 18$ and $a_6 = 1458$.

Geometric Series

A **geometric series** is the sum of the terms of a geometric sequence.

The sum S_n of the first n terms of a geometric sequence with first term a_1 and common ratio r is

$$S_n = \frac{a_1(r^n - 1)}{r - 1} \qquad (r \neq 1)$$

42. Find the sum of the first six terms of the geometric sequence with first term 8 and common ratio –2.

43. Find the sum of the first 10 terms of the geometric sequence $5, 10, 20, 40 \ldots$

Infinite Geometric Series

The sum of the terms of certain infinite geometric series can be found. If the absolute value of the common ratio, $|r|$, is less than 1, the sum S of the infinite geometric series is $S = \dfrac{a_1}{1 - r}$.

44. Find the sum of the infinite geometric sequence with $a_1 = 5$ and $r = \dfrac{1}{4}$.

45. Find a fraction equivalent to the repeated decimal $0.454545\ldots$

Factorials

Definitions
$n! = n(n-1)(n-2) \ldots (2)(1)$
    for n any positive integer
$0! = 1$

Evaluate.

46. $\dfrac{4!}{0!}$

47. $\dfrac{8!}{3!5!}$

48. $\dfrac{6!}{6!0!}$

Binomial coefficient

The **binomial coefficient** $\dbinom{n}{r}$ is defined by

$$\binom{n}{r} = \frac{n!}{r!\,(n-r)!}$$

Evaluate.

49. $\dbinom{4}{0}$

50. $\dbinom{4}{1}$

51. $\dbinom{4}{2}$

52. $\dbinom{4}{3}$

53. $\dbinom{4}{4}$

The Binomial Theorem

For x and y real numbers and n a positive integer,
$$(x + y)^n = \binom{n}{0}x^n y^0 + \binom{n}{1}x^{n-1}y^1 + \binom{n}{2}x^{n-2}y^2$$
$$+ \ldots + \binom{n}{n}x^0 y^n$$
Take note of the following to help you remember the theorem:
1. The exponents on x start with n and decrease by 1 for each term.
2. The exponents on y start with 0 and increase by 1 for each term.
3. The binomial coefficient for each term is $\dbinom{n}{r}$ where r is the exponent on the second term.

54. Expand $(x + y)^4$.

55. Expand $(2a + 3)^5$.

Pascal's Triangle

The coefficients of the terms in the expansion of $(x + y)^n$ form an interesting pattern. When these coefficients are placed in rows, the display is called Pascal's Triangle:

1	Coefficients in $(x+y)^0$
1 1	Coefficients in $(x+y)^1$
1 2 1	Coefficients in $(x+y)^2$
1 3 3 1	Coefficients in $(x+y)^3$
1 4 6 4 1	Coefficients in $(x+y)^4$

Each row begins and ends with 1 and each term inside is the sum of the two terms above it.

56. Use Pascal's Triangle to expand $(x+y)^6$

Finding the kth Term in a Binomial Expansion

The kth term in the expansion of $(x+y)^n$ is

$$\binom{n}{k-1} x^{n-(k-1)} y^{k-1}$$

57. Find the fifth term in the expansion of $(x+y)^9$.

58. Find the eighth term in the expansion of $(3a-2b)^{12}$.

Check Yourself

1. $1.3, 6, 9, 12, \mathbf{15, 18, 21}$ These numbers are successive multiples of 3. (**Sequences**)

2. $\dfrac{1}{3}, \dfrac{1}{5}, \dfrac{1}{7}, \dfrac{1}{9}, \dfrac{\mathbf{1}}{\mathbf{11}}, \dfrac{\mathbf{1}}{\mathbf{13}}, \dfrac{\mathbf{1}}{\mathbf{15}}$ The denominators are successive odd numbers. (**Sequences**)

3. $1, -\dfrac{1}{2}, \dfrac{1}{4}, -\dfrac{1}{8}, \dfrac{1}{16}, -\dfrac{\mathbf{1}}{\mathbf{32}}, \dfrac{\mathbf{1}}{\mathbf{64}}, -\dfrac{\mathbf{1}}{\mathbf{128}}$ The denominators are successive powers of 2. The signs of the fractions alternate $+, -, +, - \dots$ (**Sequences**)

4. $a_n = 2n - 1$

   $a_1 = 2(1) - 1 = 2 - 1 = 1$ Replace n with 1, 2, 3, 4, and 5.

   $a_2 = 2(2) - 1 = 4 - 1 = 3$

   $a_3 = 2(3) - 1 = 6 - 1 = 5$

   $a_4 = 2(4) - 1 = 8 - 1 = 7$

   $a_5 = 2(5) - 1 = 10 - 1 = 9$ (**Sequences**)

5. $a_n = \dfrac{n+1}{n}$

   $a_1 = \dfrac{1+1}{1} = \dfrac{2}{1}$ Replace n with 1, 2, 3, 4, and 5.

   $a_2 = \dfrac{2+1}{2} = \dfrac{3}{2}$

   $a_3 = \dfrac{3+1}{3} = \dfrac{4}{3}$

$$a_4 = \frac{4+1}{4} = \frac{5}{4}$$

$$a_5 = \frac{5+1}{5} = \frac{6}{5} \text{ (Sequences)}$$

6. $a_n = n^2$

$a_1 = 1^2 = 1$	Replace n with 1, 2, 3, 4, and 5.

$$a_2 = 2^2 = 4$$

$$a_3 = 3^2 = 9$$

$$a_4 = 4^2 = 16$$

$$a_5 = 5^2 = 25 \text{ (Sequences)}$$

7. $a_n = (-1)^{n+1}$

$a_1 = (-1)^{1+1} = (-1)^2 = 1$	Replace n with 1, 2, 3, 4, and 5.

$$a_2 = (-1)^{2+1} = (-1)^3 = -1$$

$$a_3 = (-1)^{3+1} = (-1)^4 = 1$$

$$a_4 = (-1)^{4+1} = (-1)^5 = -1$$

$$a_5 = (-1)^{5+1} = (-1)^6 = 1 \text{ (Sequences)}$$

8.

$a_1 = 5$	a_1 is the first term.
$a_n = -2a_{n-1}$	This is the recursion formula.
$a_2 = -2a_{2-1}$	Substitute $n = 2$.
$\quad = -2a_1$	Simplify.
$\quad = -2(5)$	Replace a_1 with 5.
$\quad = -10$	a_2 is the second term.
$a_3 = -2\,a_{3-1}$	Substitute $n = 3$.
$\quad = -2a_2$	Simplify.
$\quad = -2(-10)$	Replace a_2 with -10.
$\quad = 20$	This is the third term.
$a_4 = -2a_{4-1}$	Substitute $n = 4$.

$= -2a_3$	Simplify.
$= -2(20)$	Replace a_3 with 20.
$= -40$	This is the fourth term.
$a_5 = -2a_{5-1}$	Substitute $n = 5$.
$= -2a_4$	Simplify.
$= -2(-40)$	Replace a_4 with -40.
$= 80$	This is the fifth term. **(Sequences)**

9.

$a_1 = 1$	a_1 is the first term.
$a_n = 3a_{n-1} + 1$	This is the recursion formula.
$a_2 = 3a_{2-1} + 1$	Substitute $n = 2$.
$= 3a_1 + 1$	Simplify.
$= 3(1) + 1$	Replace a_1 with 1.
$= 4$	This is the second term.
$a_3 = 3a_{3-1} + 1$	Substitute $n = 3$.
$= 3a_2 + 1$	Simplify.
$= 3(4) + 1$	Replace a_2 with 4.
$= 13$	This is the third term.
$a_4 = 3a_{4-1} + 1$	Substitute $n = 4$.
$= 3a_3 + 1$	Simplify.
$= 3(13) + 1$	Replace a_3 with 13.
$= 40$	This is the fourth term.
$a_5 = 3a_{5-1} + 1$	Substitute $n = 5$.
$= 3a_4 + 1$	Simplify.
$= 3(40) + 1$	Replace a_4 with 40.
$= 121$	This is the fifth term. **(Sequences)**

10.

$a_1 = 2$	
$a_n = 2a_{n-1} + n$	This is the recursion formula.
$a_2 = 2a_{2-1} + 2$	Substitute $n = 2$.
$= 2a_1 + 2$	Simplify.

$$= 2(2) + 2$$ Replace a_1 with 2.

$$= 6$$ This is the second term.

$$a_3 = 2a_{3-1} + 3$$ Substitute $n = 3$.

$$= 2a_2 + 3$$ Simplify.

$$= 2(6) + 3$$ Replace a_2 with 6.

$$= 15$$ This is the third term.

$$a_4 = 2a_{4-1} + 4$$ Substitute $n = 4$.

$$= 2a_3 + 4$$ Simplify.

$$= 2(15) + 4$$ Replace a_3 with 15.

$$= 34$$ This is the fourth term.

$$a_5 = 2a_{5-1} + 5$$ Substitute $n = 5$.

$$= 2a_4 + 5$$ Simplify.

$$= 2(34) + 5$$ Replace a_4 with 34.

$$= 73$$ This is the fifth term. (**Sequences**)

11. $\displaystyle\sum_{i=1}^{5} (2i - 3)$

$$= [2(1) - 3] + [2(2) - 3] + [2(3) - 3] + [2(4) - 3] + [2(5) - 3]$$ Replace i, starting with 1 and ending with 5.

$$= -1 + 1 + 3 + 5 + 7$$ Simplify each expression in brackets.

$$= 15$$ Add. (**Series**)

12. $\displaystyle\sum_{i=2}^{6} i^2 = (2)^2 + (3)^2 + (4)^2 + (5)^2 + (6)^2$ Replace i, starting with 2 and ending with 6.

$$= 4 + 9 + 16 + 25 + 36$$ Simplify each term.

$$= 90$$ Add. (**Series**)

13. $\displaystyle\sum_{i=1}^{4} (i - 1)(i + 4) = (1 - 1)(1 + 4) + (2 - 1)(2 + 4) + (3 - 1)(3 + 4) + (4 - 1)(4 + 4)$

Replace i, starting with 1 and ending with 4.

$$= 0(5) + 1(6) + 2(7) + 3(8)$$ Simplify.

$$= 0 + 6 + 14 + 24$$

$$= 44$$ Add. (**Series**)

14. $\displaystyle\sum_{i=3}^{5} \frac{i^2-3}{2i+4} = \frac{3^2-3}{2\,(3)+4} + \frac{4^2-3}{2\,(4)+4} + \frac{5^2-3}{2\,(5)+4}$ Replace i, starting with 3 and ending with 5.

$\displaystyle = \frac{9-3}{6+4} + \frac{16-3}{8+4} + \frac{25-3}{10+4}$ Perform multiplications.

$\displaystyle = \frac{6}{10} + \frac{13}{12} + \frac{22}{14}$ Simplify numerators and denominators.

$\displaystyle = \frac{3}{5} + \frac{13}{12} + \frac{11}{7}$ Reduce.

$\displaystyle = \frac{3}{5}\cdot\frac{84}{84} + \frac{13}{12}\cdot\frac{35}{35} + \frac{11}{7}\cdot\frac{60}{60}$ Use $5 \cdot 12 \cdot 7 = 420$ as the LCD.

$\displaystyle = \frac{252}{420} + \frac{455}{420} + \frac{660}{420} = \frac{1367}{420}$ Add. **(Series)**

15. $\displaystyle\sum_{i=1}^{4} (2x-i) = (2x-1)+(2x-2)+(2x-3)+(2x-4)$ Replace i starting with 1 and ending with 4.

Stop here since the directions only want the terms written out. **(Series)**

16. $\displaystyle\sum_{i=2}^{4} x^i = x^2 + x^3 + x^4$ Replace i, starting with 2 and ending with 4. **(Series)**

17. $\displaystyle\sum_{i=1}^{3} \frac{i^2}{x^2} = \frac{1^2}{x^2} + \frac{2^2}{x^2} + \frac{3^2}{x^2}$ Replace i, starting with 1 and ending with 3.

$\displaystyle = \frac{1}{x^2} + \frac{4}{x^2} + \frac{9}{x^2}$ Simplify the numerators. **(Series)**

18. $9 + 16 + 25 + 36 + 49 =$ These are perfect squares.

$\displaystyle 3^2 + 4^2 + 5^2 + 6^2 + 7^2 = \sum_{i=3}^{7} i^2$ **(Series)**

19. $1 + 8 + 27 + 64 =$ These are perfect cubes.

$\displaystyle 1^3 + 2^3 + 3^3 + 4^3 = \sum_{i=1}^{4} i^3$ **(Series)**

20. $\displaystyle\frac{1}{2} + \frac{2}{3} + \frac{3}{4} + \frac{4}{5} + \frac{5}{6} = \sum_{i=1}^{5} \frac{i}{i+1}$ The numerators are successive integers, starting at 1. Each denominator is one more than the numerator. **(Series)**

21. $2, 4, 6, 8, 10 \ldots$

$10 - 8 = 2$ Subtract any term from the term that follows it.

$8 - 6 = 2$

$d = 2$

(Arithmetic sequences and series)

22. $6, 11, 16, 21, 26 \ldots$

$26 - 21 = 5$

$d = 5$

(Arithmetic sequences and series)

23. $4, 1, -2, -5, -8 \ldots$

$-5 - (-2) = -5 + 2 = -3$ Subtract any term from the term that follows it.

$d = -3$

(Arithmetic sequences and series)

24. $1, \dfrac{4}{3}, \dfrac{5}{3}, 2, \dfrac{7}{3}, \dfrac{8}{3}, 3 \ldots$

$\dfrac{5}{3} - \dfrac{4}{3} = \dfrac{1}{3}$

$d = \dfrac{1}{3}$

(Arithmetic sequences and series)

25. $a_1 = 5$ $d = 4.$

$a_2 = 5 + 4 = 9$ Add 4 to a_1.

$a_3 = 9 + 4 = 13$ Add 4 to a_2.

$a_4 = 13 + 4 = 17$ Add 4 to a_3.

$a_5 = 17 + 4 = 21$ Add 4 to a_4.

(Arithmetic sequences and series)

26. $a_1 = 7$ $d = -2.$

$a_2 = 7 + (-2) = 5$ Add -2 to a_1.

$a_3 = 5 + (-2) = 3$ Add -2 to a_2.

$a_4 = 3 + (-2) = 1$ Add -2 to a_3.

$a_5 = 1 + (-2) = -1$ Add -2 to a_4.

27. $a_1 = 6$ $d = -4.$

$a_n = a_1 + (n - 1)d$ Write the formula.

$a_{23} = a_1 + (23 - 1)d$ $n = 23$

$a_{23} = a_1 + 22d$ Simplify.

$a_{23} = 6 + 22(-4)$ Substitute $a_1 = 6$ and $d = -4$.

$a_{23} = 6 - 88$ Multiply.

$a_{23} = -82$ Subtract.

(Arithmetic sequences and series)

28. In $4, -2, -8, -14 \ldots,$ $a_1 = 4$ and $d = -8 - (-2) = -8 + 2 = -6$

$a_n = a_1 + (n-1)d$ Write the formula.

$a_{100} = a_1 + (100 - 1)d$ $n = 100$

$a_{100} = a_1 + 99d$ Simplify.

$a_{100} = 4 + 99(-6)$ Substitute $a_1 = 4$ and $d = -6$.

$a_{100} = 4 - 594$ Multiply.

$a_{100} = -590$ Subtract.

(Arithmetic sequences and series)

29. $a_{11} = a_1 + (11 - 1)d$ Write the formula for a_n with $n = 11$ since a_{11} is given.

$a_{11} = a_1 + 10d$ Simplify.

$35 = a_1 + 10d$ Substitute $a_{11} = 35$.

$a_{15} = a_1 + (15 - 1)d$ Write the formula for a_n with $n = 15$.

$a_{15} = a_1 + 14d$ Simplify.

$47 = a_1 + 14d$ Substitute $a_{15} = 47$.

Now solve the resulting system of equations:

$a_1 + 10d = 35$ Solve the system of equations.

$a_1 + 14d = 47$

$-a_1 - 10d = -35$ Multiply the first equation by -1.

$a_1 + 14d = 47$

$4d = 12$ Add the equations.

$d = 3$ Divide by 4.

$a_1 + 10(3) = 35$ Substitute $d = 3$ into the first equation.

$a_1 + 30 = 35$

$a_1 = 5$

$a_{24} = a_1 + (24 - 1)d$ Use the formula for a_n with $n = 24$.

$a_{24} = 5 + 23(3)$ Substitute $a_1 = 5$ and $d = 3$.

$a_{24} = 5 + 69 = 74$

(Arithmetic sequences and series)

30. $a_1 = 1, a_{10} = 10$ Here $n = 10$ so find a_1 and a_{10}.

$S_n = \dfrac{n(a_1 + a_n)}{2}$ Write the formula for S_n.

$S_{10} = \dfrac{10(1 + 10)}{2}$ Substitute $n = 10, a_1 = 1$, and $a_{10} = 10$.

$S_{10} = \dfrac{10(11)}{2} = 55$ Simplify.

or use the alternate formula:

$S_n = \dfrac{n}{2}[2a_1 + (n - 1)d]$ Write the formula for S_n.

$S_{10} = \dfrac{10}{2}[2(1) + (10 - 1)1]$ Substitute $n = 10, a_1 = 1, d = 1$.

$S_{10} = \dfrac{10}{2}[2 + 9]$

$S_{10} = 5(11) = S_{10} = 55$

(Arithmetic sequences and series)

31. We know $n = 9$, but do not know a_9, the ninth term. We could find a_9 but it is more convenient to use the formula $S_n = \dfrac{n}{2}[2a_1 + (n - 1)d]$.

$S_9 = \dfrac{9}{2}[2(3) + (9 - 1)8]$ Substitute $n = 9, a_1 = 3, d = 8$.

$S_9 = \dfrac{9}{2}[2(3) + 8(8)]$ Simplify.

$S_9 = \dfrac{9}{2}[6 + 64]$ Multiply.

$S_9 = \dfrac{9}{2}(70)$ Add.

$S_9 = 315$

(Arithmetic sequences and series)

32. $3, 6, 12, 24, 48, \ldots$

   $\dfrac{24}{12} = 2$ Divide any term by the term preceding it.

   $\dfrac{12}{6} = 2$

   $r = 2$ **(Geometric sequences and series)**

33. $2, -6, 18, -54, 162, \ldots$

   $\dfrac{-54}{18} = -3$ Divide any term by the term preceding it.

   $\dfrac{18}{-6} = -3$

   $r = -3$ **(Geometric sequences and series)**

34. $4, 1, \dfrac{1}{4}, \dfrac{1}{16}, \dfrac{1}{64}, \ldots$

   $1 \div 4 = \dfrac{1}{4}$ Divide any term by the term preceding it.

   $\dfrac{1}{16} \div \dfrac{1}{4} = \dfrac{1}{16} \cdot \dfrac{4}{1} = \dfrac{1}{4}$

   $r = \dfrac{1}{4}$ **(Geometric sequences and series)**

35. $a_1 = 5, r = -2$

   $a_2 = a_1 r = 5(-2) = -10$ Multiply a_1 by $r = -2$.

   $a_3 = a_2 r = -10(-2) = 20$ Multiply a_2 by $r = -2$.

   $a_4 = a_3 r = 20(-2) = -40$ Multiply a_3 by $r = -2$.

   $a_5 = a_4 r = -40(-2) = 80$ Multiply a_4 by $r = -2$.

   (Geometric sequences and series)

36. $a_1 = 1, r = \dfrac{1}{2}$

   $a_2 = a_1 r = 1\left(\dfrac{1}{2}\right) = \dfrac{1}{2}$ Multiply a_1 by $r = \dfrac{1}{2}$.

   $a_3 = a_2 r = \dfrac{1}{2}\left(\dfrac{1}{2}\right) = \dfrac{1}{4}$ Multiply a_2 by $r = \dfrac{1}{2}$.

$a_4 = \dfrac{1}{4}\left(\dfrac{1}{2}\right) = \dfrac{1}{8}$ Multiply a_3 by $r = \dfrac{1}{2}$.

$a_5 = \dfrac{1}{8}\left(\dfrac{1}{2}\right) = \dfrac{1}{16}$ Multiply a_4 by $r = \dfrac{1}{2}$.

(Geometric sequences and series)

37. $a_n = a_1 r^{n-1}$ Write the formula.

$a_4 = a_1 r^{4-1}$ Substitute $n = 4$.

$a_4 = a_1 r^3$ Simplify.

$a_4 = 5(-3)^3$ Substitute $a_1 = 5$ and $r = -3$.

$a_4 = 5(-27) = -135$

(Geometric sequences and series)

38. In the geometric sequence $1, 3, 9, 27, 81 \ldots, a_1 = 1$ and $r = \dfrac{3}{1} = 3$ (divide any term by the preceding term).

$a_n = a_1 r^{n-1}$ Write the formula.

$a_{10} = a_1 r^{10-1}$ Substitute $n = 10$.

$a_{10} = a_1 r^9$ Simplify.

$a_{10} = 1(3)^9$ Substitute $a_1 = 1$ and $r = 3$.

$a_{10} = 3^9 = 19{,}683$

(Geometric sequences and series)

39. In $3, 6, 12, 24 \ldots,\quad a_1 = 3, r = \dfrac{6}{3} = 2$

$a_n = a_1 r^{n-1}$ Write the formula.

$a_n = 3(2)^{n-1}$ Substitute $a_1 = 3, r = 2$.

(Geometric sequences and series)

40. In $\dfrac{1}{4}, \dfrac{3}{4}, \dfrac{9}{4}, \dfrac{27}{4} \ldots,\quad a_1 = \dfrac{1}{4}, r = \dfrac{3}{4} \div \dfrac{1}{4} = \dfrac{3}{4} \cdot \dfrac{4}{1} = 3$

$a_n = a_1 r^{n-1}$ Write the formula.

$a_n = \dfrac{1}{4}(3)^{n-1}$

(Geometric sequences and series)

41. $a_n = a_1 r^{n-1}$ Write the formula.

$a_2 = a_1 r^{2-1}$ Substitute $n = 2$ since a_2 is given.

$a_2 = a_1 r^1$ Simplify.

$18 = a_1 r^1$ Substitute $a_2 = 18$.

$a_n = a_1 r^{n-1}$ Write the formula.

$a_6 = a_1 r^{6-1}$ Substitute $n = 6$ since a_6 is given.

$a_6 = a_1 r^5$ Simplify.

$1458 = a_1 r^5$ Substitute $a_6 = 1458$.

Now set up a ratio using the two equalities.

$$\frac{a_1 r^5}{a_1 r^1} = \frac{1458}{18}$$

$r^4 = 81$ Simplify each side.

$r = 3$ Take the fourth root of both sides.

Now find a_1:

$18 = a_1 r^1$

$18 = a_1 (3)^1$

$6 = a_1$

So the general term a_n is

$a_n = 6(3)^{n-1}$.

(Geometric sequences and series)

42. $S_6 = \dfrac{a_1 (r^6 - 1)}{r - 1}$ Use $n = 6$.

$S_6 = \dfrac{8\,[\,(-2)^6 - 1]}{(-2) - 1}$ Substitute $a_1 = 8$ and $r = -2$.

$S_6 = \dfrac{8\,(64 - 1)}{-3}$ Simplify.

$S_6 = -168$ Reduce.

(Geometric sequences and series)

43. In the sequence $5, 10, 20, 40 \ldots, a_1 = 5, r = \dfrac{10}{5} = 2$.

$S_{10} = \dfrac{a_1 (r^{10} - 1)}{r - 1}$ Use $n = 10$.

$S_{10} = \dfrac{5 (2^{10} - 1)}{2 - 1}$ Substitute $a_1 = 5, r = 2$.

$S_{10} = 5(2^{10} - 1)$ Simplify.

(Geometric sequences and series)

44. $S = \dfrac{a_1}{1 - r}$ Write the formula.

$S = \dfrac{5}{1 - \dfrac{1}{4}}$ Substitute $a_1 = 5, r = \dfrac{1}{4}$.

$= \dfrac{5}{\dfrac{3}{4}}$ Simplify.

$= 5 \cdot \dfrac{4}{3} = \dfrac{20}{3}$ Invert and multiply.

(Geometric sequences and series)

45. Write the decimal as a sum: $0.454545\ldots = 0.45 + 0.0045 + 0.000045 + \ldots$

The terms in the sum are a geometric sequence with $a_1 = 0.45$ and $r = \dfrac{0.0045}{0.45} = 0.01$.

$S = \dfrac{a_1}{1 - r}$ Write the formula.

$S = \dfrac{0.45}{1 - 0.01}$ Substitute $a_1 = 0.45$ and $r = 0.01$.

$S = \dfrac{0.45}{0.99}$ Simplify.

$S = \dfrac{45}{99} = \dfrac{5}{11}$ Reduce. If one carries out the division of 5 by 11 one produces the
repeated decimal $0.454545\ldots$

(Geometric sequences and series)

46. $\dfrac{4!}{0!} = \dfrac{4 \cdot 3 \cdot 2 \cdot 1}{1} = 24$ Use the definition of $n!$ and $0!$ **(Factorials)**

47. $\dfrac{8!}{3!5!} = \dfrac{8 \cdot 7 \cdot 6 \cdot 5 \cdot 4 \cdot 3 \cdot 2 \cdot 1}{3 \cdot 2 \cdot 1 \cdot 5 \cdot 4 \cdot 3 \cdot 2 \cdot 1} = 56$ Use the definition of $n!$ **(Factorials)**

48. $\dfrac{6!}{6!0!} = \dfrac{6 \cdot 5 \cdot 4 \cdot 3 \cdot 2 \cdot 1}{6 \cdot 5 \cdot 4 \cdot 3 \cdot 2 \cdot 1 \cdot 1} = 1$ Use the definition of $n!$ and $0!$ **(Factorials)**

49. $\dbinom{4}{0} = \dfrac{4!}{0!\,(4-0)!}$ Use the definition of $\dbinom{n}{r}$ with $n = 4, r = 0$.

    $= \dfrac{4!}{0!4!}$ Simplify.

    $= \dfrac{4!}{1 \cdot 4!} = 1$ Use $\dfrac{n!}{n!} = 1$. **(Binomial coefficient)**

50. $\dbinom{4}{1} = \dfrac{4!}{1!\,(4-1)!}$ Use the definition of $\dbinom{n}{r}$ with $n = 4, r = 1$.

    $= \dfrac{4!}{1!3!}$ Simplify.

    $= \dfrac{4 \cdot 3 \cdot 2 \cdot 1}{1 \cdot 3 \cdot 2 \cdot 1} = 4$ Use the definition of $n!$ **(Binomial coefficient)**

51. $\dbinom{4}{2} = \dfrac{4!}{2!\,(4-2)!}$ Use the definition of $\dbinom{n}{r}$ with $n = 4, r = 2$.

    $= \dfrac{4!}{2!2!}$ Simplify.

    $= \dfrac{4 \cdot 3 \cdot 2 \cdot 1}{2 \cdot 1 \cdot 2 \cdot 1} = 6$ Use the definition of $n!$ **(Binomial coefficient)**

52. $\dbinom{4}{3} = \dfrac{4!}{3!\,(4-3)!}$ Use the definition of $\dbinom{n}{r}$ with $n = 4, r = 3$.

    $= \dfrac{4!}{3!1!} = \dfrac{4 \cdot 3 \cdot 2 \cdot 1}{3 \cdot 2 \cdot 1 \cdot 1} = 4$ Use the definition of $n!$ **(Binomial coefficient)**

53. $\dbinom{4}{4} = \dfrac{4!}{4!\,(4-4)!}$ Use the definition of $\dbinom{n}{r}$ with $n = 4, r = 4$.

    $= \dfrac{4!}{4!0!} = \dfrac{4!}{4!1} = 1$ Use $0! = 1$. **(Binomial coefficient)**

54. Use the formula with $n = 4$:

    $\dbinom{4}{0} x^4 y^0 + \dbinom{4}{1} x^3 y^1 + \dbinom{4}{2} x^2 y^2 + \dbinom{4}{3} x^1 y^3 + \dbinom{4}{4} x^0 y^4$ Note that the coefficients were computed in #49 - #53.

    $= 1x^4 + 4x^3 y + 6x^2 y^2 + 4xy^3 + 1y^4$ **(The binomial theorem)**

55. Use the formula with $n = 5$ substituting $2a$ for x and 3 for y:

$$\binom{5}{0}(2a)^5(3)^0 + \binom{5}{1}(2a)^4(3)^1 + \binom{5}{2}(2a)^3(3)^2 + \binom{5}{3}(2a)^2(3)^3 + \binom{5}{4}(2a)^1(3)^4 + \binom{5}{5}(2a)^0(3)^5$$

$$= \frac{5!}{0!\,(5-0)!}32a^5 + \frac{5!}{1!\,(5-1)!}(16a^4)(3) + \frac{5!}{2!\,(5-2)!}(8a^3)(9) + \frac{5!}{3!\,(5-3)!}(4a^2)(27) +$$

$$\frac{5!}{4!\,(5-4)!}(2a)(81) + \frac{5!}{5!\,(5-5)!}(1)(243)$$

$$= (1)32a^5 + 5(16a^4)(3) + 10(8a^3)(9) + 10(4a^2)(27) + 5(2a)(81) + 1(1)(243)$$

$$= 32a^5 + 240a^4 + 720a^3 + 1080a^2 + 810a + 243$$

(The binomial theorem)

56. Building on the fourth row, we have

$$1 \quad 4 \quad 6 \quad 4 \quad 1$$

$$1 \quad 5 \quad 10 \quad 10 \quad 5 \quad 1$$

$$1 \quad 6 \quad 15 \quad 20 \quad 15 \quad 6 \quad 1$$

so,

$$(x + y)^6 = 1x^6 + 6x^5y + 15x^4y^2 + 20x^3y^3 + 15x^2y^4 + 6xy^5 + 1y^6$$

(The binomial theorem)

57 $\binom{n}{k-1}x^{n-(k-1)}y^{k-1}$ Write the formula.

$\binom{9}{5-1}x^{9-(5-1)}y^{5-1}$ Substitute $n = 9, k = 5$.

$\binom{9}{4}x^5y^4$ Simplify.

$\dfrac{9!}{4!\,(9-4)!}x^5y^4$ Find the coefficient.

$\dfrac{9 \cdot 8 \cdot 7 \cdot 6 \cdot 5!}{4 \cdot 3 \cdot 2 \cdot 1 \cdot 5!}x^5y^4$ Simplify.

$= 126x^5y^4$

(The binomial theorem)

58 $\binom{n}{k-1}x^{n-(k-1)}y^{k-1}$ Write the formula.

$\binom{12}{8-1}(3a)^{12-(8-1)}(-2b)^{8-1}$ Substitute $n = 12, k = 8, x = 3a$ and $y = -2b$.

$\binom{12}{7}(3a)^5(-2b)^7$ Simplify.

$$\frac{12!}{7!\,(12-7)!}\,(243a^5)(-128b^7) = \frac{12\cdot 11\cdot 10\cdot 9\cdot 8\cdot 7!}{7!5!}\,(-31104a^5b^7) = 792(-31104a^5b^7)$$

$$= -24{,}634{,}368a^5b^7$$

(The binomial theorem)

Grade Yourself

Circle the question numbers that you had incorrect. Then indicate the number of questions you missed. If you answered more than three questions incorrectly, you need to focus on that topic. (If a topic has less than three questions and you had at least one wrong, we suggest you study that topic also. Read your textbook, a review book, or ask your teacher for help.)

Subject: Sequences and Series

Topic	Question Numbers	Number Incorrect
Sequences	1, 2, 3, 4, 5, 6, 7, 8, 9, 10	
Series	11, 12, 13, 14, 15, 16, 17, 18, 19, 20	
Arithmetic sequences and series	21, 22, 23, 24, 25, 26, 27, 28, 29, 30, 31	
Geometric sequences and series	32, 33, 34, 35, 36, 37, 38, 39, 40, 41, 42, 43, 44, 45	
Factorials	46, 47, 48	
Binomial coefficient	49, 50, 51, 52, 53	
Binomial theorem	54, 55, 56, 57, 58	